Cell and Tissue Culture
for Medical Research

Cell and Tissue Culture for Medical Research

Edited by
ALAN DOYLE
The Wellcome Trust
London, UK
and
J. BRYAN GRIFFITHS
Scientific Consultancy and Publishing
Porton, Salisbury, UK

JOHN WILEY & SONS, LTD

Chichester • New York • Weinheim • Brisbane • Singapore • Toronto

Other Wiley Editorial Offices

John Wiley & Sons, Inc., 605 Third Avenue,
New York, NY 10158–0012, USA

WILEY-VCH Verlag GmbH, Pappelallee 3,
D-69469 Weinheim, Germany

Jacaranda Wiley Ltd, 33 Park Road, Milton,
Queensland 4064, Australia

John Wiley & Sons (Asia) Pte Ltd, 2 Clementi Loop #02–01,
Jin Xing Distripark, Singapore 129809

John Wiley & Sons (Canada) Ltd, 22 Worcester Road,
Rexdale, Ontario M9W 1L1, Canada

Library of Congress Cataloging-in-Publication Data
Cell and tissue culture for medical research /
    edited by Alan Doyle and J. Bryan Griffiths.
        p. cm.
    Includes bibliographical references and index.
    ISBN 0-471-85213-9 (alk. paper)
        1. Tissue culture–Laboratory manuals. 2. Cell culture–
Laboratory manuals. I. Doyle, Alan. II. Griffiths, Bryan.
R853.T58 C45   2000
616.07′56–dc21   99–089625

British Library Cataloguing in Publication Data
A catalogue record for this book is available from the British Library

Cover photograph: Phase contrast photomicrograph of a developing CML-LTMC stromal layer
(week 4) with several cobblestone areas or areas of active hemopoiesis. Reproduced from Human
Hemopoietic Stem Cell Culture: Long-Term Marrow Culture by Catriona Jamieson and Armand
Keating in Cell & Tissue Culture : Laboratory Procedures, A Doyle, JB Griffiths and DG Newell
(eds) (1993), with permission of the authors and John Wiley & Sons Ltd.

ISBN 0 471 85213 9

Typeset in 10/12pt Times by Florence Production Ltd, Stoodleigh, Devon
Printed and bound in Great Britain by Biddles Ltd, Guildford, UK.
This book is printed on acid-free paper responsibly manufactured from sustainable forestry, in which
at least two trees are planted for each one used for paper production.

Contents

Contributors

Colin F. Arlett
MRC Cell Mutation Unit, University of Sussex,
Falmer, Brighton BN1 9RR, UK
Section: 4.4C

John Armitage
Division of Ophthalmology, University of Bristol,
Bristol Eye Hospital, Bristol BS1 2LX, UK
Section: 5.3

Bryan Bolton
EACC, CAMR, Salisbury, Wiltshire SP4 0JG, UK
Section: 4.4A

Gérard Brugal
Equipe RFMQ, Laboratoire TIMC-CNRS-IMAG (UMR 5525),
Institut Albert Bonniot Faculté de Médicine – Université J. Fourier,
F38706 La Tronche Cedex, France
Sections: 3.2 & 3.3

J.V. Castell
Unidad de Hepatología Experimental, Centro de Investigation, Hospital
Universitario La Fe, Avda, Campanar 21, E-46009 Valencia, Spain
Section: 5.6

N. Randall Chu
Kennedy Institute of Rheumatology, Charing Cross Sunley Research Centre
Trust, 1 Lurgan Avenue, Hammersmith, London W6 8LW, UK
Section: 4.4C

Stuart Clark
Microgen Bioproducts Ltd, 1 Admiralty Way, Camberley, Surrey GU15 3DT, UK
Section: 5.2

Catherine Clarke
LICR/UCL Breast Cancer Laboratory, Department of Surgery, Royal Free and
University College Medical School, Charles Bell House, 67–73 Riding House
Street, London W1P 7LD, UK
Section: 3.4

Thomas G. Cotter
Head of Department of Biochemistry, University College of Cork,
Lee Maltings, Prospect Row, Cork, Ireland
Section: 3.7

Irene Cour
Section: 2.3

Steven M. D'Ambrosio
The Ohio State University, Division of Radiobiology,
Department of Radiology, Columbus, OH 43210, USA
Section: 4.10

John L. Darling
University Department of Neurosurgery, Institute of Neurology,
The National Hospital for Neurology and Neurosurgery, Queen Square,
London WC1N 3BG, UK
Section: 4.15

Alan Doyle
The Wellcome Trust, 183 Euston Road, London NW1 2BE, UK
Sections: 1.1, 1.2, 2.2A, 2.6, 3.1, 4.1 & 5.4

Hans G. Drexler
DSMZ, German Collection of Micro-organisms & Cell Cultures,
Department of Human & Animal Cell Cultures, Mascheroder Weg 18,
D-3300 Braunschweig, Germany
Section: 4.5

Marie-Jeanne Drian
Etudes Pratique des Hautes Etudes, Laboratory of Quantitative Cellular
Neurobiology, Institut National de la Santé et de la Recherche Médicale,
INSERM U 336, Université Montpellier 2 – CASE 106, Place E. Bataillon,
34095, Montpellier Cedex 5, France
Section: 4.14

Marc Feldmann
Kennedy Institute of Rheumatology, Charing Cross Sunley Research Centre Trust,
1 Lurgan Avenue, Hammersmith, London W6 8LW, UK
Section: 4.4C

Margaret Fitchett
Oxford Medical Genetics Laboratories, Oxford Radcliffe Hospital NHS Trust,
The Churchill, Headington, Oxford OX3 7LJ, UK
Section: 4.12

Ruth. E. Gibson-D'Ambrosio
The Ohio State University, Department of Radiology,
Division of Radiobiology, 103 Wiseman Hall, 400 W. 12th Avenue,
Columbus, OH 43210, USA
Section: 4.10

M.J. Gomez-Lechón
Unidad de Hepatología Experimental, Centro de Investigación, Hospital La Fe,
Avda. Campanar 21, E-46009-Valencia, Spain
Section: 5.6

James Gray
Public Health Laboratory Service, Addenbrooke's Hospital, Level 6,
Hills Road, Cambridge CB32 2QW, UK
Section: 1.4

Bryan Griffiths
Scientific Consultancy & Publishing, 5 Bourne Gardens, Porton, Salisbury,
Wiltshire, SP4 0NU, UK
Sections: 2.1, 2.7, 3.1 & 4.1

Richard L. Guerrant
Division of Geographic and International Medicine,
University of Virginia School of Medicine, PO Box 801379, Charlottesville,
VA 22903, USA
Section: 4.9

Raymonde Guillot
Faculté de Medecine Paris Ouest, Laboratorie D'Histo-Embryologie,
45 rue des Saints-Péres, 75006, Paris, France
Section: 4.13

Angela Hague
University of Bristol, Department of Pathology & Microbiology,
Medical Sciences Building, University Walk, Bristol BS8 1TD, UK
Section: 4.16

Robert J. Hay
Director, Cell Biology, American Type Culture Collection,
10801 University Blvd, Manassas, VA 201110, USA
Section: 2.3

Leonard Hayflick
University of California San Francisco, School of Medicine,
36991 Greencroft Close, PO Box 89, The Sea Ranch,
CA 95497, USA
Section: 4.6

Laurie Haynes
University of Bristol, Department of Zoology, Woodland Road,
Bristol BS8 1UG, UK
Section: 4.3

Gisele Hodges
Sections: 4.1 & 5.4

Zhen-Bo Hu
German Collection of Micro-organisms & Cell Cultures,
Department of Human & Animal Cell Cultures, Mascheroder Weg 18,
D-3300 Braunschweig, Germany
Section: 4.5

Rowena James
Pharmagene Laboratories Ltd, 2A Orchard Road, Royston,
Herts SG8 5HD, UK
Section: 1.3

Siän Jones
CRC Laboratories, University of Birmingham, The Medical School,
Department of Cancer Studies, Vincent Drive, Birmingham B15 2TJ, UK
Section: 5.5B

Norbert König
EPHE Biol. Cell. Qmant, INSERM U 336, Université Montpellier 2/CC 106,
Place E. Bataillon, 34095 Montpellier Cedex 5, France
Section: 4.14

Eero Lehtonen
University of Helsinki, Department of Pathology, PO Box 21,
(Haartmaninkatu 3), FIN-00014, Helsinki, Finland
Section: 4.11

Seamus V. Lennon
Department of Science, Institute of Technology Sligo, Ballinooe, Sligo, Ireland
Section: 3.7

Claire Linge
Raft Institute of Plastic Surgery, Mount Vernon Hospital, Northwood,
Middlesex, HA6 2RN, UK
Section: 2.5

Charles D Little
Section: 4.9

Marco Londei
Kennedy Institute of Rheumatology, Charing Cross Sunley Research Centre
Trust, 1 Lurgan Avenue, Hammersmith, London W6 8LW, UK
Section: 4.4C

Caroline MacDonald
The University of Paisley, Department of Biology, High Street,
Paisley, PA1 2BE, UK
Sections: 5.1 & 5.5A

Roderick A.F. MacLeod
DSMZ, Deutsche Sammlung von Mikrooganismen und Zelkulturen GmbH,
Mascheroder Weg 1B, D-38124 Braunschweig, Germany
Section: 4.7

Louis M. Mansky
McArdle Laboratory for Cancer Research, 1400 University Avenue,
University of Wisconsin-Madison, Madison, Wisconsin 53706, USA
Sections: 5.5C & 5.5D

Barry J. Marshall
Section: 4.9

Richard W. McCallum
Section: 4.9

John S. McLean
Safety Co-ordinator, Department of Biological Sciences, University of Paisley,
PA1 2BE, UK
Sections: 5.5B, 5.5C & 5.5D

Stella Margaret McLean
Department of Biological Sciences, University of Paisley, PA1 2BE, UK
Section: 5.5B

Paul Monaghan
Institute for Animal Health, Purbright Laboratory, Ash Road,
Purbright, Woking, Surrey GU24 0NF, UK
Section: 3.4

Christopher B. Morris
The Wellcome Trust Centre for Human Genetics, Windmill Road, Headington,
Oxford OX3 7BN, UK
Sections: 1.2, 1.5 & 1.6

Jon Mowles
Stream Cottage, Market Lavington, Devizes,
Wiltshire SN10 4AN, UK
Sections: 2.2A & 2.4

J. Greg Murison
Kennedy Institute of Rheumatology, Charing Cross Sunley Research Centre
Trust, 1 Lurgan Avenue, Hammersmith, London W6 8LW, UK
Section: 4.4C

Diane G. Newell
Silver Birches, Longparish Road, Wherwell, Andover, Hants SP11 7AW, UK
Sections 1.5 & 3.1

Mike O'Donovan
AstraZeneca R & D Charnwood, Bakewell Road, Loughborough,
Leicestershire LE11 5RH, UK
Section: 4.4C

Michael G. Ormerod
34 Wray Park Road, Reigate, Surrey RH2 0DE, UK
Section: 3.5

Christos Paraskeva
University of Bristol, Department of Pathology & Microbiology,
Medical Sciences Building, University Walk, Bristol BS8 1TD, UK
Section: 4.16

Mariann Pedrotti-Krueger
John Hopkins University, School of Medicine, 720 Rutland Avenue,
Baltimore, Maryland 21205, USA
Section: 4.4B

Margaret Penno
John Hopkins University, School of Medicine, 720 Rutland Avenue,
Baltimore, Maryland 21205, USA
Section: 4.4B

David A. Peura
University of Virginia, Health Sciences Center, Department of Medicine,
PO Box 145, Charlottesville, VA 22908, USA
Section: 4.9

X. Ponsoda
Departament de Parasitologia i Biologia Cel·lular, Facultat de Biològiques,
Universitat de València, Avda Dr, Moliner 50, E-46100-Burjassot, Spain
Section: 5.6

Emma Ramsdale
University of Oxford, Kingman's Laboratory,
Department of Biochemistry, South Parks Road, Oxford OX1 3QU, UK
Section: 5.5B

Tanya Ray
John Hopkins University, School of Medicine, 720 Rutland Avenue, Baltimore,
Maryland 21205, USA
Section: 4.4B

Marcelle Regnier
L'Oreal, Centre de Recherche Charles Zviak, 90 rue du General Roguet,
92583 Clichy Cedex, France
Section: 4.8

Maria Soledad Santisteban-Otegui
Université Joseph Fourier, Equipe RFMQ, Laboratoire TIMC-CNRS-JMAG
(UMR 5525), Institut Albert Bonniot, Faculté de Médecine,
F38706 La Tronche, Cedex, France (postage address USA)
Section: 3.3

Jerzy Sarosiek
University of Virginia, Health Sciences Center, Department of Medicine,
PO Box 145, Charlottesville, VA 22908, USA
Section: 4.9

Lauri Saxén
University of Helsinki, Department of Pathology,
PO Box 21 (Haartmaninkatu 3), FIN-00014, Helsinki, Finland
Section: 4.11

Glyn Stacey
NIBSC, Blanche Lane, South Mimms, Potters Bar, EN6 3QG, UK
Section: 2.2B & 2.6

Anja Tuomi
University of Helsinki, Department of Pathology, PO Box 21
(Haartmaninkatu 3), FIN-00014, Helsinki, Finland
Section: 4.11

Sally Warburton
ECACC, CAMR, Salisbury, Wiltshire SP4 0JG, UK
Section: 1.3

T.H. Ward
Drug Development and Imaging, Paterson Institute for Cancer Research,
Christie Hospital NHS Trust, Wilmslow Road, Manchester M20 4BX, UK
Section: 4.2

Ashley Wilson
CCTR Cryotech, Department of Biology, PO Box 373, York YO1 5YW, UK
Section: 3.6

John Worthington
Section: 1.5

Preface

The comprehensive manual *Cell and Tissue Culture: Laboratory Procedures* edited by A. Doyle, J.B. Griffiths and D.G. Newell, was first published in 1993, with quarterly additions and updates up to 1998. The initial publication was well received by the scientific community. Numerous requests have been received from a range of people, saying: 'When will a series of subset volumes be produced?' In response to this demand we decided to look afresh at the wealth of material available in the main publication and adapt from this 'highlights' which we believe will be of particular value to targeted users. The first of these was a subset for the biotechnologist; this current volume presents techniques for medical research. This is partly in response to the ever-increasing interest and application of cell culture in the new medicinal applications of cell and tissue engineering, and cell therapy. Many of the contributions have been updated from the original for this publication but some are published here for the first time. It is certainly not our intention to reproduce all of the manual in this fashion but to provide core procedures for many of the specialist groups that can be identified as benefiting from them. We aim to appeal to scientists who may be new to cell culture and require the practical guidance that *Cell and Tissue Culture: Laboratory Procedures* has to offer. There is also the added benefit that valuable technical information is made available without the major investment in the whole publication. We believe that these subsets fulfil a need and constitute a major development on the original publication, and we look forward to offering further titles in this series.

A. Doyle and J. B. Griffiths
Managing Editors

Safety

Neither the editors, contributors nor John Wiley & Sons Ltd accept any responsibility or liability for loss or damage occasioned to any person or property through using the materials, instructions, methods, or ideas contained herein, or acting or refraining from acting as a result of such use. While the editors, contributors and publisher believe that the data, recipes, practical procedures and other information, as set forth in this book, are in accord with current recommendations and practice at the time of publication, they accept no legal responsibility for any errors or omissions, and make no warranty, express or implied, with respect to material contained herein. Attention to safety aspects is an integral part of all laboratory procedures and national legislations impose legal requirements on those persons planning or carrying out such procedures. It remains the responsibility of the reader to ensure that the procedures which are followed are carried out in a safe manner and that all necessary safety instructions and national regulations are implemented.

In view of ongoing research, equipment modifications and changes in governmental regulations, the reader is urged to review and evaluate the information provided by the manufacturer, for each reagent, piece of equipment or device, for any changes in the instructions or usage and for added warnings and precautions.

CHAPTER 1

BASIC LABORATORY SET-UP AND PROCEDURES

1.1 GENERAL INTRODUCTION

This section covers some of the essential requirements before tissue/cell culture can be carried out efficiently, effectively and safely. These techniques are common regardless of the long-term aims of the culture. In addition it should be stressed that great consideration should be given to the location of the work and the necessary specialist basic equipment required to carry out the work (see Doyle & Morris 1993; Morris & Jones 1993).

The medical research worker has a wealth of systems to choose from before the decision has to be taken to establish a cell line for a specific purpose *de novo*. Naturally, the requirements (and not least the Intellectual Property considerations) will dictate the 'utility' of existing material. Even so, if the cells for exploitation or the basic tools to create the *in vitro* systems already exist, then authenticated sources of starting material are essential. This can mean obtaining cell stocks from culture collections such as ECACC, ATCC, DSMZ, and the Riken Cell Bank. The advantage of basing a project on existing, well-characterized, authenticated and quality-controlled stocks cannot be over-emphasized. The added advantage is that much of the material available from collections is free of constraints on exploitation. However, should a new source of cells be needed then the Overview to Chapter 4 deals comprehensively with establishing new cell lines.

The standards for the cryopreservation, storage and routine quality control of cell stocks are widely recognized (Stacey *et al.* 1995). Cryopreservation of a well-characterized, dependable, high-viability (achieved by controlled-rate freezing), microbial-contaminant-free cell stock is fundamental to both the academic researcher as well as the commercial producer. Existing legislation provides for a set of international standards and the US FDA/CBER 'Points to Consider' are seen as the benchmark in this field (CBER 1993).

A further consideration is the safety aspect of handling cell lines. The minimum standard to be applied in any cell culture laboratory is Category 2 containment although this is not suggesting any inherent hazard. Even though a risk assessment is made (see Section 4.1), it is a fact that in most cases the extent of characterization and thus information on the potential hazard of handling particular material is incomplete. This is especially true with respect to the presence of adventitious agents (e.g. viruses) in cell lines. There may be particular concern in the handling of patient material with regard to hepatitis/HIV/HTLV status and a balanced view on risk has to be taken. The topic is a large one (Stacey *et al.* 1998) and suffice to say that once minimum standards are set they can be all-embracing for every cell type handled.

Of fundamental importance in the routine handling of cell cultures is mycoplasma contamination status. If present, the concentration of mycoplasmas in the culture supernatant can be in the region of 10^6–10^8 mycoplasmas ml^{-1}. Unlike

Cell and Tissue Culture for Medical Research, edited by A. Doyle and J.B. Griffiths.
© 2000 John Wiley & Sons, Ltd

bacterial and fungal contaminants, they do not necessarily manifest themselves in terms of pH change and/or turbidity and they can be present in low numbers and often masked by blanket use of antibiotics. Mycoplasmas elicit numerous deleterious effects and their presence is incompatible with standardized systems. Routinely, broth and agar culture or Hoechst DNA stain are the methods of choice for detection, although increasingly polymerase chain reaction (PCR) methods are becoming available (Doyle & Bolton 1994). Tests have to be part of a regular routine and not just seen as 'one-off' procedures at the start of a piece of work. Elimination of contamination is possible but costly in time and resources, and is not always successful, so it is better to check early rather than later; this re-emphasizes the importance of authenticated cell banks to return to in case of contamination.

A further point that is often overlooked in the routine use of differentiated cell types is the need to set defined limits to the number of population doublings that cell should undergo *in vitro* before returning to the original stored stocks. Genetic drift/cell selection can lead to loss of differentiated functions and some preliminary studies may be necessary at the start of a project to set sensible and practical limits.

Finally, it must be emphasized that no amount of testing can replace the day to-day vigilance of laboratory workers routinely handling cells. Any alteration in normal growth pattern or morphology should not be ignored because this may well indicate a fundamental problem well in advance of other more formal testing parameters.

REFERENCES

Centre for Biologics Evaluation and Research (CBER) (1993) *Points to Consider in Characterization of Cell Lines Used to Produce Biologics.* US Food and Drugs Administration, Bethesda, USA.

Doyle A & Bolton BJ (1994) The quality control of cell lines and the prevention, detection and cure of contamination. In: *Basic Cell Culture: a Practical Approach,* pp. 243–271 IRL Press, Oxford, UK.

Doyle A & Morris CB (1993) Introduction to setting up a tissue culture laboratory. In: Doyle A, Griffiths JB & Newell DG (eds) *Cell & Tissue Culture: Laboratory Procedures,* pp. 1A:1.1–3. Wiley, Chichester.

Morris CB & Jones B (1993) Safety Aspects (Handling, Equipment.and Containment) In: Doyle A, Griffiths JB & Newell DG (eds) *Cell & Tissue Culture: Laboratory Procedures,* pp. 1A:2.1–15. Wiley, Chichester.

Stacey, G, Doyle A & Hambleton P (eds) (1998) *Safety in Tissue Culture.* Kluwer, London.

Stacey, GN, Parodi, B & Doyle A (1995) The European Tissue Culture Society (ETCS) initiative on quality control of cell lines. *Experiments in Clinical Cancer Research* 4: 210–211.

1.2 CRYOPRESERVATION: BASIC TECHNIQUES AND STORAGE

BASIC TECHNIQUES

A major achievement permitting the development of animal and human cell technology has been the determination of the parameters for routine cell cryopreservation. The importance of a stable, reliable, secure supply of material held at temperatures below –130°C cannot be overstated. At its most simple the technology is based on slow freeze and fast thaw, together with high protein concentration and the presence of an agent which increases membrane permeability. The other, perhaps obvious, ingredient is 'healthy' cells.

Preliminary considerations

The first consideration is whether glass or plastic ampoules should be used; both have their advantages and disadvantages. A correctly sealed glass ampoule will not permit entry of liquid nitrogen which might be the case with a screw-capped plastic ampoule. This is important in two respects: the presence of liquid nitrogen during rapid warming could risk the danger of explosion due to the rapid expansion of gas; liquefied gas is not sterile and could carry contaminating objects into the ampoule. However, glass ampoules are comparatively difficult to seal, and have to be thoroughly tested before freezing. An additional problem with glass is labelling; ceramic ink is needed which requires specialized equipment and a certain degree of expertise. For most users plastic ampoules are more convenient as they are pre-sterilized and will label with marker pens. In addition there is a commercially available heat-shrinking sheath which can be placed over the ampoules to minimize the risk of entry of liquid.

The means of freezing at a controlled rate is another major consideration. A controlled-rate freezing apparatus is considered by some as an expensive luxury. Nevertheless, successful freezing is virtually guaranteed (although it is always advisable to perform a viability check before discarding the live culture) and a chart recorder/printout of each freeze run can be kept. In those cases where the expense of specialized equipment cannot be justified, the alternative methods can be used, but some pre-validation of the methodology to be employed is strongly recommended.

Cell and Tissue Culture for Medical Research, edited by A. Doyle and J.B. Griffiths.
© 2000 John Wiley & Sons, Ltd

PROCEDURE: CRYOPRESERVATION USING CONTROLLED-RATE FREEZING APPARATUS

Materials and equipment

- Freeze medium, i.e. growth medium with 20% foetal bovine serum (FBS) and 9% dimethylsulfoxide (DMSO) – in some cases 5–10% glycerol may be used
- Glass or plastic ampoules with a capacity of 1.5–2.0 ml (Nunc, Corning, Falcon)
- Pipettes (1 ml and 10 ml) or a dispensing syringe (for larger volumes)
- Centrifuge tubes (10 ml)
- Dispensing vessel
- Sealing torch for glass ampoules
- Racks to hold the ampoules (preferably drilled aluminum)
- Programmable freezer (Planer Products Ltd, Sunbury-on-Thames, UK), two-stage freezer (Taylor-Wharton) or well-insulated polystyrene box
- Liquid nitrogen refrigerator
- Protective goggles, face mask and gloves

Note: Cell cultures must be healthy (confirmed by observation with an inverted microscope) and in log phase of growth. This can be ensured by using pre-confluent cultures at below maximum cell density and performing a medium change 24 h prior to freezing.

1. Resuspend adherent cells by the routine subculture method (see Section 1.6). Suspension cells can be taken directly from culture. It is essential to ensure that any trypsin used for removal of adherent cells is neutralized by the addition of medium containing serum to a volume at least equal to that of the enzyme.
2. Perform a viability count by Trypan Blue exclusion (see Section 1.3). Viability should be higher than 95% – a figure below 80% is not usually acceptable and such cultures should be rejected.
3. Centrifuge the cells with the minimum of *g* force necessary to form a pellet (e.g. 100 *g* for 5–10 min).
4. Decant the supernatant and resuspend the cells in the cryoprotectant solution at a final concentration of 4–10×10^6 cells ml^{-1}. Dispense the cells into the ampoules using a pipette. For larger volumes an automatic syringe may be used but in this case the cell suspension should be placed on a magnetic stirrer in order to prevent settling.

Note: If large volumes of cell suspension are to be frozen a problem can arise with bicarbonate-buffered medium due to increased alkalinity. It may be necessary to gas the mixture with 5–10% CO_2/air to prevent cell death at pH values above 7.7. Alternatively a medium formulation without bicarbonate can be used (e.g. Leibovitz's L-15). The use of zwitterionic buffers (e.g. HEPES) is not recommended as this may cause toxicity problems.

5. Seal glass ampoules (if used) using a propane/oxygen or acetylene/oxygen torch. Unless automated equipment is used, practice will be necessary to produce a perfect seal. After sealing, test the ampoules by immersion in a solution of 0.05% methylene blue at 4°C and holding for 40 min. Discard any ampoules containing blue dye. This holding period has some benefit in that it allows equilibration.

6. Hold plastic ampoules at 4°C for up to an hour before freezing if several batches are to be prepared. The equilibration time can aid cell survival.
7. Transfer the ampoules for cryopreservation to the freezing chamber of a programmable freezer. The freezer should be programmed to the manufacturer's instructions.

The following programme is offered as one suitable for a wide variety of cell types. However, it is essential to test several combinations of cooling rate to optimize the viability on resuscitation. The main consideration is to minimize the effect of the latent heat of fusion during phase transition, i.e. when liquid freezes (see Table 1.2.1).

Ampoules should be held at +4°C for 10 min in the cooling chamber, prior to starting the freeze run. This allows them to equilibrate with the reference ampoule.

It may be necessary to adjust the temperature range and rate of cooling of Ramp 2 to reduce heat generated during phase transition. The freeze medium will normally undergo this transition between –4°C and –8°C. Once the ampoules have reached –180°C transfer to a nitrogen storage vessel. When the ampoules reach –60°C, the cooling rate can be increased to –10°C min^{-1} until –180°C is achieved.

LONG-TERM STORAGE

Long-term storage of cell stocks requires good record-keeping and maintenance of an inventory with each removal logged to ensure adequate stock levels. This can best be achieved by the use of a microcomputer with appropriate software.

A wide range of liquid nitrogen refrigerators are available. Whatever the choice they must be alarmed to prevent accidental drying out. In addition the ampoules must be kept in more than one location. The safe-deposit facilities provided by culture collections are highly recommended.

The decision whether to store ampoules in the liquid or gaseous phase of nitrogen will depend on the need to guarantee absolute sterility of the ampoule contents or to maintain a constant temperature. Ampoules held permanently in liquid will maintain viability indefinitely. Those stored in the gaseous phase will be subject to temperature variations as the liquid level changes, and the refrigerator is opened for access. The best solution is to maintain stocks in both phases by maintaining refrigerators half full with liquid.

Table 1.2.1 Temperature range and cooling rate

Ramp	Starting temperature (°C)	Upper temperature (°C)	Rate of climb (°C min^{-1})
Ramp 1	Ambient	–10	–2
Ramp 2	–10	–30	–12
Ramp 3	–30	–60	–2
Ramp 4	–60	–180	–10

In the routine handling of cell stocks it is essential to avoid wide fluctuations in storage temperature. Frequent removal of the inventory system from the refrigerator with the associated variation in storage temperature will reduce viability. It is recommended that stocks are periodically checked for maintenance of viability (e.g. every 2–3 years).

SUPPLEMENTARY PROCEDURE: THAWING

1. Achieve rapid thawing by transferring ampoules directly to a 37°C water-bath. If an ampoule contains potentially hazardous material it is advisable to add 1–2% (w/v) chloramine-T to the water-bath.
2. Take care not to submerge the cap of the plastic ampoule in order to prevent contaminated water entering the ampoule. A simple method is to use a plastic rack designed to hold tubes of the same diameter (e.g. Nalgene plastics 500 series, Rochester, New York) and place in the correct depth of water. Alternatively a piece of foam with holes in it for the ampoules will float on the surface.
3. Once thawed, transfer to the sterile work area. Soak glass ampoules in 70% ethanol and allow to dry. A pre-scored ampoule can be snapped open using a tissue soaked in 70% ethanol. Unscored types must be cut with a diamond and then they can be soaked with 70% ethanol before opening. Plastic ampoules can be opened while gripped with a tissue soaked with 70% ethanol.
4. Using a pipette transfer the ampoule contents into either a flask or a centrifuge tube containing pre-warmed growth medium. Slow addition is recommended. If it is essential to remove the cryoprotectant at this stage, centrifugation should be at the lowest speed necessary to form a cell pellet. For cultures started directly from an ampoule the first medium change should be at 24 h unless recommended otherwise.

    Note: Start-up cell density is of critical importance. The cells must be encouraged to condition the medium as rapidly as possible and prevent a protracted lay phase. In the case of most suspension cultures a cell density of $3-6 \times 10^5$ cells/ml is recommended. For adherent cells a density of $2-5 \times 10^4$ cells/ml is normally appropriate.
5. Examine cells daily using an inverted microscope and subculture as soon as confluency or maximum cell density is reached. This is frequently indicated by a medium colour change to orange or orange–red, i.e. acid pH.

ALTERNATIVE PROCEDURE: MECHANICAL FREEZING AND TWO-STAGE FREEZING

Mechanical freezing

Cryopreservation using a –80°C or –130°C mechanical freezer is probably the least reliable of methods. However, low-temperature freezers are widely used to cryopreserve cell lines.

To achieve slow cooling, ampoules must be heavily insulated. A block of polystyrene with individual holes large enough to take ampoules should be prepared.

The polystyrene must be 1–2 cm deep all around the ampoule, with no air spaces. The insulation required will depend on the final temperature of the freezer. The ampoules are placed in the block, which is then located near the middle of the freezer and left overnight (16–24 h) before being transferred to a liquid nitrogen storage vessel.

Before using this method on a regular basis a series of tests should be made to check the cell viability after freezing. If it varies significantly from the viability of the cells prior to freezing, i.e. a drop of more than 15–20%, the insulation will have to be modified. This method is not recommended for preparing master or reference cell stocks.

Two-stage freezing

Ampoules are placed at a holding temperature of –20 to –40°C for up to 24–48 h and then transferred directly to –196°C. Similar results can be achieved for a small number of ampoules by using a small device which holds ampoules in liquid nitrogen vapour in the neck of a Dewar flask. This device has a mechanism which allows the ampoules to be held at adjustable heights above the nitrogen, thus providing a variable cooling rate. After a holding time of 10–20 min the ampoules are plunged into liquid nitrogen and then transferred to their final storage location.

Note: Both of these alternatives require preliminary experimental work to determine the optimum conditions.

In all cases the cells should be examined for viability following cryopreservation.

DISCUSSION

Safety considerations

During the handling of frozen ampoules a full face mask and gloves must be worn. Ampoules containing pathogens should be thawed on an outer sealed container. In order to prevent explosion of ampoules on removal from the liquid phase it is recommended that transfer to the gaseous phase is carried out 24 h prior to thawing. To reduce the risk of injury from exploding ampoules during thawing it is advisable to place ampoules in a perforated container before placing in a water bath.

Additional considerations

For certain specialized cell types it may be considered necessary to include partic- ular essential growth factors (e.g. epidermal growth factor (EGF) or interleukins) to the freeze medium in order to maintain surface receptor stability during cryo- preservation. Hybridomas often cause problems on revival due to instability over the initial culture period resulting in an increasing proportion of dead cells. When the level of cell debris becomes high, further cell death may be induced by the accumulation of toxic products. Starting cultures on macrophage feeder layers in high serum concentration (e.g. 20%) medium will provide rapidly conditioned medium and the macrophages will endocytose cell debris. BALB/c peritoneal macrophages taken from a single mouse will provide sufficient cells for several 24-well plates at a concentration of 2×10^4 cells well^{-1}.

The problem of pH stability can be further exacerbated during thawing by the tendency of CO_2 to dissociate from solution. At a time when membranes are at their most fragile it is important to avoid exposure to alkaline pH. An alternative approach is to use whole serum with the cryoprotectant. In addition to greater pH stability the higher protein concentration is protective. Finally, it is essential to use freshly prepared freezing mixture every time. DMSO should be obtained from a supplier who is able to offer the most recently prepared stocks available. Once received this should be filter sterilized (0.2 μm) through a filter specifically designed for DMSO (Nalgene) and stored in glass at –20°C. This will ensure that there are no problems due to oxidation of DMSO. In those cases where DMSO causes differentiation (e.g. the human promyelocytic leukaemia cell line HL60) glycerol should be used.

Continuity of supply

To ensure the continuity of supply of cells it is essential to calculate future needs. This can be done by estimating the number of ampoules required to complete a particular investigation.

A simple rule of thumb is to allow a frozen ampoule of cells to supply up to 30 passages and then replace these cells with a fresh ampoule. If it is assumed that passaging will occur twice a week, one ampoule can supply cells for about 3–4 months.

The reasons for not continuing a cell culture indefinitely are to prevent the possibility of microbial contamination, including mycoplasma, and to reduce genotypic and/or phenotypic drift from the original material.

To prepare adequate cell stocks ('cell banks') the original material should be cultured free of all antibiotics and then tested for microbial and mycoplasma infection (see Section 2.2). Cultures that are free from contamination should be cultured to provide enough cells to freeze between 5 and 10 ampoules. This is the 'seed stock' or 'master cell bank'. One ampoule from this bank should then be resuscitated, tested for viability and microbial contamination, and expanded to produce a 20–50-ampoule 'working cell bank'.

The number of passages from the original source should always be recorded where known.

Cultures for use in the laboratory should always be started from the working cell bank. When this bank is exhausted a new bank should be produced from a master bank ampoule. Detailed and accurate records of the production and usage of ampoules must be maintained.

LITERATURE REVIEW

After 200 years of experimentation with preservation of sperm it was Polge *et al.* (1949) who made the observation that glycerol enhanced the survival of fowl semen at –79°C. The technique was then transferred to several other cell types, including erythrocytes (Smith 1950). The next step was the discovery of the cryoprotective properties of DMSO (Lovelock & Bishop 1959). It was not until the work of Mazur (1970) that the mechanism of cryopreservation was better understood. He showed that cells cooled slowly in the presence of cryoprotectant shrink and do not contain intracellular ice.

Each cell has its own optimum cooling rate, which is determined by its water permeability, and the role of DMSO is to increase that membrane permeability. The reversal of intracellular ice formation, and consequent intracellular damage on thawing, is essential to successful cryopreservation (Mazur 1977).

The cell cycle is also important in cryopreservation, as shown in synchronized cultures of Chinese hamster cells (Koch *et al.* 1970) and Hela cells (Terasima & Yasukawa 1977). Cells frozen in late S phase recovered considerably faster than those in the G_2 phase.

REFERENCES

Ashwood-Smith MJ & Farrant J (eds) (1980) *Low Temperature Preservation in Medicine and Biology*. Pitman Medical, London.

Doyle A, Morris CB & Armitage WJ (1988) Cryopreservation of animal cells. In: Mizrahi A (ed.) *Advances in Biotechnological Processes*, vol. 7, pp. 1–17. Alan R Liss, New York.

Koch GJ, Kruuv J & Bruckschwaiger CW (1970) Survival of synchronised Chinese Hamster cells following freezing in liquid nitrogen. *Experimental Cell Research* 63: 476–477.

Lovelock JE & Bishop MWH (1959) Prevention of freezing damage to living cells. *Nature* 183: 1394–1395.

Mazur P (1970) The freezing of biological systems. *Science* 168: 939–949.

Mazur P (1977) The role of intracellular freezing in the death of cells at supraoptimal rates. *Cryobiology* 14: 251–272.

Polge C, Smith AU & Parkes AS (1949) Revival of spermatozoa after vitrification and dehydration at low temperatures. *Nature* 164: 1394–666.

Smith AU (1950) Prevention of haemolysis during freezing and thawing of red blood cells. *Lancet* II: 910–911.

Terasima T & Yasukawa M (1977) Dependence of freeze–thaw damage on growth phase and cell cycle of cultured mammalian cells. *Cryobiology* 14: 379–381.

1.3 HAEMOCYTOMETER CELL COUNTS AND VIABILITY STUDIES

In order to ensure that cell cultures have reached the optimum level of growth before routine subculture or freezing, it is helpful to obtain an accurate cell count and a measure of the percentage viability of the cell population. The most common routine method for cell counting which is efficient and accurate is with the use of a haemocytometer. There are several types on the market of which the improved Neubauer has proved most popular.

A thick, flat counting chamber coverslip rests on the counting chamber at a distance of 0.1 mm above the base of the slide. The base has rulings accurately engraved on it, comprising 1 mm squares, some of which are further divided into smaller squares.

When cell suspensions are allowed to fill the chamber, they can be observed under a microscope and the cells counted in a chosen number of ruled squares. From these counts, the cell count per ml of suspension can be calculated. Hybridoma cells and others which grow in suspension may be counted directly. Cell lines which are attached will need to be removed from the tissue culture flask by trypsinization. Since accuracy of counting requires a minimum of approximately 10^5 cells ml^{-1} it may be necessary to resuspend the cells in a smaller volume of medium.

To ensure that a cell culture is growing exponentially it is useful to know the percentage viability and percentage of dead cells and hence the stage of growth of the cells. This can be estimated by their appearance under the microscope, as live healthy cells are usually round, refractile and relatively small in comparison to dead cells, which can appear larger, crenated and non-refractile when in suspension. The use of viability stains such as Trypan blue ensures a more quantitative analysis of the condition of the culture. Trypan blue is a stain which will only enter across the membranes of dead/non-viable cells.

When a cell suspension is diluted with Trypan blue, viable cells stay small, round and refractile. Non-viable cells become swollen, larger and dark blue. Both the total count of cells/ml and percentage of viable cells can be determined.

PROCEDURE: HAEMOCYTOMETER CELL COUNT

Materials and equipment

- 0.4 g Trypan blue in 100 ml physiological saline. Pass through a 0.22 μm filter to remove any debris.

Cell and Tissue Culture for Medical Research, edited by A. Doyle and J.B. Griffiths.
© 2000 John Wiley & Sons, Ltd

- Haemocytometer with coverslip – Improved Neubauer British Standard for Haemocytometer Counting Chambers, BS 748:1963 (Figure 4B:1.1)
- Hand-held counter
- Microscope – low power, ×40–×100 magnification

⚠ Trypan blue is harmful if ingested or inhaled. It is irritating to the eyes, harmful by skin contact and has been found to cause cancer in laboratory animals. Appropriate precautions should be taken when handling Trypan blue and the use of an extraction hood and gloves is advised.

1. Thoroughly clean the haemocytometer and coverslip and wipe both with 70% alcohol before use.
2. Moisten the edges of the coverslip or breathe on the chamber to provide moisture before placing the cover slip centrally over the counting area and across the grooves.
3. Gently move the coverslip back and forth over the chamber until Newton's rings (rainbow-like interference patterns) appear, indicating that the coverslip is in the correct position to allow accurate counting, i.e. the depth of the counting chamber is now 0.1 mm.
4. Mix the cell suspension gently and add an aliquot to the Trypan blue solution (see Table 1.3.1). The dilution will depend on the cell concentration and may need to be adjusted to achieve the appropriate range of cells to be counted (see Step 6). Draw a sample into a Pasteur pipette after mixing thoroughly and

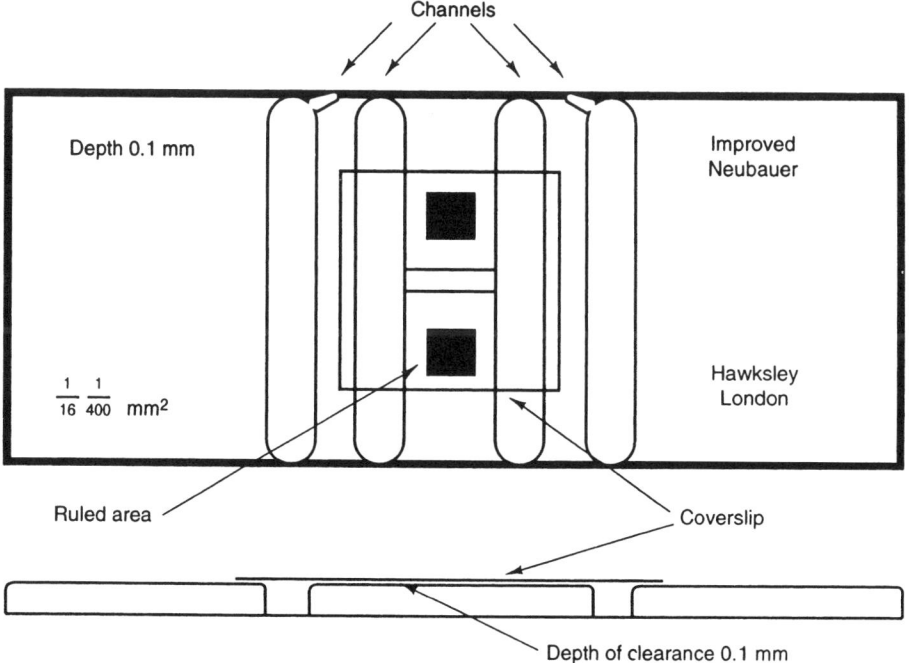

Figure 1.3.1 Improved Neubauer haemocytometer with coverslip.

allow the tip of the pipette to rest at the junction between the counting chamber and the coverslip. Draw the cell suspension in to fill the chamber. No pressure is required since the fluid will be drawn into the chamber by capillarity. Both halves of the chamber should be filled to allow for counting in duplicate.
5. Using a light microscope at low power, focus on the counting chamber.

Table 1.3.1 Examples of suitable dilutions

Cells	Trypan blue	Dilution
0.1 ml	0.1 ml	2-fold dilution
0.1 ml	0.3 ml	4-fold dilution
0.1 ml	0.9 ml	10-fold dilution

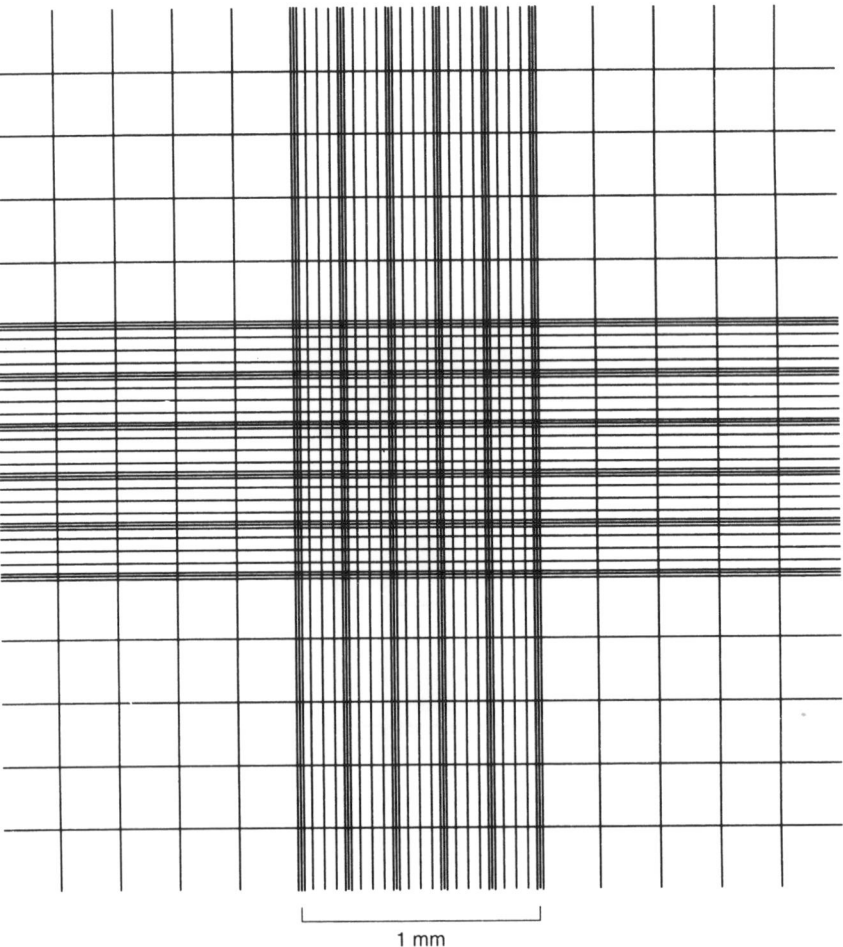

1 mm

Figure 1.3.2 Magnified view of the cell counting chamber grid. The central 1 mm square area N is divided into 25 smaller squares each 1/25 mm^2. These are enclosed by triple ruled lines and are further subdivided into 16 squares, each 1/400 mm^2 to ensure accurate counting of very small cells (e.g. erythrocytes).

6. Count the number of cells (stained and unstained separately) in 1 mm² areas (see Figure 1.3.2) until at least 200 unstained cells have been counted. As a rule, the cells in the left-hand and top grid markings should be included in a square, and those in the right-hand and bottom markings excluded (see Figure 1.3.3).
7. Count the viable and non-viable cells in both halves of the chamber.
8. Calculations.
    i. Total number of viable cells.
        $A \times B \times C \times 10^4$.
    ii. Total dead cell count.
        $A \times B \times D \times 10^4$.
    iii.To give a total cell count = viable cell count + dead cell count
    iv. % viability $= \dfrac{\text{Viable cell count}}{\text{Total cell count}} \times 100\%$

    where A = volume of cells
            B = dilution factor in Trypan blue (4B:1.1)
            C = mean number of unstained cells (i.e. unstained count no. of areas counted

    D is the mean number of dead/stained cells.
    10^4 is the conversion of 0.1 mm³ to ml.

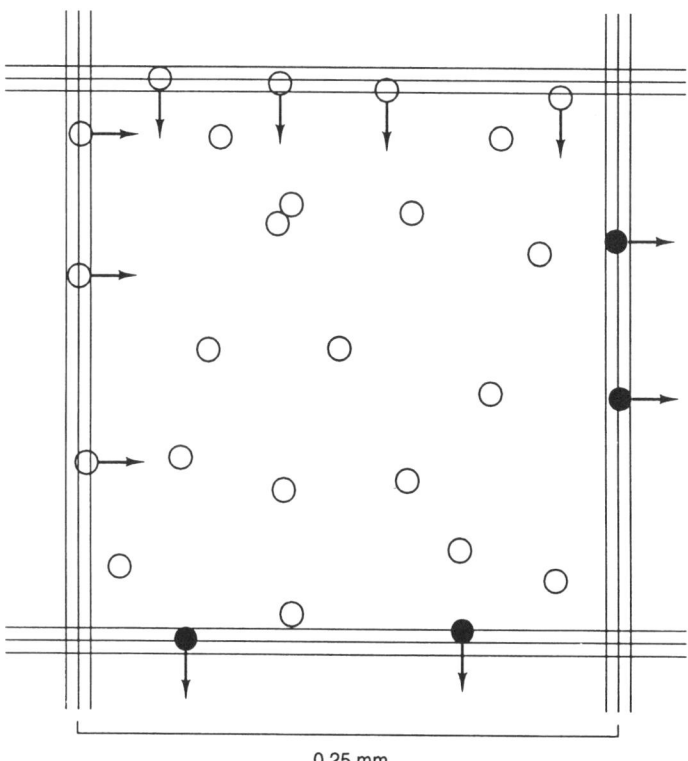

0.25 mm

Figure 1.3.3 Diagram indicating which cells are to be counted: ○, count; ●, count on next grid.

DISCUSSION

Troubleshooting

The counting chamber and coverslip should be scrupulously clean and without scratches. The clarity of the rulings on the chamber is vital and metallized glass improves both clarity and visibility.

Errors can be caused during sampling, dilution, mixing, and filling the chamber and by inaccurate counting.

The following can cause inaccuracy of the haemocytometer method by affecting the volume of the chamber.

- Overflowing the chamber and allowing sample to run into the channels/moat
- Incompletely filling the chamber
- Air bubbles or debris in the chamber

There is an inherent error in the method due to random distribution of the cells in the chamber. The error is reduced by increasing the number of cells to be counted and duplicating the counts.

Time considerations

It is essential to examine the cells within 5 min of immersing them in Trypan blue.

1.4 PROPHYLACTIC USE OF ANTIBIOTICS IN CELLS AND TISSUES WITH A HIGH RISK OF MICROBIAL CONTAMINATION

The inadvertent introduction of bacteria, fungi or mycoplasma species into cell or tissue cultures will result in the rapid overgrowth of the contaminating organism and the eventual destruction of the cultures. The prophylactic use of antibiotics where appropriate in conjunction with good aseptic technique can reduce considerably the chance contamination and subsequent overgrowth of bacteria, fungi and mycoplasmas.

> **Antimicrobial prophylaxis with antibiotics should be instigated only when there is a clear risk of microbial contamination. The establishment of cell lines from surgical or post mortem material and the use of cell lines for the isolation or viruses and other microorganisms from clinical samples are procedures which carry a high risk of microbial contamination and would therefore benefit from antimicrobial prophylaxis. Alternatively, avoiding microbial contamination during the routine subculture of established cell lines can best be achieved by the instigation of good aseptic techniques. An effective antibiotic regimen should not only control microbial contamination but also limit the indiscriminate use of antibiotics in the cell culture laboratory. This approach should limit the development of resistant bacteria and mycoplasma species.**

The majority of antibiotics used for controlling microbial contamination of cell or tissue cultures are effective against both Gram-positive and Gram-negative bacteria. Gentamicin and ciprofloxacin exhibit activity against mycoplasma species, and the polyenes, amphotericin B and nystatin are useful in preventing contamination with yeasts and fungi (See table 1.4.1). See Wiedemann & Atkinson (1991) for a comprehensive review of antimicrobial susceptibility.

The combination of penicillin G at a concentration of 10^5 U l^{-1}, streptomycin sulphate (100 mg l^{-1}) and amphotericin B (5 mg l^{-1}) has found wide application in preventing bacterial and fungal contamination in cell and tissue cultures. More recently gentamicin sulphate at a concentration of 50 mg/l has been used as an alternative to streptomycin sulphate because of its activity against *Pseudomonas* species. This combination has been particularly useful in reducing the incidence

Cell and Tissue Culture for Medical Research, edited by A. Doyle and J.B. Griffiths.
© 2000 John Wiley & Sons, Ltd

Table 1.4.1 Prophylactic use of antibiotics in cell and tissue cultures

Antibiotic	Antibiotic spectrum	Recommended concentration*
AMINOGLYCOSIDES		
Amikacin	Gram +ve and −ve bacteria	100.0 mg l^{-1}
Dihydrostreptomycin sulphate	Gram +ve and −ve bacteria	100.0 mg l^{-1}
Gentamicin sulphate	Gram +ve and −ve bacteria mycoplasmas	50.0 mg l^{-1}
Kanamycin sulphate	Gram +ve and −ve bacteria	100.0 mg l^{-1}
Neomycin sulphate	Gram +ve and −ve bacteria	50.0 mg l^{-1}
Streptomycin sulphate	Gram +ve and −ve bacteria	100.0 mg l^{-1}
Tobramycin sulphate	Gram +ve and −ve bacteria	50.0 mg l^{-1}
CEPHALOSPORINS		
Ceftazidime	Gram +ve and −ve bacteria	50.0 mg l^{-1}
Cephalothin	Gram +ve and −ve bacteria	100.0 mg l^{-1}
CHLORAMPHENICOL	Gram −ve bacteria	5.0 mg l^{-1}
GLYCOPEPTIDES		
Vancomycin hydrochloride	Gram +ve bacteria	50.0 mg l^{-1}
MACROLIDES		
Erythromycin	Gram +ve bacteria	100.0 mg l^{-1}
PENICILLINS		
Ampicillin	Gram +ve and −ve bacteria	100.0 mg l^{-1}
Potassium benzyl-penicillin	Gram +ve bacteria	10^5 U l^{-1}
Potassium phenoxy-methyl penicillin	Gram +ve bacteria	10^5 U l^{-1}
POLYENES		
Amphotericin B (deoxycholate complex)	Fungi and yeasts	2.5 mg l^{-1}
Amphotericin B (methyl ester)	Fungi and yeasts	10.0 mg l^{-1}
Nystatin	Fungi and yeasts	50.0 mg l^{-1}
POLYMIXINS		
Polymixin B sulphate	Gram −ve bacteria	50.0 mg l^{-1}
4-QUINOLONES		
Ciprofloxacin	Gram +ve and −ve bacteria mycoplasmas	10.0 mg l^{-1}
TETRACYCLINES		
7-Chlortetracycline hydrochloride	Gram +ve and −ve bacteria	10.0 mg l^{-1}
6-Dimethyl-7-chlortetracycline hydrochloride	Gram +ve and −ve bacteria	5.0 mg l^{-1}
5-Hydroxytetracycline	Gram +ve and −ve bacteria	5.0 mg l^{-1}
Tetracycline hydrochloride	Gram +ve and −ve bacteria	10.0 mg l^{-1}

* Paul (1975), Schanffner (1979), Gray & Brenwald (1991).

of bacterial contamination in cell cultures used for isolating viruses from clinical samples. However the use of gentamicin is not to be encouraged.

Several factors should be considered when determining the most appropriate antibiotic or combination of antibiotics for use in cell or tissue culture. The antibiotic's cytotoxicity and stability in cell culture medium and its spectrum of antimicrobial activity should be determined. Also, the types of contaminants that may be encountered and interactions with other antibiotics should be considered.

PROCEDURE: DETERMINING THE CYTOTOXICITY OF ANTIBIOTICS IN CELL AND TISSUE CULTURE

Antibiotics may be cytotoxic at concentrations approaching their effective antimicrobial levels. As cytotoxicity is cell dependent, it is necessary to determine the toxic concentration of any proposed antibiotic in the different cell types in current use.

The toxic effects of antibiotics reduces the uptake of the vital stain neutral red by the cells. When a dye-uptake assay is performed in microtitre plates, the dye can be eluted and its concentration measured colorimetrically (Gray & Brenwald 1991). The reduction in the uptake of neutral red by the antibiotic-treated cells, when compared to control cells grown in antibiotic-free medium, is a measure of the cytotoxicity of the antibiotic (see table 1.4.2).

Reagents and solutions

Neutral red phosphate-buffered saline

0.15% Neutral red in 0.1 M phosphate-buffered saline (PBS) adjusted to pH 5.5.

Table 1.4.2 Cytotoxic effects of streptomycin, gentamicin, vancomycin and ciprofloxacin on vero cell cultures determined in a dye-uptake assay

Antibiotic concentration (mg/l)	Streptomycin	Mean optical density Gentamicin	(510 nm) values Vancomycin	Ciprofloxacin
0.0	0.528	0.562	0.488	0.533
9.8	0.620	0.617	0.535	0.538
19.5	0.612	0.657	0.568	0.531
39.0	0.607	0.678	0.567	0.498
78.1	0.632	0.636	0.568	0.395
156.2	0.636	0.675	0.565	0.325
312.5	0.603	0.641	0.606	0.344
625.0	0.577	0.563	0.584	0.284
1250.0	0.497	0.412	0.657	NT
2500.0	0.206	0.183	0.682	NT
5000.0	0.136	0.157	0.389	NT
−3 SD from cell control mean	0.428	0.528	0.463	0.478

NT: not tested

0.1 M PBS adjusted to pH 6.5

Phosphate ethanol buffer
10% ethanol in 0.1 M PBS adjusted to pH 4.2.

Materials and equipment

- Neutral red
- PBS pH 7.3
- Antibiotic-free growth medium
- Spectrophotometer with 510 nm wavelength filter
- 96-well flat-bottomed tissue culture grade microtitre tray
- Ethanol

1. Add 100 µl cell suspension, at a concentration that will produce confluent monolayers of cells after 3 days, to each well of a 96-well flat-bottomed microtitre tray.
2. Prepare a fresh solution of the antibiotic at 100 times the proposed concentration in growth medium.
3. Perform doubling dilutions of the antibiotic from neat to 1 in 1024 and add 100 µl of each dilution to each of the eight wells in columns 1 to 11 of the microtitre tray.
4. Add 100 µl antibiotic-free growth medium to each of the eight wells in column 12 to act as cell controls.
5. Incubate the microtitre tray in an atmosphere of 5% CO_2 for 3 days.
6. After incubation carefully aspirate the medium.
7. Add 50 µl 0.15% neutral red in PBS pH 5.5 to each well of the microtitre tray.
8. Incubate the tray at 37°C for 60 min.
9. Carefully aspirate the stain.
10. Wash each well two times with PBS pH 6.5.
11. Add 100 µl phosphate/ethanol buffer to each well of the microtitre tray.
12. Shake for 30 s and read the optical density (OD) of each well at 510 nm wavelength in a spectrophotometer equipped to take microtitre trays.
13. Calculate the arithmetic mean and standard deviation (SD) of the ODS of the cell controls.
14. Calculate the mean OD at each concentration of antibiotic.

The antibiotic is cytotoxic at concentrations whose mean OD values show a reduction of > 3 SD from the mean OD of the cell controls (see table 1.4.2).

ALTERNATIVE PROCEDURE: CYTOTOXICITY TEST IN LYMPHOBLASTOID CELL CULTURES

The dye-uptake assay for determining cytotoxicity is difficult to perform with cells grown in suspension, e.g. lymphoblastoid cells. Determining the number of viable cells, growing in medium containing different concentrations of the proposed antibiotic, a haemocytometer chamber may be used as an alternative procedure.

Additional materials

- 200 ml cell suspension
- Trypan blue solution (0.4%)
- Haemocytometer chamber

1. Aliquot 200 ml cell suspension, grown for two passages in antibiotic-free growth medium, into 20 sterile universal bottles.
2. Centrifuge the cells at 1000 rev/min for 5 min and discard the supernatant.
3. Resuspend each of four aliquots of cells in 10 ml growth medium containing either no antibiotic, half the proposed concentration of antibiotic, the proposed concentration, five times the proposed concentration and ten times the proposed concentration and transfer to culture flasks.
4. Incubate the cell cultures for 3 days at 37°C in an atmosphere of 5% CO_2.
5. After incubation count the number of viable cells in each flask and record the mean value for each antibiotic concentration.
   Perform a viability count using a haemocytometer. See Section 1.3.
6. Resuspend the cells, discard 5 ml of the cell suspension and allow the remaining cells to settle.
7. Decant the medium and replenish with 10 ml fresh growth medium containing the appropriate antibiotic concentration.
8. Incubate the cell cultures for 3 days at 37°C in an atmosphere of 5% CO_2.
9. Continue with Steps 5, 6, 7 and 8 for ten passages or until a significant reduction in the number of viable cells in the presence of the antibiotic is detected.

A reduction in the mean viable cell count obtained from cells grown in the presence of the antibiotic by 3 SD or more from the mean viable cell count obtained from cells grown in antibiotic-free medium is significant.

STABILITY OF ANTIBIOTICS IN CELL AND TISSUE CULTURE MEDIUM

The effective concentration of any antibiotic used to prevent bacterial or fungal contamination must be maintained throughout the time the medium is in contact with the cells. The stability of antibiotics in culture medium, at temperatures required for cell growth, is variable. The antibacterial activity of the cephalosporin, ceftazidime, is reduced by 50% after incubation at 37°C for 1½ days whereas the aminoglycoside, gentamicin, and the 4-quinolone, ciprofloxacin, remain stable over 5 days. Therefore, it is important to determine the stability of any proposed antibiotic in culture medium at the optimum temperature for cell growth. Also, if antibiotic solutions are prepared in bulk for storage at –20°C, their stability should be determined at regular intervals to ensure their effectiveness.

ANTIBIOTIC ASSAYS

The effective concentration of an antibiotic in cell or tissue culture medium can be determined in a diffusion assay. The culture medium to be assayed as well as standard concentrations of the antibiotic are added to wells cut in an agar plate seeded with a bacterial species susceptible to the antibiotic under test. After incubation, zones of bacterial growth inhibition obtained with the standard antibiotic concentrations are used to construct a dose-response curve from which the concentration of the antibiotic under test can be determined (Reeves & Wise 1978).

PRELIMINARY PROCEDURE: ANTIBIOTIC SOLUTION

1. Prepare cell culture medium containing the antibiotic at the proposed working concentration.
2. Incubate the medium at 37°C for 5 days and assay four aliquots for antibacterial activity each day.

PRELIMINARY PROCEDURE: PREPARATION OF ASSAY PLATES

• Diagnostic sensitivity test (DST) agar (Oxoid)

1. Pour 225 ml DST agar, cooled to 50°C and adjusted to the correct pH (see Table 1.4.3), onto the 25×25 cm plate, ensuring the plate is flat and horizontal.
2. Dry the surface of the plate for 30 min at 37°C.

PROCEDURE: DIFFUSION ASSAY

Materials and equipment

• Susceptible reference organism (see Table 1.4.3)
• Diagnostic sensitivity test (DST) agar (Oxoid)
• Glass-bottomed assay plate (25×25 cm)
• Sterile cork borer (6–9 mm)

1. Flood the plate with a suspension of the cultured reference organism, diluted to produce semi-confluent growth when the assay is read.
   If the cell culture medium to be assayed contains more than one antibiotic the reference organism should be susceptible to the antibiotic under test and resistant to all other antibiotics contained in the cell culture medium.
2. Dry the surface of the plate for 30 min at 37°C.
3. Prepare standard concentrations of the test antibiotic in culture medium.
   These standards are best made up fresh and should be prepared from powders of known potency and not from therapeutic preparations. Standards should cover a range of concentrations from the lower limit of sensitivity of the diffusion assay to just above the original concentration of the test antibiotic.
4. Cut and remove equally spaced 6–9 mm diameter agar plugs from the plate with a sterile cork borer, taking care not to lift the surrounding agar.
   A maximum of 64 wells can be cut in a 25×25 cm plate.

Table 1.4.3 Stability of antibiotics in cell and tissue culture medium

Antibiotic	Organism for use in antibiotic assays	pH of antibiotic assay	Stability in cell culture medium (days)
AMINOGLYCOSIDES			
Amikacin	*B. subtilis*	7.8	≥5
Dihydrostreptomycin	NCTC* 8236 or		≥5
Gentamicin sulphate	*Kl. edwardsii*		≥5
Kanamycin sulphate	NCTC 10896		≥5
Neomycin sulphate			≥5
Streptomycin sulphate			3
Tobramycin sulphate			≥5
CEPHALOSPORINS			
Ceftazidime	*B. subtilis*	6.8	1
Cephalothin	NCTC 8236 or		3
	S. lutea		
	NCTC 8340 or		
	Staph. aureus		
	NCTC 6571		
CHLORAMPHENICOL	*S. lutea*	7.4	≥5
	NCTC 8340		
GLYCOPEPTIDES			
Vancomycin	*B. subtilis*	7.8	≥5
	NCTC 8236		
	Staph. aureus		
	NCTC 6571		
MACROLIDES			
Erythromycin	*B. subtilis*	7.8	3
	NCTC 8236 or		
	Staph. aureus		
	NCTC 6571		
PENICILLINS			
Ampicillin	*B. subtilis*	6.8	3
Potassium benzyl-penicillin	NCTC 8236 or *S. lutea*		3
Potassium phenoxy-methyl penicillin	NCTC 8340 or *Staph. aureus* NCTC 6571		3
POLYENES			
Amphotericin B (deoxycholate complex)	*Sacch. cerevisia* NCTC 10716	7.2	3
Amphotericin B (methyl ester)	*C. albicans* NCTC 3242		1
Nystatin			3
POLYMIXINS			
Polymixin B sulphate	*E. coli*	7.4	≥5
	NCTC* 10418		
4-QUINOLONES			
Ciprofloxacin	*E. coli*	7.4	≥5
	NCTC 10418		

Table 1.4.3 Continued

Antibiotic	Organism for use in antibiotic assays	pH of antibiotic assay	Stability in cell culture medium (days)
TETRACYCLINES			
7-Chlortetracycline hydrochloride	*B. cereus* NCTC 10320	6.6	1
6-Dimethyl-7- chlortetracycline hydrochloride	or *Staph. aureus* NCTC 6571		≥5
5-Hydroxytetracycline			3
Tetracycline hydrochloride			4

*NCTC – National Collection of Type Cultures, Central Public Health Laboratory, Colindale, UK.

5. Fill each of four wells to the brim with either the standard antibiotic solutions or the test samples, placing the standards and samples at random across the plate.
6. Incubate the plate for 18 h at 37°C.
7. After incubation, measure the diameter of each zone of growth inhibition and determine the mean of the four aliquots of each standard antibiotic concentration.
8. Construct a dose–response curve on semi-logarithmic graph paper, plotting the antibiotic concentration on the log scale and the mean diameter of the zones of inhibition on the arithmetic scale.
9. Calculate the mean diameter of the zones of inhibition obtained with the four test sample aliquots and determine the concentration of antibiotic by interpolation from the dose–response curve.

DISCUSSION

Alternative procedures

Non-microbiological methods such as enzyme immunoassay (EMIT, Syva) and fluoroimmunoassay (TDX, Abbott) are available commercially for assaying antibiotics. Although these assays are generally rapid and specific they require dedicated equipment and specialized reagents.

Special considerations

Although gentamicin has found wide-spread use in the control of *Pseudomonas* species in cell cultures used for virus isolation, a significant number of gentamicin-resistant organisms have been isolated from hospital patients. This is particularly true of patients who are immunosuppressed and require long-term antibiotic therapy. Gentamicin-resistant organisms such as coryneforms, faecal streptococci, coagulase-negative staphylococci and Gram-negative bacilli such as

Pseudomonas acidovorans, Xanthomonas maltophilia and *Enterobacter cloacae* have been shown to contaminate cell cultures grown in medium containing 50 mg l^{-1} gentamicin and inoculated with samples collected from immunosuppressed patients.

Antibiotics such as ceftazidime (50 mg l^{-1}) and ciprofloxacin (10 mg l^{-1}), which have broad-spectrum activity against Gram-negative bacteria, and vancomycin (50 mg l^{-1}), which inhibits most Gram-positive organisms, should be considered when cell cultures are to be inoculated with samples collected from immunosuppressed patients. Also, the replacement of penicillin and gentamicin by vancomycin and amikacin has been shown to reduce bacterial contamination in cell cultures used for virus isolation (Lo *et al.* 1996).

The combined action of two or more antibiotics can be different from that seen when the drugs are used singly. As well as synergy, when the antimicrobial action of two drugs exceeds that of one alone, and antagonism, when the activity of one drug is diminished in the presence of another, the cytotoxic effects of combined antibiotics may occur at a lower concentration than that determined for the individual drugs. Therefore, when combinations of antibiotics are used to prevent contamination of cell or tissue cultures the cytotoxicity, stability and effectiveness of these combinations must also be determined.

Antibiotics should be used in cell and tissue cultures to prevent the overgrowth of contaminating bacteria, fungi and mycoplasmas and should only be used to decontaminate cell or tissue cultures when they are irreplaceable. The use of antibiotics, which have the ability to penetrate cell membranes, should be avoided where cell lines are used to study bacterial pathogenesis, and performing one subculture of the cell line in antibiotic-free medium will not be sufficient to remove all traces of the antibiotic. Antibiotics have been reported to have inhibitory or stimulatory effects on the biochemistry processes of some cell types and their effects should be determined before these processes are studied (Martinez-Liarte *et al.* 1995 and Suzuki *et al.* 1997). It is essential that an effective antibiotic regimen is established so that the indiscriminate use of antibiotics, under ill-defined conditions, does not lead to the emergence of multiply resistant strains of bacteria, fungi or mycoplasmas.

REFERENCES

Gray JJ & Brenwald NP (1991) The use of antibiotics to control bacterial overgrowth of cell cultures used for diagnostic virology. *Journal of Virological Methods* 32: 163–170.

Lo JY, Lim WW, Tam BK & Lai MY (1996) Vancomycin and amikacin in cell cultures for virus isolation. *Pathology* 28: 366–369.

Martinez-Liarte JH, Solano F & Lozano JA (1995) Effect of penicillin-streptomycin and other antibiotics on melanogenic parameters in cultured B16/F10 melanoma cells. *Pigment Cell Research* 8: 83–88.

Paul J (1975) *Cell and Tissue Culture*, pp. 140–143. Churchill Livingstone, Edinburgh.

Reeves DS & Wise R (1978) Antibiotic assay in clinical microbiology. In: Reeves DS, Phillips I, Williams JD & Wise R (eds) *Laboratory Methods in Antimicrobial Chemotherapy*, pp. 137–143, Churchill Livingstone, Edinburgh.

Schaffner CP (1979) Animal and plant tissue culture decontamination. In: Maramorosh

K & Hirumi H (eds) *Practical Tissue Culture Applications*, pp. 203–214. Academic Press, New York.

Suzuki H, Shimomura A, Ikeda K, Furukawa M, Oshima T & Takasaka T (1997) Inhibitory effects of macrolides on inter-leukin-8 secretion from cultured human nasal epithelial cells. *Laryngoscope* 107: 1661–1666.

Wiedemann B & Atkinson BA (1991) Susceptibility to antibiotics: species incidence and trends. In: Lorian V (ed.) *Antibiotics in Laboratory Medicine*, pp. 962–1150. Williams & Wilkins, Baltimore.

1.5 ENZYMIC TECHNIQUES FOR TISSUE DISSOCIATION

The procedures used to isolate cells from tissue must be designed to maximize the yield of functionally viable and dissociated cells remaining after digestion. Cultures can be initiated from a variety of tissues, i.e. embryonic, normal and tumour. When a choice is possible, embryonic tissue is often the simplest source from which cell cultures can be derived. Alternatively, biopsy specimens may be used from animals and humans.

It may be necessary to perform some mechanical separation prior to the use of enzymes. This can be achieved by first finely cutting up tissues using sterile scalpels. Ideally pieces of tissue should be 1 mm^3 or smaller if possible. This maximizes the surface area available for enzyme digestion. Tissues that are to be dissected should be placed in serum-free medium to provide an isotonic bathing solution.

When cell cultures are produced from tissues of the respiratory, digestive and genitourinary systems, addition of antibiotics is advisable to prevent infection during the preliminary stages.

Before preparing cell cultures the following list of parameters should be addressed to decide the suitability of the prepared material.

1. Type of tissue.
2. Species of origin.
3. Age of the animal.
4. Dissociation medium used.
5. Enzyme(s) used.
6. Impurities in any crude enzyme preparation used.
7. Concentration(s) of enzyme(s) used.
8. Temperature.
9. Incubation times.

To achieve effective results the last six variable conditions must be defined empirically. Despite the widespread use of enzymes for tissue culture over the years, the mechanisms of action are poorly understood, so that the choice of one technique over another is often arbitrary.

Variations in methodologies are due largely to the complex and dynamic nature of the extracellular matrix and to the use of relatively crude and ill-defined enzyme preparations. Recent advances in understanding the constituents of the extracellular matrix and the purity and characterization of available enzymes allow a more directed approach to the dissociation of cells.

Cell and Tissue Culture for Medical Research, edited by A. Doyle and J.B. Griffiths.
© 2000 John Wiley & Sons, Ltd

TISSUE TYPES

The extracellular matrix is composed of a wide variety of proteins, glycoproteins, lipids and glycolipids, all of which can differ in abundance from species to species, tissue to tissue and with developmental age.

Epithelial tissue

In the adult, epithelium forms such tissues as the epidermis, the glandular appendages of skin, the outer layer of the cornea, the lining of the alimentary and reproductive tracts, peritoneal and serous cavities, and blood and lymph vessels (where it is usually referred to as 'endothelium'). Structures derived from extensions of the primitive gut, including portions of the liver, pancreas, pituitary, gastric and intestinal glands, are also composed of epithelial tissue.

Epithelial cells are usually packed so closely together that there is very little intercellular material between them. An extremely tight bond exists between adjacent cells, making dissociation of epithelium a difficult process.

Four types of intercellular bond are found on the lateral surfaces of adjacent cells, i.e. zonula occludens, zonula adherens, macula adherens and nexus. The first is more commonly termed the 'tight junction' because of the large number of fusion sites between adjacent cells.

The integrity of these bonds can be broken in the presence of the enzymes collagenase, trypsin and hyaluronidase; they can either be used simply or in combination. Addition of chelating agents, i.e. EDTA and EGTA, can enhance enzyme activity by depleting calcium in desmosomes (macula adherens).

Connective tissue

Connective tissue develops from mesenchymal cells and forms the dermis of the skin, the capsules and stroma of several organs, the sheaths of neural and muscular cells and bundles, mucous and serous membranes, cartilage, bone, tendons, ligaments and adipose tissue.

Connective tissue is composed of cells and extracellular fibres embedded in an amorphous ground substance and is classified as loose or dense, depending upon the relative abundance of the fibres. The cells, which may be either fixed or wandering, include fibroblasts, adipocytes, histiocytes, lymphocytes, monocytes, eosinophils, neutrophils, macrophages, mast cells, and mesenchymal cells.

There are three types of fibres – collagenous, reticular and elastic – although there is evidence that the former two may simply be different morphological forms of the same basic protein. The proportion of cells, fibres and ground substance varies greatly in different tissues and changes markedly during the course of development.

Collagen is a major component of connective tissue, e.g. skin, tendon, blood vessels. It is composed of rod-like fibres composed of repeat units of triple-helical polypeptides. Extensive cross-linking of the fibres makes collagen rather insoluble and therefore resistant to hydrolytic attack by most proteases.

Reticular and elastic fibres are generally less abundant, but can be more widely distributed in the tissues of young animals. Like collagenase they have a complex structure of repeating fibre subunits, and cross-linking in the case of elastic fibres. The three fibre types are embedded in a complex mixture of glycoproteins, e.g. hyaluronic acid, chrondroitin and keratin sulphate.

The extensive cross-linking of fibres and complexity of the surrounding matrix makes connective tissue very insoluble. To dissociate the tissue a combination of enzymes is necessary, including collagenase, pronase, hyaluronidase and elastase.

A brief description of these and other useful enzymes is given below, followed by a detailed methodology.

ENZYMES

Collagenase

Collagenase can be derived from mammalian tissue or certain bacteria, e.g. *Clostridium histolyticum*. Mammalian collagenases split collagen's triple-helical conformation at specific sites to yield uncoiled fibre fragments, i.e. gelatin. Collagenase from *Clostridium histolyticum* has a similar mode of action. This enzyme is more accurately referred to as clostridiopeptidase A.

Purified clostridiopeptidase A alone is usually inefficient in dissociating tissue due to incomplete hydrolysis of all collagenous polypeptides and its limited activity against high concentrations of non-collagen proteins and other macromolecules found in the extracellular matrix. The collagenase most commonly used for tissue dissociation is a crude preparation containing clostridiopeptidase A in addition to a number of other proteases, polysaccharidases and lipases. Crude collagenase is well suited for tissue dissociation since it contains the enzyme required to attack native collagen and reticular fibres in addition to the enzymes which hydrolyse the other proteins, polysaccharides and lipids in the extracellular matrix of connective and epithelial tissues.

Four basic types of collagenase are now identified and available commercially.

- **Type 1** contains average amounts of assayed activities (collagenase, caseinase, clostripain, and tryptic activities). It is generally recommended for fat cells, adrenal cells and liver cells.
- **Type 2** contains greater clostripain activity. It is generally used for heart, bone, muscle, thyroid, cartilage and liver.
- **Type 3** is selected for low tryptic activity. It is usually used for mammary cells.
- **Type 4** is selected because of low tryptic activity. It is commonly used for islets and other applications where receptor integrity is crucial.

Correlations between type and effectiveness with different tissues have been good, but not perfect, due in part to variable parameters of use.

Trypsin

Trypsin is a pancreatic serine protease with a specificity for peptide bonds involving the carboxyl group of the basic amino acids arginine and lysine. Trypsin is one of the most highly specific proteases known, although it also exhibits some esterase and amidase activity.

Purified trypsin alone is usually ineffective for tissue dissociation since it shows little selectivity for extracellular proteins. Combinations of purified trypsin and other enzymes such as elastase have proven effective for dissociation.

'Trypsin' is also the name commercial suppliers have given to pancreatin, a crude mixture of proteases, polysaccharidases, nucleases and lipases extracted from porcine pancreas. Crude 'trypsins' (like National Formulary (NF) 1 : 250 and 1 : 300) (a commonly used trypsin preparation) are widely used for dissociating tissues, perhaps because the tryptic and contaminating proteolytic and polysac-charidase activities do bring about a preferential attack of the extracellular matrix. It appears, however, that crude trypsin and crude collagenase dissociate tissues by different mechanisms, and difficulties are often encountered when using NF 1 : 250 preparations – the most common being incomplete solubility, lot-to-lot variability, and cell toxicity.

In tissue culture laboratories researchers used purified trypsin to release cells into suspension from monolayers growing on the interior surfaces of culture vessels.

Elastase

Pancreatic elastase is a serine protease with a specificity for peptide bonds adja-cent to neutral amino acids. It also exhibits esterase and amidase activity. While elastase will hydrolyse a wide variety of protein substrates, it is unique among proteases in its ability to hydrolyse native elastin, a substrate not attacked by trypsin, chymotrypsin or pepsin. It is produced in the pancreas as an inactive zymogen, proelastase, and activated in the duodenum by trypsin. Elastase is also found in blood components and bacteria.

Because elastin is found in highest concentrations in the elastic fibres of connec-tive tissues, elastase is frequently used to dissociate tissues which contain extensive intercellular fibre networks. For this purpose, it is usually used with other enzymes such as collagenase, trypsin and chymotrypsin. Elastase is the enzyme of choice for the isolation of Type II cells from the lung.

Hyaluronidase

Hyaluronidase is a polysaccharidase with a specificity for endo-N-acetylhexo-saminic bonds between 2-acetoamido-2-deoxy-β-D-glucose and D-glucoronate. These bonds are common in hyaluronic acid and chrondroitin sulphate A and C. Because these substances are found in high concentrations in the ground substance of virtually all connective tissues, hyaluronidase is often used for the dissociation of tissues, usually in combination with crude protease such as collagenase.

Papain

Papain is a sulphydryl protease from *Carica papaya* latex. It has wide specificity and it will degrade most protein substrates more extensively than the pancreatic proteases. It also exhibits esterase activity.

With some tissues papain has proved less damaging and more effective than other proteases. Huettner & Baughman (1986) described a method using papain

to obtain high yields of viable, morphologically intact cortical neurons from post-
natal rats.

Deoxyribonuclease I

Often, as a result of cell damage, nucleic acid leaks into the dissociation medium,
increasing viscosity and causing handling problems. Deoxyribonuclease will digest
the nucleic acids without damaging the intact cells.

Pronase

This protease is a bacterial enzyme derived from *Streptomyces frisens*. In addition
to gentle disaggregation of tissues it is often used as an alternative to trypsin for
subculturing cell lines. Because of its mild action, pronase may have to be used
in conjunction with another enzyme, e.g. collagenase.

Dispase

This is another bacterial enzyme, but obtained from *Bacillus polymyxa*. It is a
neutral metalloenzyme requiring calcium for activity. The enzyme is useful for
tissues of low hyaluronic acid content and is usually used in conjunction with other
enzymes for complete digestion.

EDTA

Ethylenediaminotetraacetic acid at a concentration of 0.02% (w/v) in calcium- and
magnesium-free phosphate-buffered saline (PBS) can be used in conjunction with
trypsin to enhance tissue disaggregation. Acting as a chelating agent it dissociates
intercellular links such as tight junctions. For this reason it is not suitable for use
with dispase.

Trypsin inhibitor (soybean)

The trypsin inhibitor from soybean inactivates trypsin on an equimolar basis.
However, it exhibits no effects on the esterolytic, proteolytic or elastolytic activities of porcine elastase.

PREPARATION AND STORAGE OF ENZYMES

Once diluted with medium or buffer, proteolytic enzymes can undergo autolysis.
Dissolve enzymes immediately before use. Special care must be taken with the
deoxyribonuclease. This product is very prone to shear denaturation. Mix gently.
   Reconstituted enzymes should not be stored at 2–8°C. If necessary they can be
aliquoted and frozen at –20°C. Avoid repeated freeze–thaw cycles.
   All enzymes, upon reconstitution, can be sterile filtered through a 0.22 μm
membrane. Generally all the enzymes, except trypsin, can be directly dissolved in

a balanced salt solution or buffer of choice. A stock solution of trypsin should be made initially by reconstituting the enzyme in 0.001 M HCl. This solution can be diluted into the digestion balanced salt solution (BSS) or buffer immediately prior to use.

A compilation of standard balanced salt solutions is given in Table 1.5.1. A review of the References can be helpful in selecting an appropriate dissociation solution.

DISCUSSION: OPTIMIZATION OF CELL DISSOCIATION

Although optimization of a cell isolation procedure for a particular cell type is dependent upon the adequate recovery of cells having various required characteristics, some guidelines can be established.

There is a complex relationship between cell yield and viability. In general there is an area of optimized recovery balanced between yield and viability; working near the middle of this range will reduce variability in the results of the cell isolation procedure.

The relative digestive powers of the commonly used enzymes are given in Figure 1.5.1.

Troubleshooting

• *Low yield/low viability* Over/under-dissociation, cellular damage. Change to less digestive enzyme and/or decrease working concentration (e.g. from trypsin to collagenase/from Type 2 collagenase to Type 1).

Table 1.5.1 Composition of selected balanced salt solutions[a,b]

	Ringer[c]	Tyrode[d,e]	Gey[f]	Earle[g]	Puck[h]	Hanks[i]	Dulbecco (PBS)[j,k]
NaCl	9.00	8.00	7.00	6.80	8.00	8.00	8.00
KCl	0.42	0.20	0.37	0.40	0.40	0.40	0.20
$CaCl_2$	0.25	0.20	0.17	0.20	0.012	0.14	0.10
$MgCl_2.6H_2O$		0.10	0.21			0.10	0.10
$MgSO_4.7H_2O$			0.07	0.10	0.154	0.10	
$Na_2HPO_4.12H_2O$			0.30		0.39	0.12	2.31
$NaH_2PO_4.H_2O$		0.05		0.125			
KH_2PO_4			0.03		0.15	0.06	0.20
$NaHCO_3$		1.00	2.27	2.20		0.35	
Glucose		1.00	1.00	1.00	1.10	1.00	
Phenol red				0.05	0.005	0.02	
Atmosphere	Air	Air	95% air/5% CO_2	95% air/5% CO_2	Air	Air	Air

[a]Amounts are given as grams per litre of solution. [b]In some instances the values given represent calculations from data presented by the authors to account for the use of hydrated or anhydrous salts. Ringer (1895). [d]Tyrode (1910). [e]Parker (1961). [f]Gey & Gey (1936). [g]Earle (1943). [h]Puck *et al.* (1958). [i]Hanks & Wallace (1949). [j]Phosphate-buffered saline. [k]Dulbecco & Vogt (1954).

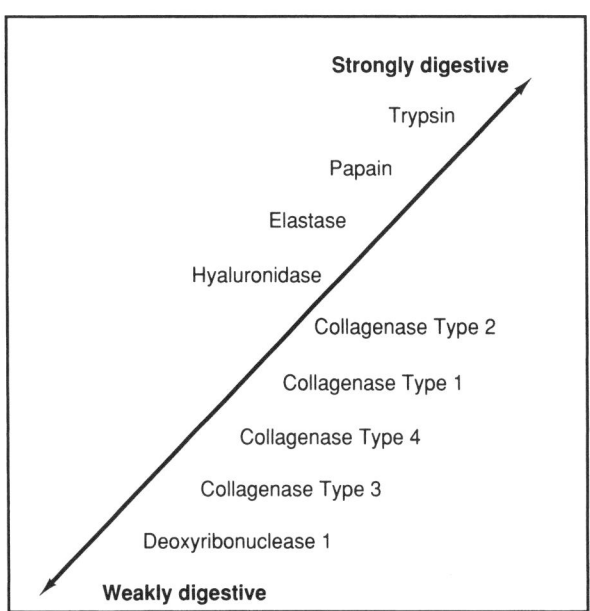

Figure 1.5.1 Enzyme digestion scale.

- *Low yield/high viability* Under-dissociation. Increase enzyme concentration and/or incubation time and monitor both yield and viability response. If yield remains poor, evaluate a more digestive enzyme and/or the addition of a secondary enzyme(s).
- *High yield/low viability* Good dissociation, cellular damage. Enzyme overly digestive and/or at too high a working concentration. Reduce concentration and/or incubation time and monitor yield and viability response. Try diluting the proteolytic action by adding bovine serum albumin (BSA) (0.1–0.5% w v^{-1}) or soybean trypsin inhibitor (0.01–0.1% w v^{-1}) to the dissociation. Try using less proteolytic enzyme although yield may be affected and should be monitored.
- *High yield/high viability* The place to be. Consider evaluating the effect of dissociation parameters to learn their limitations for future reference.

Optimization strategy

Set up proposed preliminary dissociation conditions based on known properties of enzymes, their working concentrations and buffer conditions. If the use of more than one enzyme is considered, optimize the concentration of the primary enzyme (the one at the highest relative concentration) before adding the secondary enzyme(s). For example, if considering collagenase 0.1% with DNase 0.01% or collagenase 0.075% with hyalurodinase 0.025%, optimize the collagenase concentration empirically before evaluating the effects of either the hyaluronidase or the deoxyribonuclease.

After optimizing the primary enzyme's concentration and incubation conditions, evaluate any secondary enzyme(s).

Initially vary the concentration of the primary enzyme approximately 50% relative to the referenced procedure(s). The above example of collagenase concentrations 0.1% and 0.075% suggests the evaluation of enzyme concentrations between 0.025% and 0.15%. The concentration increment should be evenly distributed to cover this entire range. As a result, incremental concentrations of 0.025%, 0.05%, 0.075%, 0.10%, 0.125% and 0.15% would be indicated. To simplify the initial screening the middle of the range can be selected and, after evaluation of yield and viability results, a decision can be made regarding the need for further studies. In this case initial collagenase concentrations evaluated may be 0.05%, 0.075%, 0.105 and 0.125%.

Note: Historically, most tissue dissociation and cell isolation procedures have cited the enzyme concentration used in terms of weight per unit volume (w v^{-1}). More recently, however, some researchers have begun to use the enzymes on an activity basis, that is, units per millilitre (u ml^{-1}). Use either method but consider the advantages and disadvantages of each.

- The traditional weight per unit volume method most probably resulted from the use of cruder, partially purified mixtures of enzymes and is used independently of any specific or contaminating activities which may be present. With some of these crude preparations the lot-to-lot variation can be significant, resulting in up to a two-fold difference in the amount of enzymic activity added on a weight basis.
- Adding by activity can result in a possible two-fold difference in the amount of weight added to a dissociation.

Both methods ignore the relative contaminant activity levels. Upon establishing a basic method, consider presampling different lots of enzyme(s) to evaluate these factors and to select a lot of enzyme which has minimal effect upon the critical parameters of a specific application.

Important: For accurate evaluation of a particular procedure's performance, cell yield and viability should be quantitated and compared. After optimizing basic dissociation and isolation conditions, the specific application parameters such as metabolic function(s) or receptor binding capability should also be evaluated. Based upon these results the method may be judged suitable for use or re-optimized for higher retention of native cellular characteristics.

Cell separation

In addition to the use of enzymes for establishing cell cultures, additional separation techniques may be necessary to obtain cells with the desired characteristic. Some of the more commonly used methods are briefly described below.

Sedimentation

The simplest method of separating different cell types is the use of a gradient at unit gravity. Most mammalian cells have a density around 1.06 g cm^{-3}. Elution from serum or Ficoll gradients has been used successfully to separate a wide variety of cell types, including bone marrow and smooth muscle (Wells 1982). Advantages of sedimentation are that inexpensive equipment can be used, although more expensive and sophisticated systems are available, and separation efficiency is high. However, this system has a limited capacity, i.e. 10^6–10^8 cells, and requires several hours to achieve maximum separation, during which time cell death can occur.

Centrifugation

The process of separation can be accelerated if a density gradient is used in conjunction with isopycnic (buoyant) centrifugation (Boyum 1984). A variety of commercially available non-viscous, non-toxic media can be used, e.g. Ficoll, Lymphoprep. Nycodenz, Metrizamide and Percoll. They are based on either dextran or silica. Sterile gradients are prepared in centrifuge tubes, usually within the range 1.015–1.120 g ml^{-1}; for example, mammalian lymphocytes separate at 1.086 g ml^{-1}. Separations are normally performed at 100–1000 *g.*

Using velocity centrifugation, good separations can be achieved of cells whose buoyant density lies at the extreme of the normal range, e.g. mast cells. This method has been widely used to separate leukocyte populations. Its main limitations are the degree of overlap between cell buoyancies and the relatively small numbers that can be processed.

This last limitation can be overcome by using a special rotor for centrifugal elutriation (Conkie 1986). The rotor is designed to accept a continual flow of cells for separation which are pumped through a separation chamber in which a gradient of flow rates is generated. Cells reach an equilibrium point in the chamber depending on their density. This can be seen through a viewing window. The cells are then collected by increasing the flow rate. Examples of cells separated by this method are embryonic rat (Schengrund & Repman 1979) and alveolar cells (Greenleaf *et al.* 1979). The main advantages of this method are speed and large cell capacity, i.e. up to 5×10^9 cells run^{-1}.

Fluorescence-activated Cell Sorter (FACS)

This technique utilizes the ability of cells to be separated according to their surface charge (Herzenberg *et al.* 1976). The desired population of cells is labelled with a specific marker, e.g. monoclonal antibody, to which is bound a fluorescent molecule. Fluorescent cells are detected by a laser beam and the information is then displayed on a screen, as a two- or three-dimensional graph. After graph coordinates have been set, labeled cells are collected by passing all the cells between two oppositely charged plates and collecting similarly charged cells as single drops, each containing one cell.

This system is in use in laboratories throughout the world, especially those specializing in studies of the immune system. It allows a very high efficiency of separation and cloned cells can be produced directly from the outflow. Its main limitations are that the maximum number of cells that can be separated is currently 3×10^4 cells s^{-1}, and cells which clump together cannot be used with a flow cytometer.

Magnetic beads

Specific antibodies can be bound to the surface of iron-containing beads using ligands, i.e. Dynabeads (Dynal). When antibodies against cell surface receptors, e.g. lectins, are used it is possible to selectively bind cell populations from disaggregated tissue. The beads with the attached cells can then be separated from the other cells by use of a magnet. The required population can then be obtained directly if a specific antibody is available, or by elimination if other cell types are removed.

Selective killing

Selection of the required cell type can sometimes be achieved by elimination of the remaining cells. Antibodies to specific cell surface markers, preferably IgM, are added to the cell mixture, followed by complement, which induces cell lysis. The method is limited in the availability of specific antibodies and the type of complement suitable for each cell type; for example, rabbit complement is often toxic to human cells.

Cloning

Once a purified or partially purified population of cells is established, a final step towards obtaining a homogeneous culture is cloning. The simplest method is limiting dilution (see hybridoma production), i.e. aliquots of medium contain single cells (McCullough & Spier 1990). Individual cells can be selected by the use of a micromanipulator. This is particularly useful with mixed populations of suspension cells in which different types can be identified by size or morphology. For attached cells a small metal ring, i.e. a cloning ring, can be placed around individual cells or small colonies to allow trypsinization of the desired cells (Freshney 1987). It is not possible to clone all cell types due to a specific requirement for growth factors.

REFERENCES

Boyum A (1984) Separation of lymphocytes, granulocytes, and monocytes from human blood using iodinated density gradient media. *Methods in Enzymology* 108: 88.

Conkie D (1986) Separation of viable cells by centrifugal elutriation. *Animal Cell Culture: A Practical Approach*. IRL Press, Oxford.

Dulbecco R & Vogt M (1954) Plaque formation and isolation of pure lines with poliomyelitus viruses. *Journal of Experimental Medicine* 199: 167–182.

Earle WR (1943) Production of malignancy in vitro. IV. The mouse fibroblast cultures and changes seen in the living cells. *Journals of the National Cancer Institute* 4: 165–212.

Freshney RI (1987) *Culture of Animal Cells: A Manual of Basic Techniques*. A.R. Liss, New York.

Gey GO & Gey MK (1936) *American Journal of Cancer* 27: 55.

Greenleaf RD, Mason RJ & Williams MC (1979) Isolation of alveolar type II cells by centrifugal elutriation. *In Vitro* 15: 673.

Hanks JH & Wallace RE (1949) Relation of oxygen and temperature in the preservation of tissues by refrigeration. *Proceedings of the Society for Experimental Biology and Medicine* 71: 196–200.

Herzenberg LA, Sweet RG & Herzenberg LA (1976) Fluorescence-activated cell sorting. *Scientific American* 234: 108.

Huettner JE & Baughman RW (1986) Primary culture of identified neurons from the visual cortex of postnatal rats. *Journal of Neuroscience* 6: 3044.

McCullough KC & Spier RE (1990) *Monoclonal Antibodies in Biotechnology: Theoretical and Practical Aspects.* Cambridge University Press, Cambridge.

Parker RC (1961) *Methods of Tissue Culture*, 3rd edn, p. 57. Harper, New York.

Puck TT, Cieciura SJ & Robinson A (1958) Long-term cultivation of euploid cells from human and animal subjects. *Journal of Experimental Medicine* 108: 945–955.

Ringer S (1895) *Journal of Physiology* (*London*) 18: 425.

Schengrund CL & Repman MA (1979) Differential enrichment of cells from embryonic rat cerebra by centrifugal elutriation. *Journal of Neurochemistry* 33: 283.

Tyrode MV (1910) *Archives Internationales de Pharmacodynamie et de Therapie* 20: 205.

Wells (1982) In Cell Separation (ed Pretlow TG and Pretlow TP) Vol 1, 169. Academic Press, London.

1.6 ROUTINE SUBCULTURING

When an attached cell line has either covered the surface available for growth or depletes the nutrients in the surrounding medium, it must be subcultured into new vessels, e.g. flasks or dishes. Attached cells will exhibit strong to very light adherence, depending on cell type. For example, a confluent monolayer of the dog kidney cell line, MDCK, adheres to surfaces very strongly and requires both a protease (i.e. trypsin) and a chelating agent (i.e. EDTA) to detach the cells. Conversely the Chinese hamster cell line (CHO) can be loosened by gentle tapping of the culture vessel, or pipetting off the monolayer. However, the majority of attached cell lines will require at least the addition of a protease to dissociate the cells from the growth surface.

Cells in suspension usually have very low levels of internal structural proteins, e.g. actin and vimentin, and lose some of their differentiated functions. However, suspension cell lines originally derived from attached cells will generally re-attach if allowed to settle, returning to their original morphology and structure.

Many tumour cell lines once established in culture do not produce attachment factors, and remain in suspension either as single cells or clumps. Examples of this cell type are lymphoblasts (e.g. Raji and hybridomas). These are termed static suspension cultures.

GENERAL

Each cell line must be assessed according to its risk of infection, i.e. biohazard risk. In the UK, hazardous substances, which include biological material, are covered by the Control of Substances Hazardous to Health (COSHH) regulations (1988).

To prevent either microbial infection or cross-contamination of cell lines, it is essential to maintain separate reagent bottles for each cell line, i.e. PBS, trypsin, medium. If only small quantities are required, aliquot the reagent into individual bottles. Mark each bottle in use with the cell line name and date of use.

Additionally, each cell line must be handled separately, always working with those which have been characterized first. Allow a 10–15 min 'clearance time' between working with different cell lines, making sure that the flow cabinet has been thoroughly decontaminated by wiping all internal surfaces, including the sides and glass screen, and that all reagents for the previous cell line are removed. If gloves are worn, these must also be changed. Hands must be washed after each subculture.

Animal cell lines which have not been fully characterized should be handled under at least containment level 2, unless a higher level is required, e.g. level 3 for hepatitis, HIV and SIV infected cell lines. For further information on handling of equipment and Containment see Section 1.1.

Cell and Tissue Culture for Medical Research, edited by A. Doyle and J.B. Griffiths.
© 2000 John Wiley & Sons, Ltd

PROCEDURE: SUBCULTURE-ATTACHED CELLS

Materials and equipment

- Trypsin 0.25% or trypsin 0.25% with 1 mM EDTA (pH 7.5–7.8)
- Trypan blue 0.4% (w/v) in phosphate-buffered saline (PBS)

1. Prewarm the cell culture medium and PBS to the usual culture temperature. Allow the trypsin or trypsin–EDTA solution to reach room temperature. Note that repeated warming will inactivate trypsin.
2. Examine the cultures microscopically for their level of confluence, morphological appearance and any signs of microbial contamination.
3. Decant the culture medium from the cells by either pouring or pipetting it off. If the medium is poured off (in the case of flasks) invert the culture so that the medium runs down the side of the flask opposite to the cells. Note that decanting rather than pipetting off medium can, however, increase the risk of contamination. It is essential, whenever possible, to delete antibiotics from the medium, in order to detect any infections at an early stage.
4. Wash the cells with a volume of Ca^{2+}/Mg^{2+}-free PBS equal to at least half the volume of growth medium. Decant the PBS. Repeat, if the cells are known to be strongly adherent.
5. Add trypsin or trypsin-EDTA to the cells, 1–2 ml/25 cm^2 flask, and gently spread over the entire surface by tilting the vessel. Leave inside the cabinet for 15–60 s, at room temperature, and then decant or pipette off most of the solution.
6. Place the cultures in an incubator at their normal growing temperature, e.g. 24°C for insect cell lines and 37°C for mammalian cell lines. The cells should detach after 2–10 min as seen by gently tilting the vessel and observing whether the cell layer moves. Gently tap the side of the vessel to increase detachment. Detached cells will be seen as a 'milky' suspension.
7. Microscopically examine the cultures to ensure that all the cells have detached. Reincubate if necessary.
8. Collect the cells by pipetting culture medium over the surface of the vessel, i.e. 2–5 ml/25 cm^2. Mix carefully to disperse the cells into a single-cell suspension. Measure the total suspension volume. Note that trypsin activity is neutralized by serum proteins. It is important to ensure that the volume of serum in the culture medium of the new culture vessels is equal to, or exceeds, the volume of trypsin in the vessel subcultured. Cells will not attach if the trypsin is not neutralized.
9. At this point the cells can be counted. When culturing a new cell line it is advisable to make a viable cell count to estimate the maximum cell density achieved, i.e. cell cm^{-2}, and the expected viability.
10. Transfer the cell suspension into prepared culture vessels by inoculating the appropriate volume of the cell suspension. For routine subculture a split ratio is used (1 : 2, 1 : 3 etc.). This is calculated as the ratio of surface areas; for example, a 75 cm^2 flask subcultured into 225 cm^2 has a split ratio of 1 into 3 (1 : 3), i.e. three 75 cm^2 or one 225 cm^2 Roux flasks. Alternatively, incubate the cells at a specific density, e.g. 2×10^4 cells cm^{-2}. First calculate the total

number of cells required, which is total surface area (cm²) × cell density required; and then the volume of cell suspension needed to inoculate the new vessels, which is total cells required/(cells/ml). For example, to inoculate a 75 cm² flask at 2×10^4 cells cm⁻² will require:

$$75 \times 2 \times 10^4 = 150 \times 10^4 \text{ cells } (= 1.5 \times 10^6)$$

If the suspension has 0.75×10^6 cells/ml, each new 75 cm² flask will require:

$$\frac{1.5 \times 10^6}{0.75 \times 10^6} = 2 \text{ ml cell suspension}$$

11. Place the inoculated vessels in an incubator at the correct temperature. If a CO_2 incubator is used leave culture flask caps loose.

Check the cultures after a few hours for cell attachment and pH. If the cultures are too acid (yellow), or alkaline (purple), the CO_2 concentration must be immediately checked and adjusted. To prevent the loss of cultures due to a failure in the supply of CO_2, keep the flask caps tightened after the cultures have equilibrated, i.e. 1–2 h.

The duration between subcultures will depend on the incubation temperature and cell type. The majority of mammalian cell lines require subculturing every 3–7 days. If the duration is longer than 5 days, change the medium of the cultures every 3–4 days.

ALTERNATIVE PROCEDURE: SERUM-FREE CULTURES

Any trypsin present at the time of subculture must be removed or neutralized when cells are grown in serum-free medium.

1. After trypsinizing cultures (Steps 1–7 above) collect the cells with fresh culture medium, i.e. 3–5 ml/25 cm² and transfer into a screw-top centrifuge tube.
2. Centrifuge the cells at 80–100 g for 5 min.
3. Remove the supernatant and resuspend the cells in fresh medium.
4. Recentrifuge the cells at 80–100 g for 5 min.
5. Remove the supernatant, resuspend the cells in a small volume of fresh medium and distribute into new vessels.

Alternatively, soybean trypsin inhibitor (Sigma) can be added to the trypsinized cells at a concentration of 1 mg/1–3 mg trypsin. After 5 min at room temperature the cells are centrifuged, the supernatant removed and the cells distributed into new vessels.

ALTERNATIVE PROCEDURE: MECHANICAL REMOVAL OF CELLS

Certain types of cell line may be harmed if exposed to proteases, e.g. NCTC 2071, L-M. Alternatively, some studies of surface membrane components or membrane transport, e.g. endocytosis, cannot be performed in the presence of proteolytic enzymes. In such cases the collection or subculture of the cells requires mechanical dislodgement from the culture vessel surface. The simplest method is to use a metal or glass rod bent at right angles and covered in a piece of silicon rubber tubing (rubber policeman). Sterilize by autoclaving. Rubbing the flat edge of the angle across the cell sheet will remove the cells in clumps; inevitably some cells will be destroyed due to mechanical stress.

When the cells are to be subcultured they should be collected in fresh medium and gently centrifuged. Resuspend the pellet in a small volume of medium and mix with a fine-tipped pipette before inoculating new vessels.

DISCUSSION: ALTERNATIVE PROTEASES

When cell detachment is slow or incomplete, or cell viability is low after detachment, it may be advisable to use an alternative protease to trypsin, e.g. pronase and dispase (Boehringer Mannheim Biochemicals). Detachment is usually very rapid, and frequently with less damage to the cells. However, the enzymes must be removed by centrifugation as they cannot be neutralized with serum. As all enzymes have an optimum pH range for their reaction with specific substrates, a small sample of each batch should be tested before use.

Slow cell detachment and aggregation

Trypsin in solution gradually loses activity at 4°C. Repeated warming and cooling increases the loss of activity. Cell aggregation or 'clumping' is due to surface membrane adhesion of carbohydrate components produced during protease activity. To obtain a mono-dispersed culture resuspend the cells in the smallest volume possible and mix with a fine-tipped pipette (glass Pasteur or sterile yellow Eppendorf tip). If this is not successful, use EDTA ($0.2 \, gl^{-1}$) dissolved in 0.85 g NaCI/1 or pronase to detach the cells.

Note: When EDTA is added to trypsin the pH should be checked and readjusted to pH 7.5–7.8 using either 0. 1 M NAOH or 0. 1 M HCI.

Split ratios and population doublings

For convenience most attached cell lines are subcultured by dividing the contents of a vessel into an exact number of new vessels. However, if it is necessary to record the population doublings (PDL) this can he calculated as follows:

2-fold split (1: 2) = 1 PDL: 4-fold split (1: 4) = 2 PDL: 8-fold split (1: 8) = 3 PDL: 16-fold split (1: 16) = 4 PDL

This estimate is based on new cultures reaching the same cell density as the culture from which they were derived.

Cell lines vary in their plating efficiency, i.e. the number of cells in a viable population which will attach and grow. If the plating efficiency of a cell line is below 75% care should be taken not to use high split ratios which may reduce the optimum growth rate, due to a prolonged lag phase. The plating efficiency of cells in serum-free culture will usually be lower than those in serum-containing medium.

If a more accurate record of PDL is required this can be calculated using the following equation:

$$PDL = \frac{\log_{10} \text{ cell count at harvest} - Lo_{10} \text{ cell count at inoculation}}{0.301}$$

For example, if the number of cells inoculated $= 4 \times 10^6$ and the number of cells harvested $= 18.5 \times 10^6$ then:

$$PDL = \frac{1.267 - 0.602 = 2.209}{0.301}$$

Note: It is necessary to adjust the cell counts to the same power of ten, i.e. in the example above both counts are expressed at 10^6 and not 4×10^6 and 1.85×10^7.

PROCEDURE: SUBCULTURE – SUSPENSION CELLS

Subculture

1. Visually examine cultures for signs of high-density growth, i.e. cell turbidity and acid pH of the medium (orange-yellow when phenol red is included in the medium). Static cultures should also be examined microscopically. Healthy cells will appear refractile, whereas dying cultures show cell lysis and shrunken cells. Check for signs of microbial contamination.
2. Remove a sample of cells by pipette for a viable cell count, ensuring that the culture is thoroughly mixed. If the cells have a tendency to clump, mix the sample again after removal using a fine-tipped pipette. To obtain a statistically accurate cell count a minimum of 200–300 cells should be removed for counting. The accuracy is increased slightly if 34 times this number are counted.
3. Perform a viable cell count (see Section 1.3).
4. Calculate the volume of new culture(s) required and prepare the culture vessels with medium. Where appropriate gas with 5% CO_2.
5. Centrifuge (80–100 g for 5 min) the number of cells needed to inoculate the new vessels i.e. number of cells/ml × number of ml of culture medium. For example, if the inoculating density is to be 2×10^5 cells ml^{-1} and the new culture volume 200 ml, then the number of cells needed is $2 \times 10^5 \times 200 = 400 \times 10^5$ ($= 40 \times 10^6$).
6. Calculate the volume of culture to be centrifuged by dividing the total number of cells required by the number of cells/ml (culture density); for example, if the culture density is 9.5×10^5 cells/ml the volume of culture required for 40×10^6 cells is:

$$\frac{400 \times 10^5}{9.5 \times 10^5} = 42.1 \text{ ml}$$

7. After centrifugation decant the supernatant, resuspend the cell pellet in a small volume of medium and inoculate into the new vessels. The majority of suspension cells can be maintained in the concentration range 105–106 cells ml^{-1}. However, when working with a new cell line keep the cultures within a narrower range, e.g. 2–8 × 105 cells ml^{-1}, until their optimum density range can be determined. Once a growth curve is established, maintain a density range which allows subculturing every 35 days. Shorter periods are inconvenient, and longer times between subculturing may result in nutrient exhaustion. Exceptions are slow-growing cells which must be maintained at lower densities, e.g. the mouse T-cell (CTLL).

8. Regas culture vessels with 5% CO_2. This is especially important when using mechanical stirring vessels as the pH of the medium has a tendency to rise in this type of vessel.

9. Set the stirring speed of stirring vessels so as to just maintain the cells in suspension (20–100 rev min^{-1} for Techne vessels and 30–150 rev min^{-1} for Bellco vessels). Speed is increased with volume.

ALTERNATIVE PROCEDURE: SUBCULTURE BY DILUTION

It is possible to maintain many suspension cell lines by simple dilution, i.e. removing the required number of cells directly from the culture and adding them to fresh medium in a new culture vessel. This has the advantage that some conditioned medium is carried over and can reduce the lag phase of new cultures. This method is especially convenient when working with large volumes.

However, after a few subcultures the level of exhausted medium may increase, due to increasing carry-over of old medium at each passage. This will be evidenced by a steady decline in cell viability. If subculturing by dilution continues, cell viability will eventually decline to a point where dead cells exceed viable cells.

When the cells are over-diluted to overcome this problem, they may die due to an inability to rapidly condition the medium. Alternatively, surviving cells may undergo selection pressures, resulting in altered characteristics. The solution to this is to remove the cells from the exhausted medium by centrifugation, and follow the subculture procedure in Steps 2–9 above.

DISCUSSION: SERUM-FREE CULTURES

If low-serum or serum-free medium is used with mechanically stirred cultures, it is especially important to control the culture pH, since the buffering capacity of such medium is considerably reduced. Bioreactors usually include this facility but with simple vessels continued monitoring and addition of CO_2 is not always possible.

Check cultures daily and regas when necessary. Reduction of the bicarbonate in the medium may facilitate cell growth as may the addition of zwitterionic buffers, e.g. HEPES.

REFERENCE

Health and Safety Executive (1988) *Control of Substances Hazardous to Health. Approved Codes of Practice*, HMSO, London.

CHAPTER 2

SPECIALIZED TECHNIQUES

2.1 CELL QUANTIFICATION – AN OVERVIEW

To measure performance, to achieve reproducibility, or to make comparative studies, a means of quantifying the cell population is needed. The range of methods allow one to obtain quantitative measures of viable and non-viable cells (haemocytometer), viable and dividing cells only (colony-forming efficiency), viable (MTT and Neutral Red), and loss of viability (LDH). The full range of methods are well documented in books on cell culture (e.g. Doyle *et al.* 1993; Doyle & Griffiths 1998; see also Background reading). Classically, direct counts of cell numbers using a microscopic counting chamber (haemocytometer), usually in conjunction with a vital stain (e.g. Trypan blue) to distinguish viable and non-viable cells, is used. However, all vital stains are subjective and cannot give absolute values, and cell numbers take no account of differences in cell size/mass. The method is simple, quick and cheap, and requires only a small fraction of the total cells from a cell suspension.

Automation of cell counting (total cells) is possible with electronic counters, especially for non-clumping single-suspension cells. A method which is widely used for total cell numbers is counting cell nuclei after dissolving the cytoplasm (Sanford *et al.* 1951). This is particularly useful for large clumps of cells, where cells are inaccessible (e.g. in matrices), or where cells are difficult to trypsinize off substrates (e.g. microcarriers). If determination of viability is the prime consideration then the most accurate method is to count the cells which have the ability to divide by the mitotic index method.

If cell mass, rather than number, is the important factor then a cell constituent has to be measured. The options are dry weight (large numbers required for accuracy), protein (probably the best parameter), DNA, protein nitrogen, or lipids. DNA is widely measured with spectrophotometric techniques based on the Hoechst 33258 stain (Doyle *et al.* 1993).

It may not be possible to perform a direct cell count, e.g. during a monolayer culture or for cells immobilized in matrices, so an indirect measurement has to be carried out. Examples are glucose or oxygen consumption rates which can be converted to cell numbers from predetermined standard curves. Other metabolites include lactic and pyruvic acid and carbon dioxide. Again there are reservations about such methods as these metabolic rates are not constant throughout the growth cycle of a specific cell and may also be influenced by changes in the culture which are the subject of the investigation. However, they do provide a good indication of the viable cell mass in a culture, if not an absolute value, and are widely used to give growth yields or specific utilization rates. The advantage is, of course, that the method is non-destructive and many of the parameters have

Cell and Tissue Culture for Medical Research, edited by A. Doyle and J.B. Griffiths.
© 2000 John Wiley & Sons, Ltd

been developed into on-line measurements for cell growth in bioreactors. A useful parameter being increasingly used is lactate dehydrogenase activity released into the medium. This is released by dying/dead cells and therefore gives a quantitative measurement of the non-viable population. This method can be modified to measure viable cell mass by a reverse titration procedure; that is by controlled lysis of the cells and measuring the increase in enzyme activity.

These biochemical measurements are best used over a time-course in culture so that successive readings will show a definite trend in the culture dynamics, e.g. stationary, growing or dying. The rule is not to rely on one method alone, but to combine two or more. Mention should be made of the use of radioisotopes to measure rates of synthesis (e.g. protein, DNA and RNA synthesis). However, these are usually kept for specific situations rather than general use.

There are many specific methods available which are suitable for a certain set of conditions, or a particular experimental aim. These include radiochromium labelling (radiochromium binds to viable protein and is released on cell damage/death), cellular ATP, intracellular volume (by ratios of radiochemicals), specific stains (e.g. MTT (3-(4,5-dimethylthiazol-2-yl)-2,5-diphenyl tetrazolium bromide), triphenyltetrazolium chloride, neutral red) redox potential, turbidity, calorimetry (Griffiths 1985), and biomass monitors (e.g. Aber Instruments Ltd).

HAEMOCYTOMETER CELL COUNTS

The most common routine method for cell counting which is both efficient and accurate is to use a haemocytometer. This is a counting chamber with 0.1 mm depth and squared graduations for ease of counting. Using a vital stain, such as Trypan blue, a viable and non-viable count can be made. Accurate sampling, diluting and filling the chamber are essential. Overfilling, inclusion of air bubbles and incomplete cleaning of the chamber cause errors. Inherent statistical errors are reduced by counting enough cells in duplicate determinations. However haemocytometer counting is the simplest and most versatile (viable and non-viable) method, with the advantage of giving a direct measurement (i.e. actual cells).

COLONY-FORMING EFFICIENCY

This is a true test of viability in that it depends upon a single cell being able to divide enough times to form a visible, countable colony, usually after a minimum of five or six doublings to give colonies of at least 50 cells (Doyle *et al.* 1993; Doyle & Griffiths 1998). The ability to form colonies is a measure of reproductive integrity and is measured as colony-forming efficiency (CFE). This is often referred to as Plating Efficiency (PE) when colonies are grown on plastic culture devices. As well as measuring the proliferative capability of a population the assay is widely used to determine the effect of cytotoxic agents The method can be applied to colony formation in soft agar and on culture surfaces with or without feeder cells. A problem to be overcome is that dilution to isolate single cells is stressful to the cell, thus additional feeding strategies such as cell feeder layers may have to be used. Also it takes many days to complete the assay.

MTT ASSAY

This viability assay depends upon the activity of an enzyme which can be measured colorimetrically (Doyle *et al.* 1993; Doyle & Griffiths 1998). It is quick, sensitive, accurate and large numbers of samples can be tested as it is an automatable method with a scanning spectrophotometer. It measures viable cells whether dividing or non-dividing, and also metabolic activation or inhibition of cells. The assay is based on the capacity of mitochondrial dehydrogenase enzymes to convert the yellow water-soluble substrate MTT into dark-blue formazan product which is insoluble in water. The amount of formazan product is directly proportional to the viable cell number in a variety of cell types (Mosmann 1983). The results are consistent with those obtained from the tritiated [3H]-thymidine uptake assay. Care is needed with preparing standard curves.

LDH ACTIVITY

This method is based on the concept that dead cells release LDH into the medium which can be measured in a spectrophotometric method. It is thus a quantitative measure for the loss of cell viability and is a response to cell viability and not cell metabolism (Wagner *et al.* 1992). It can also be used as a measure of total cells in a reverse titration procedure (lyse all the cells artificially and measure total LDH) (Racher *et al.* 1990). Problems are that stability of LDH is variable depending upon cell type, and release from cells is also variable depending upon the extent of membrane damage through to complete cell lysis. Culture conditions also affect levels of intracellular LDH.

NEUTRAL RED (NR) ASSAY

Neutral Red is a supravital dye that accumulates in lysosomes of viable cells, and on extraction can be measured spectrophotometrically (Babich & Borenfreund 1992; Doyle *et al.* 1993). The method is principally aimed towards cytotoxicity assay as it is simple, fast, sensitive, economical and highly reproducible. It is best to minimize or avoid serum in the medium.

CHOICE OF METHOD

In conclusion, there is no ideal method and the nature of the study often determines the parameter to be measured. It is best not to rely on one method alone as a combination of several methods will give a far more accurate picture (e.g. cell counts for numbers, cell protein for mass will show growth even if no division is occurring). Also account must be taken of the fact that all viability determinations will stress the cell (e.g. dilution, mixing, time factor, trypsinization etc). However, as most determination are to show relative changes, or specific metabolic activities, the inaccuracies should be remembered rather than be a concern, so that results are not over-interpreted.

REFERENCES

Babich H & Borenfreund E (1992) Neutral red assay for toxicology *in vitro* In: Watson RR (ed.) *In Vitro Meth*ods *in Toxicology*, Chapter 17, pp. 237–251. CRC Press, Boca Raton, Florida.

Doyle A, Griffiths JB & Newell D (eds) (1993*) Cell and Tissue Culture: Laboratory Procedures.* Wiley, Chichester.

Doyle A & Griffiths JB (eds) (1998) *Cell and Tissue Culture: Laboratory Procedures in Biotechnology.* Wiley, Chichester.

Griffiths J B (1985) Cell biology; experimental aspects. In: Spier RE & Griffiths JB (eds) *Animal Cell Biotechnology*, Vol 1, pp. 49–8. Academic Press, Orlando, FL.

Mosmann T (1983) Rapid colorimetric assay for cellular growth and survival: application to proliferation and cytotoxicity assays. *Journal of Immunological Methods* 65: 55–59.

Racher AJ, Looby D & Griffiths JB (1990) Studies on monoclonal antibody production by a hybridoma cell line (CIE3) immobilised in a fixed bed, porosphere culture system. *Journal of Biotechnology* 15: 129–146.

Sanford KK, Earle WR & Evans VJ (1951) The measurement of proliferation in tissue cultures by enumeration of cell nuclei. *Journal of the National Cancer Institute* 11: 773–795.

Wagner A, Marc A, Engasser JM & Einsele A (1992) The use of lactic dehydrogenase (LDH) release kinetics for the evaluation of death and growth of mammalian cells in perfusion reactors. *Biotechnology and Bioengineering* 39: 320–325.

BACKGROUND READING

Cook JA & Mitchell JB (1989) *Analytical Biochemistry* 179:1–7

Doyle A & Griffiths JB (eds) (1997) *Mammalian Cell Culture: Essential Techniques.* Wiley (and Bios Scientific Publishers Ltd), Chichester.

Freshney RI (1994) *Culture of Animal Cells: A manual of basic technique.* Wiley-Liss, New York.

Patterson M K (1979) Measurement of growth and viability of cells in culture. In: Jakoby W B & Paston I H (eds) *Cell Culture*, Vol. 11, pp. 141–151. Academic Press, New York.

2.2 MYCOPLASMA

A Detection of Mycoplasma

Mycoplasma is a generic term given to organisms of the order Mycoplasmatales that can infect cell cultures. Those that belong to the families Mycoplasmataceae (*Mycoplasma*) and Acholeplasmataceae (*Acholeplasma*) are of particular interest.

The first observation of mycoplasma infection of cell cultures was by Robinson *et al.* (1956). The incidence of such infection has since been found to vary from laboratory to laboratory. At present 12% of cell lines received by the ECACC are infected but this may be an uncharacteristically low figure because many lines are screened for mycoplasma prior to deposition.

Mycoplasmas differ from other prokaryotes by their lack of a cell wall. They are unable to produce even precursors of bacterial cell wall polymers, unlike L-forms of bacteria that can do so under the right environmental conditions. Their size is another distinguishing feature; they are the smallest self-replicating prokaryotes, with coccoid forms of only 0.3 μm diameter capable of reproduction. Their genome size is approximately one-sixth that of *Escherichia coli*.

The importance of mycoplasma detection in cell cultures should not be underestimated. The concentration of mycoplasmas in the supernatant can be typically in the region of 10^6–10^8 mycoplasmas ml^{-1}. Additionally, mycoplasmas will cytadsorb to the host cells. They do not necessarily manifest themselves in the manner of most bacterial or fungal contaminants, e.g. pH change or culture turbidity. It is important therefore to adopt an active routine detection procedure. Mycoplasmas have been shown to elicit various effects, including the following:

- Induction of chromosome aberrations (Aula & Nichols 1967)
- Induction of morphological alterations, including cytopathology (Butler & Leach 1964)
- Interference in the rate of growth of cells (McGarrity *et al.* 1980)
- Influence of nucleic acid (Levine *et al.* 1968) and amino acid (Stanbridge *et al.* 1971) metabolism
- Induction of membrane alteration (Wise *et al.* 1978) and even cell transformation (MacPherson & Russel 1966)

Mycoplasma contamination is usually caused by any of five common species. The organisms and their natural hosts are *M. hyorhinis* (pig), *M. arginini* (cow), *M. orale* (humans), *A. laidlawii* (cow), and *M. fermentans* (humans).

Acholeplasma alone has no sterol requirement.

A range of assay techniques are available for the detection of mycoplasma contamination. These include staining, culture, DNA probes and co-cultivation.

Cell and Tissue Culture for Medical Research, edited by A. Doyle and J.B. Griffiths.
© 2000 John Wiley & Sons, Ltd

To remove the risk of false positives and false negatives, two methods at least should be employed. The ECACC recommends the use of enrichment broth and agar culture and Hoechst 33258 DNA staining.

PROCEDURE: DNA STAIN

 This stain is heat and light sensitive. The toxic properties of Hoechst 33258 are unknown, and therefore gloves should be worn when handling the powder or solution.

Reagents and solutions

Use analytical grade (Analar) reagents and distilled water.

Carnoy's fixative

For each preparation:

- Methanol, 3 ml
- Acetic acid (glacial), 1 ml
- Prepare fresh as required

Hoechst stain stock solution (100 ml)

bisbenzimide-Hoechst 33258:

- Add 10 mg of Hoechst 33258 to 100 ml of water
- Filter-sterilize using a 0.2 μm filter
- Wrap the container in aluminium foil and store in the dark at 4°C

Hoechst stain working solution (50 ml)

Prepare fresh each time as necessary by adding 50 μl of stock solution to 50 ml of water.

Mountant

Take 22.2 ml of citric acid (0.1 M) and 27.8 ml of disodium phosphate (0.2 M) and:

- Autoclave
- Mix with 50 ml of glycerol
- Adjust pH to 5.5
- Filter-sterilize and store at 4°C

Agar media

Mycoplasma agar base (80 ml) (Oxoid, Basingstoke, Hants, UK): add 2.8 g of agar base to 80 ml of water, mix and autoclave at 15 psi for 15 min. Prepare fresh as necessary.

Pig serum (10 ml)

Dispense in universal containers in 10 ml aliquots. Heat to 56°C for 45 min and store at −30°C.

Yeast extract (10 ml) (Oxoid, Basingstoke, Hants, UK)

Add 7 g to 100 ml of water, mix and autoclave at 15 psi for 15 min. Dispense in universal containers in 10 ml aliquots and store at 4°C.

Agar preparation

Prepared medium must be used within 10 days.

1. Autoclave agar at 15 psi for 15 min. Cool to 50°C and mix with the other constituents (which have been warmed to 50°C).
2. Dispense 8 ml per 5 cm diameter Petri dish. Store at 4°C in sealed plastic bags.

Broth media

Mycoplasma both base (70 ml) (Oxoid, Basingstoke, Hants, UK): Add 2 g of broth base to 70 ml of water, mix and autoclave at 15 psi for 15 min.

Horse serum (Tissue Culture Services, Botolph Claydon, Buckingham, UK)

Dispense in universal containers in 20 ml aliquots, do not heat inactivate and store at –30°C.

Yeast extract (Oxoid, Basingstoke, Hants, UK)

Prepare as for agar media.

Broth preparation

Mix constituents, dispense in glass vials in 1.8 ml aliquots and store at 40°C. Complete medium may be stored without deterioration for several weeks.

Materials and equipment

- Carnoy's fixative
- Methanol
- Acetic acid
- Hoechst stain stock solution
- Hoechst stain working solution (prepare fresh as required)
- Mountant
- Fluorescence microscope equipped with epifluorescence (e.g. Zeiss) and 340–380 nm excitation filter and 430 nm suppression filter

Note: Both methods used for the detection of mycoplasma have the following general principles in common. The cells to be tested should, before testing, complete at least two passages in antibiotic-free media. Infection may be hidden by the presence of antibiotics. Cell cultures tested from frozen ampoules should undergo at least two passages because cryoprotectants may also mask infection.

1. Using a routine method of subculture, harvest adherent cells, e.g. trypsin, and resuspend in the original culture medium to a concentration of about 5×10^5 cells ml^{-1}.

2. Test the suspension lines direct from the culture at about 5×10^5 cells ml^{-1}.
   Note: Experience of working with any particular cell line should remove the absolute necessity for an accurate cell count. An adequate number of cells should be added to dishes so that a semi-confluent spread of cells on the coverslip is obtained at the time of observation (at 1 day and 3 days post-incubation). Cell overgrowth would make interpretation difficult. If it is thought that after 3 days growth the media would be exhausted, the cells should be resuspended in a mixture of their original medium and fresh culture media.
3. Add 2–3 ml test cells to each of two tissue culture dishes containing glass coverslips.
4. Incubate at $36 \pm 1°C$ in a humidified 5% CO_2/95% air atmosphere for 12–24 h.
5. Remove one dish and leave the remaining dish for a further 48 h.
6. Before fixing the cells, examine them under an inverted microscope for bacterial and fungal contamination at $\times 100$ magnification.
7. Fix the cells by adding approximately 2 ml of Carnoy's fixative dropwise to the edge of the dish and leave for 3 min at room temperature. It is particularly important for suspension cultures to add fixative in this way because it avoids cells being swept to one side of the culture dish.
8. Decant the fixative to a waste bottle (avoiding fumes) for careful disposal, add a further 2 ml aliquot of fixative to the dish and leave for 3 min at room temperature.
9. Decant the fixative to a waste bottle.
10. Allow the coverslip to air-dry. Invert the lid of the dish and use forceps to rest the coverslip against the lid for about 30 min.
11. Add 2 ml of Hoechst stain (wearing gloves) to the coverslip and leave for 5 min at room temperature. Shield the coverslip from direct light at this point.
12. Decant the stain to a waste bottle.
13. Add one drop of mountant to a labelled slide. Place the coverslip cell-side down, on a slide.
14. Under UV epifluorescence examine the slide at $\times 100$ magnification with oil immersion.

ALTERNATIVE PROCEDURE: USE OF INDICATOR CELL LINES

An alternative to the described procedure is the use of an indicator cell line onto which the test cells are inoculated. The advantages are: standardization of the system; increased surface area of cytoplasm to reveal mycoplasma; and mycoplasma screening of serum and other cell culture reagents that may be inoculated onto the indicator line. Positive and negative controls may be included in each assay. However, it may prove impracticable for many laboratories to grow mycoplasma cultures to use as positive controls. The disadvantages of using an indicator line include the time and effort required for preparation.

A recommended indicator line is the Vero African green monkey kidney cell line, which has a high cytoplasm/nucleus ratio. The indicator cells are added at a concentration of approximately 1×10^4 cells ml^{-1} to sterile coverslips in tissue culture dishes 10–24 h before inoculation. The test sample should be added at a concentration of approximately (5×10^5 cells ml^{-1} to give a semi-confluent monolayer at the time of observation (at 1 and 3 days' post-inoculation of the sample). Slides are prepared in the same way as those prepared by the direct method.

SUPPLEMENTARY PROCEDURE: MICROBIOLOGICAL CULTURE

1. Using a sterile swab, harvest adherent cells. Resuspend them in the culture medium to a concentration of about 5×10^5 cells ml^{-1}.
2. Test the suspension cells direct at about 5×10^5 cells ml^{-1}.
3. Inoculate agar plate with 0.1 ml of the test sample.
4. Inoculate broth with 0.2 ml of the test sample.
5. Incubate test plate under anaerobic conditions (Gaspak, BBL Microbiology Systems, Cockeysville. MD, USA) at $36 \pm 1°$C for 21 days.
6. At approximately 7 and 14 days' post-inoculation, subculture test broth onto agar plates and incubate anaerobically at $36 \pm 1°$C.
7. After 7, 14 and 21 days' incubation, agar plates are examined under $\times 40$ or $\times 100$ magnification using an inverted microscope.

Culture of mycoplasma

1. Each new batch of media ingredients should be subject to quality control before agar or broth preparation. It is especially important to show that new batches of pig and horse serum can support the growth of a representative sample of species found infecting cell cultures, e.g. any two of *M. orale*, *M. hominis*, *M. fermentans*, *M. arginini*, *M, hyorhinis* or *A. laidlawii*. The National Collection of Type Cultures (Colindale, London, UK) or the American Type Culture Collection (Manassas, Virginia, USA) may supply type strains, or wild-type strains may be used. Stock positive control cultures may be kept frozen at –70°C in mycoplasma broth.
2. Quality control of each batch of complete broth media should be performed before use by adding at least two different strains of positive control mycoplasmas. A non-inoculated broth should be incubated as a negative control.
3. Mycoplasma agar medium should be shown to be able to support mycoplasma growth. The test can be performed at the time of the test sample inoculation. A non-inoculated plate should also be included as a negative control.
4. It is necessary to be able to distinguish 'pseudocolonies' and cell aggregates from mycoplasma colonies on agar. Pseudocolonies are caused by crystal formation and may possibly increase in size. They are distinguished from genuine mycoplasma colonies by using Dienes stain, which does not stain pseudocolonies but stains mycoplasma colonies blue. Most fungal and bacterial colonies also appear colourless. Cell aggregates, however, do not increase in size and

thus are most easily distinguished. An indicator of genuine mycoplasma colonies is their typical 'fried egg' appearance on agar. This, however, is not always apparent in primary isolates. By using a sterile bacteriological loop, cells may be disrupted, leaving the agar surface free of aggregates, but mycoplasma colonies, because of the nature of their growth will leave a central core embedded in the agar.

SUPPLEMENTARY PROCEDURE: ELIMINATION OF CONTAMINATION

Removal of broken cells debris before Hoechst staining

Cell debris in a culture often gives a microscopic image similar to that of mycoplasma contamination after the cells are stained with Hoechst 33258. To avoid possible 'false-positive' results of mycoplasma contamination, samples (culture supernatant of a test cell line) should be filtered aseptically using a membrane filter with 0.8 μm pore size to remove cell debris.

Contaminating mycoplasmas in animal cell culture usually grow up to a concentration of 10^5–10^8 ml^{-1}. Therefore, although 90%, at most, of mycoplasmas will be trapped on the filter (Ohno & Takeuchi 1990), the mycoplasmas that pass through the membrane are detectable by direct Hoechst staining on a slide-glass or by staining after propagation in a host cell culture.

DISCUSSION

DNA stain

The fluorochrome dye Hoechst 33258 binds specifically to DNA. Cultures infected with mycoplasma are seen under fluorescence microscopy as fluorescing nuclei with extranuclear mycoplasmal DNA (Plate 2.2A1, A, B) whereas uninfected cell cultures contain fluorescing nuclei against a negative background (Plate 2.2A.1, B, C).

Mycoplasmas may appear as: filamentous forms, some of which may branch, indicating a culture in logarithmic growth: or as cocci, which is typical of an aged mycoplasma culture. Some slide preparations may contain extranuclear fluorescence produced by disintegrating nuclei and these should not be confused with mycoplasmas. Normally, fluorescence of this type is not uniform in size and is too large to be mycoplasma. Contaminating fungi or bacteria will also stain using this technique but will appear much brighter and larger than mycoplasmas.

The main advantages of the staining method are that results are obtained speedily and the, as yet, non-cultivable *M. hyorhinis* strains that have been detected in cell cultures can be observed.

Both positive and negative slides may be kept in the dark for several weeks without deterioration.

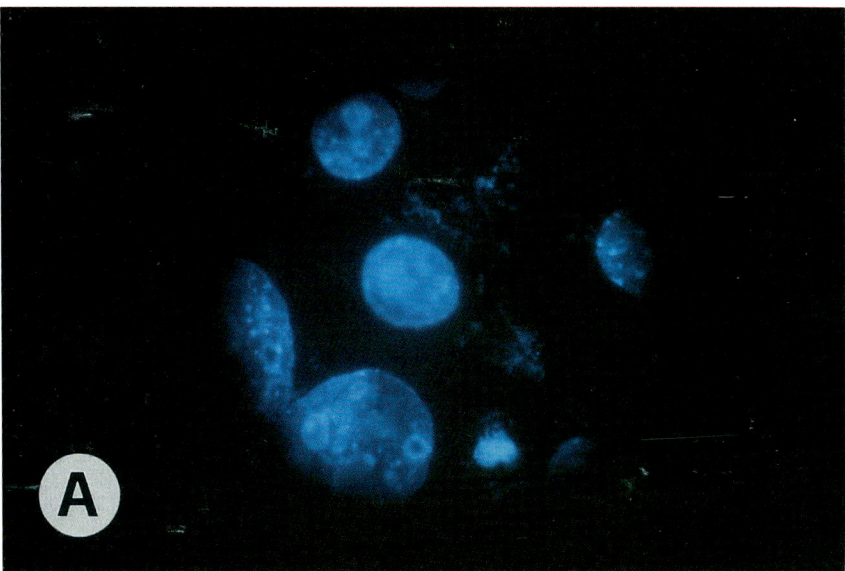

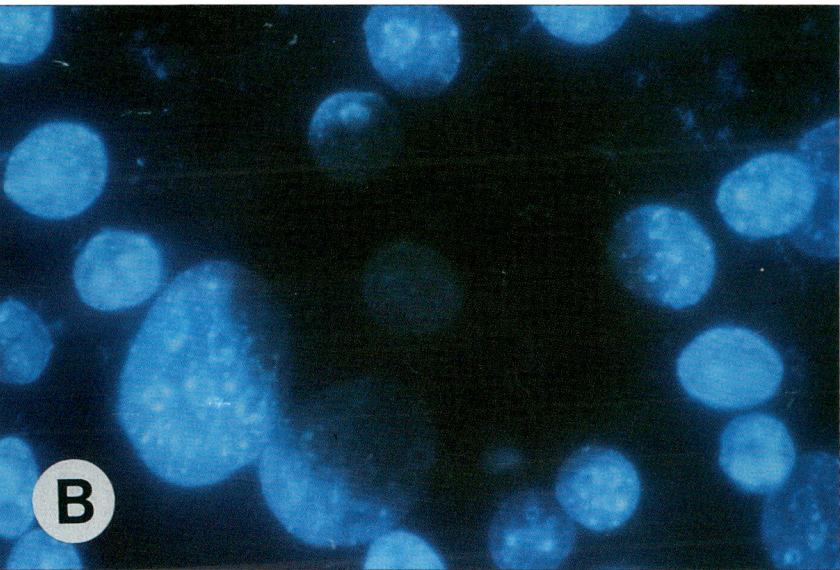

Plate 2.2A.1 Mycoplasma-infected cell cultures stained with Hoechst 33258 and mounted in plain buffered glycerol (A, B) or in the presence of 0.01% 1,4-*p*-phenylenediamine (PPD) (C, D). The field diaphragm of the microscope was partially closed and photographs of the central area of each field (A, plain; B, 0.01% PPD) were taken immediately. The central areas delimited by the partially closed diaphragm were exposed to ultraviolet light for 2 min, and then the whole fields were immediately photographed after opening the diaphragm (B, plain; D, 0.01% PPD). In the central area of the culture mounted in plain buffered glycerol (B, central area) fluorescence is virtually abolished, whereas the staining intensity of the culture mounted in the presence of 0.01% PPD is unaffected (D). Each series of photographs was taken with the same exposure time, using a ×100/NA 1.3 oil-immersion fluorite objective.

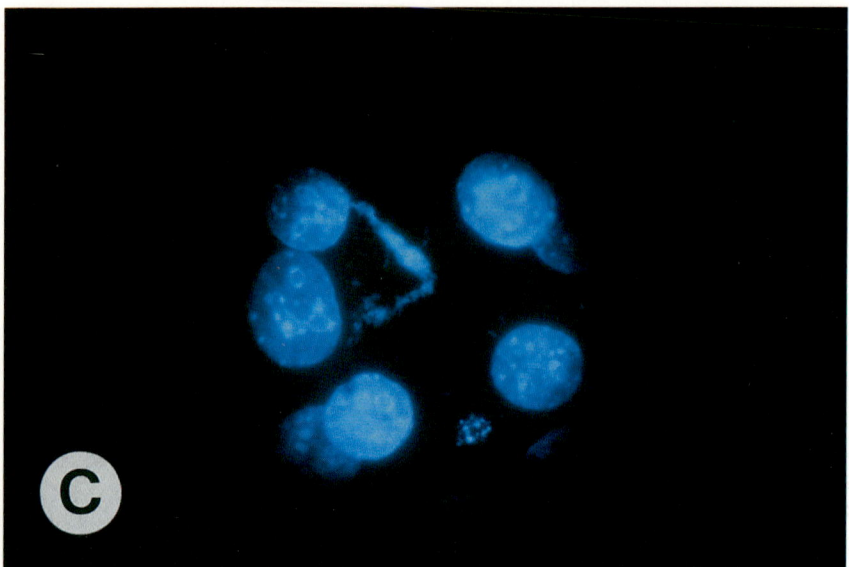

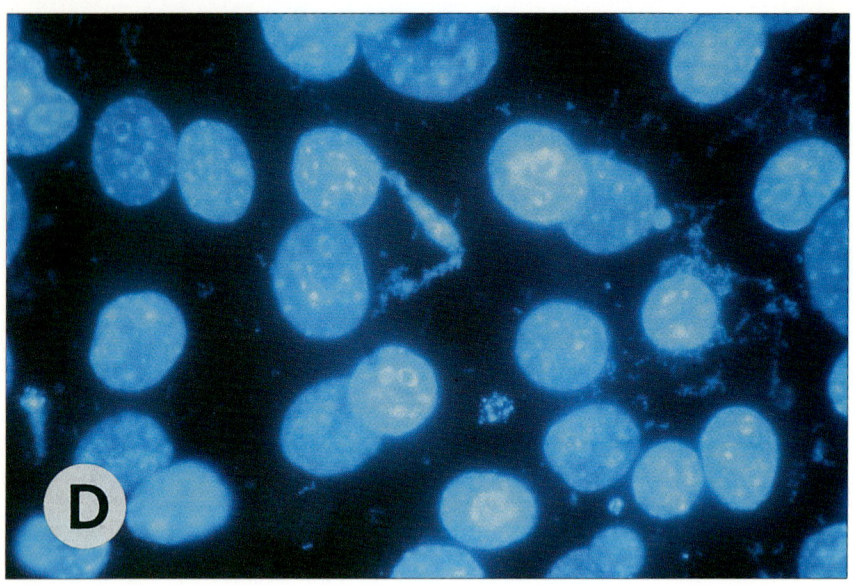

REFERENCES

American Type Culture Collection, 10801 University Boulevard, Manassas, Virginia, 20110 2209, USA.

Aula P & Nichols WW (1967) The cytogenetic effects of mycoplasma in human leucocyte cultures. *Journal of Cell Physiology* 70: 281–290.

Butler M & Leach RH (1964) A mycoplasma which induces acidity and cytopathic effect in tissue culture. *Journal of General Microbiology* 34: 285–294.

European Collection of Cell Cultures, CAMR, Porton Down, Salisbury, Wiltshire, SP4 0JG, UK.

Levine EM, Thomas L, McGregor D, Hayflick L & Eagle M (1968) Altered nucleic acid metabolism in human cell cultures infected with mycoplasma. *Proceedings of the National Academy of Sciences of the USA* 60: 583–589.

MacPherson I & Russel W (1966) Transformations in hamster cells mediated by mycoplasmas. *Nature* 210: 1343–1345.

McGarrity GJ, Phillips D & Vaidya A (1980) Mycoplasmal infection of lymphocyte cultures: Infection with *M. salivarium*, *In Vitro* 16: 346–356.

National Collection of Type Cultures, Central Public Health Laboratory, 61 Colindale Avenue, Colindale, London NW9 5HT, UK.

Ohno T & Takeuchi M (1990) Test for mycoplasma contamination. Standardized protocols for quality control of animal cell lines. Report from JTCA Cell Bank Committee, *Tissue Culture Research Communications* 9: Suppl. 9–11.

Robinson LB, Wichelhausen RB & Roizman B (1956) Contamination of human cell cultures by pleuro-pneumonia-like organisms. *Science* 124: 1147–1148.

Stanbridge EJ, Hayflick L & Perkins FT (1971) Modification of amino acid concentrations induced by mycoplasmas in cell culture medium. *Nature (London) New Biology* 232: 242–244.

Wise KS, Cassell GH & Action RT (1978) Selective association of murine T lymphoblastoid cell surface alloantigens with mycoplasma hyorhinis. *Proceedings of the National Academy of Sciences of the USA* 75: 4479–4483.

B Detection of Mycoplasma by DNA Amplification (PCR)

Mycoplasma contamination of cell cultures is a serious issue for in vitro cell culture due to the possible effects of such contamination on the characteristics of cell cultures (e.g. chromosome abnormalities, transformation, altered growth rate, competition for nutrients) (Rotem & Naot 1998). Thus the presence of mycoplasma in experimental cell cultures is undesirable as it may compromise the quality of research data. The presence of mycoplasma in cell lines used in the manufacture of medical therapeutic products is not acceptable since some mycoplasma species are pathogenic or may produce biologically active substances. Furthermore, the presence of any mycoplasma strain raises the risk of contamination of cell culture derived products with mycoplasma antigens and/or DNA. All of these concerns are addressed by regulatory bodies in guidelines dealing with the acceptability of cell substrates for the production of biologicals (WHO 1998; ICH 1997) that require such cell lines to be free of mycoplasma. The problem of mycoplasma contamination is wide spread due to a number of factors: 1) resistance to most antibiotics used in tissue culture, 2) mycoplasma contamination is not generally evident macroscopically or by standard light microscopy, 3) mycoplasma appear to survive in aerosols and in contaminated cell culture reagents. As a consequence mycoplasma contamination can spread rapidly between cultures in the laboratory and routine screening of cultures is essential to avoid this.

Many species of mycoplasma have been discovered as contaminants of cell cultures and any detection system must be able to identify contamination with any one of these. The methods used should also have adequate sensitivity to reveal low level contamination that may result from inhibition of mycoplasma growth by some antibiotics. It is advisable to use more than one technique for detection to provide confidence in negative results. This is particularly important in the testing of cryopreserved stocks of cells. PCR is a particularly useful technique since it can be applied routinely and is potentially very sensitive. Numerous groups have reported the use of PCR for routine detection of mycoplasma in cell cultures (e.g. Pruckler *et al.* 1995; Nissen *et al.* 1996; Kobayashi *et al.* 1995) and a straightforward method based on the detection of 16S rRNA genomic sequences of mycoplasma and acholeplasma is given below (van Kuppeveld *et al.* 1994).

Materials and equipment

- Programmable PCR temperature cycler
- Gel electrophoresis equipment
- UV transilluminator
- Gel photography equipment (e.g. Polaroid MP4 using polaroid film 667 and an orange filter for use with ethidium bromide)

 Avoid exposure to uv light and always wear uv protective safety glasses

Reagents and solutions

- Taq DNA polymerase
- 50 mM magnesium chloride
- 10X buffer supplied by the manufacturer (e.g. Gibco BRL Cat. No. 10342–020)
- Nucleotides (dATP, dTTP, dGTP, dCTP)
- DEPC-treated water
- Primers (forward primer GPO-3, 5'-GGGAGCAAACAGGATTAGATAC-CCT-3'; reverse primer MGSO (5'-TGCACCATCTGTCACTCTGTTAAC-CTC-3')

- Controls: uninfected cells, positive cell supernatant and distilled water. The positive control should be well characterized and standardized to ensure that the same type and level of material is used for successive analyses.
- 10XTBE buffer (20 mM Tris base, 20 mM sodium chloride, 2 mM disodium EDTA, 0.25% SDS, pH 7.5)
- Ethidium bromide solution (10 mg ml^{-1})
- 6X gel loading buffer.

PROCEDURE: AMPLIFICATION

Preparation of samples for PCR amplification

1. Pour off supernatant from cell culture to be tested. Label 1.5 ml eppendorf tubes with test code for each culture to be tested. Transfer 1ml of culture supernatant to the appropriate tube and close the cap tightly.
2. Fill a glass beaker with tap water and bring to the boil on a hotplate. When boiling, add the microtubes containing the test supernatants to the beaker and leave in boiling water for five minutes.
3. Remove tubes from the water and cool them to room temperature. The supernatants can be used immediately or stored frozen at or below –20° C.

Preparation of PCR reactions, amplification and analysis

1. Label 0.2 ml microtubes including positive and negative controls.
2. To each tube add, 10 μl 10 × buffer (as supplied by the Taq enzyme manufacturer), 1.5 mM magnesium chloride, 50 pmol each primer, each deoxynucleotide triphosphate at a concentration of 0.2 mM, 0.2 U (0.5 μl) *Taq*. Make up to 90 μl with DEPC-treated water. All reagents should be kept on ice throughout.
3. Add 10 μl of treated supernatant to the appropriate tube. Vortex tubes to mix then centrifuge to bring the liquid to the bottom of the tubes.
4. For traditional non-peltier cycler machines (e.g. Omnigene, Hybaid, UK) use 0.5 ml tubes and overlay the PCR reaction mixtures with 50 μl of mineral oil. In such machines the following amplification cycles should be appropriate: 94°C for 1 min, 55°C for 1 min, 72°C for 2 mins. This cycle is carried out 40 times. When using peltier devices (e.g. Thermus, MWG Biotech, Germany) the following cycles should be appropriate: 94°C for 1 min, 53°C for 1 min, adjust the ramp so that it goes up to 72°C at 1°C/second, 72°C for 2 mins. This cycle is carried out 40 times.

5. The PCR products can then be visualized by running on a 2% agarose gel at
   100 volts for 45 mins and staining the gel with 10 µl of ethidium bromide in
   100 ml of 1X TBE for 20 mins. The product amplified in a positive sample
   is 280 bp.

- Preparation of amplification reactions should be carried out in an area separated from the amplification procedure and stored PCR products. This will
  reduce the opportunity for contamination and false positive reactions
- Reaction reagents should be prepared, aliquotted and stored in the reaction
  preparation area
- Use positive displacement pipettes and pipette tips with filters to prevent contamination

DISCUSSION

The method described is straightforward and has been shown to identify contamination with the mycoplasma species and Acholeplasma found in cell cultures.
Sample preparation is very simple and a result can be obtained in one day. PCR
is reported to perform well when compared with other detection techniques
(Hopert et al. 1993; van Kuppeveld et al. 1994) and nested PCR may provide
methods with potentially very high sensitivity (Prickler et al. 1995). A commercial PCR kit is also now available from Stratagene Europe (Amsterdam Zuidoost)
and this uses a method that enables detection and subsequent species identification (Nissen et al. 1996).

It is important to remember that a PCR method can only be as good as the
operator's technique. It is therefore important to adopt the approaches to isolation of different stages of the PCR process as described above and place special
emphasis on careful training of staff. In addition it is important to remember that
an important variable in the PCR process is the PCR cycler and a method developed on one type of cycler may need to be adapted by altering the incubation
times and possibly the cooling/warming rates for another make of machine.

PCR testing should be standardized by preparing an aliquotted positive control
that can be used in parallel with each set of samples. This should be used at high
concentration and at a low concentration (close to the limit of detection) to assure
reproducible quality of testing. When analysing samples from a wide variety of
cell cultures it may be necessary to check for inhibitory substances that may cause
false negatives. This can be done by performing each test twice and spiking one
of the samples with positive control material. When both the spiked sample and
the unspiked sample are negative then inhibition of the PCR may have occurred
and appropriate retesting should be performed.

PCR is likely to become the most common technique for the detection of
mycoplasma in cell culture. However, it is important that for cell banks and other
important samples another non-specific technique (e.g. culture, Hoechst stain) is
used to help exclude false results that could arise in routine testing by PCR.

REFERENCES

Hopert A, Uphoff CC, Wirth M, Hauser H & Drexler HG (1993) Specificity and sensitivity of polymerase chain reaction (PCR) in comparison with other methods for the detection of mycoplasma contamination in cell lines. *J. Immunol. Methods,* 164: 91–100.

Human Medicines Evaluation Unit (1997) ICH Topic Q 5 D – Quality of biotechnological products: Derivation and characterisation of cell substrates used for production of biotechnological/biological products. European Agency for the Evaluation of Medicinal Products, ICH – Technical Co-ordination, London, UK.

Nissen E, Vollenbroich D & Pauli G (1996) Comparison of PCR methods for mycoplasma in cell cultures. In Vitro Cell. *Dev. Biol.-Animal* 32: 463–464.

Pruckler JM, Pruckler JM & Ades EW (1995) Detection by polymerase chain reaction of all common mycoplasma in a cell culture facility. *Pathobiology* 63: 9–11.

Rottem S and Naot Y (1998) Subversion and exploitation of host cells by mycoplasmas. *Trends Microbiol.* 6: 436–440.

van Kuppeveld FJ, Johansson KE, Galama JM, Kissing J, Bolske G, van der Logt JT & Melchers WJ (1994) Detection of mycoplasma in cell cultures by a mycoplasma group-specific PCR. *Appl. Environ. Microbiol.* 60: 149–152.

WHO expert committee on biological standardisation and executive board (1998) Requirements for the use of animal cells as in vitro substrates for the production of biologicals. World Health Organisation, Geneva.

2.3 BACTERIA AND FUNGI

For any reasonable quality control programme involving cell lines it is absolutely necessary to include routine culture tests for bacteria and fungi. Most often the presence of such contaminants will be obvious even to the novice. This is especially true when the cultures are set up in the absence of all antibiotics; a practice that is strongly recommended. However, while overt microbial contamination may be more common, many organisms will grow slowly in the usual cell cultures, sometimes obscured by animal cell debris, and remain unnoticed. The steps outlined below will reveal all the most usual bacterial or fungal organisms that might be expected to thrive in conventional cell culture systems.

PROCEDURE: DETECTION OF BACTERIA AND FUNGI IN CELL CULTURES

Materials and equipment

- Dehydrated media for bacterial/fungal cultivation
- Bacto Sabouraud dextrose broth (Difco)
- Bacto thioglycollate medium (Difco)
- Trypticase soy broth powder (Baltimore Biological Laboratory)
- Bacto brain heart infusion agar (Difco)
- Bacto blood agar base (Difco)
- YM broth base (Difco)
- Nutrient broth base (Difco)
- Bacto yeast extract (Difco)
- Reference microbial cultures for positive controls
- Hotplates and stirring apparatus
- GasPak Anaerobic System (Baltimore Biological Laboratory)

To examine cell cultures or suspect media for bacterial or fungal contaminants, proceed as follows:

1. Using an inverted microscope, equipped with phase contrast optics if possible, examine cell culture vessels individually. Scrutiny should be especially rigorous in cases in which large-scale production is involved. Check each culture first using low power. The suppliers listed provide the specific media required but other suitable vendors exist. Batches of media should be tested for optimal growth promotion before use in cell culture quality control.
2. After moving the cultures to a suitable isolated area, remove aliquots of fluid from cultures that are suspect and retain these for further examination. Alternatively, autoclave and discard all such cultures.

Cell and Tissue Culture for Medical Research, edited by A. Doyle and J.B. Griffiths.
© 2000 John Wiley & Sons, Ltd

3. Prepare wet mounts using drops of the test fluids and observe under high power.
4. Prepare smears, heat-fix and stain by any conventional method (e.g. Wright's stain), and examine under oil immersion.
5. Consult Cour *et al.* (1979) and Freshney (1987) for photomicrographs of representative contaminants and further details.

Microscopic examination is only sufficient for detection of gross contaminations and even some of these cannot be readily detected by simple observation. Therefore, an extensive series of culture tests is also required to provide reasonable assurance that a cell line stock or medium is free of fungi and bacteria.

To perform these on standard cell cultures or stocks of frozen cells:

1. Pool and mix the contents of about 5% of the ampoules from each freeze lot prepared using narrow-width pipettes. Aliquots from cultures to be tested should generally include some of the monolayer or cell suspension. It is commonly recommended that antibiotics be omitted from media used to cultivate and preserve stock cell populations. If antibiotics are used, the pooled suspension should be centrifuged at 2000 g for 20 min and the pellet resuspended in antibiotic-free medium. A series of three such washes with antibiotic-free medium prior to testing will reduce the concentration of antibiotics which could obscure contamination.
2. From each pool, inoculate each of the test media listed in Table 2.3.1 with a minimum of 0.3 ml of the test cell suspension and incubate under the conditions indicated. Include positive and negative controls comprising a suitable range of bacteria and fungi which might be anticipated. A recommended grouping consists of *Pseudomonas aeruginosa*, *Micrococcus salivarius*, *Escherichia coli*, *Bacteroides distasonis*, *Penicillium notatum*, *Aspergillus niger* and *Candida albicans*.

Table 2.3.1 Regimen for detecting bacterial or fungal contamination in cell cultures

Test medium	Temperature (°C)	Aerobic state	Observation time (days)
Blood agar with:			
fresh defibrinated	37	Aerobic	14
rabbit blood (5%)	37	Anaerobic	
Thioglycollate broth	37	Aerobic	14
	26		
Trypticase soy broth	37	Aerobic	14
	26		
Brain heart infusion broth	37	Aerobic	14
	26		
Sabouraud broth	37	Aerobic	21
	26		
YM broth	37	Aerobic	21
	26		
Nutrient broth with			
2% yeast	37	Aerobic	21
yeast	26		

For further details see Cour *et al.* (1979).

3. Observe as suggested for 14–21 days before concluding that the test is negative. Contamination is indicated if colonies appear on solid media or if any of the liquid media become turbid.

DISCUSSION

While this procedure will permit detection of most common bacterial and fungal organisms that grow in cell cultures, it has been noted that at least one, very fastidious, bacterial strain initially escaped observation. This was present in nine different cultures from a single clinical laboratory in the USA submitted for testing and expansion under a government contract. The organism grew extremely slowly but could be detected after 3 weeks' incubation with cell cultures that had no antibiotics and had no fluid changes. Samples so developed were inoculated to sheep blood agar plates and New York City broth (ATCC (American Type Culture Collection) medium 1685). The organism could be observed during a subsequent 6-week incubation period at 37°C.

The Bacteriology Department at ATCC determined the appropriate culture conditions for this microorganism, and tentatively identified it as a *Corynebacterium*. Antibiotic sensitivity tests revealed bacteriostasis with some compounds but no bactericidal antibiotics have yet been found.

This incident emphasizes the critical importance of diligent testing of cell cultures for contaminant microorganisms. By combining procedures such as those described here with procedures included elsewhere in this volume (e.g. fluorescent or nucleic acid probes for mycoplasma and viruses) one can be more certain that clean cell cultures are available for experimentation.

REFERENCES

Cour I, Maxwell G & Hay RJ (1979) Tests for bacterial and fungal contaminants in cell cultures as applied at the ATCC. *Tissue Culture Association Manual* 5: 1157–1160.

Freshney RI (1987) *Culture of Animal Cells – A Manual of Basic Techniques*, 2nd edn. Alan R Liss Inc., New York.

2.4 ELIMINATION OF CONTAMINATION

In the event of cell cultures becoming contaminated with bacteria, fungi or mycoplasmas, the best course of action is to discard the culture, check cell culture reagents for contamination, thoroughly disinfect all safety cabinets and work surfaces and resuscitate a fresh culture from previously frozen stock. In the case of contamination with a spore-forming organism, and where such facilities exist, room fumigation may also be advisable. However, in the case of irreplaceable stocks this course of action may not be possible and antibiotic treatment may be necessary to eliminate the contamination.

PROCEDURE: ERADICATION

Materials

- Antibiotic of choice (see Tables 2.4.1–2.4.3)
- Appropriate cell culture growth medium

1. Culture cells in the presence of the chosen antibiotic for 10–14 days. Each passage should be performed at the highest dilution of cells at which growth occurs. If the contaminant is still detectable after this time, it is unlikely that the antibiotic used will prove successful, and another antibiotic should be tried.
2. During the course of antibiotic treatment and also 5–7 days after treatment in the case of bacteria and fungi, and 25–30 days after treatment for mycoplasma-contaminated lines, re-test the culture by an appropriate method. Regular routine testing should, however, occur after this period. This is particularly important in the case of mycoplasmas, where low levels of infection may persist after antibiotic treatment.

DISCUSSION

The antibiotic of choice used to eliminate any contaminant is dependent upon the sensitivity of the organism in question. If facilities are available to identify the organism and perform antibiotic sensitivity assays an appropriate antibiotic may be chosen for elimination and used at minimum concentration.

In the case of bacteria and fungi a Gram stain would give an indication of which antibiotic to use (see Tables 2.4.1 and 2.4.2).

Cell and Tissue Culture for Medical Research, edited by A. Doyle and J.B. Griffiths.
© 2000 John Wiley & Sons, Ltd

Table 2.4.1 Antibiotics active against bacteria

Antibiotic	Working concentration	Gram-positive bacteria	Gram-negative bacteria
Ampicillin	100 mg l^{-1}	✓	✓
Cephalothin	100 mg l^{-1}	✓	✓
Gentamicin	50 mg l^{-1}	✓	✓
Kanamycin	100 mg l^{-1}	✓	✓
Neomycin	50 mg l^{-1}	✓	✓
Penicillin V	100 mg l^{-1}	✓	
Polymyxin B	50 mg l^{-1}		✓
Streptomycin	100 mg l^{-1}	✓	✓
Tetracycline	10 mg l^{-1}	✓	✓

Antibiotics available from Sigma.

Table 2.4.2 Antibiotics active against fungi

Amphotericin B (Sigma)	2.5 mg l^{-1}
Ketaconazole (Sigma)	10 mg l^{-1}
Nystatin (Sigma)	50 mg l^{-1}

Unlike bacterial or fungal contamination, mycoplasma infection is not always detectable in a cell culture by the usual microscopic methods.

Various antibiotics have been investigated for their ability to eradicate mycoplasmas from cell lines. Historically the antibiotics of choice were either minocycline or tiamulin. However, many cell culture mycoplasmas are now resistant to these antibiotics. Today the most effective are the quinalone antibiotics ciprofloxacin (Mowles 1988) and MRA (see Table 2.4.3).

It is extremely important that after antibiotic treatment the cells are maintained in antibiotic-free media. Lack of evidence of mycoplasma infection does not necessarily indicate that the culture is free of such infection, as the level of infection may be below the limit of detection. For this reason it is suggested that antibiotic-free growth be continued to allow any residual infection to reach levels which are detectable. If, after this period, no mycoplasmas are detected, the line may be considered to be mycoplasma-free.

Other methods of mycoplasma elimination

For most laboratories antibiotic treatment is the most convenient method of eradication.

Table 2.4.3 Antibiotics active against mycoplasmas

Ciprofloxacin (Bayer)	20 mg l^{-1}
MRA (Mycoplasma Removal Agent – ICN–Flow)	0.5 mg l^{-1}

Other methods are available which vary in their degree of success. These include passage in athymic nude mice (Van Diggelen *et al.* 1977), growth of cells in rabbit or guinea pig serum (Nair 1985), and the use of nucleic acid analogues (Marcus *et al.* 1980).

REFERENCES

Marcus M, Lavi U, Nattenberg A, Rottem S & Markowitz D (1980) Selective killing of mycoplasma from contaminated mammalian cells in cell cultures. *Nature* 285: 659–699.

Mowles JM (1988) The use of Ciprofloxacin for the elimination of mycoplasmas from naturally infected cell lines. *Cytotechnology* 1: 355–358.

Nair CN (1985) Elimination of mycoplasma contaminants from cell cultures with animal serum. *Proceedings of the Society for Experimental Biology and Medicine* 179: 254–258.

Van Diggelen OP, Shin S & Phillips DM (1977) Reduction in cellular tumour-genicity after mycoplasma infection and elimination of mycoplasma from infected cultures by passage in nude mice. *Cancer Research* 37: 2680–2687.

2.5 REMOVAL OF FIBROBLASTS

For many cell types, deriving primary cultures can prove difficult due to domination of the cultures by fibroblasts. The fibroblasts have a proliferative advantage over many more specialized cell types, such as epithelia, tending to outgrow these cells even when the initial level of fibroblast contamination is low. This overgrowth renders long-term culture, or work requiring pure populations of non-fibroblastic cells, impracticable in most cases.

CHOICE OF METHOD

A variety of methods to control the extent of fibroblast contamination have been reported; however, most of these cannot be applied universally. The successful removal of fibroblasts from a particular culture system is critically dependent on the choice of technique.

The use of monoclonal antibodies has allowed the development of highly specific techniques which target fibroblasts for destruction or removal. They are widely applicable, and two methods are given in detail.

The first and most effective technique is an immunoadsorption method termed 'panning'. In contrast to the majority of fibroblast-control/removal techniques, 'panning' involves the treatment of cell suspensions rather than adherent cultures. A single 'panning' treatment usually results in complete elimination of the contaminating fibroblasts, whereas most other techniques often require repetitive treatment over a number of subsequent passages. The reason for this difference in efficiency is that a significant number of fibroblasts are located beneath the colonies of specialized cells. The protected position of these fibroblasts probably makes them inaccessible to adherent culture treatments, whereas this is clearly not a problem when using the 'panning' technique. 'Panning' is also comparatively simple and inexpensive. The method relies on the presence of the cell surface glycoprotein, Thy-1, on human fibroblasts but not on many specialized human cell types (such as keratinocytes).

The second immunological method involves the use of fibroblast-specific antibody and complement. Although this method is only applicable to adherent cultures and may not kill inaccessible fibroblasts it still remains an effective and simple treatment especially where cell suspensions are not available, e.g. explant cultures. Nevertheless, a small degree of non-specific destruction of the specialized cells can take place.

Both the 'panning' and the complement methods may be used with a number of antibodies which are specific to antigens present on the cell surface of fibroblasts. It is obviously important to establish that the antibody being used does not bind

Cell and Tissue Culture for Medical Research, edited by A. Doyle and J.B. Griffiths.
© 2000 John Wiley & Sons, Ltd

to the cells of interest. The choice of antibody to use with complement is some-what more limited, since antibodies must be tested for their ability to kill when used with complement.

Finally, the last method, which is given in detail, is the simplest and most rapid, involving a brief treatment with EDTA and trituration. Although this technique is not as thorough at removing fibroblasts as the others, it is still useful for the removal of large areas of fibroblastic outgrowth, and can be used to remove the majority of fibroblasts before other techniques are used. The suitability of this method depends on the relative insensitivity of more specialized cells to EDTA and is most commonly used on epithelial cell cultures, such as keratinocytes.

Each of these techniques can be used repeatedly on several consecutive subcul-tures, although this is not usually necessary when using the 'panning' method correctly.

A number of other methods are given in outline only, since these are some-what less effective and in some cases involve time-consuming preparation or extremely toxic reagents.

PRELIMINARY PROCEDURE: PREPARATION OF DISHES FOR 'PANNING'

Materials and equipment

- Affinity purified goat anti-mouse-IgG antibody
- 0.05 M Tris buffer (pH 9.5)
- 5% foetal bovine serum (FBS) in phosphate-buffered saline (PBS)
- Bacteriological-grade Petri dishes (50 mm)

1. Place 1 ml Tris buffer into each bacteriological-grade Petri dish and add goat anti-mouse-IgG antibody to give a final concentration of 10 µg ml^{-1}.
2. Incubate at room temperature for 40 min.
3. Carefully wash dishes once with PBS and twice with 5% FBS in PBS.

PRELIMINARY PROCEDURE: PREPARATION OF CELLS FOR 'PANNING'

Materials and equipment

- Mouse monoclonal antibody F15-42-1 in the form of undiluted ascites (specific to human Thy-1)
- Trypsin 0.25%
- 5% FBS in PBS

1. Trypsinize cells as normal to obtain a single cell suspension.
2. Wash cells in 5% FBS in PBS (to inactivate trypsin) and estimate cell number using a haemocytometer.
3. Resuspend cells in a 1 : 1000 dilution of F15-42-1 in 5% FBS/PBS so that there are 10^6 cells per 100 µl diluted antibody.

4. Incubate at 4°C for 1 h and then wash cells twice with 5% FBS in PBS.
5. Resuspend cells in 5% FBS in PBS at a concentration of 4×10^5 cells ml^{-1}.

PROCEDURE: 'PANNING'

Materials

- 5% FBS in PBS

1. Gently add 1 ml of treated cell suspension to each treated dish.
2. Incubate at 4°C on a level surface.
3. After 40 min gently swirl dishes and incubate for a further 30 min.
4. Gently pour cell suspensions from each dish.
5. Carefully wash dishes twice with 5% FBS in PBS (by pouring washes onto the side/lip of the dish) taking care not to dislodge bound cells.
6. The washes can then be pooled along with the cell suspension (the Thy-1 negative cells of interest) and can be placed back into culture or used immediately whilst the fibroblasts should have remained on the dish if sufficient care was taken with the washing.

ALTERNATIVE PROCEDURE: COMPLEMENT AND ANTIBODIES

Materials

- Monoclonal antibody LICR LON/FIB86 (fibroblast specific) ascites
- Rabbit complement (whole rabbit serum)
- Treatment medium (serum-free, HEPES, buffered medium)

1. Remove normal growth medium from cell or explant cultures.
2. Wash cultures with cold treatment medium.
3. Place cold treatment medium onto the culture, containing 1 : 500 dilution of antibody LICR LON/FIB86.
4. Incubate at 4°C for 1 h on a slowly rocking platform.
5. Wash cultures three times with cold treatment platform.
6. Place cold treatment medium containing a 1 : 11 dilution of rabbit complement onto cells (although concentration can vary depending on toxicity of serum).
7. Bring cultures quickly to 37°C and gently rock for 2–4 h.
8. Wash cultures with treatment medium three times and replace with normal growth medium. Fibroblast death should be detectable by phase contrast microscopy on completion of treatment.

RAPID PROCEDURE: EDTA/TRITURATION

Materials and equipment

- 0.02% of EDTA in PBS

1. Remove normal growth medium from cultures and wash with EDTA solution, which is removed immediately.
2. Add fresh EDTA solution, enough to cover the surface area of the flask or Petri dish, and incubate at 37°C for 2–10 min (until contaminating fibroblasts have started to round up, whereas specialized cells have not).
3. Using a pipette, take up EDTA solution and expel it with moderate force over the dish/flask surface (making sure that all the surface has been treated in this manner) until fibroblasts detach, leaving behind more specialized cells.
4. Remove EDTA solution, wash once with normal growth medium and replace with fresh growth medium.

SUPPLEMENTARY PROCEDURE: OTHER TECHNIQUES

Scraping

This involves the manual scraping of distinct areas of fibroblast outgrowth and is usually used on explant cultures. The disadvantages of this technique are that it is extremely tedious and time-consuming and has to be repeated every 2–3 days. This technique is of limited use in cases where fibroblasts have intermingled with the cells of interest.

Differential trypsinization

This involves a brief incubation at 37°C of cultures with a solution of 0.05% trypsin in 0.02% EDTA and relies on a difference in trypsin sensitivity between fibroblasts and the cells of interest.

Cloning

This depends on the cells of interest having a high proliferative capacity when in a cloning environment, in order to allow the production of a clone sufficiently large for subsequent experimentation. Although this is not often possible, the proliferative potential of cells may be enhanced by the use of either conditioned medium or a feeder layer of proliferation-blocked cells.

Supplemented media

A number of compounds have been shown to slow the proliferation of fibroblasts in culture, but not that of specialized cells; these include using high levels of cholera toxin (10 nM), cis-hydroxyproline (100 μg ml^{-1}), or substituting D-valine for the L-valine which is usually present within culture medium.

Geneticin (G418 sulphate)

The use of this antibiotic can be most effective, but depends on a difference in growth rate between the cells of interest and fibroblasts. Geneticin has a cytotoxic effect after 1–2 cell divisions, with the most rapidly growing cells being killed in the shortest interval. The effects of geneticin on the specialized cells can be reduced or eliminated by reversibly slowing their proliferation rate, while maintaining that of the fibroblasts. This can be achieved by removing a suitable medium supplement or growth factor (e.g. phorbol esters from melanocyte cultures). The geneticin is then added to the cultures at a sufficient concentration (which varies with batch but is usually between 50 and 300 μg ml⁻¹) to kill proliferating cells. The geneticin-containing medium is removed after 2–3 days, the cultures are washed thoroughly and the normal growth medium is added. Fibroblast detachment should take place by 1–2 weeks after treatment. This treatment can be repeated upon subculture.

SUPPLEMENTARY PROCEDURE: IMMUNOLOGICAL TECHNIQUES

Ricin-conjugated antibodies

This technique is most useful for fibroblast-contaminated explant cultures and involves conjugating the highly toxic agent ricin to specific antibodies.

 The extreme care needed when using ricin cannot be overemphasized and for this reason this procedure cannot be recommended except in the last resort.

1. Isolate, using gel filtration, the component of the conjugated products that corresponds to one molecule of ricin conjugated to one molecule of antibody.
2. Dilute antibody-ricin conjugate to a final concentration of 1 μg ricin ml⁻¹ in standard growth medium containing 100 mM galactose. The cultures are incubated at 37°C for 1 h in this medium.
3. Wash cultures three times in standard medium.

Fibroblasts should detach from the culture surface by 1–2 weeks after treatment. The use of galactose in the treatment medium prevents nonspecific attachment of the conjugate through its ricin moiety to galactose residues on the cell surface. Although the non-specific killing of specialized cells is avoided in this way, this method is more technically demanding and time-consuming.

Cytofluorimetry

This technique involves the use of cell suspensions which have been treated with a fibroblast-specific antibody and a secondary fluorochrome-conjugated antibody. The fluroescence-activated cell sorter (FACS) can detect and separate the fluorescence-negative cells (the cells of interest) from the fluorescence-positive cells (fibroblasts). The main disadvantage to this technique is the time taken for the cells to be sorted (2–7 h), which can lead to cell damage. The results obtained with this technique are usually no better than those obtained with 'panning'.

Immunomagnetic separation

This technique involves the use of magnetic polystyrene beads, such as Dynabeads. These beads are coated with fibroblast-specific antibody, which when incubated with a cell suspension binds the fibroblasts. The beads and therefore the fibroblasts can be removed using magnetic force.

LITERATURE REVIEW

The earliest procedures established for the removal of fibroblasts are the most basic, involving a brief treatment with trypsin (Owens & Hackett 1972; mouse mammary tumors) or EDTA (Rheinwald & Green 1975; human epidermal keratinocytes) to detach fibroblasts preferentially. Fibroblastic overgrowth was shown to be inhibited by the use of mitomycin-C-treated or irradiated mouse 3T3 fibroblasts as feeder layers (Rheinwald & Green 1975; human epidermal keratinocytes). Fibroblast proliferation rates have been shown to be selectively decreased by the use of a number of compounds such as: cholera toxin (Eisinger & Marko 1982; human melanocytes); cis-hydroxyproline (Whei-Yang Kao & Prockop 1977; neuroblastoma and epidermoid carcinoma); and substituting D-valine for L-valine (Gilbert & Migeon 1975; fetal lung and kidney). An alternative approach developed by Halaban & Alfano (1984; human melanocytes) was to use the rapid proliferation of fibroblasts to advantage, together with the cell-cycle-dependent cytotoxic antibiotic, geneticin. More recently, with the availability of monoclonal antibodies, techniques have become more specific. Edwards *et al.* (1980) (human squamous cell carcinoma) and Singer *et al.* (1989) (human thymic epithelial cells) are two groups which have used antibodies along with complement-induced cytolysis. A third group (Paraskeva *et al.* 1985; human colorectal carcinoma) have used antibodies conjugated to ricin, which resulted in fibroblast destruction. Finally monoclonal antibodies have been used with a solid-phase immunoadsorption technique, 'panning', to separate fibroblasts from specialized cells. This technique was used with fibroblast-specific antibody to remove fibroblasts from human keratinocyte cultures (Linge *et al.* 1989). In addition, 'panning' was used together with antibody specific to a mouse adhesion molecule (N-CAM) to separate mouse myogenic cells and fibroblasts (Jones *et al.* 1990).

REFERENCES

Edwards PW, Easty DM & Foster CS (1980) Selective culture of epithelioid cells from a human squamous carcinoma using a monoclonal antibody to kill fibroblasts. *Cell Biology International Reports* 4: 917–922.

Eisinger M & Marko O (1982) Selective proliferation of normal human melanocytes *in vitro* in the presence of phorbol ester and cholera toxin. *Proceedings of the National Academy of Sciences of the USA* 79: 2018–2022.

Gilbert SF & Migeon BR (1975) D-valine as a selective agent for normal human and rodent epithelial cells in culture. *Cell* 5: 11–17.

Halaban R & Alfano FD (1984) Selective elimination of fibroblasts from cultures of human melanocytes. *In Vitro* 20: 447–450.

Jones GE, Murphy SJ & Watt DJ (1990) Segregation of the myogenic cell lineage in mouse muscle development. *Journal of Cell Science* 97: 659–667.

Linge C, Green MR & Brooks RF (1989) A method for the removal of fibroblasts from human tissue culture systems. *Experimental Cell Research* 185: 519–528.

Owens RB & Hackett AJ (1972) Tissue culture studies of mouse mammary tumor cells and associated viruses. *Journal of National Cancer Institute* 49: 1321–1332.

Paraskeva C, Buckle BG & Thorpe PE (1985) Selective killing of contaminating human fibroblasts in epithelial cultures derived from colorectal tumours using an anti Thy-1 antibody-ricin conjugate. *British Journal of Cancer* 51: 131–134.

Rheinwald JG & Green H (1975) Serial cultivation of strains of human epidermal keratinocytes: the formation of keratinizing colonies from single cells. *Cell* 6: 331–344.

Singer KH, Scearce RM, Tuck DT, Whichard LP, Denning SM & Haynes BF (1989) Removal of fibroblasts from human epithelia cell cultures with the use of a complement fixing monoclonal antibody reactive with human fibroblasts and monocyte/macrophages. *Journal of Investigative Dermatology* 92: 166–170.

Whei-Yang Kao W & Prockop DJ (1977) Proline analogue removes fibroblasts from cultured mixed cell populations. *Nature* 266: 63–64.

2.6 IDENTITY TESTING – AN OVERVIEW

The consequences of using a cell line whose identity is not that claimed by the provider are clearly very serious. Failure readily to identify an occurrence of cell line cross-contamination or switching during tissue culture procedures could completely invalidate a body of research or completely abort a production process. Cell line authentication is therefore an essential requirement for all cell culture laboratories and should be carried out at both the earliest passages of cultures and at regular intervals thereafter (Hay 1988). The occurrence of cross-contamination is not merely anecdotal; documented cases have been widely reported (Nelson-Rees *et al.* 1981; van Helden *et al.* 1988). Some earlier reports indicated that levels of cross-contamination may exceed 30% of cultures tested (Halton *et al.* 1983). Preferential inclusion of suspect cultures in these reports means that this figure probably represented an overestimation of the problem. Nevertheless, experience at the ECACC in the authentication of cell stocks from a wide range of laboratories indicates that cross-contamination is a neglected problem. The classic example is that of Hela contamination (Nelson-Rees *et al.* 1981). Early studies used conventional cytogenetic analysis in association with isoenzyme analysis to verify the species of origin. This is particularly easy in the case of the HeLa cell line because it has characteristic cytogenetic markers, and for isoenzyme analysis type B rather than the more usual type A glucose-6-phosphate dehydrogenase is expressed. A large number of isoenzyme tests are usually required for specific identification. A summary of the standard authentication techniques is given in Table 2.6.1, but this list is by no means exhaustive.

CYTOGENETIC ANALYSIS

Microscopic examination of the chromosomal content of a cell line provides a direct method of confirming the species of origin and allows the detection of gross aberrations in chromosome number and/or morphology. Cytogenetic analysis is very useful for specific identification of cell lines with unique chromosome markers and has proved useful to differentiate cell lines that were apparently identical by isoenzyme analysis. In one study of 47 cell lines reported by O'Brien *et al.* (1980), two cell lines could not be differentiated by eight separate enzyme tests but were readily distinguished by karyology. However, it should be borne in mind that very careful interpretation in the light of considerable experience would be required to differentiate cell lines of normal karyotype beyond the level of species. In addition, many species have not been studied extensively using karyology and

Cell and Tissue Culture for Medical Research, edited by A. Doyle and J.B. Griffiths.

Table 2.6.1 Comparison of key features of authentication techniques

	Cytogenetic analysis	Isoenzyme analysis	DNA fingerprinting/ profiling
Determination of species	✓[a]	✓	
Rapid identification of a cell line			✓
Detection of cell line variation	✓		✓
Recognized by regulatory authorities	✓	✓	?[b]

[a]✓ = useful applications.
[b]? = under consideration.

therefore the prevalence of chromosome markers in natural populations, and hence in cell lines, may be unknown. Cytogenetic analysis is time consuming and, to some extent, subjective. In most routine and research laboratories handling a wide range of cell lines, full characterization and regular monitoring of identity using the above technique would be impracticable. However, developments in flow cytometric analysis of chromosomes may enable routine chromosome analysis in the future.

ISOENZYME ANALYSIS

Isoenzyme analysis is useful for the speciation of cell lines and for the detection of contamination of one cell line with another. The method utilizes the property of isoenzymes having similar substrate specificity but different molecular structures. This affects their electrophoretic mobility. Thus each species will have a characteristic mobility pattern of isoenzymes. While the species of origin of a cell line can usually be determined with only two isoenzyme tests (lactate dehydrogenase and glucose-6-phosphate dehydrogenase), specific identification of a cell line would require a larger battery of tests (Halton *et al.* 1983). A system in which eight isoenzyme activities were investigated offers a useful compromise (O'Brien *et al.* 1980). This procedure retained the advantage of rapid testing while also giving a useful level of specificity for identification purposes. Certain proteins characterized in isoenzyme analysis may undergo post-translational modification, which varies with the cell type. This offers additional characteristics for identification that can also act as markers for tissue of origin or developmental stage (Wright *et al.* 1981). However, in analysis of enzyme systems with multi-allelic products, low-level expression of some components may lead to unreliable detection and hence mis-identification. In conclusion, the use of a wide range of isoenzyme tests for accurate identification of cell lines requires a detailed knowledge of each isoenzyme system in each species to be studied.

DNA FINGERPRINTING AND DNA PROFILING

Multilocus DNA probes such as those of Jeffreys *et al.* (1985), oligonucleotides homologous to very simple sequence repeats (Ali *et al.* 1986) and the DNA sequence of the M13 protein III gene (Ryskov *et al.* 1988) interact with a widely dispersed range of loci throughout the genomes of many organisms and produce unique fingerprints in Southern blot analysis of cell line DNA. An alternative approach is to analyse single loci bearing variable numbers of tandem repeats (VNTRs), which produce simple DNA profiles (as opposed to fingerprints) comprising one or two bands. The key advantages of this technique in identification are, firstly, the apparent simplicity of scoring such profiles and, secondly, that methods based on the polymerase chain reaction are available (Tautz 1989; Jeffreys *et al.* 1991). Probes for single loci are used under highly specific conditions, and consequently, each one may be useful in only a single species, unlike multi-locus fingerprinting. In visualizing many loci, DNA fingerprinting techniques enable recognition of changes in a cell line that might go unnoticed when using single-locus methods. Multilocus DNA fingerprinting can therefore identify cell line variation (Thacker *et al.* 1988) and cross-contamination (van Helden *et al.* 1988), and can be used to confirm stability and consistent quality in cell stocks from diverse species. Furthermore, the development of a library of fingerprints from original documented material may help in the validation of suspect cell banks by comparison against the definitive version.

The standard techniques for cell authentication (cytogenetics and isoenzyme analysis) have been invaluable to cell biologists wishing to establish cell identity and to exclude cross-contamination. However, the availability of an ever-increasing number and diversity of cell lines indicates the need for rapid and simple techniques that also enable highly specific cell line identification. In a routine quality control setting where diverse cell lines are involved, current cytogenetic methods are impractical, whereas simplified isoenzyme procedures can offer at least species identification. DNA fingerprinting and profiling are important and timely additions to the cell biologists' battery of authentication techniques. In particular, multilocus DNA fingerprinting in a single test provides specific cell line identification for many species, combined with a screen for cross-contaminated or switched cultures. All the authentication techniques described should be viewed as complementary and each one will have areas in which it is especially useful. However, molecular genetic techniques are proving extremely powerful and sensitive for routine checks on the consistency of cell cultures.

REFERENCES

Ali S, Muller CR & Epplen JT (1986) DNA fingerprinting by oligonucleotide probes specific for simple repeats. *Human Genetics* 74: 239–243.

Halton DM, Peterson WD & Hukku B (1983) Cell culture quality control by rapid isoenzymatic characterisation. *In Vitro* 19: 16–24.

Hay RJ (1988) The seed stock concept and quality control for cell lines. *Analytical Biochemistry* 171: 225–237.

Jeffreys AJ, Wilson V & Thein SL (1985) Individual specific 'fingerprints' of human DNA. *Nature* 316: 76–79.

Jeffreys AJ, McLeod A, Tamaki K, Neil DL & Monckton DG (1991) Minisatellite repeat coding as a digital approach to DNA typing. *Nature* 354: 204–210.

Nelson-Rees WA, Daniels DW & Flandermeyer RR (1981) Cross-contamination of cells in culture. *Science* 212: 446–452.

O'Brien SJ, Shannon JE & Gail MH (1980) A molecular approach to the identification and individualization of human and animal cells in culture: isoenzyme and alloenzyme genetic signatures. *In Vitro* 16: 119–135.

Ryskov AP, Jincharadze AG, Prosnyak AG, Ivanov PL & Limborskaya (1988) M13 phage as a universal marker for DNA fingerprinting of animals, plants and microorganisms. *FEBS Letters* 233: 388–392.

Tautz D (1989) Hypervariability of simple sequences as a general source of polymorphic DNA markers. *Nucleic Acids Research* 17: 6469–6471.

Thacker J, Webb MBT & Debenham PG (1988) Fingerprinting cell lines: use of human hypervariable DNA probes to characterize mammalian cell cultures. *Somatic Cell and Molecular Genetics* 14: 519–525.

van Helden PD, Wild IJF, Albrecht CF, Theron E, Thornley AL & Hoal-van Helden EG (1988) Cross-contamination of human oesophageal squamous carcinoma cell lines detected by DNA fingerprinting analysis. *Cancer Research* 48: 5660–5662.

Wright WC, Daniels WP & Fogh J (1981) Distinction of seventy one cultured human tumour cell lines by polymorphic enzyme analysis. *Journal of the National Cancer Institute* 66: 239–247.

2.7 SCALE-UP OF LABORATORY CULTURES

The three most used laboratory scale-up methods are described, that is, the roller bottle (for attached and suspension cells), spinner flask (for suspension cells), and microcarrier culture (for attached cells in spinner flasks). An alternative to roller culture for attached cells is the Cell Factory (Nunc) which is a multiple-layered polystyrene plate system in units of 600, 1200, 6000, and 24 000 cm^2. Other methods which are not described but which can be considered are the simple perfusion systems, mainly developed for hybridoma cells, such as the Technomouse (Intregra Biosciences –supports 5×10^8 cells and capable of producing up to 1.5 g MAB/ month), or the MiniPERM (Hereaus – 15–30×10^6 cells ml^{-1} producing 0.5–1.5 mg MAB ml^{-1} d^{-1}). More advanced systems are available such as the Acusyst hollow fibre cultures (Cellex Biosciences) and the CellCube (Costar), a stacked polystyrene plate system giving units of 21 250, 42 500 and 85 000 cm^2.

PROCEDURE: ROLLER BOTTLE CULTURE

Roller bottle culture is considered the first scale-up step for anchorage-dependent cells from stationary flasks or bottles but it is also suitable for suspension cells such as hybridomas. This is achieved by using all the internal surface for cell growth, rather than just the bottom of a bottle. The added advantages are that a smaller volume of medium and thus a higher product titre can be achieved; the cells are more efficiently oxygenated due to alternative exposure to medium and the gas phase; and dynamic systems usually generate higher unit cell densities than stationary systems.

The basis of the roller culture system is to place multiple cylindrical bottles into an apparatus that will rotate the bottles evenly at set rotational speeds (between 5 and 60 rph). Apparatus is available that will accommodate four to hundreds of bottles, and in fact this method is used for the large-scale production of vaccines using 29 000 bottles (1 l volume) per batch. Any cylindrical bottle of the correct tissue culture grade surface (borosilicate glass or polystyrene) can be used and there are many available commercially between 250 and 2000 cm^2 in capacity.

Materials and equipment

- Standard tissue culture media and reagents
- Roller bottles (e.g. 23×12 cm plastic bottle with surface area of 1400 cm^2 from, for example, Bibby Sterilin, Corning, Costar)

Cell and Tissue Culture for Medical Research, edited by A. Doyle and J.B. Griffiths.
© 2000 John Wiley & Sons, Ltd

- Roller bottle apparatus with speed control between 5 and 60 rph (e.g. Wheaton, Bellco, New Brunswick, Integra Biosciences)

1. Add approximately 1 ml of complete culture medium at 37°C per 5 cm^2 culture area (to give a medium:air volume ratio of 1 : 5 to 1 : 10).
2. Add 5% CO_2 in air to headspace.
3. Add $1-2 \times 10^4$ cells cm^2 ($0.5-1.0 \times 10^5$ suspension cells ml^{-1}).
4. Seal bottle and place in roller culture apparatus.
5. Rotate bottle at 12–24 rph for the initial attachment phase (2–8 h, depending upon cell type). This faster speed is to get an even distribution of cells but should be reduced for cells with low attachment efficiency.
6. Reduce revolution rate to 5–10 per hour when culture is growing.
7. Cell growth can be monitored initially under an inverted microscope and later in the culture period by visual inspection.
8. A medium change can be carried out after 4–5 days if the pH becomes acid and/or maximum cell densities are required. This can also be done to change to a production medium, or when infecting cells, and a lower medium volume can be added to get a higher product concentration.
9. Harvest cells when confluent (5–6 days) by removing the medium and trypsinizing in the conventional way. After adding trypsin, roll the bottles as speeds of 20–60 rph until cells detach. Cell yields will be similar to or up to 2-fold higher than in stationary cultures, and multilayering (non-diploid cell lines) will occur.

Comment

Successful roller bottle culture depends a great deal on solving various logistic problems. Larger roller bottles (1500 cm^2) are unwieldy to use and cannot be manipulated in many tissue culture cabinets. Special Class II cabinets are suitable; otherwise use the smaller bottles. Also, do not underestimate the time involved in harvesting cells from roller bottles. If a large number of bottles are being used, trypsinize four at a time and get the cells in fresh medium with serum before beginning the next batch. Cells deteriorate very rapidly in trypsin, and also when being held at high concentrations in medium, so do not store while harvesting multiple batches of roller bottles, but put them in a new culture as soon as possible. Unless your laboratory is geared up for this type of work (multiple tissue culture cabinets and staff), limit the number of roller bottles to 8–16 at a time. It is very difficult to remove completely all the cells from a roller bottle without repeated washing (increases contamination risk, standing time of cells and centrifugation volume) and it is wise to assume that a 10–20% loss may occur. This is particularly important when judging how many bottles to set up as an inoculum for a larger culture.

SUPPLEMENTARY PROCEDURE: EXTENDED SURFACE AREA ROLLER BOTTLE (ESRB)

The roller bottle system is a multiple process requiring considerable staff time for repeated manipulations. Thus many modifications have been introduced that increase the efficiency or surface area capacity of roller bottles. Some examples are as follows:

1. Extended Surface Area Roller Bottle (ESRB) (Bibby Sterilin Ltd). This polystyrene bottle has a ribbed or corrugated surface, doubling the surface area of the 850 cm^2 flask to 1700 cm^2.
2. ImmobaSil (Ashby Scientific Ltd/Integra Biosciences) – polystyrene flasks and roller bottles with a silicone rubber matrix that increases the surface area several-fold.
3. If hundreds of roller bottles need to be handled then a robotic culture system Cellmate (The Automation Partnership, Melbourn, UK) could be considered, which reduces manning by 10-fold as well as giving improved consistency and reproducibility.

Discussion

The roller bottle technique is a well-established and successful culture method widely used for the production of cells and products. Also, some cell lines (particularly epithelial) may not be as successfully grown in roller bottles as in stationary bottles. Common problems are streaking, clumping or inadequate spreading over the total surface (e.g. non-locomotory cell lines). An alternative scale-up route is to use multisurface stationary systems such as the Nunclon Cell Factory (Nunc) or the CellCube (Costar).

SUPPLEMENTARY PROCEDURE: SPINNER FLASK CULTURE

The first scale-up step for cells growing in suspension either naturally or after adaptation, or anchorage-dependent cells on microcarriers, is the spinner flask. This technique originated by placing a silicone or Teflon-coated magnet with a central ring into a glass vessel that is placed on a magnetic stirrer. Although this simplistic approach is still used, it is preferable to use a specially constructed flask where the magnet is suspended just above the bottom of the flask. A typical spinner flask is shown in Figure 2.7.1a, and has a stirrer shaft containing the magnet fitted to the bottle top. In addition, side arms are fitted in order to add cells, change medium or gas with oxygen or CO_2-enriched air (in which cases filters should be fitted). It is preferable to have a slightly convex surface under the stirrer bar to aid mixing. There are many variations on this theme from different suppliers, mostly with regard to size, number and positioning of the stirrer bar.

The spinner flask is a glass vessel, usually intended to be used for replicate cultures, on a multibased magnetic stirrer. Sizes range from a few millilitres to 20 l, but for ease of handling, safety and physicochemical reasons it is advisable to consider 10 l as a maximum practical size. The units are sterilized by autoclaving.

Reagents and solutions

Growth medium:
Standard tissue culture media can be used with various supplements. Serum at 10% can cause foaming at fast stirring speeds, so the serum level is usually reduced

to as low a level as possible (0.5–5%). The use of HEPES buffer to stabilize the pH during the setting-up procedure is beneficial.

Medium supplements:
Pluronic F-68 (polyglycol) (BASF, Wyandot) can be used at 0.1% to protect cells against mechanical damage, especially at reduced serum concentrations. Carboxymethyl cellulose (CMC) (15–20 cP) can also be used to protect cells from mechanical damage at 0.1% concentration. Antifoam (6 ppm) (Dow Chemical Co.) is recommended at serum concentrations above 2–3%.

Materials and equipment

- Spinner flask

1. Add 200 ml of growth medium to a 1-l spinner flask, gas with 5% CO_2 and warm to 37°C.
2. Add cells from a logarithmically growing seed culture at $1–2 \times 10^5$ cells ml^{-1}.
3. Place spinner flask on a magnetic stirrer at 100–250 rpm (this is variable depending upon the cell type and geometry of the stirrer bar and vessel – a guideline is to use the minimum speed, which gives, by visual examination, complete homogeneity).
4. Monitor cell growth at least daily by taking a small sample from the side arm (remove flask to a tissue culture cabinet) and carrying out a cell count (Trypan blue stain and a haemocytometer).
5. Monitor the pH. In closed systems (i.e. all ports closed with no filters) the pH will become acidic and need adjusting by days 2–3. Remove the vessel to a tissue culture cabinet and regas with 5% CO_2 in air; add sodium bicarbonate (5.5%); or allow cells to settle, remove around 50% of the medium and replace with fresh pre-warmed medium.
6. After 3–5 days the culture density will typically reach $1–2 \times 10^6$ ml^{-1} (for many hybridomas, $0.8–1.5 \times 10^6$ ml^{-1} only) and can be harvested, or the culture prolonged for extra cell growth with twice-daily 50% medium changes. Suspension cells tend to have only a limited stationary phase and it is thus Important to monitor growth more closely than in monolayer culture to ensure that healthy, rather than dying/dead, cells are harvested.

If cells are prone to clumping, or adhering to surfaces, then a medium with reduced Ca^{2+} and Mg^{2+} concentration should be used (e.g. suspension MEM). Additionally it is good practice to siliconize the glass vessels (see below).

There is a tendency to subculture suspension cells by dilution into fresh medium. However, some cells do not grow optimally under this regime and should be centrifuged (800 *g* for 5 min) and resuspended in new medium.

There is a wide range of spinner vessels available:

- Conventional vessels (Figure 2.7.1(a)) from Bellco or Wheaton.
- Conventional vessels but with large-bladed paddles attached to the magnet or base of the stirrer shaft (Bellco and Cellon μ spinner flasks). These vessels give more efficient stirring with better mixing at the lower speeds and are thus primarily aimed at microcarrier culture.

- The Techne system uses a radial stirring action, which is designed for good mixing at low stirring speeds and to minimize mechanical stress to cells (Figure 2.7.1(b)).
- Radial (pendulum) oscillatory stirring system similar to Techne but with dual stirrers (Cellspin – Integra Biosciences).
- The Techne BR-06 Bioreactor uses a floating impeller, which again gives a gentle stirring action. An advantage of this system is that the vessel can be used equally effectively at various working volumes from 500 ml to 3 l. This allows an *in situ* build-up of cell seed by just adding extra medium as cell growth occurs.

If both side arms are fitted with filters, then continuous aeration through the headspace can be carried out, giving better control of pH and higher cell densities.

SUPPLEMENTARY PROCEDURE: MICROCARRIER CULTURE

Microcarrier culture, i.e. the growth of anchorage-dependent cells on small particles (usually spheres) 100–300 μM in size suspended in stirred culture medium,

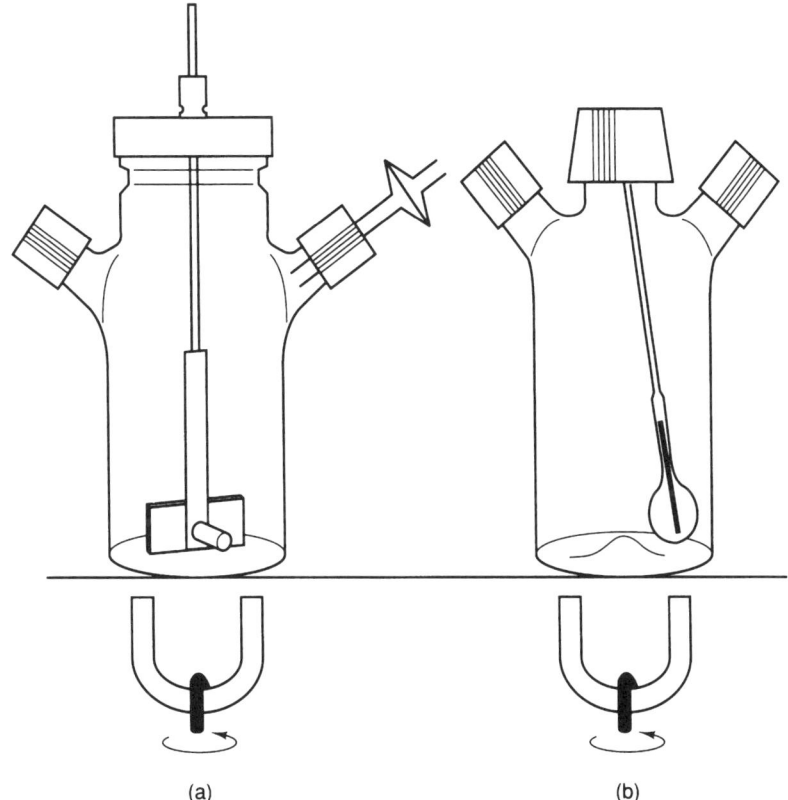

(a)   (b)

Figure 2.7.1 (a) A typical spinner flask with magnetic spinner bar. (b) A Techne flask with asymmetric stirring rod.

has made a tremendous impact on upstream processes (Griffiths 1991). During the 1960s, the availability of human diploid cell (HDC) lines allowed a rapid expansion in the manufacture of human vaccines. However, large-scale production was restricted to using many replicate small cultures (flask and roller bottles) because HDC lines were anchorage-dependent. This restriction did not apply to veterinary vaccines, which were being produced in large scaleable fermenter processes (a suspension cell line, BHK, was licensed for the production of veterinary vaccines). A similar unit scale-up process was a goal being sought for surface-growing cells; hence the large range of process bioreactors that have been developed (Griffiths 1988). The opportunity came when Van Wezel (1967) showed that cells would grow on dextran beads in stirred bioreactors. However, the chromatography-grade dextran (Sephadex A-50, Pharmacia) being used was unsuitable for consistent and reliable growth of cell lines, particularly HDC. After considerable developmental work by van Wezel and Pharmacia, a range of suitable microcarriers became available (the Cytodex range, Pharmacia). The first industrial process using microcarriers was described by Meignier et al. (1980) for foot and mouth disease virus (FMDV) vaccine production. Subsequently a whole range of microcarriers based on gelatin, collagen, polystyrene, glass, cellulose, silicone rubber, polyacrylamide and silica have been manufactured to meet all situations (Table 2.7.1). The key criteria were to get the surface chemically and electrostatically correct for cell attachment, spreading and growth. To put this development in full perspective the following facts illustrate the sheer scale of opportunity that this method gives the cell culturist:

1. 1 g of Cytodex microcarrier has a surface area of 6000 cm^2 and, used at a modest concentration of 2 g l^{-1} gives 12 000 cm^2 l^{-1}. This is equivalent to 8 large or 15 small roller bottles.

Table 2.7.1 Examples of microcarriers

Trade name	Manufacturer	Material	SG	Diam (μm)	Area (cm^2/g)
Biosilon	Nunc	Polystyrene	1.05	160–300	255
Bioglas	Solohill Eng.	Glass[a]	1.03	150–210	350
Bioplas	Solohill Eng.	Polystyrene[a]	1.04	150–210	350
Biospheres	Solohill Eng.	Collagen[a]	1.02	150–210	350
Cytodex 1	Pharmacia	DEAE Sephadex	1.03	160–230	6000
Cytodex 2	Pharmacia	DEAE Sephadex	1.04	115–200	5500
Cytodex 3	Pharmacia	Collagen	1.04	130–210	4600
Cytosphere	Lux	Polystyrene	1.04	160–230	250
Dormacell	Pfeifer & Langen	Dextran	1.05	140–240	7000
DE-53	Whatman	Cellulose	1.03	Fibres	4000
Gelibead	Hazelton Lab.	Gelatin	1.04	115–235	3800
Microdex	Dextran Prod.	DEAE Dextran	1.03	150	250
Ventreglas	Ventrex	Glass	1.03	90–210	300
Ventregel	Ventrex	Gelatin	1.03	150–250	4300

[a]Biospheres (glass, plastic, collagen) available at SG of 1.02 or 1.04 and diameters of 150–210 or 90–150 μm (manufactured by Solohill Eng. and distributed by Whatman and Cellon).

2. Suspension culture systems, unlike bottles, etc., can be environmentally controlled and have been optimized to a minimum 100 l capacity, i.e. a 1×100 l fermenter was equivalent to 800–1500 roller bottles. The scale-up potential is 10 000l (for suspension cells), although currently only 4000l has been used for microcarrier culture.

The availability of microcarriers has not only opened up industrial opportunities but has also allowed the laboratory worker to produce substantial quantities of cells and cell products for research and development purposes.

General principles

1. Cells differ in their attachment requirements, and thus a range of microcarriers should be assessed for suitability (see Table 2.7.1).
2. It is critical to have suitable culture equipment. Spinner flasks should have a special impeller (e.g. Bellco, Wheaton, Techne and Cellspin flasks are specifically designed for microcarriers) because a magnetic bar is unsuitable. The shape of the vessel is important for homogeneous mixing, as is the stirring unit, which must operate smoothly without vibration at low revolutions (15–100 rpm). Culture vessels must be siliconized to prevent microcarriers sticking to the glass, as well as the bottles used for preparing and storing microcarriers (see Preliminary Procedure below).
3. Microcarriers can be purchased as dry powders or sterilized solutions. The dry powders are swelled in Ca^{2+}/Mg^{2+}-free phosphate-buffered saline (PBS) (3 h), the supernatant is decanted and discarded and then the residual powder is washed and sterilized in the same buffer (50 ml g^{-1}) by autoclaving (15 1b in^2, 15 min, 115°C). *Warning:* do not overheat. The powder microcarriers are 'softer' than glass and plastic and can be used in higher concentrations, thus giving vastly greater unit surface areas (e.g. Cytodex, 6000 cm^2 g^{-1}, glass/polystyrene, 300 cm^2 g^{-1}). Some microcarriers are available at different specific gravities (usually between 1.01 and 1.05), offering a choice depending upon the stirring (or mixing) system – roller bottles, shake flasks or airlift bioreactors can be used – and process requirements (e.g. if multiple medium changes are to be made, the heavier beads settle out more quickly and efficiently).
4. Consideration may have to be given to using a supplemented medium during the initial stages of culture to aid cell attachment and to offset the effects of low cell density, which will be more critical in microcarrier than stationary culture. The supplementation may be simple, e.g. non-essential amino acids, pyruvate (0.1 mg ml^{-1}), adenine (10 pg ml^{-1}), hypoxanthine (3 pg ml^{-1}) and thymidine (10 pg ml^{-1}). Other supplements include tryptose phosphate broth (1 mg ml^{-1}), HEPES (5 mM), transferrin (10 mg l^{-1}) and fibronectin (2 pg ml^{-1}). Serum, unless serum factors are added, may have to be used at 5–10% initially before being reduced after 1–2 days of culture.
5. The conventional cell-counting procedure of trypsinizing cells and then counting are suspended in 0.1 M citric acid containing 0.1% crystal violet in a haemocytometer with Trypan blue can be tedious or even inaccurate (due to problems

in removing cells from the microcarrier sludge). A better method is to use the nuclei-counting method developed by Sanford *et al.* (1951). The microcarriers are suspended in 0.1 M citric acid containing 0.1% crystal violet. The mixture can be vortex mixed or left at 37°C for at least 1 h. An advantage is that samples can be stored for long periods (at least 1 week) at 4°C before being counted. However, it is a total, not viable, cell count. Cells can also be fixed and stained on microcarriers for microscopic examination using standard procedures. Haematoxylin is the most widely used stain.

6. It is possible to re-use some microcarriers using good washing procedures and re-equilibration in PBS. Except for glass microcarriers this is not recommended, especially for more than one recycle, because performance drops significantly. There are reports of *in situ* recolonization of microcarriers (Crespi & Thilly 1981), which is feasible but once again reduces the efficiency of the process (lower yields and heterogeneity in cell numbers between carriers).

7. Enzymatic harvesting (e.g. trypsin) of some cells from microcarriers can be difficult or damaging. Consideration can be given to digesting some microcarriers (e.g. by dextranase, collagenase) because this gives a far quicker and less damaging means of getting a single-cell suspension.

8. Microcarrier culture requires more critical preparation than non-dynamic culture systems. To ensure success it is very important that all experimental details are carried out optimally. Of particular importance is the quality of the cell inoculum. This should be rapidly dividing, not stationary, and cells should be in a good condition (i.e. not trypsin-damaged) and as near a single-cell suspension as possible. It is good practice to feed the seed culture 24 h before use. Also ensure that medium is prewarmed and pre-equilibrated for pH (shifts in pH during attachment are extremely damaging). Due to the number of manipulations in the process, extreme care with aseptic technique should be taken, and as many as possible of the steps carried out in a tissue culture cabinet. Do not inoculate below the minimum number; the culture may not initiate or, if it does, will not reach maximum cell density.

PRELIMINARY PROCEDURE: SILICONIZATION

Materials and equipment

• Dimethyldichlorosilane (Merck or Hopkin and Williams (as Repelcote))

1. Add a small volume of siliconizing fluid to clean glassware (spinner flask, bottles and pipettes used for handling microcarriers) and wet all surfaces.
2. Drain off excess fluid and allow glassware to dry.
3. Wash glassware in distilled water (combination of three washes/prolonged immersion).
4. Autoclave.

This siliconization process will allow glassware to be used through many repeat processes but glassware should be retreated at least annually.

PROCEDURE: MICROCARRIER CULTURE

Materials and equipment

- Eagle's minimum essential medium (MEM) (or equivalent alternative) with 10% foetal bovine serum (FBS), Eagle's non-essential amino acids and other supplements as necessary (see 'General principles', point 4).
- Anchorage-dependent cell line (e.g. MRC-5)
- Microcarrier, e.g. Cytodex 3 (Pharmacia)
- Ca^{2+}/Mg^{2+}-free PBS
- Spinner vessel adapted for microcarrier culture (e.g. Bellco or Techne)

1. Add complete medium to spinner flask (200 ml in 1 l flask), gas with 5% CO_2 and allow to equilibrate.
2. Decant PBS from sterilized stock solution of Cytodex 3 and replace with growth medium (1 g to 30–50 ml). Add Cytodex 3 to spinner vessel to give a final concentration of 2 g l^{-1} (1–3 g l^{-1}).
3. Put spinner on magnetic stirrer at 37°C and allow temperature and physiological conditions to equilibrate (minimum 1 h).
4. Add cell inoculum obtained by trypsinization of late log phase cells (pre-stationary phase) at over five cells per bead (optimum to ensure cells on all beads is seven); inoculate at the same density per square centimetre as with other culture types, e.g. $5–10 \times 10^4$ cm^{-2}. Cytodex 3 at 2 g l^{-1} inoculated at six cells per bead gives:

$$8 \times 10^6 \text{ microcarriers} = 4.8 \times 10^7 \text{ cells } l^{-1}$$

$$9500 \text{ cm}^2 = 5 \times 10^4 \text{ cells cm}^{-2}$$

$$200 \text{ ml} = 2.5 \times 10^5 \text{ cells ml}^{-1}$$

5. Place spinner flask on magnetic stirrer and stir at the minimum speed to ensure that all cells and carriers are in suspension (usually 20–30 rpm). It is advantageous if cells and carriers are limited to the lower 60–70% of the culture for this purpose. Alternatively either:

   - Inoculate in 50% of the final medium volume and add the rest of the medium after 4–8 h, or
   - Stir intermittently (for 1 min every 20 min) for the first 4–8 h. However, only use this option for cells with very poor plating efficiency because it causes clumping and uneven distribution of cells per bead

6. When the cells have attached (expect 90% attachment) the stirring speed can be increased to allow complete homogeneity (40–60 rpm).
7. Monitor progress of culture by taking 1 ml samples at least daily; observe microscopically and carry out cell (nuclei) counts.
8. As the cell density increases there is often a tendency for microcarriers to begin clumping. This can be avoided by increasing the stirring speed to 75–90 rpm.
9. After 3–4 days the culture will become acid. Remove the spinner flask and re-gas the headspace and/or add sodium bicarbonate (5.5% stock solution).

With some cells, or at microcarrier densities of 3 g l^{-1} or more, partial medium changes should be carried out. Allow culture to settle (5–10 min), siphon off at least 50% (usually 70%) of the medium and replace with prewarmed fresh medium (serum can be reduced or omitted at this stage). Replace spinner flask on stirrer.

10. After 4–5 days cells reach a maximum cell density (confluency) at the same level as in static cultures, although multilayering is not so prevalent. Thus a cell yield of 1–2 × 10^6 ml^{-1} (2–4 × 10^5 cm^{-2}) can be expected.

11. Cells can be harvested in the following way:

   - Allow culture to settle (10 min)
   - Decant off as much medium as possible (> 90%)
   - Add warm Ca^{2+}/Mg^{2+} -free PBS (or EDTA in PBS) and mix
   - Allow culture to settle and decant off as much. PBS as possible
   - Add 0.25% trypsin (30 ml) and stir at 75–100 rpm for 10–20 min at 37°C
   - Allow the beads to settle out (2 min). Either decant trypsin plus cells or pour mixture through a sterile, coarse-sinter, glass filter or a specially designed filter such as the Cellector (Bellco)
   - Centrifuge cells (800 *g* for 5 min) and resuspend in fresh medium (with serum or trypsin inhibitor, e.g. soybean inhibitor at 0.5 mg ml^{-1})

As a guide to calculating settled volume and medium entrapment by micro-carriers, 1 g of Cytodex, for example, has a volume of 15–18 ml.

DISCUSSION

The basic principles for using microcarrier culture are described, together with many notes on how to avoid problems and get the most out of this very powerful and useful technology. The description is suited to small-scale processes based on spinner flasks (i.e. 200 ml to 5 l) at levels that do not need special adaptations for perfusion, etc. This does not mean that scale-up is not possible; in fact, it has been volumetrically scaled up to 4000 l for the production of interferon and viral vaccines. It has also been scaled up in density by the use of spin-filters (Griffiths *et al.* 1987) to allow continuous perfusion of the culture and thus operation at microcarrier concentrations up to 15 g l^{-1} and cell densities over 10^7 ml^{-1}.

An advance on the solid microcarriers described above are the porous micro-carriers which allow cells to grow within the fibrous matrix of the microcarrier. This gives an enormously increased surface area for cell attachment and very high unit cell density, as well as protecting the cells from shear force damage. Porous microcarrier technology was pioneered by the Verax Corporation (now Cellex Biosciences Inc.). Turnkey units were available from 16 ml to 24 L fluidized beds (Runstadler *et al.* 1989). The smallest system in the range, Verax System One, is a benchtop continuous perfusion fluidized bioreactor suitable for process assess-ment and development, and also for laboratory-scale production of MABs. Cells are immobilized in porous collagen microspheres, weighted to give a specific gravity of 1.6, which allowed high recycle flow rates (typically 75 cm min^{-1}) to give effi-cient fluidization. The microspheres have a sponge-like structure with a pore size of 20–40 μm and internal pore volume of 85% allowing immobilization of cells to densities of 1 to 4×10^8 ml^{-1}. They are fluidized in the form of a slurry.

An alternative commercial system is the Cytopilot (Pharmacia) which is a fluidized system using polyethylene carriers (Cytoline) and supports 12×10^7 cells ml^{-1} carrier (Valle *et al.* 1998). It is available as the Cytopilot-Mini (400 ml bed) for laboratory-scale operation as well as sizes up to 25 L.

There is a range of porous microcarriers available (Table 2.7.2), which allow the design of one's own fluidized system, or alternatively some of the carriers are designed for stirred cultures.

Table 2.7.2 Porous microcarriers

Trade name	Supplier	Material	Diam (μm)	Culture mode[a]
Cultispher G	HyClone	Gelatin	170–270	F
			300–500	
Cytocell	Pharmacia	Cellulose	180–210	S
Cellsnow	Kirin Ltd	Cellulose	800–1000	S
Cytoline 1, 2	Pharmacia	Polyethylene	1200–1500	F
ImmobaSil	Ashby Scientific	Silicone rubber	1000	S
Microsphere	Cellex Biosciences	Collagen	500–600	F
Siran	Schott Glasswerke	Glass	600–1000	F
			4000–6000	X
Microcarrier	Asahi Chem. Ind. Co.	Cellulose	300	S

[a] S, stirred; F, fluidized; X, fixed bed.

REFERENCES

Crespi CL & Thilly WG (1981) Continuous cell propagation using low-charge microcarriers. *Biotechnology and Bioengineering* 23: 983–993.

Griffiths JB (1988) Overview of cell culture systems and their scale-up. In: Spier RE & Griffiths JB (eds) *Animal Cell Biotechnology*, Vol. 3, pp. 179–220. Academic Press, London.

Griffiths JB (1991) Cultural revolutions. *Chemistry and Industry* 18: 682–684.

Griffiths JB, Cameron DR & Looby D (1987) A comparison of unit process systems for anchorage dependent cells. *Developments in Biological Standardization* 66: 331–338.

Meignier B, Mougeot H & Favre H (1980) Foot and mouth disease virus production on microcarrier-grown cells. *Developments in Biological Standardization* 46: 249–256.

Runstadler PW Jr, Tung AS, Hayman EG, Ray NG, Sample JG & DeLucia DE (1989). Continuous culture with macroporous matrix, fluidised bed systems. In: Lubiniecki A S (ed.) *Large Scale Mammalian Cell Culture Technology*, pp. 363–391. Marcel Dekker, New York.

Sanford KK, Earle WR & Evans VJ (1951) The measurement of proliferation in tissue cultures by enumeration of cell nuclei. *Journal of the National Cancer Institute* 11: 773–795.

Valle MA, Kaufmann J, Bentley WE & Shiloach J (1998). In: Merten O-W, Perrin P & Griffiths JB (eds) *New Developments and New Applications in Animal Cell Technology*, pp. 381. Kluwer Academic Publishers, Dordrecht.

Van Wezel AL (1967) The growth of cell strains and primary cells on microcarriers in homogeneous culture. *Nature (London)* 216: 64–65.

BACKGROUND READING

Griffiths JB (1992) Scaling-up of animal cell
cultures In: Freshney RL (ed.) *Animal Cell
Culture: a Practical Approach,* pp. 47–93.
IRL Press, Oxford.

CHAPTER 3

CELL CHARACTERIZATION AND ANALYSIS

3.1 TECHNIQUES FOR CELL CHARACTERIZATION AND ANALYSIS – AN OVERVIEW

Tissue culture cells utilized in medical research need to be characterized for a variety of reasons. All cell populations are to some extent heterogeneous. This heterogeneity is most commonly observed as variations in morphology, biochemical or biological properties, expression of cell surface components or even position in the cell cycle. Generally, some form of detailed characterization is performed to establish the relationship of newly derived cell lines to the cells of origin, usually in terms of cell type and degree of differentiation.

A battery of technologies and procedures is available for characterization studies. This is particularly important when the cells have been subcultured for some time since the original characterization; such treatment can easily result in selection of a cell line with undesirable properties, or, worse, enhance the chance of contamination with other, faster-growing cell lines. A strategy for the characterization of newly isolated cell lines is given in Table 3.1.1.

This chapter is primarily concerned with a variety of specialist, but generally readily available, techniques which define cell populations in terms of their morphological and biochemical characteristics. In particular these techniques determine the heterogeneity/homogeneity, the cell kinetics and the cellular interactions. Not surprisingly the majority of such techniques are morphologically based, performed on living and/or fixed cells. The limitations of such techniques in terms of resolution or artifacts should always be kept in mind. However, morphology alone cannot always define properties, and functional markers utilizing techniques like cytochemistry or immunological probes are useful adjuncts. The increasing availability of image-processing and flow cytometry equipment has allowed such procedures to become quantitative and, in the case of the latter, preparative.

A further characteristic of any cell population is its kinetics or rate of cell turnover. Direct cell counting can, at best, only indicate the proportion of viable cells and is useful for studying total cell populations. In comparison, cell kinetics determines the number of actively dividing cells.

The procedures described in this chapter aim to contribute to establishing the functional properties of the cell as these criteria are essential prerequisites for experimental procedures involving cell cultures.

Cell and Tissue Culture for Medical Research, edited by A. Doyle and J.B. Griffiths.
© 2000 John Wiley & Sons, Ltd

Table 3.1.1 Cell characterization

Morphology
- Whole cell
  1. Cytochemistry
- Sections
  1. Light microscopy stains
  2. Transmission electron microscopy
- Unfixed
  1. Phase contrast
  2. Confocal

Biochemistry
- Cytochemistry
- Isoenzymes
- DNA fingerprinting (see relevant chapter)
- Product expression – type and kinetics (see relevant chapter)
- Nutrient requirements and metabolic coefficients (see relevant chapter)

Differentiation markers
- Unfixed cells
  1. Light microscopy – fluorescent probes
- Fixed cells
  1. Scanning electron microscopy – gold or latex probes
  2. Transmission electron microscopy/sequential stained sections
- Fixed, permeabilized cells
  1. Light microscopy/transmission electron microscopy – gold- or enzyme-linked probes
- Stained sections
  1. Light microscopy/transmission electron microscopy – gold- or enzyme-linked probes

Kinetics
- Viable counts
- Cloning
- Nucleic acid concentration
- Total protein (etc.)
- ^{13}H-Thymidine incorporation
- BUdR assay (etc.)

3.2 IMAGE CYTOMETRY – AN OVERVIEW

In biological research and medical analyses the development and application of exploratory methods in cytology and histology have led to ever-increasing needs in the field of computerized microscopic image processing. These needs and the instrumentation now becoming available to address them will be reviewed here.

A variety of imaging modes, preparation techniques and analytical methods is available for the investigation of the diverse range of specimens arising from research using cell and tissue culture.

Each imaging code (often used in conjunction with a specific preparation technique) will provide a unique piece of microstructural information about the specimen. Consequently, an integrated approach is often necessary to derive the exact breadth of information needed.

The methods listed below have been applied to a range of specimen types outlined in Table 3.2.1. Those techniques highlighted in bold are considered to be especially appropriate for the application.

GOALS OF MEDICAL MICROSCOPIC IMAGING

The observation of microscopic images as an element in the medical decision process involves:

- The screening of precancerous and cancerous lesions, which can become systematic for easily accessible organs such as the uterine cervix, blood, urinary cells exfoliating from the bladder, and the upper aerodigestive tracts
- The diagnosis of all benign or malignant tumours on biopsies of affected organs
- The prognosis based on an evaluation of the severity of cellular or tissue lesions
- The follow-up of patients and evaluation of anti-cancer therapies on the basis of biopsies of the target organs after treatment

All severe pathologies (except in cardiology) are confirmed or rejected on the basis of microscopic examination of affected cells or tissues.

DIVERSITY OF MICROSCOPIC IMAGING

Beyond the medical use of microscopes, these instruments are an established tool for biological investigations involving several imaging technologies covering an

Cell and Tissue Culture for Medical Research, edited by A. Doyle and J.B. Griffiths.
© 2000 John Wiley & Sons, Ltd

Table 3.2.1 Imaging/instrumentation

Transmission Electron Microscopy (TEM)

T1. Bright Field imaging
T2. Dark Field imaging
T3. Electron Diffraction analysis
T4. High Resolution/Field Emission Imaging

Scanning Electron Microscopy (SEM)

S5. Secondary Electron Imaging
S6. Backscatter Electron Imaging
S7. Cathodoluminescence imaging

Preparation Methods

Preparation Techniques for Transmission Electron Microscopy

PT1. Primary Fixatives
PT2. Secondary Fixatives
PT3. Observation of Whole Mounts by Positive Staining
PT4. Observation of Whole Mounts by Negative Staining
PT5. Embedding and Ultrathin Sectioning
PT6. Staining and Ultrastructural Cytochemistry
PT7. Ambient Temperature Replication and Shadowing Methods
PT8. Cryo-preparation Techniques for Observation of Hydrated Cells and
          Macromolecules

Preparation Techniques for Scanning Electron Microscopy

PS1. Preparation of Dehydrated Cells and Macromolecules
PS2. Cryo-preparation of Hydrated Cells and Macromolecules

Analytical

Additional Analytical Techniques

A1. Electron Probe X-ray Microanalysis
A2. Energy Loss and Energy Filtering Spectroscopy and Imaging
A3. EM Autoradiography
A4. Immunocytochemistry
A5. Image Analysis
A6. Image Reconstruction

Investigation/Tissue/Cell	Appropriate EM Techniques and Associated Analytical Method
Cell Quantification	**T1, S5,** PT1, PT2, PT3, PT5, PT6, PT8, PS1, PS2, A1, A2, **A5, A6.**
Detection of Mycoplasma	T1, **T4,** S5, PT1, PT2, PT4, PT5, PT6, PT8, PS1, PS2, A4, A5, A6.
Bacteria and Fungi	**T1,** T2, T4, **S5,** PT1, PT2, PT4, PT5, PT6, PT8, PS1, PS2, A4, A5, A6.
Elimination of Contamination	**T1,** T2, PT1, PT2, PT4, PT5, PT6, PT8, A4.
Fibroblasts	**T1,** T2, T4, **S5,** PT1, PT2, PT4, PT5, PT6, PT8, PS1, PS2, A4, A5, A6.
Cell Characterization and Analysis	**T1,** T4, **S5,** PT1, PT2, PT4, PT5, PT6, PT8, PS1, PS2, **A4, A5, A6.**

Table 3.2.1 Continued

Investigation/Tissue/Cell	Appropriate EM Techniques and Associated Analytical Method
Cell Death in Culture Systems	**T1,** T4, S5, PT1, PT2, PT4, PT5, PT6, PT8, PS1, PS2, **A4,** A5, A6.
Cells Lines for Medical Research	**T1,** T4, **S5,** PT1, PT2, PT4, PT5, PT6, PT8, PS1, PS2, A4, A5, A6.
Explant Cultures	**T1,** T4, **S5,** PT1, PT2, PT4, PT5, PT6, PT8, PS1, PS2, A4, A5, A6.
Haemopoetic Cells in Culture	T1, T4, S5, PT1, PT2, PT4, PT5, PT8, PS1, PS2, A4, A5.
EBV transformation	T1, T2, T4, S5, PT1, PT2, PT4, PT5, PT6, PT8, PS1, PS2 A4, A5.
Cryopreservation of human blood for b-cell immortalization	T1, T4, S5, PT1, PT2, PT5, PT6, **PT8,** PS1, **PS2,** A4, A5.
T-Cells and Thymocytese	**T1,** T4, **S5,** PT1, PT2, PT4, PT5, PT6, PT8, PS1, PS2, A4, A5.
Establishment of Human Leukaemia Cell Lines	T1, T4, S5, PT1, PT2, PT4, PT5, PT6, PT8, PS1, PS2, **A4,** A5.
Derivation and subculturing of human diploid cell strains	T1, T4, S5, PT1, PT2, PT4, PT5, PT6, PT8, PS1, PS2, A4, A5.
Initiation of adherent monolayer primary cell cultures	T1, T4, **S5,** PT1, PT2, PT4, PT5, PT6, **PT8,** PS1, **PS2,** A4, A5.
Human keratinocytes	**T1,** T4, **S5,** PT1, PT2, PT4, PT5, PT6, PT8, PS1, PS2, A4, A5, A6.
Human Gastric Epithelial Cells	T1, T4, S5, PT1, PT2, PT4, PT5, PT6, PS1, PS2, A4, A5.
Human Hepatocytes	T1, T4, S5, PT1, PT2, PT4, PT5, PT6, PT8, PS1, PS2, A4, A5.
Embryonic kidney in organ culture	T1, T4, S5, PT1, PT2, PT4, PT5, PT6, PT8, PS1, PS2, A4, A5.
Culture of Amniocytes, Chorionic Villi and Human Foetal and Placental Tissue	T1, T4, S5, PT1, PT2, PT4, PT5, PT6, PT8, PS1, PS2, A4, A5.
Endothelial Cells	T1, T4, S5, PT1, PT2, PT4, PT5, PT6, PT8, PS1, PS2, A4, A5, A6.
Nerve Cells	T1, T4, S5, PT1, PT2, PT4, PT5, PT6, PT8, PS1, PS2, A4, A5, A6.
Neuronal and Glial Tumours	T1, T4, S5, PT1, PT2, PT4, PT5, PT6, PT8, PS1, PS2, **A4,** A5.
Colon Adenocarcinoma Cells	T1, T4, S5, PT1, PT2, PT4, PT5, PT6, PT8, PS1, PS2, **A4,** A5.
Tissue Banking	T1, T4, S5, PT1, PT2, PT4, PT5, PT6, **PT8,** PS1, **PS2,** A4, A5.
Cell and Tissue Engineering	T1, T4, S5, PT1, PT2, PT4, PT5, PT6, PT8, PS1, PS2, A4, A5, A6.
Gene Therapy & Recombinant Cells	T1, T4, PT1, PT2, PT4, PT5, PT6, A4, A5.
Toxicity Testing	T1, T4, S5, PT1, PT2, PT4, PT5, PT6, PT8, PS1, PS2, **A1, A2,** A4, A5.

extensive field of applications but which have a variety of requirements, constraints and expectations related to:

- The nature of the sample to be analysed, which itself depends on the biological material observed (blood cells, cultured cells, organ biopsy, etc.); the type of preparation (histological section, smear, imprint, etc.); the revelatory techniques (qualitative staining, stoichiometric staining, autoradiography, immunocytochemistry, immunohistochemistry); and the final quality of the sample obtained
- The nature of the questions relating to the samples, such as: the improvement in the image to obtain greater detail; the distribution of certain quantitative features (DNA content, cell or nuclear size, cellular density, labelling intensity); the identification and counting of the cell types (blasts, atypical cells, malignant cells, mitoses); the diagnosis of a sample on the basis of either cell identification and/or counting, or tissue-architecture analysis (cancer screening by means of exfoliative cytology, immunotoxicity tests, allergologic tests of basophilic degranulation); the understanding of the biological mechanisms based on a comparison between several samples (cell proliferation, differential genetic expression, malignant transformation, viral transformation, *in situ* enzyme kinetics, metabolic pathways, etc.)

LABORATORY REQUIREMENTS

The laboratory requirements related to microscopic imaging are expressed in terms of application needs such as:

- The effective representation of the information contained in the microscopic image (minimal size of detectable elements, range of measurable optical densities, sensitivity to fluorescence, etc.)
- The quantitative description of cells and tissues (size, shape, colour, texture, etc.) complementing qualitative observations
- The objective identification of the observed objects (malignant cells, hypertrophied lobules, mitoses, blasts, etc.) to complement the subjective observations
- The analytic approach to biological phenomena as conveyed by their images (cell proliferation, ploidy, dysplasia, differentiation, etc.) to confirm or reject the usual intuitive judgements

Thus the system for quantitative microscopy, and its attendant methodologies capable of satisfying such requirements, must be able not only to process images by exploiting all the technological possibilities, but also to make many different kinds of measurements and evaluate their statistical significance. Above all such a system must permit a *direct comparison* between the sample and the derived data set, and do so at any time during the analysis. All these functions must be presented in a manner which is directly comprehensible to the computer-illiterate user.

MAJOR ERGONOMIC WEAKNESS

Viewing at the microscope does not appear to be easily changeable into viewing at a monitor screen (even HDTV, large-format colour display) for many reasons:

- The video image format that provides the highest resolution at the pixel level does not permit the entire microscope field to be observed. This prevents the observer from using the image contextual information that is often required to evaluate the extent to which the cell of interest deviates from normal.
- The optical depth of the microscope objective is very limited so that the video image contains only a restricted part of the microscope image, while the observer usually scans the microscope image forward and backward by small and frequent changes of focus.

In practice the microscopist using an image analysis system is thus obliged to continually alternate between the microscope and the computer display(s). This is time-consuming and very rapidly becomes both tedious and tiring.

One possible solution is to project the computer images, graphics and alpha-numeric commands onto the microscopic optical image itself in order to achieve a proper ergonomy. The monitor(s) and the keyboard can thus be suppressed and merely replaced by a 'mouse' whose pointer is also projected into the microscope viewing field. This dramatically improves the observer-to-computer relationship. Using such an interface, called 'highly optimized microscope environment' (HOME), the observer sees the conventional image with a superimposed menu selection line and dialogue boxes.

The microscope field is addressed in the image frame of the computer. Thus when human observation is interactively combined with the measuring facilities offered by the computer, the cells to be measured (usually a few cells among hundreds) can be selected by just clicking on them. The segmentation mask of that cell can be projected on its optical image for verification purposes and the parameters measured on that cell can be both displayed in the microscope field and/or stored in a file, etc. Statistical calculations and morphological functions are computed in real time by the resident software and are available at any time during the application as well as afterwards in the form of hard copy.

COMMENT

In conclusion, computer-assisted microscopy has been making steady and exciting progress in research and medical imagery since its inception 30 years ago. But sophisticated image processing will not penetrate routine laboratory procedures until, and unless, the systems are orientated to the way the human observer thinks and works. Indeed, assisting the human experts to improve their diagnostic reliability requires two basic components: powerful numerical image processing and analysis; and human-to-computer interfacing compatible with the microscopist's behaviour.

The biologist will never be supplanted by an image analyser, however sophisticated. But a properly designed, ergonomic system will extend the range of the human observer's heretofore qualitative perception to include quantitative characteristics of the cells, thus bridging the gap between cellular and molecular biology.

3.3 FLUORESCENT IMAGE CYTOMETRY

Fluorescent image cytometry is based on the measurement, point by point, of the fluorescence emitted by cells under adequate excitation wavelength. The images are digitized by a high-resolution analogue-to-digital converter and then analysed by different computer-resident algorithms. The measured fluorescence can be intrinsic or result from the staining of different cellular compounds (Brugal 1984). Various types of fluorescence can be measured on the same cell provided that adequate block filters are used to excite and to recover the fluorescence (De Biasio *et al.* 1987). This technique applies to fixed cells as well as to living cells. In the latter case an extremely sensitive detector should be used because living cells can tolerate only very weak excitation light.

An epi-fluorescence microscope fitted with objectives of different magnifications and a light source with a wide range of wavelengths are required, and also a sensitive detector which, depending on the application, may be a video camera or a cooled charged device (CCD) camera. A control monitor, an image analyser, a computer-driven microscope and an image-analysis system are also needed. Different resident programs may be available for statistical analysis as well.

Applications involve the quantitation of different cell constituents, e.g. DNA, RNA, individual or total proteins (Santisteban *et al.* 1992; Leger *et al.* 1990), as well as their topographic distribution (Humbert *et al.* 1990). In the case of DNA this topographic distribution is very informative since chromatin structure has been shown to be related to the major cell functions; replication, transcription and chromosome condensation. At the molecular level fluorescence image cytometry makes it possible to detect and localize individual genes or gene products in relation to cell differentiation or proliferation, assessed by the DNA content (Du Manoir *et al.* 1991), or proliferation antigen content. Nevertheless the spatial relationships of the particular genes or gene products, with respect to other nuclear or cellular structures, are more efficiently studied using three-dimensional fluorescence techniques, by means of confocal laser scanning microscopy (CLSM). The objects are illuminated point by point within an optical section, avoiding two of the problems encountered with conventional fluorescence microscopes, i.e. background fluorescence and scatter. The CLSM makes it possible to obtain optical sections every 0.5 µm on the z-axis through objects as thick as 500 µm. Although the photomultiplier (single or double) used as the detector provides both linearity and high sensitivity, the use of this microscope for three-dimensional quantitative measurements is still very limited and concerns only DNA (Rigaut *et al.* 1991).

Cell and Tissue Culture for Medical Research, edited by A. Doyle and J.B. Griffiths.
© 2000 John Wiley & Sons, Ltd

PROCEDURE: IMAGE ACQUISITION

Equipment

- Microscope
- Detector

The commonly used specific fluorochromes have excitation wavelengths from UV to red. Mercury (50–100 W) or xenon (75 W) arc lamps are most often used because they have illumination lines throughout the spectrum. They are used in combination with block filters containing band pass and high pass or low pass filters for excitation and emission respectively, as well as dichroic filters to prevent excitation light from entering the detectors. Alternatively, lasers may be used that provide excitation wavelengths from 356 to 665 nm. They are increasingly used in laser scanning microscopy (LSM) (Wilke 1982). In this instrument the microscopic field is scanned by vibrating mirrors or rotating polygons that shield the laser light on the specimen. Very low microscope objective magnification can be used, making this the instrument of choice when large areas of microscopic samples are to be observed.

Objectives that combine fluorescence detection with conventional modes of observation, such as phase contrast, are preferable in order to search for the objects without exciting the sample and thus fading the fluorescent light. Since the spatial resolution depends on the numerical aperture of the objective, a high numerical aperture (NA) must be used when high resolution is needed (0.25 μm for objectives $\times$ 100 with 1.3 NA). Low-magnification dry- or oil-immersion lenses are preferable for applications where only measurements of substance content are desired, since they permit the analysis of large areas of the sample and concentrate less excitation power on each point of the image.

A *good* detector should combine the following characteristics: high dynamic range, good linearity and high spatial resolution. The dynamic range refers to the number of grey levels that can be discriminated between black and white. The linearity implies that the proportionality of the measurement is respected for every point of the dynamic range.

Photomultiplier detectors, used in laser scan microscopy, provide a good dynamic range and linearity, but since the mirrors are used to scan the image, LSM is much slower than conventional microscopy using video cameras or CCD arrays. A new generation of CCD cameras is now available; high-resolution cooled CCD arrays (Arndt-Jovin & Jovin 1989) that are able to detect low fluorescent light intensity at low speed. When video frequencies are used, the sensitivity of the video cameras needs to be increased through the use of a silicon-intensified target (SIT). The normal CCD arrays also require an amplified signal. This amplification results in a decrease in spatial resolution and a greater heterogeneity of the target (shading). Generally, the video cameras, as well as CCD arrays, have to be checked for linearity and heterogeneity and corrections made if necessary. To check the linearity of the detector, repeated measurements of an empty microscopic field must be made with different incident intensities varying from 1% to 100% of the maximum saturation signal for a given gain. Discrete variations on the incident light may be obtained with a series of neutral filters (Wratten KODAK gelatin ND 509) of known density in the light path.

Heterogeneity

Different factors contribute to the heterogeneity of a fluorescent image:

* The heterogeneity of the excitation light
* The irregularity of the sample because of the differences in its thickness and/or the mounting medium
* The geometric and/or chromatic aberrations of the optics
* The heterogeneity of response of the channels of the intensifier
* The heterogeneity of camera target ('shading')
* The response of the analogue-to-digital converter

In order to correct the digital image, two types of corrections may be used:

1. To correct the shading, the black current of the camera is subtracted, point by point, from the digitized cell image (Inoue 1986).
2. Each point of the image is divided by the image of a homogeneous field where the points of highest fluorescence intensity give a signal at 255. The resulting digital image is multiplied by 255 to preserve the dynamics of the signal. This method has the advantage of taking into account not only the heterogeneity associated with the camera but all those related to the sample – the optics and the electronics. Fluorescent phosphor crystals and uranyl glass can be used for this type of correction. Moreover, the standard can be made by the user (Camus *et al.* 1989).

Before starting the measurements the lamp, as well as the camera, must be warm enough to deliver a stable signal. The lamp power source should be connected to a stabilizer to avoid variations due to the alterations in the electric current.

Acquisition

1. Search for objects to be measured in phase contrast light and place them in the centre of the field.
2. Open and close shutters to allow fluorescent light to impinge on the sample for a given time. The duration of the excitation must be determined for each fluorochrome used since the rate in decrease of their fluorescent intensity ('fading') is different for each of them. Simple repetitive quantitative measurements of the fluorescence intensity can be carried out to determine the maximum acceptable excitation before significant fading occurs.
3. Digitize the images by an analogue-to-digital converter and input to the analyser.
4. A resident algorithm then carries out the pretreatment operations (subtraction or division) on the image.

Image segmentation – object labelling and featuring

Image processing sometimes requires that the image be improved to enhance structures or weak staining, especially on living cells. The signal-to-noise ratio may be enhanced by averaging the image (cumulation of various digitized images).

Segmentation (Wells *et al.* 1992) of the image is the most difficult step in image processing. In this step, cells and cellular organelles of interest to the user are isolated from the background by means of resident programs supplied by the manufacturer. For specimens having irregular staining, gradient-type algorithms are used. Segmentation is even further improved by contributions from the modelling of visual perception.

As a result of the segmentation a binary image is obtained and the labels for each object are calculated on this image. Different tools exist to eliminate the non-interesting objects and to separate coupled objects.

Parameters are calculated for each object on the grey-level image. The parameters belong to different families as listed below:

- Morphological parameters for the shape and dimensions of the objects
- Densitometric parameters for the integrated fluorescence intensity and the distribution of the fluorescence intensities throughout the object
- Textual parameters for the spatial relations between the fluorescence intensities of the objects

Depending on the application, 1–30 parameters can be calculated on each object. Because fluorometric readings are relative values, the integrated fluorescence of one specimen cannot be compared with the values obtained from another unless a standard is used to calibrate the instrument. Two types of standards can be used: those that standardize the microscope and detector as already discussed above; and those that serve as reference for a particular fluorochrome. The second type of standard is provided by particles of microscopic dimensions for which the fluorochrome quantity is well known, such as Sepharose beads or micro-droplets of known diameter. Also, for the DNA content measurements, cells with a known diploid DNA content such as lymphocytes may be used as a reference for the calculation of the DNA content of other cells.

DISCUSSION: CRITICAL PARAMETERS

The main sources of error in fluorescence image cytometry are fading (decrease of the fluorescence intensity under excitation light), bleaching (the loss of fluorescence due to the transformation of the fluorescent dye molecule by the light), and reabsorption of fluorescent light. The best way to avoid fading is to limit the power and the duration of the excitation. To this end, electronic shutters and attenuating filters, or polarizers, are used. Oxygen scavengers are included in the mounting medium to reduce bleaching, but they are usually incompatible with vital staining.

REFERENCES

Arndt-Jovin DJ & Jovin TM (1989) Fluorescence labeling and microscopy of DNA. *Methods in Cell Biology* 30: 417–448.

Brugal G (1984) Image analysis of microscopic preparations. In: Jasmin G &

Proschek L (eds) *Methods and Achievements in Experimental Pathology*, pp. 1–33. Karger S, Basel.

Camus E, Santisteban Otegui MS, Monet JD & Brugal G (1989) Quantification de la fluorescence en cytology par videomicrofluroimetrie. *Innovations in Technical and Biological Medicine* 11: 96–106.

DeBiasio R, Bright GR, Ernst LA, Waggoner AS & Taylor DL (1987) Five-parameter fluorescence imaging: wound healing of living Swiss 3T3 cells. *Journal of Cell Biology* 105: 1613–1622.

Du Manoir S, Guillaud P, Camus E, Seigneurin D & Brugal G (1991) Ki-67 labeling in postmitotic cells defines different Ki-67 pathways within the 2c compartment. *Cytometry* 12: 455–463.

Humbert C, Giroud F & Brugal G (1990) Detection of S cells and evaluation of DNA denaturation protocols by image cytometry of fluorescent BrdUrd labeling. *Cytometry* 11: 481–489.

Inoue S (1986) *Video Microscopy*. Plenum Press, New York.

Leger I, Giroud F & Brugal G (1990) Quantitative analysis of cytoskeletal proteins throughout the cell cycle of the MRC-5 fibroblastic cell line. *Analytical and Quantitative Cytology and Histology* 12: 321–326.

Rigaut JP, Vassy J, Herlin P, Duigou F, Masson E, Briane D, Foucrier J, Carvajal-Gonzalez S, Downs A & Mandard AM (1991) Three-dimensional DNA image cytometry by confocal scanning laser microscopy in thick tissue blocks. *Cytometry* 12: 511–524.

Santisteban MS, Montmasson MP, Giroud F, Ronot X & Brugal G (1992) Fluorescence image cytometry of nuclear DNA content versus chromatin pattern: a comparative study of ten fluorochromes. *Journal of Histochemistry and Cytochemistry* 40: 1789–1797.

Wells WA, Rainer RO & Memoli VA (1992) Basic principles of image processing. *American Journal of Clinical Pathology* 98: 493–501.

Wilke V (1982) Laser scanning in microscopy. *Proceedings of the Royal Microscopy Society* 17: 4–21.

3.4 IMMUNOCYTOCHEMICAL LABELLING AND ANALYSIS FOR LIGHT MICROSCOPY

Antibody markers have become invaluable tools in the analysis of *in vitro* cultures. They can be used not only in the fundamental characterization of cells but also in the localization of endogenous and transfected proteins. The simplest systems to consider are monolayer cultures, and the methods to characterize these will form the basic procedure. The methods are also applicable to cells grown in suspension. More complex multilayered cultures or organ cultures are amenable to a similar approach, but the analysis of the labelled cultures may be more difficult.

Unless the primary antibody is directly linked to some form of marker, an indirect immunological staining procedure is normally used. A wide variety of detection systems is available for visualization of the bound antibodies, and these include fluorescent markers, enzyme-based markers and colloidal gold which can be detected by silver enhancement (see Figure 3.4.1). Whichever detection system is chosen, the basic methodology remains the same. Detection of cytoplasmic antigens requires that the cells are fixed and made permeable prior to labelling, whereas cell surface antigen labelling may be undertaken on live cells. Alternatively, the presence of a specific antigen may need to be correlated with cellular morphology and this will often be easier to achieve with an enzyme-based detection system rather than a fluorescence detection method. Each of these alternatives is given in the procedures below.

CELL CULTURE

Cells should be grown on tissue culture chamber slides (Life Technologies Ltd, Paisley, Scotland) in which the cells are grown directly on a slide. The cells may be labelled *in situ,* after which the chamber is detached to allow a coverslip to be applied. Alternatively, cells may be grown on glass coverslips in multiwell tissue culture trays. An ideal combination is 13 mm coverslips in 16 mm wells. The cells can then be labelled *in situ* using the minimum quantity of reagents. Cells grown in 25 cm^2 flasks require unrealistic quantities of reagents to label *in situ*. However, methods for labelling such cells are described below.

Cell and Tissue Culture for Medical Research, edited by A. Doyle and J.B. Griffiths.
© 2000 John Wiley & Sons, Ltd

DETECTION METHOD

Having bound a primary antibody, there is a variety of methods available to visualize the antigen. The simplest and most versatile approach is to apply a second antibody recognizing the species in which the primary antibody was raised. Thus for a mouse monoclonal the second antibody could be sheep anti-mouse immunoglobulins. This second antibody has a choice of conjugated markers and these include:

- Fluorescent markers – commonly fluorescein isothiocyanate (FITC) or rhodamine although a wide range of fluorescent dyes are now available (e.g. Molecular Probes: Cambridge Bioscience, Cambridge)
- Enzymes which will produce an insoluble reaction product when incubated with a suitable substrate
- Colloidal gold, which is visualized for light microscopy by deposition of silver on the bound colloidal gold by a chemical reaction known as silver enhancement

The choice of detection system may be determined as much by familiarity with a particular methodology as for any other reason. Fluorescence methods are the simplest as they do not require substrate incubations or silver enhancement, and they are extremely sensitive, particularly when examined with a conventional fluorescence microscope equipped with a sensitive CCD camera or a confocal scanning laser microscope. The main disadvantage with the technique is that

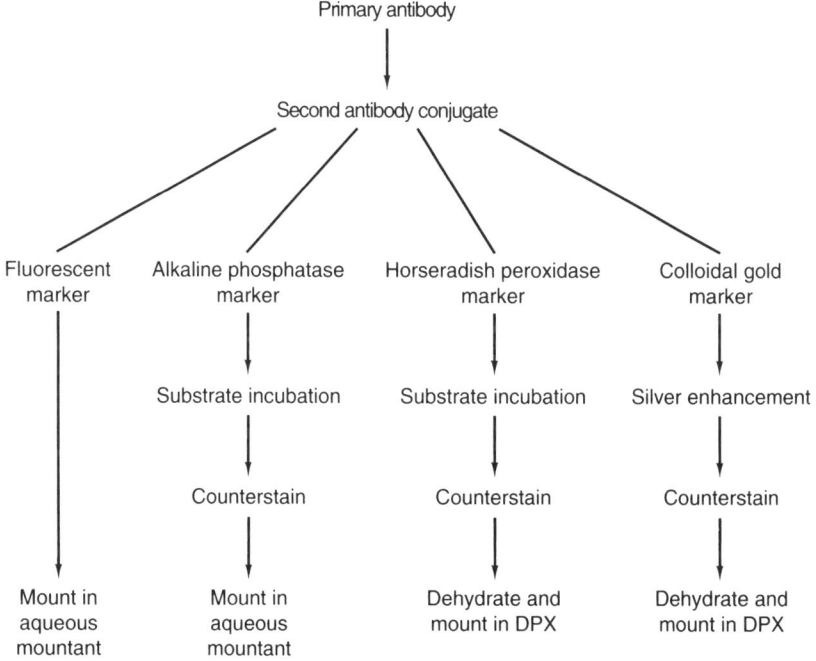

Figure 3.4.1 The different methods of immunolocalization for light microscopy.

cellular morphology is not easy to image although parallel fluorescence and phase contrast images can be recorded.

Following the second antibody – conjugate incubation, fluorescent conjugates require only to be mounted before viewing. Enzyme and colloidal gold conjugates require a further visualization step but these methods allow the cells to be counterstained and viewed by bright field microscopy, allowing correlation of antigen location and cellular morphology.

Suitable substrates for enzyme-linked second antibodies deposit a coloured reaction product at the site of the bound antibody. For peroxidase conjugates, a brown reaction product is formed by incubation in 3,3'-diaminobenzidine tetrachloride, although other colours can be obtained with different substrates. Alkaline phosphatase conjugates produce a red reaction product with naphthol AS-TR and Fast Red.

Colloidal gold conjugates are available in a variety of sizes, but the highest sensitivity is obtained with the smaller 1 or 5 nm particles. These particles are visualized by silver enhancement. While the reagents are simple to prepare, early methods were light sensitive (Danscher 1981). Commercial enhancement reagents are not light sensitive and are relatively inexpensive. Care must be exercised to wash the cells to reduce the number of ions available, otherwise a high background will appear. Ultra-pure 18MΩ cm water is ideal although glass double-distilled water will suffice.

The silver-enhanced gold is seen as small black particles and the longer the silver enhancement time, the larger they become. While they can be seen as black dots in transmitted light, they are more easily seen by using epi-polarized light microscopy. With this imaging mode, the gold particles are seen as bright blue specks against a black background. A careful double exposure of epi-polarized and transmitted light gives the morphology and immunolocalization on one image.

The indirect immunolabelling methods described will suffice for the majority of applications, but there a number of variations which offer potentially higher sensitivity. These include the peroxidase/anti-peroxidase (PAP) and biotin/avidin detection systems.

Indirect immunolabelling of cells on glass coverslips

Reagents and solutions

NAS substrate
The substrate solution should be made up fresh prior to use. Suspend 5 mg naphthol AS-TR phosphoric acid sodium salt in a drop of dimethylformamide. Mix 10 ml veronal acetate buffer (30 mM, pH 9.2) containing 1 mM levamisole as an inhibitor of endogenous alkaline phosphatase and 5 mg Fast Red TR. Mix the two solutions immediately before use and filter.

Veronal acetate buffer
247.5 ml distilled water, 0.972 g sodium acetate trihydrate, 1.472 g sodium barbitone. Adjust pH to 9.2 using 0.1 M hydrochloric acid (approximately 2.5 ml).

Mayer's hemalum
Dissolve 1 g hematoxylin, 50 g aluminium potassium sulphate and 0.2 g sodium iodate in 1000 ml distilled water. Leave overnight to dissolve. Add 50 g chloral hydrate and 1 g citric acid. Mix and boil for 5 min. Cool and filter. Avoid other haemalum recipes (e.g. Ehrlich's) which require differentiation in acid alcohol, as this will remove the reaction product.

Glycerin jelly
Dissolve 30 g gelatin in 180 ml distilled water by heating. Add 210 ml glycerin, and 0.75 g phenol.

DAB substrate
Immediately prior to use make up a solution of 100 mg 3,3'-diaminobenzidine tetrachloride (DAB) in 100 ml Tris-HCl (0.1 M, pH 7.2) plus 100 ml distilled water and 66 μl hydrogen peroxide.

Materials and equipment

Fixation is required to stabilize the cell structure and to allow reagents to penetrate the cell membrane. There are a number of alternatives, but the two choices here will cover most situations. Paraformaldehyde will give the better cellular preservation and is the fixative of choice. Some antigens are affected by aldehyde fixation and cold methanol fixation will then be a good alternative.

4% paraformaldehyde in PBS
0.05% Triton X-100 in PBS
Methanol or acetone at 4°C
Phosphate-buffered saline plus 0.5% (v/w) bovine serum albumin (PBS/BSA)
Primary antibody
Second antibody–conjugate
Microscope slides
Nail varnish

Ensure that the refrigerator is sparkproof.

1. Wash the cells with PBS
2. Fix with 4% paraformaldehyde for 30 min
3. Wash with PBS
4. Permeabilize with 0.05% Triton X-100 for 15 min
5. Wash with PBS
6. Incubate in primary antibody diluted as appropriate in PBS/BSA for 60 min. If correct dilution is unknown, 0.1–20 μg/ml is the broad range of concentration, depending upon the reagent; 200–250 μl is adequate for 16 mm wells.
7. Wash (3 × 5 min) in PBS.
8. Incubate in species-specific secondary antibody conjugated to the marker of choice diluted as appropriate in PBS/BSA for 60 min.
9. Wash (3 × 5 min) in PBS.

Alternative method

Fix cells for 5 min in cold methanol or acetone
Wash in PBS
Continue from step 6, above
At this point the methods for the different markers diverge.

Fluorescent conjugates

Additional materials

Aqueous base mounting medium, e.g. Hydromount (National Diagnostics, Hull, UK)

Anti-quenching agent, e.g. Citifluor (Citifluor Ltd, Canterbury, UK)

Following the wash (step 9 in basic procedure), mount the coverslips on a microscope slide (cells downwards). Any aqueous base mounting medium will be satisfactory, but FITC can fade under illumination and the addition of an antiquenching agent (e.g. Citifluor) will give more stable samples. Cells may have grown not only on the top of the coverslip but also on the underside. These can be removed by carefully wiping what was the underside of the coverslip with a tissue before mounting. The coverslip can be sealed in place by running a ring of clear nail varnish around the edge of the coverslip and allowing to dry. Store stained specimens at 4°C wrapped in aluminium foil to keep dark.

Nuclear counterstain

It is often helpful to have the nucleus labelled to give an idea of the overall cell structure.

Additional materials

DAPI made up at 1 μg/ml in aqueous mounting medium
Pancreatic RNAse solution.

1. Dissolve 10 mg ml^{-1} pancreatic RNAse 1A (Sigma, Poole, UK) in 0.01 M sodium acetate pH 5.2
2. Heat to 100°C for 15 min to inactivate any DNAse
3. Cool to room temperature and adjust pH to 7.4 using 1 M Tris HCl buffer
4. Aliquot and freeze

If using a conventional fluorescence microscope, include DAPI (1 μg ml^{-1}) in the aqueous mounting medium. Alternatively, this is commercially available ready made up (e.g. Vectashield, Vector Laboratories, Peterborough, UK). DAPI is excited at 360 nm and many confocal microscopes do not have suitable lasers for this excitation. Propidium Iodide (PI) will label nuclei red and contrast well with FITC labelling. If the cells are methanol fixed, label prior to mounting by incubating for 2 min in a solution of PI (1 μg ml^{-1}) in PBS.

If the cells have been fixed in paraformaldehyde, the PI will bind to cytoplasmic RNA and an RNAse step is needed. Incubate the cells prior to mounting in DNAse-free RNAse. Thaw an aliquot of RNAse and dilute 1 : 200 in PBS. Incubate cells at 37°C for 15 min followed by PI as for methanol fixed cells.

Alkaline phosphatase conjugates

Additional materials

NAS substrate
Veronal buffer
Mayer's haemalum
Aqueous base mountant, e.g. Aquamount (National Diagnostics, Hull, UK), or glycerin jelly

Following the wash (step 9 in basic procedure) make up the substrate.

1. Wash (2×2 min) in distilled water.
2. Incubate in substrate for 45–60 min, watching for colour development.
3. Wash (2×2 min) in distilled water.
4. Counterstain with Mayer's haemalum.
5. Mount on a microscope slide using a water-based mountant.
6. Seal the edge of the coverslip with nail varnish.

Horseradish peroxidase conjugates

Additional materials

DAB substrate
Histo-clear (National Diagnostics, Aylesbury, UK)

Following the wash (step 9 in basic procedure)

1. Wash (2×2 min) in distilled water.
2. Incubate for 5 min in the DAB solution.
3. Wash (2×2 min) in distilled water.
4. Counterstain with Mayer's haemalum.
5. Dehydrate in ethanol 70% for 1 min, ethanol 100% for 1 min $\times 2$ and Histo-clear for 1 min.
6. Mount in DPX.

Colloidal gold conjugates

Additional materials

Silver enhancer (e.g. IntenSE from ApBiotech, Aylesbury, UK)
High-purity double distilled water or Elga UHQ water

Following the wash (step 9 in basic procedure):

1. Wash (2×5 min) in high-purity water.
2. *Silver enhance:* Times will vary with source of the reagents and the intensity of labelling and may vary from 15 to 30 min.
3. Wash (2×5 min) in high-purity water.
4. Counterstain in Mayer's haemalum.
5. Dehydrate in ethanol (see 'Horseradish peroxidase conjugates').
6. Mount in DPX.
7. Image with light microscope – preferably with an epi-polarized light attachment

Labelling live cells for surface antigens

Occasionally, it is necessary to stain live cells unfixed. This may be to compare with fixed cells in order to determine the proportions of the antigen present on the cell membrane and in the cytoplasm. Alternatively, it may be useful to label live cells and incubate for a short time before final fixation and analysis.

Throughout the procedure, all reagents and equipment coming into contact with the cells (e.g. pipette tips) must be precooled and kept on ice in order to stabilize the cell membrane or to prevent antigen internalization.

Additional materials

L-15 Leibovitz medium (Gibco BRL, Life Technologies Ltd, Paisley, Scotland) or PBS/BSA

1. Precool the cells on ice for 20 min.
2. Wash (3×5 min) with Leibovitz medium, L-15 + 10% foetal bovine serum (FBS) or PBS/BSA at 4°C.
3. Incubate cells for 60 min with primary antibody diluted in L-15/FBS or PBS/BSA as appropriate.
4. Wash (3×5 min) in medium.
5. Incubate for 60 min with second antibody diluted in medium/FBS or PBS/BSA
6. Wash (3×5 min) in medium.
7. Fix the cells on the coverslips for 15 min.

The choice of fixative is not critical. Aldehyde fixatives offer improved morphology over methanol as used in the basic procedure. Formal saline, or 2% glutaraldehyde in phosphate buffer, as for electron microscopy, are commonly used. Following fixation the coverslips should be treated as in the basic procedure, depending upon the choice of detection system.

Multilayered cultures or organ cultures

Multilayered cell cultures must be fixed and made permeable before staining as in the basic procedure. The difficulty encountered with these cultures is in analysing the results. While it may be possible to analyse two cell layers that have been labelled with fluorescent markers, the use of a confocal laser scanning microscope will provide unequivocal discrimination between the two layers.

Organ cultures and many-layered cell cultures will need to be sectioned (either after fixation and embedding or as frozen sections) and labelled by routine light microscopy methods.

Cells grown in suspension

Cells are labelled in suspension by gently centrifuging the cells (500 g) to a pellet, discarding the supernatant, and resuspending in the next reagent. Following labelling, the cells must be attached to slides by cytocentrifugation at 1200 g for 5–10 min, or by incubating a solution of approximately 10×5 cells ml^{-1} on a slide previously coated in a solution of 1 mg ml^{-1} polylysine.

Alternative approaches include either attaching the cells to a microscope slide as above, and labelling the cells on the slide or preparing a cell pellet by centrifugation (500 g) which is then sectioned for light microscope immunocytochemistry. If the antigen is fixation-stable, the pellet can be fixed, embedded in wax and sectioned. Fixation-labile antigens will mean the cells must be frozen and have frozen sections prepared. The sections are then labelled as for routine microscopy.

Cells grown in culture flasks

The quantity of reagent needed to label cells grown in culture flasks precludes staining *in situ*. Cells must be removed from the flask prior to staining.

Cells grown in 25 cm^2 flasks can be removed by trypsinization or, if the antigen is trypsin labile, the cells should be carefully scraped from the culture vessel. Cells may be labelled in suspension by centrifuging and resuspending at each step. This method is usually only employed when labeling membrane antigens. The cells can be treated as for suspension-grown cells.

Multiple antigens

If more than one antigen is to be detected in a culture, two different labelling methods must be chosen. If a suitably equipped fluorescence microscope is available, the two antigens can be detected with different fluorochromes such as FITC and rhodamine or Texas Red. Alternatively, a peroxidase-linked conjugate for one antigen and an alkaline-phosphatase-linked conjugate for the second (Monaghan *et al.* 1990), or a mixture of enzyme and colloidal-gold-linked conjugates, can be employed.

Care must be taken to ensure there is no cross-reactivity between the two primary antibodies. If they are raised in two different species, then the species-specific second antibodies should not cross-react. Indeed, it is possible to mix the primary antibodies for the first incubation, and mix the conjugates for the second incubation. The enzyme substrate incubations will need to be undertaken sequentially. For some combinations, the order of substrate incubation is important, and will need to be checked.

Where both primary antibodies are monoclonals, but of different subclass, subclass-specific second antibodies may be available. If not, the first antigen will need to be detected, followed by a blocking procedure to mask any remaining primary antibody before proceeding to the second antigen detection. For example, with mouse monoclonals, incubation with unlabelled sheep anti-mouse immunoglobulins should mask any primary antibody. Strict controls are especially necessary for such experiments.

ALTERNATIVE DETECTION METHODS

The methods described have been restricted to indirect immunolocalization techniques. Variations on this method include protein-A- or protein-G-conjugated secondary reagents and the use of PAP and avidin/biotin methods.

CONTROLS

A negative control should always be included in any labelling experiment. The simplest form is a 'no first antibody' where the primary antibody is omitted from the first incubation. Where the purified antigen is available, a polyclonal antiserum can be incubated with the antigen prior to adding to the sample. If the staining is removed by the pre-incubation, then the antiserum is behaving correctly. This is known as an 'absorption control'. For monoclonal antibodies, this control is not worth while, as it is certain that the antigen will bind to the antibody and preclude staining of the cells.

To be completely certain that a positive immunolabelling result is due to the presence of a particular antigen, it is necessary to undertake Western blotting to check that the antigen being detected is of the correct molecular weight.

REFERENCES

Bullock GR & Petrusz P (1983) *Techniques in Immunocytochemistry*, Vols 1 and 2. Academic Press, London.

Danscher G (1981) Localisation of gold in biological tissue. A photochemical method for light and electron microscopy. *Histochemistry* 71: 81–88.

Harlow E & Lane D (1988) *Antibodies: a Laboratory Manual.* Cold Spring Harbor Laboratory, New York.

Monaghan P, Ormerod MG & O'Hare MJ (1990) Epidermal growth factor receptors and EGF responsiveness of the human breast carcinoma cell line PMC42. *International Journal of Cancer* 46: 935–943.

3.5 FLOW CYTOMETRY – AN OVERVIEW

Flow cytometry, as the name implies, is the measurement of cells in a flow system. The purpose of the flow system is to deliver cells singly to a point of measurement at which a beam of light (usually from a laser) is focused. Fluorescence from selectively labelled cells and scattered light are collected and recorded.

The advantage of flow cytometry lies in the ability to measure several parameters on tens of thousands of individual cells within a few minutes, thereby accurately defining subpopulations. Its disadvantage is that a preparation of single particles (cells, nuclei, chromosomes) is required, so that tissue architecture is destroyed and spatial information lost. The large amount of data recorded must be processed by a well-programmed computer.

Several recent books have given overviews of flow cytometry (Shapiro 1995; Melamed *et al.* 1990; Watson 1991; Givan 1992; Bauer *et al.* 1993) and detailed descriptions of many of the methods in common use. (Darzynkiewicz *et al.* 1994; Macey 1994; Jaroszeski & Heller 1998; Ormerod 2000; Radbruch 1992; Robinson *et al.* 1993).

INSTRUMENTATION

The basic instrument consists of a source of light, flow cell, optical components to focus light of different colours onto the detectors, electronics to amplify and process the resulting signals and a computer (Figure 3.5.1). The simpler instruments, usually supplied as benchtop models, only analyse cells while more complex and expensive machines can also physically sort cells.

The flow cell

The purpose of the flow cell is to deliver cells singly to a specific point on which the source of light is focused. The sample is injected into the centre of a stream of liquid called the sheath fluid (normally either water or saline solution). The cell is designed so that the sample stream is hydrodynamically focused, delivering the cells to the point of detection with an accuracy of ±1 μm or better.

There are three types of flow cell found in commercial instruments: quartz cuvettes, 'sense-in-air' and those based on a channel cut in a solid block. In the first, light is focused onto the sample within the flow cell; in the second, the sheath plus sample emerges into the air from a nozzle, just below which the laser beam is focused; in the third, an epi-illumination system is used, similar to a fluorescence

Cell and Tissue Culture for Medical Research, edited by A. Doyle and J.B. Griffiths.
© 2000 John Wiley & Sons, Ltd

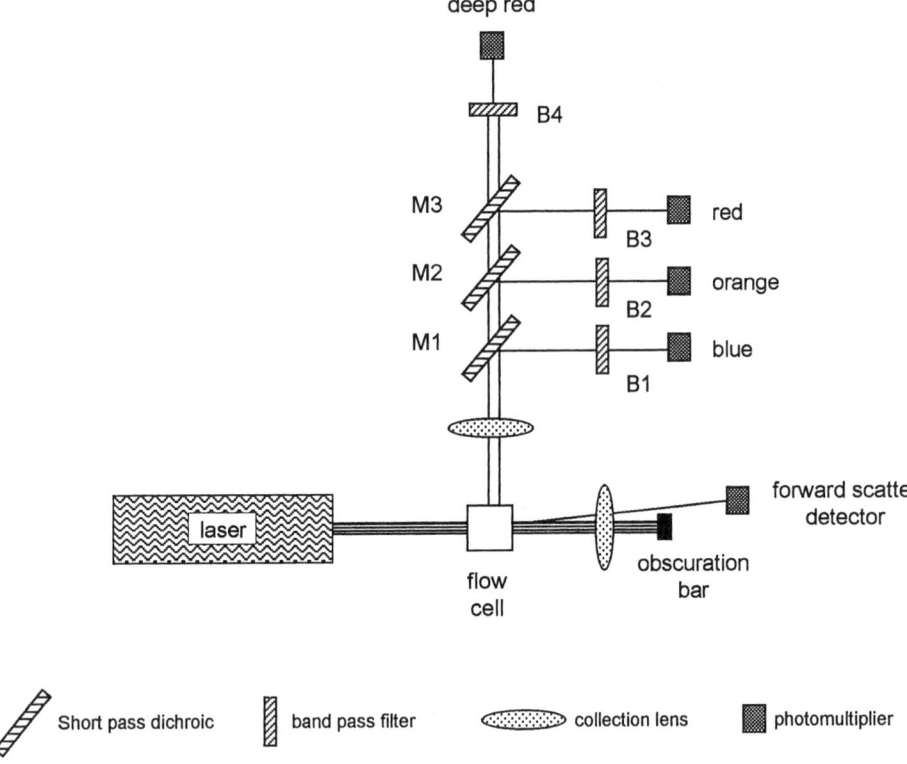

Figure 3.5.1 The layout of a typical flow cytometer. Bi-B4, barrier filters; M1-M4, dichroic mirrors; D1-D5, detectors. Typical properties of the filters and mirrors are given in Table 3.5.1.

microscope with the flow cell replacing the microscope slide. The 'stream-in-air' system is used in cell sorters.

Light sources

The light source is normally either a laser or an arc lamp; the former is preferred because it produces monochromatic light with a small 'spot' size, which can be focused into a small volume. Argon-ion tuned to 488 nm (blue) is the commonest laser used.

Optics

The light beam is focused onto the sample stream, usually to give an elliptical beam or even a narrow slit of light (used for DNA analysis, for further explanation see Ormerod 2000). Scattered and fluorescent light of different wavelengths from the cells is separated by means of a series of dichroic mirrors and barrier filters. Table 3.5.1 gives the properties of filters and mirrors that could be used in

Table 3.5.1 Dichroic mirrors and filters that might be used in a flow cytometer

Filter		Parameter detected
Long pass dichroic	M1 : 500 nm	
	M2: 560 nm	
	M3: 600 nm	
Band pass filter	B1 : 488/10 nm	Scattered light
	B2: 530/30 nm	Fluorescein
	B3: 580/30 nm	Phycoerythrin
Long pass filter	B4: 610 nm	Per CP

The letters and numbers refer to the layout shown in Figure 3.5.1. A long pass dichroic mirror reflects light below the given wavelength passing light of longer wavelength. The numbers for the band pass filters give the wavelength of transmission/the 50% band width.

the layout shown in Figure 3.5.1. This configuration would be suitable for measuring scattered light and green (fluorescein), orange (phycoerythrin) and red (peridinin-chlorophyll, PerCP) fluorescences.

Fluorochromes have wide emission spectra and, when multiple fluorescences are measured, the filters will not completely separate the different fluorescences. A correction for this spectral overlap must be applied either electronically or in the computer software.

Detectors

A photodiode diode is sufficient for the measurement of forward-scattered light. Photomultipliers are used for measuring fluorescences and orthogonal scatter.

Signal processing

The signal from a photomultiplier is processed to give a signal proportional to the fluorescence of the cell. Together with the integrated area, the width and peak of the pulse may also be recorded for analysis to distinguish between single cells and doublets (Ormerod 2000; Watson 1991; Bauer *et al.* 1992). There is a choice between linear and logarithmic amplification; the former should be used for DNA measurement, while the latter is often used for immunofluorescence.

DATA ANALYSIS

The raw data can be written in a continuous stream onto a disk (so-called listed data). The advantage of storing raw data is that it can be re-analysed off-line.

During analysis, 'gating' is employed. Data from one or two parameters are displayed, and regions of interest (gates) are defined to select certain populations of cells for display of further parameters. The different parameters are displayed by the computer as either uni- or bivariate histograms (cytograms). The latter are shown as 'dot plots', contour plots or pseudo-three-dimensional isometric plots (Figure 3.5.2).

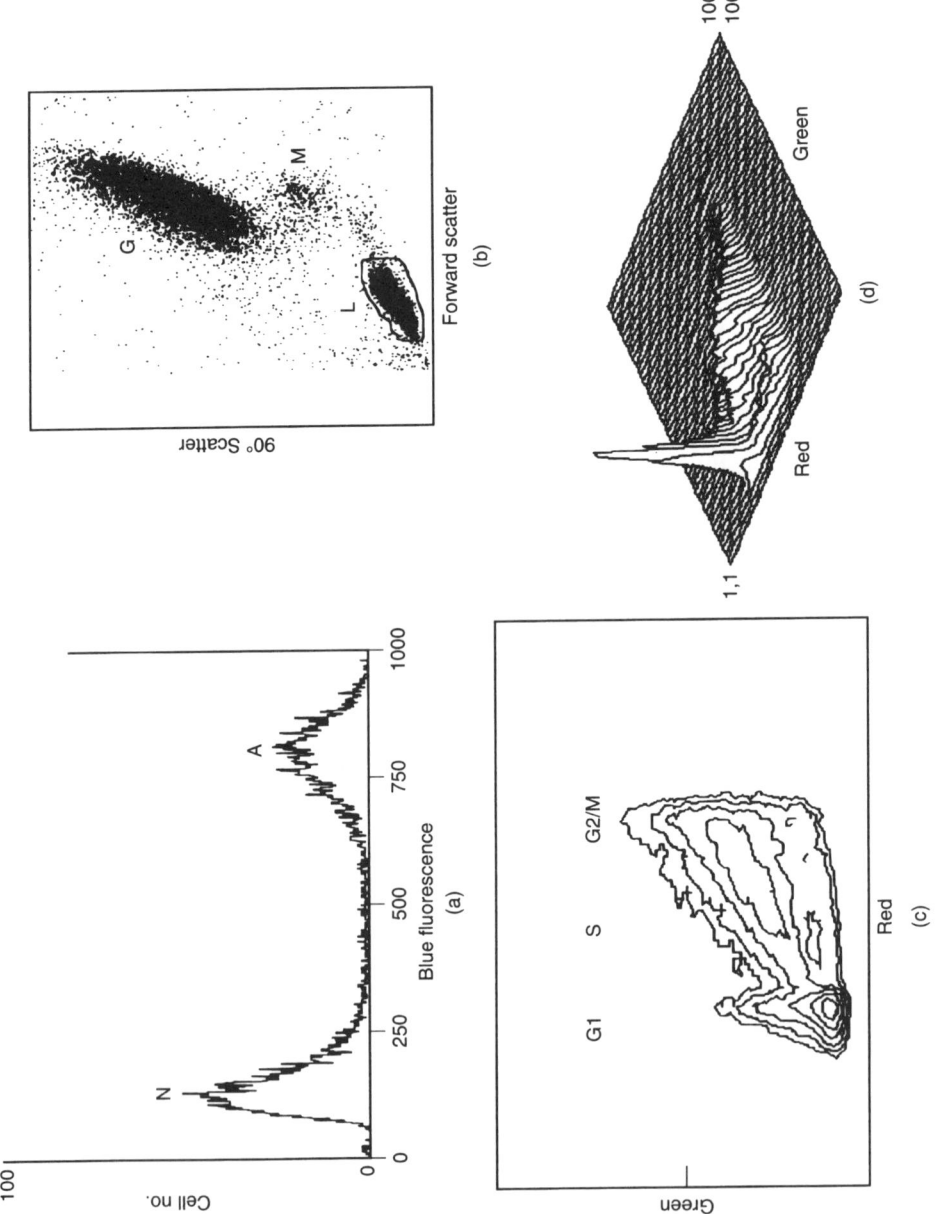

CELL SORTING

The commonest method of sorting cells is by electrostatic deflection of charged droplets. A conductive sheath fluid is used (buffered saline). The flow cell is vibrated vertically by means of a piezoelectric transducer, causing the fluid from the exit nozzle to break up into droplets. The flow cell is charged at the moment a cell of interest is inside the droplet about to be created. The stream of droplets passes through a pair of charged plates so that the charged droplets are deflected and collected.

Under ideal conditions, cells can be sorted with a purity of > 98% although the total number of cells obtained is low compared to many other methods of cell separation.

APPLICATIONS

Immunofluorescence

The detection of up to four compounds fluorescing at different wavelengths permits multi-parametric analysis of cells. Routinely, fluorescein is used as the first label, phycoerythrin as a second label and phycoerythrin-Texas Red conjugate, phyco-erythrin-cyanine5 or peridinin-chlorophyll as the third. A wide range of directly labelled monoclonal antibodies and labelled anti-Igs are available from several manufacturers.

DNA analysis

The second most common application is measurement of DNA to give a picture of the cell cycle and, in the case of clinical samples, to measure ploidy. DNA analysis can be combined with the measurement of antigen (see Figure 3.5.3).

Figure 3.5.2 Displays recorded from a flow cytometer. (a) A univariate histogram of cell number versus blue fluorescence. Unfixed cells from a murine haemopoietic cell line were stained with the bisbenzimidazole Hoechst 33342 and excited with UV light. Apoptosis had been induced by 18 h earlier withdrawing a growth factor, interleukin-3, required by these cells for growth. The normal cells (N) showed low, and the apoptotic cells (A) high fluorescence. Dead cells had been excluded from the analysis by staining with propidium iodide (PI), which is excluded by cells with an intact plasma membrane. (b) A bivariate histogram (cytogram) of orthogonal versus forward light scatter from human peripheral blood leukocytes. The clusters show lymphocytes (L), monocytes (M) and granulocytes (G). (c, d) A bivariate histogram of green (fluorescein/cyclin B1) versus red (PI/DNA) fluorescence. Cells from a human lymphoblastoid cell line (W1L2) were harvested, fixed in 70% ethanol, and washed and suspended in a buffer containing 0.1% Triton X-100. The cells were then stained with a monoclonal murine antibody against human cyclin B1 followed by fluorescein-anti-mouse Ig. PI was added at 10 μg ml^{-1}. The variation of cyclin B1 through the cell cycle is shown. (c) is a contour plot and (d) a pseudo-three-dimensional isometric plot.

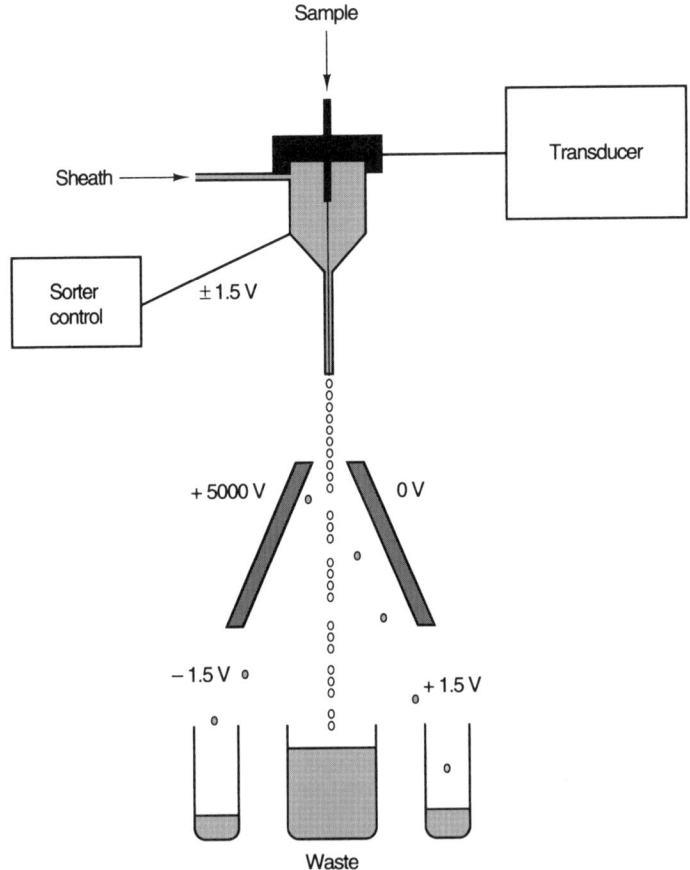

Figure 3.5.3 The major components of a cell sorter.

Other applications

There is a wide variety of other applications, which include:

- Measurement of RNA content
- Measurement of protein content
- Kinetic analysis of intracellular enzymes (particularly esterases) – estimation of cell viability
- Measurement of membrane potential
- Following changes in membrane permeability
- Measurement of the production of intracellular oxidative species
- Tracking cells *in vivo*
- Monitoring electropermeabilization
- Measurement of drug uptake
- Monitoring endocytosis
- Measurement of intracellular pH
- Measurement of intracellular calcium ions

- Estimation of the number of apoptotic cells
- Measurement of intracellular glutathione
- Chromosome analysis and sorting
- Monitoring fusion of cells

CHOICE OF INSTRUMENT

The two important points to consider in the choice of an instrument are (a) whether a cell sorter is needed (this will double the cost) and (b) whether a second laser is needed. A list of manufacturers of flow cytometers and cell sorters is given below. (Most of these companies have offices in more than one country. The address of their main office is given here.)

Beckman Coulter, Inc., PO Box 169015, Miami, FL 33116–9015, USA. +1 800 635 3497. Web site: www.beckmancoulter.com

Becton Dickinson Immunocytometry Systems, 2350 Qume Drive, San Jose, CA 95131–1807, USA. +1 800 223 8226. Web site: www.bdfacs.com

Cytomation, Fort Collins, Colorado, USA. +1 800 822 9902. Fax: +1 970 226 0107.

E-mail: sales@cytomation.com. Web site: www.cytomation.com

Optoflow AS, PO Box 70, Bogerud, N-0621 Oslo, Norway. +47 22 62 7080. Fax: +47 22 62 72 75. E-mail: gjelsnes@optflow.com.

Partec GmbH, Otto-Hahn-Strasse 32, D-48161 Münster, Germany. +49 2534 8008–0. Fax: +49 2534 8008–90. E-mail: info@partec.de. Web site: www.partec.de

FURTHER READING

Bauer, KD, Duque, RE & Shankey, TV (eds) (1992) *Clinical Flow Cytometry Principles and Applications* Williams and Wilkins, Baltimore.

Darzynkiewicz, Z, Robinson, JP & Crissman, HA (eds) (1994) *Flow Cytometry.* Methods in Cell Biology, 41 & 42 Academic Press, San Diego.

Givan, AL (1992) *Flow Cytometry First Principles.* Wiley-Liss, New York.

Jaroszeski, MJ & Heller, R (eds) (1998) *Flow Cytometry Protocols.* Methods in Molecular Biology, 91, Humana Press, Towowa, NJ.

Macey, MG (ed) (1994) *Flow Cytometry: Clinical Applications.* Blackwell Scientific, Oxford.

Melamed, MR, Lindmo, T & Mendelsohn, MI (eds) (1990) *Flow Cytometry and Sorting,* 2nd edn. Wiley-Liss, New York.

Ormerod, MG (ed.) (2000) *Flow Cytometry. A Practical Approach,* 3rd edn. IRL Press at Oxford University Press, Oxford.

Radbruch, A (ed.) (1992) *Flow Cytometry and Cell Sorting.* Springer-Verlag, Berlin.

Robinson, JP (ed.) (1993) *Handbook of Flow Cytometry Methods.* Wiley-Liss, New York.

Shapiro, HM (1995) *Practical Flow Cytometry,* 3rd edn. Alan R Liss, New York.

Watson, JV (1991) *Introduction to Flow Cytometry.* Cambridge University Press, Cambridge.

3.6 ELECTRON MICROSCOPY – AN OVERVIEW

The resolving power of any microscope (even a simple glass lens) has been defined as the smallest distance between two adjacent points located on the specimen being observed, where they can still be distinguished as two separately identifiable points. The best resolving power of any type of microscope equates to about one half the wavelength of the illuminating beam. For visible light using glass lenses, this is about 200 nm. However, by recognizing the wave nature of electrons we can employ them as a *source* of illumination. By focusing the electrons using electromagnetic lenses, we are able to improve the resolving power by at least a 1000-fold.

However, because the mean free path of electrons in air is only a few micrometres, the optics of the microscope need to be enclosed within a high vacuum chamber. At an operating vacuum of 10^{-4} Torr, electrons will travel about 2500 mm before encountering and being scattered by a gas molecule. In addition, because the mean free path of electrons within a solid specimen is even smaller, the section thickness (in the case of the transmission electron microscope) needs to be in the range 50–120 nm for the electrons to penetrate the specimen, interact with it and achieve maximum resolution.

In the scanning electron microscope resolving power is dictated, among other parameters, by the diameter of the electron probe focused onto the specimen surface.

INSTRUMENTATION

Introduction

The two types of electron microscope commonly employed by researchers in cell and tissue culture are the scanning (SEM) and transmission (TEM) electron microscope. They are extremely powerful tools for probing the ultrastructure and cytochemistry of cultured cells and tissues and have the potential to image microstructures down to the molecular level of resolution.

Commercially manufactured transmission electron microscopes (TEMs) have been available now for about 60 years. The first scanning electron microscope (SEM) appeared in the late 1960s with the advent of the highly successful Cambridge 'Stereoscan' series of instruments. The application of X-ray microanalysis coupled with cryopreparation techniques have enabled confident qualitative and quantitative microanalyses of inorganic components within cells and tissues.

Cell and Tissue Culture for Medical Research, edited by A. Doyle and J.B. Griffiths.
© 2000 John Wiley & Sons, Ltd

In addition to TEM and SEM, scanning transmission electron microscopy (STEM) was developed in which the electron probe is scanned across an ultra-thin section of the specimen in the form of a raster, the signal being collected the other side of the specimen electronically. The advantage of this is that, as in SEM, contrast may be enhanced electronically. STEM has found particular application to the observation and X-ray analysis of ultrathin frozen sections of tissue or thin vitreous films of cells where contrast may be introduced into the specimen electronically rather than using electron opaque heavy metals.

Transmission electron microscopy (TEM)

In its simplest form a TEM consists of an aligned stack of electromagnetic lenses above which is a source of electrons in the form of a tungsten filament (cathode) above an anode plate (Figure 3.6.1). Below the stack of electromagnets is a cathodoluminescent screen for observing the image. The whole aligned stack is contained within a vacuum chamber which is evacuated by a two-stage pumping system involving a rotary pump then an oil-diffusion or ion-getter or turbo-molecular pumping system. The specimen is placed at a plane within the objective lens of the microscope. The specimen has to be thin enough for electrons to either pass through it or interact with it, and thus generate an image at the other side.

Imaging modes

- *Bright field* In transmission electron microscopy all the electrons scattered by their interaction with the specimen can contribute to the final image. In bright field imaging an objective aperture of a particular diameter is centred around the transmitted beam. The aperture thus blocks out some, or all, of the scattered electrons, depending upon its diameter.
- *Dark field* It is also possible to centre the objective aperture around a scattered or diffracted beam thus producing a dark field image of the specimen in the way to cause scattering. The background onto which the specimen image is projected is therefore dark (Otten 1991). This form of observation is particularly useful for imaging isolated suspensions of macromolecules.

    There are three principal applications of dark-field TEM; (1) improving contrast in biological specimens with inherently low contrast, (2) providing images with high contrast of defects in certain types of crystal lattice, (3) providing images of crystals with high contrast which have a specific orientation within the specimen.
- *Electron diffraction* All four-lens transmission electron microscopes are capable of forming electron diffraction patterns (diffractograms). The TEM with two projector lenses is essentially an electron diffraction camera of variable length (Dyson 1993). The diffraction pattern may appear on the screen as either (1) a spot pattern, or (2) a pattern of concentric rings (implying that the specimen includes many tiny individual crystals, or (3) a line or *Kikuchi* pattern where the specimen thickness is greater than optimum for selected area diffraction work (Andrews *et al.* 1971).

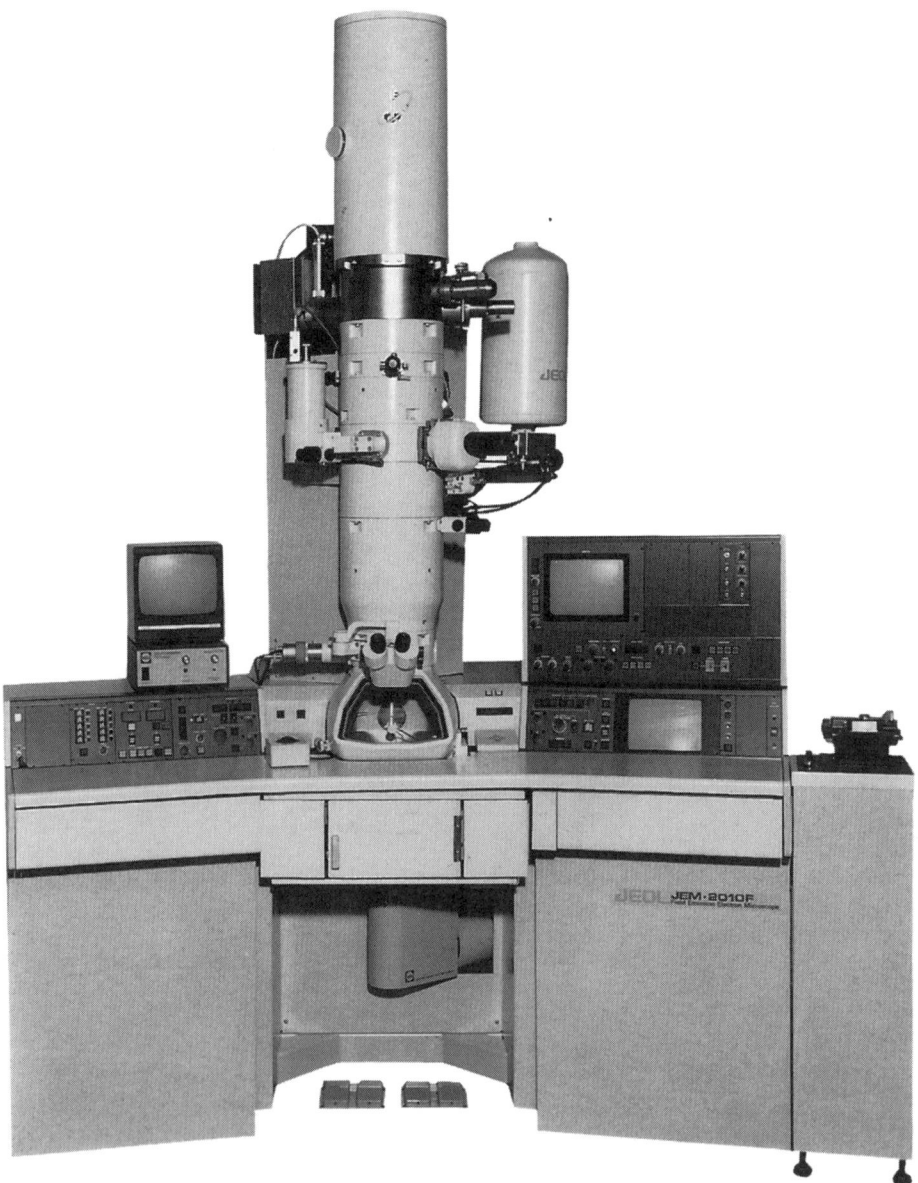

Figure 3.6.1a Field emission high-resolution analytical electron microscope, JEM-2010F. Reproduced with permission of JEOL (UK) Ltd.

Data derived from diffractograms enables the characterization of individual components of a crystalline material, providing information on dimensions of a crystal lattice and its orientation within the sample, atomic ordering, faulting, twinning and information regarding the growth of a specific phase within the specimen.

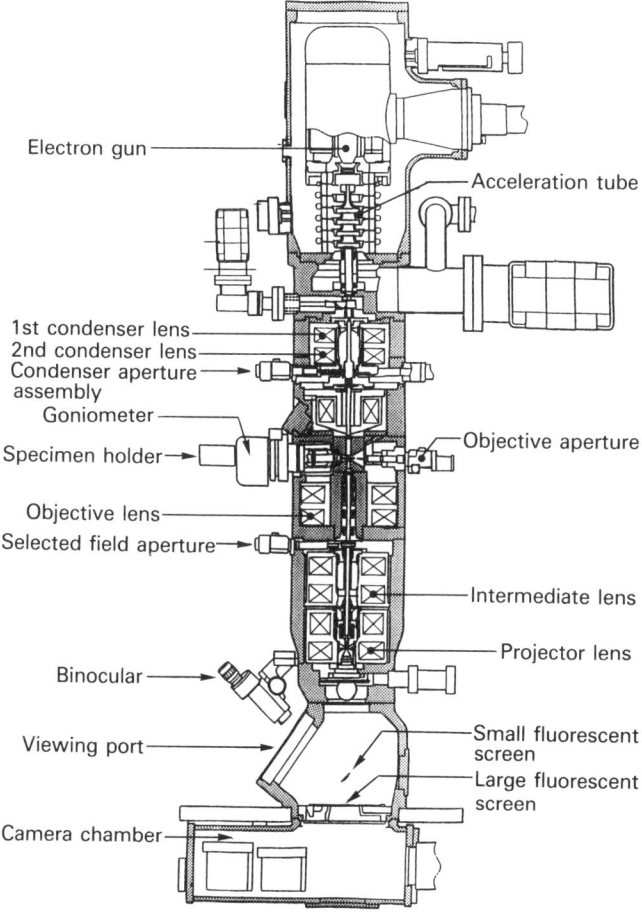

Figure 3.6.1b Cross-sectional view of column. Reproduced with permission of JEOL (UK) Ltd.

Resolution and contrast

In transmission electron microscopy, the prerequisites for obtaining images with high resolution are (1) that the microscope itself must have a high resolving power, but (2) that the specimen being observed must have sufficient contrast between the two points to be distinguished. This first condition is only satisfied in instruments with very low aberrations where they are adjusted so that the beam is optimal to the optical axis, the aperture size is optimal and the focusing is also optimal. In the second condition optimum contrast is achieved not only due to electron/specimen interaction but also to the focusing conditions of the objective lens.

Image capture and recording

In capturing images from the TEM, several through-focus recordings should be taken, centering around the optimum focus. To achieve optimum contrast (see above) the focus should be adjusted to 5–10 nm underfocus.

Traditionally, TEM images have been recorded by projecting the electron image directly onto a fine-grain EM plate. Today the image is more likely to be captured using a solid-state camera and recorded on a computer database where many tens of thousands of images may be stored digitally.

Scanning electron microscopy (SEM)

In its simplest form an SEM consists of an aligned stack of eletromagnetic lenses above which is an electron generator in the form of a tungsten filament (cathode) sitting above an anode plate. Below the stack of electromagnets is a set of scan coils which move the focused electron beam as a raster across the surface of the specimen which sits on a movable stage below. Signals from the specimen surface are captured by various detectors to one side which amplify signals to attenuate the brightness on a CRT. The electron gun, aligned electromagnetic stack and specimen are all contained with a vacuum chamber which is evacuated by a two-stage pumping system involving a rotary (roughing) pump then a high vacuum oil-diffusion or ion-getter or turbomolecular pumping system.

Accelerating voltage

In SEM the accelerating voltage employed for observation has a profound effect upon the final image and signals generated. Low accelerating voltages while generally providing low signal-to-noise ratio will provide more true *surface* information about the specimen since the depth of penetration and therefore the excitation volume is smaller. At higher kVs a stronger signal is produced due to greater energization, but more subsurface information regarding the specimen is generated. kV is important in X-ray analysis since there is an optimum excitation potential (thus delivering maximum X-ray signal) for each element to be analysed.

Signals and detectors

A variety of interactions occur between the primary electron beam and the solid specimen. These interactions result in a range of signals which can be collected and converted into valuable structural and chemical information about the specimen. Signals are of two types: (1) electron signals and (2) non-electron signals.

Two of the most important electron signals used in SEM are emission of secondary electrons (emissive mode) and reflected or backscatterred electrons. Non-electron signals include cathodoluminescent emission and X-ray emission. *Auger* emission, which is a method enabling analysis of the surface atoms of a specimen, has almost unique application in materials science.

For electron and non-electron emissions, a variety of detectors has been designed to capture a high yield of each of the specific signals:

- *Secondary electron (SE) imaging* In its simplest form (designed by Everhart & Thornley 1960) the secondary electron detector consists of a thin film of a metal deposited at the end of a light pipe. The film is maintained at a high positive potential in order to attract secondary electrons. The electrons impinging on the film cause it to scintillate and the light pipe guides the flash of light to a photomultiplier where its intensity is measured. This amplified signal modulates the brightness on the CRT used for observation. Contrast is a function of the topography of the surface being scanned. Secondary electron signals thus provide topographical information about the specimen surface.
- *Backscatter electron (BSE) imaging* The backscatter electron yield due to specimen/beam interaction is about the same as for secondary electrons but their capture needs some additional effort to achieve the same efficiency as for secondaries. Because BSEs cannot be easily deflected by magnetic or electric fields, they must be collected by a detector which physically subtends a sufficiently solid angle at the specimen to collect a high fraction of electrons. Because the BSE emission contains both compositional (Z-contrast) as well as topographical information, the two signals can be collected separately and subsequently combined (mixed) in various ratios. This is achieved by collecting the signals separately, using a four-quadrant detector. There are four types of detector in common use for collection of the BSE signal: (1) a modified *Everhart–Thornley* detector, (2) a ring scintillator detector, (3) solid state detectors and (4) electron channel plates.
- *Cathodoluminescence* When certain organic or inorganic materials are irradiated with electrons, they emit photons in the visible or infra-red portion of the spectrum (Yacobi & Holt 1990). If the magnitude of the photons is measured using a photomultiplier then the resultant signal can be used to modulate the brightness of the spot on the CRT display. Cathodoluminescence provides analysis of the chemical nature of the specimen surface but also renders information about is physical state and is of special value to the study of rocks and minerals and also fabricated solid-state structures. Since the yield of photons is very low, the efficiency of collection of the emitted photons is vital to the provision of a good signal. In addition, photon yield is far greater if the temperature of the specimen is lowered. There is thus considerable scope for combining cathodoluminescnce observation with cryo-techniques (see below).

Field emission

Resolving power is ultimately linked to gun brightness and a relatively new generation of high brightness, cold filament electron guns using *Schottky* or field emission are up to a thousand times brighter than conventional thermionic filament sources. Electrons are drawn from the cold tip of the metal filament by a very strong electric field (10^9 V m^{-1}). The actual filament tip in an FE gun is only 10 nm across compared with the 100 μm diameter tip of a conventional thermionic filament, thereby providing a smaller wavelength range and thus greater resolving power.

Image capture and recording

Traditionally, the images generated by the SEM have been recorded using a conventional 35 mm or 2¼-inch square camera which is focused on a special CRT with a low after-glow phosphor (Wergin 1993). Today the image is more likely to be captured using a CCD camera and stored on a computer database where many tens of thousands of images can be recorded digitally. The image source is usually analogue, the brightness of the sample being measured at a particular position and moment in time. The image acquisition involves the conversion of that analogue signal into a digital counterpart, i.e. a square matrix of pixels (image points) representing the field of view captured.

PREPARATION TECHNIQUES

Introduction

The ability to distinguish between two closely located points within the specimen (resolution) while clearly, first, being a function of the resolving power of the instrument, is also directly related to the degree of contrast between the two points we wish to distinguish. The techniques we use to prepare specimens for TEM often dictate how well we achieve the second.

Techniques for transmission electron microscopy

Fixatives and fixation methods

The purpose of this, the first treatment step, is to arrest metabolic activity and therefore post-mortem changes within the cells and tissues and then to preserve as well as possible, their chemistry and ultrastructure. The 'ideal' fixation should preserve ultrastructure while minimising deformation and dissolution of cellular and microstructural components (Hayat 1981).

For animal tissues to be observed for TEM a 'double fixation' regime is usually adopted employing a primary then secondary fixative. For small blocks of tissues (typically 1 mm^3) an immersion time of 1 h is frequently used for the primary, then the secondary fixative. Accordingly, there is a temptation to reduce these times when fixing cultured cells in suspension or cell monolayers as flat sheets, where it might be argued that fixation will occur much faster. In practice, the optimum fixation times are often found to be the same, even when single cells are suspended within the fixative solution.

The primary fixative solution can be exposed to the specimen either by simple immersion of tiny (1 or 0.5 mm cubes) excised tissue blocks into the solution or, preferably, by perfusion of the fixative through the vasculature etc. of the intact organ of the animal in order to aid speed of penetration. After primary fixation, tissue blocks are washed in buffer and then exposed to the secondary fixative. The use of microwave excitation during fixation in order to improve rates of penetration of the fixative into the cells is well recognised (Giberson & Demaree 1999).

- *Primary fixatives* These are fixatives which penetrate the fresh tissue (ideally as rapidly as possible) and thus provide the initial stabilization stage. While they may fix proteins well, a secondary fixation stage is often necessary to stabilize lipids. Primary fixatives commonly used are often aldehydes (which have a high crosslinking capacity). These include: glutaraldehyde (glutaric acid dialdehyde), formaldehyde and acrolein (Hayat 1981). A mixture of both glutaraldehyde and formaldehyde (freshly prepared from depolymerized paraformaldehyde) is often used for cell cultures and has the additional advantage of rapid penetration into the cell monolayer or pellet.
- *Secondary fixatives* The secondary fixative solution is normally a buffered tetroxide of either osmium or (rarely) ruthenium liquid or vapour (Takahashi 1990). Despite its rare use, ruthenium tetroxide is actually an extremely valuable fixative for membrane ultrastructure studies (Pettari 1979). For secondary fixation, osmium tetroxide is most commonly used for cells and tisues and as well as its excellent fixation characteristics, the differential deposition of osmium *within* various parts of the tissue imparts valuable contrast to those regions.

Observation of whole mounts

Provided they are sufficiently thin, dry (dried) non-sectioned samples can be mounted onto support grids and observed in the TEM. These may vary from whole cells which have been stained then critical-point dried (see below, and see Wilson 1993b) to large protein crystals such as insulin or glucagon which are mounted whole onto a grid and either positively or negatively stained. Such whole mounts can be fixed in the vapour phase of either aldehydes or osmium tetroxide.

- *Positive staining* A variety of heavy metal salts can be used to stain various components within the cell. Solutions include (most commonly) uranium and lead, but also vanadium, molybdenum, barium, tungsten, silver and ruthenium. Generally, for whole mounts, the concentration of the solution used is much lower and at a more neutral pH with shorter exposure times than for solutions used to stain ultrathin resin-embedded sections (see below).
- *Negative staining* Negative staining employs solutions of electron opaque heavy metals such as phosphotungstic acid, phosphomolybdic acid and uranyl acetate (Horne 1993). The technique is applicable to nanostructures such as bacterial components, viral particles, microsomes and specific isolated organelles, macromolecules and fibrous materials. They are mixed with the heavy metal solution then a small drop of the suspension is deposited on a coated grid. The particulates sediment then adheres to the grid surface and excess stain is blotted away and allowed to air dry. As the electron opaque stain rains down onto the specimen during drying, it tends to pile up against vertical edges of the specimen and also fills crevices on its surface. The particulates are thus surrounded by electron opaque material and negative contrast is achieved. Unlike positive contrast stains, where reactions with specific chemical groups on the specimen occur, there is no binding with negative staining. Also, because the dried electron opaque stain exists in an amorphous vitreous state (rather than micro-crystalline) very high resolution detail can be visualized.

Embedding and ultrathin sectioning

- *Dehydration* After fixation, tissue is washed free from residual fixative and it may then be dehydrated in either alcohol or acetone series (50%, 70%, 90% up to 100%). An additional staining stage with uranyl acetate may potentially be included with the 70% dehydration step.
- *Resin embedding* The specimen must next be completely infiltrated with a resin prior to it being hardened (polymerized) in order to facilitate the subsequent sectioning step. Impregnation of the tissues gives additional support to the specimen. A variety of resin formulations are available for various applications. These vary from the most commonly used epoxy resins through to hydrophilic acrylate/methacrylate formulations (useful for immunocytochemistry) and polyester and melamine resins. If the dehydrating agent (i.e. acetone or ethanol) is either poor, or immiscible with the resin formulation being used, a transition medium (i.e. a solution which is soluble in both) may have to be used as an intermediary stage. Examples of such transition media are 1,2 epoxypropane (syn. propylene oxide) or the safer CMP-30 (Robards & Wilson 1993).

    Infiltration involves the gradual replacement of the transition solvent with an embedding medium by gradually decreasing the concentration of the solvent and increasing the concentration of the resin. After the resin infiltration stage, which may take one or two days for dense tissues, the resin is hardened by thermal polymerization or some other reaction (depending upon the formulation).

Ultrathin sectioning and section mounting

The resin-embedded tissue is now sufficiently rigid to allow it to be sectioned using an ultramicrotome. The surface of the resin-infiltrated tissue is trimmed to a trapezium, then ultrathin sections in the thickness range 50–120 nm are cut, using either glass or diamond knives. Ribbons of these sections are floated off the knife edge onto water contained within a reservoir at the edge of the knife. Sections of an appropriate thickness are then collected onto 3.05 mm specimen support grids which can be fabricated from a variety of materials including (most commonly) copper and nickel or, for certain applications, platinum and (for X-ray analysis) Nylon. A thin plastic (Formvar, Butvar or Pioloform) film (Baumeister & Hahn 1978; Wilson 1993a) can be first deposited across the grid to provide additional support for the delicate sections.

Staining and ultrastructural cytochemistry

Even with osmium post-fixation, the inherent electron contrast within the tissue sections is very poor and they must be stained with a variety of heavy metals which bind to various chemical groups within the specimen and impart contrast (see above).

    While a variety of staining techniques based on heavy metals such as vanadium, tungsten, lanthanum, bismuth, and molybdenum may be employed to impart contrast into various components of the specimen (Lewis & Knight 1977) by far the most commonly used reliable solutions are uranyl acetate (Watson 1958a,b) and lead citrate (Reynolds 1963) which in combination stain a broad spectrum of cell components. In addition, the use of microwave excitation during staining in order to improve rates of penetration of the stain into the cells is well recognized (Giberson & Demaree 1999; Estrada *et al.* 1985).

A variety of cytochemical reactions may be employed to stain specific molecules within the cell. These include enzyme systems, nucleic acids, structural carbohydrates such as cellulose, lignin and pectin as well as proteins such as actin and myosin. Special care must be taken when choosing a fixative and fixation protocol for monitoring cytochemical reactions within specimens since certain fixatives may denature binding sites. For example, osmium tetroxide should not be employed since it denatures enzymes.

Cryopreparation techniques for observation of hydrated cells and macromolecules

Rapid cryofixation of tissue-cultured cells and monolayers can often provide a much more efficient initial preparation than conventional chemical fixation. Assuming cryofixation has been achieved by rapidly cooling the specimens, they can subsequently be prepared for introduction into the transmission electron microscope by a variety of routes.

- *Direct observation of frozen hydrated cells* If cells are frozen as a thin (possibly vitreous) film onto a TEM grid, they can then be mounted onto a cold stage and transferred directly for observation into the column of the TEM. If the inherent contrast of the cells is low, this can be enhanced elecronically by their observation in a scanning transmission electron microscope (STEM) (see above) where contrast may be enhanced electronically.
- *Freeze-fracture (+etching) replication* After the initial rapid freezing stage, frozen specimens are transferred to the vacuum chamber of a freeze-fracture machine where they are fractured open and the resulting surface shadowed at an angle (between 10 and 60′) prior to the replica being coated with a backing layer of carbon. If the specimen is rotated during the shadowing stage (rotary shadowing), structure in the resulting replica is far easier to understand. In addition, low angle rotary shadowing highlights high resolution detail in very small structures such as viral and protein particles. If solid particles or structures are obscured by ice, the latter may be lightly sublimed by first warming the fracture surface to an etching temperature (typically –90~–100°C) prior to replication. After fracture (etching) and shadowing and backing, the resultant 'replica' on the surface of the frozen specimen is parted from the thawing substrate, cleaned in solvents if necessary, then mounted onto a grid and observed in the TEM. This technique provides quite high-resolution information (4 nm) about the morphology of the fracture plane of the specimen.
- *Freeze-substitution and low-temperature embedding* Freeze-substitution involves the rapid cryofixation of a specimen followed by its subsequent immersion in a polar organic solvent (at low temperature) containing fixatives. The temperature of the solvent (typically –80°C) is such that ice crystal growth (and

therefore damage to cell ultrastructure) is minimized. A very broad range of solvents applicable to the 'substitution cocktail' have been identified. After substitution, and while still cold, the specimen is washed free from fixative and slowly replaced with a low-temperature resin. The latter infiltrates the specimen at low temperature and is subsequently polymerised with UV light.

Techniques for scanning electron microscopy

Preparation of dehydrated cells and macromolecules

Unless a special technique (requiring a specialist microscope) such as 'High-pressure SEM' (Shah 1990) or 'environmental SEM' (Danilatos 1993) is used, most biological cells and tissues will need to have their inherent water removed before their introduction into the SEM.

- *Fixatives and fixation methods* Most routine fixation methods used for SEM have been previously developed for TEM (see above). However, because SEM specimens tend to be a little larger than those excised and prepared for TEM, buffered formaldehyde/glutaraldehyde mixes (which penetrate quickly) are useful and therefore popular. In addition, the use of microwave excitation during fixation in order to improve rates of penetration of the fixative into the cells is well recognized (Giberson & Demaree 1999). It must be stressed, however, that there is no 'universal fixative' for SEM preparation and that for each specimen type there is a special requirement.
- *Replacement of water and solvent* This is normally performed by immersing the specimens in an ascending alcohol or acetone series i.e. 50%, 70%, 90%, 100% of the solvent in distilled water. The 100% solvent should contain an activated molecular sieve as a drying agent, which should be regenerated regularly.
- *Critical-point drying (CPD)* Of all the techniques available for preparation of cells for SEM, critical-point drying is still probably the most commonly used, although this situation may change as more laboratories buy cryo-preparation systems for their SEMs (see Cohen 1974). At the critical point (temperature and pressure) of the liquid contained within the cells to be prepared, the interfacial tension between liquid and aqueous phases of the liquid is minimal, therefore disruption of the delicate structure of the cells will also be minimal.

  The critical point for water is impractically high (22 Mpa and 374°C) but for CO_2 the critical point is more achievable (7.4 Mpa and 31°C). Since water is not very miscible with liquid CO_2, specimens are first dehydrated in a transition solvent (i.e. acetone) soluble in both. Acetone-soaked specimens are then introduced into a small pressure chamber and liquid CO_2 is introduced and slowly replaces the acetone within the specimens. After a suitable exchange period the chamber is warmed to the critical point and due to the expansion of the liquid CO_2 the pressure also rises to the critical pressure. Gas is then slowly released from the chamber and the specimens are dry.
- *Solvent drying* Cell suspensions and large macromolecules can be air dried from solvents having a very low surface tension. The specimens are first dehydrated

in alcohol or acetone series (see above) then transferred into the drying solvent such as tetramethylsilane (TMS). After complete infiltration with the drying solvent a small drop of the solvent-soaked cells are deposited onto a flat, non-porous surface such as a cover glass, and allowed to dry at ambient temperature and pressure. Because of the extremely low surface tension, the deleterious effect of these forces on the tissue is minimized.

- *Specimen mounting and coating* Dried specimens are affixed to flat-topped specimen stubs using either double-sided adhesive tape or a very thin smear of two-pack epoxy cement for every small specimens. For very tiny specimens a small glass or plastic cover slip may be mounted onto the surface of the stub first to provide a relatively flat, featureless surface. Small specimens may be attached using polylysine (Nagarajan & Bates 1981) or a small drop of the adhesive dissolved from the surface of a piece of Sellotape using chloroform or indeed a variety of different adhesives for different applications (see Blackwood 1993). The solid SEM specimen receives electrons from the primary beam onto its surface and, by way of balance, loses electrons by secondary electron generation and backscattered loss. If the two parts of the equation do not balance then the result will be accumulation of positive or negative charge on the surface resulting in 'charging'. Charging manifests itself as very bright zones on the specimen surface, immediate loss of resolution and often distortion of the image. If the specimen surface is electrically conductive or is rendered so and is made continuous with earth, then zero electrical potential exists and charging does not occur. Dry biological samples are generally poor electrical insulators and therefore must be rendered conductive by the application of a non-resolvable metallic layer over the specimen surface. This is generally achieved using a technique called sputter-coating in which a metal such as gold, palladium, chromium or platinum is deposited as a semi-coherent layer onto the specimen surface via a reactive plasma typically derived from argon gas (Echlin *et al.* 1980).

Cryopreparation of hydrated cells and macromolecules

- *Cryo-fixation* Any low-temperature preparation technique for electron microscopy involves the all-important prior step of cryofixation. This involves the rapid cooling of the specimen to a temperature at which water and liquids are frozen and at which ice crystal growth is minimal. Among other things, the rate at which heat is withdrawn from the specimen dictates the ice crystal size within the specimen and in order to minimize this, the cooling rate must be very high indeed. A variety of methods are used to cool specimens for SEM and TEM. These include rapid plunge freezing into liquid propane or ethane or subcooled nitrogen, propane jet freezing, spray freezing (for unicells and viral particles), fast slamming onto a polished silver surface and hyperbaric freezing. After the initial freezing stage, there are three principal cryomethods for examining cells and macromolecules in the SEM: (1) direct observation of frozen hydrated (etched) coated specimens, (2) freeze-drying/molecular distillation of frozen specimens, and (3) freeze-substitution and critical-point drying of frozen specimens.

- *Direct observation* This involves rapidly freezing the cell pellet or monolayer (see freezing methods above) then transferring it under vacuum on a cold stage (typically –165°C) to a cold stage in a preparation chamber where the cell pellet can be fractured using a cold knife, to reveal both the cell surface and also its internal structure. Ice can be sublimed from the fracture surface by warming the specimen briefly to –80 to –90°C (i.e. for one or two minutes). The specimen is then cooled again to –165°C and sputter coated with gold, gold palladium, chromium or platinum depending upon the application. The specimen is then transferred to a cold stage in the SEM which is generally maintained at –165°C using cooled nitrogen gas.
- *Freeze-drying* Freeze-drying (or the more controlled technique of molecular distillation) involves rapidly freezing the cell pellet or monolayer (see freezing methods above) then holding it in a vacuum chamber on a temperature-controlled cold stage. A metal surface colder than the specimens is held close to them to trap subliming water molecules from the specimen surface. Generally, freeze-drying of very tiny specimens such as small cell pellets will occur at –60°C over a period of three days at a rotary pump vacuum. Molecular distillation is performed at high (turbopump) vacuum at temperatures approaching 95°C over very much more protracted periods.
- *Freeze-substitution* This technique involves rapidly freezing the cell pellet or monolayer (see freezing methods above) then transferring the frozen pellet to a freeze-substitution 'cocktail' containing fixatives, a solvent (typically acetone for SEM) and a molecular sieve. The 'cocktail' is maintained at –80°C and the frozen specimens are gently swirled within it. Over a period of days the ice is replaced by solvent and the molecular sieve traps water molecules released from the specimen. This eliminates the collapse and shrinkage that inevitably occurs within the specimen when it is dehydrated conventionally at ambient temperature. After three or four days the specimen is warmed to about –40°C and the 'cocktail' is replaced with pure cold acetone. The specimens and acetone are then gradually warmed to ambient temperature and the dehydrated, fixed specimens are finally critical-point dried (see above) (Humbel & Muller 1986).

Additional analytical techniques

Electron probe X-ray microanalysis

The interactions of the manifold types of atoms within the specimen with the incident electron beam yields a spectrum of valuable signals for the electron microscopist. When a high-energy incident electron collides with an electron of a specific type of atom within the specimen the electron from the latter will be ejected. As a result of its electron loss, the atom now becomes ionized and the resting energy state is restored by the transition of an electron from an outer higher energy shell to fill the vacancy. The excess energy is then released as an X-ray photon whose *wavelength* and *energy* are characteristic of the atom from which it was released. The last 25 years or so have seen the evolution of more and more sensitive spectrometers and software support, to detect and measure either the wavelength or energy of the X-rays in order to provide a sensitive inorganic analysis.

Despite its inestimable value in electron microscopic analysis, it must be borne in mind that the results arising from X-ray analysis depend on five important factors: (1) the application of a specimen preparation technique which minimizes both the *loss* of inorganic elements and also their *relocation* within cells and tissues (the technique cannot distinguish between bound and free ions); (2) optimization of the operating parameters of the microscope and detector to provide the best possible signal from the specimen; (3) minimization of electron beam damage during the initial visual screening of the specimen; (4) the use of robust quantitative methodology/software; and (5) the use of reliable quantitative standards. For procedures to prepare cultured cells in suspension for X-ray microanalysis see Fernandez-Segura *et al.* (1999).

- *Wavelength dispersive (WD) X-ray analysis* All elements of the Periodic Table in theory, with the exception of H, He and Li, can be detected and analysed using a modern WD spectrometer. Because of the low kVs, long working distances and very high beam currents needed to obtain the optimum WD signal, high-resolution analysis is generally impossible.
- *Energy dispersive (ED) X-ray analysis* While not as sensitive as wavelength spectrometry, a usable signal can be delivered to the ED spectrometer at high kV and much lower beam current (therefore maximizing image signal and minimizing specimen damage).

Energy loss and energy filtering spectroscopy and imaging

Energy loss spectroscopy (EELS) can provide chemical, electronic and interatomic information about the specimen at very high spatial resolution. A conventional electron microscope fitted with an electron spectrometer can achieve this (Hunt & Williams 1991).

In addition to EELS, the electron spectrometer may be used as a filter in order to provide an image using electrons which, after interaction with the specimen, have experienced a defined range of energy loss. The energy filtering TEM (EFTEM) spectrometer is tuned such that only electrons of a specific energy, corresponding to a specific accelerating voltage, remain on the optical axis of the electron microscope after passing through the filter. The EFTEM technique thus provides images with far improved contrast.

EM autoradiography

Autoradiography involves the use of a detector layer (a silver halide photographic emulsion) which is placed next to a specimen in which a specific radio-label is localized. Macroscopic autoradiographs may be viewed with the unaided eye. LM autoradiography requires a light microscope to resolve the autoradiograph and an electron microscope in the case of EM autoradiography (Williams 1977a,b).

In EM autoradiography a radio-label is incorporated and bound into a specific region or compartment of the cell or tissue in order to spatially define the ultrastructural boundaries' compartments. Furthermore, it is possible to quantify the amount of radiolabel bound within the compartment.

Image analysis techniques (point source density function) are often used to interpret and analyse clustering patterns of grains within a specimen.

Immunocytochemistry

The antibody/antigen reaction is so specific that it can now be used reliably at the EM level to localize specific components within the cell. The success of immunocytochemical localization depends upon three criteria: (1) the availability of very specific antibodies for the antigen under investigation, (2) a good electron opaque probe of suitable size, (3) a specimen preparation technique which faithfully preserves both the antigenicity of proteins and retains their position within the cell. Monoclonal antibodies are extremely valuable since they react with only one epitope.

Antibodies are bound either directly or indirectly to a metal probe with a very high electron opacity (e.g. colloidal gold particles) (Faulk & Taylor 1971). Binding of the antibody/probe to antigens can be performed either pre- or post-embedding in resin. In the former the wet tissue is labelled, in the latter, ultra-thin sections are labelled. The availability of different sized gold particles allows multi-labelling experiments to be undertaken due to the ability to distinguish between two different antibodies.

The gold probe particles are between 1 and 50 nm in diameter and therefore beyond the resolving power of the LM. However, after localization the effective size of the gold particles can be increased by encapsulating them in a thick coat of silver (silver enhancement technique) by exposing them to silver halide in the presence of light, thus providing particles resolvable in the light microscope.

The major problem with immunocytochemistry is retaining the antigenic activity of the specimen during sample preparation. Different antigens are affected to various degrees by the different crosslinking fixatives used in electron microscopy. The embedding resins can also mask the epitopes which are exposed to bind with the antibodies. High concentrations of strongly crosslinking fixatives such as glutaraldehyde are avoided. Instead, a very low concentration of glutaraldehyde with formaldehyde may be employed (Beesley 1993). Resins with a more 'open' surface strucutre such as acrylic resins are used to increase the chance of the antigen being exposed at the surface after sectioning.

Ideally, chemical fixation should be replaced with cryo-fixation (see above) followed by the analysis of thawed crysosections. In this way, chemical fixation, dehydration and resin infiltration are avoided.

Image interpretation and analysis

In the average electron micrograph there is an overwhelming wealth of information available for potential analysis. A methodical approach to simplifying the task of interpretation has been outlined by Murphy (1993) and involves evaluation of the negative (analogue data source), production of a working micrograph and identification of cell ultrastructure. The latter may be further analysed in terms of estimation of section thickness, observation of empty spaces, evaluation of membrane structure, evaluation of organelle structure, identification of dark precipitates, identification of sectioning artefacts, identification of artefacts originating directly from the microscope, photographic artefacts and completion of interpretation.

EM image analysis is the means by which we can derive valuable numerical data from an image bearing structural information. It is especially valuable where we are examining the effects of, for example, different treatments on cells, and need to understand the statistical significance, or not, of the differences between treated cells and controls. It enables the conversion of ostensibly structural data into a numerical form which is more easily manipulated statistically.

Prior to any analysis the issue of specimen sampling must be addressed and the appropriate sample size for the application determined. This will involve determining optimum number of cell pellets, sections of pellets and cells counted per section. Collection of data and subsequent image analysis may then be performed either in the microscope or outside the microsope on micrographs. Stereology and morphometry is a specific branch of image analysis involving (in the case of cell biology) the collection of accurate data on cell and organelle size, shape and distribution (Weibel 1979; Williams 1977a,b).

Image reconstruction

Powerful computer databases now allow us to perform very sophisticated image reconstruction which involves storing two-dimensional images from serial sections then reconstructing them to form a representation of the bulk structure in three dimensions.

REFERENCES

Andrews KW, Dyson DJ, & Keown SR (1971) *Interpretation of Electron Diffraction Patterns*, 2nd edn. Adam Hilger, Bristol.

Baumeister W & Hahn M (1978) Specimen supports. In: Hayat MA (ed.), *Principles and Techniques in Electron Microscopy: Biological Applications*, Vol. 8. Van Nostrand Reinhold, New York.

Beesley J (1989) Colloidal Gold: A New Perspective for Cytochemical Marking, *Royal Microscopial Society Handbooks 17*. Oxford Univ. Press, Oxford.

Blackwood AW (1993) Mounting and storing specimens. In: Robards AW & Wilson AJ (eds), *Procedures in Electron Microscopy*, Module 10:4, pp. 10:4.1–10:4.18. Wiley, Chichester.

Cohen AL (1974) Critical-point drying. In: Hayat, MA (ed.), *Principles and Techniques of Scanning Electron Microscopy*, Vol. 1. Van Nostrand Reinhold, New York.

Danilatos GD (1993) Introduction to the ESEM instrument. *Microscopy Research and Technique* 25: 354–361.

Dyson DJ (1993) Electron diffraction techniques. In: Robards AW & Wilson AJ (eds), *Procedures in Electron Microscopy*, Module 9, pp. 9:4.1–9:4.18. Wiley, Chichester.

Echlin P, Broers AN & Gee W (1980) Improved resolution of sputter-coated metal films. *Scanning Electron Microscopy* 1: 163–170.

Estrada JC, Brinn NT & Bossen EH (1985) A rapid method of staining ultrathin sections for surgical pathology TEM with the use of the microwave oven. *Brief Scientific Reports* 83(5): 639–641.

Everhart TE & Thornley RF (1960) Wide-band detector for micro, micro-ampere low energy electron currents. *Journal of Scientific Instruments* 37: 246–248.

Faulk WP & Taylor GM (1971) An immuno-colloid method for the electron microscope. *Immunocytochemistry* 8: 1081–1083

Fernandez-Segura E, Canizares FJ, Cubero MA, Campos A & Warley A. (1999) A procedure to prepare cultured cells in suspension for electron probe X-ray microcanalysis: its application to scanning and transmission electron microscopy. *Journal of Microscopy* 196, Pt 1, 19–25.

Giberson TR & Demaree RS (1999) Microwave processing and techniques for EM. In: Hajibabheri MAN (ed.), *Electron Microscopy Methods and Protocols*. Methods in Molecular Biology Series, Vol. 117. pp 145–157.

Hashimoto H. (1993) High resolution TEM. In: Robards AW & Wilson AJ (eds), *Procedures in Electron Microscopy*, Module 9:6, pp. 9:6.1–9:6.24. Wiley, Chichester.

Hayat MA (1981) *Fixation for Electron Microscopy*. Academic Press, London.

Hofer F (1993) Energy loss spectroscopy. In: Robards AW & Wilson AJ (eds), *Procedures in Electron Microscopy*, Module 15:3, pp. 15:3.1–15:3.21. Wiley & Sons, Chichester.

Horne RW (1993) Negative staining of various biological particles. In: Robards AW & Wilson AJ (eds), *Procedures in Electron Microscopy*, Module 5:8, pp. 5:8.1–5:8.9. Wiley & Sons, Chichester.

Humbel B & Muller M (1986) Freeze-substitution and low temperature embedding. In: Muller, M, Becker, RP, Boyde, A & Wolosewick, JJ (eds), *The Science of Biological Specimen Preparation*. SEM Inc. AMF O'Hare, Chicago, IL.

Hunt JA & Williams DB (1991) Electron energy-loss spectrum imaging. *Ultramicroscopy* 38: 47–73.

Lewis PR & Knight DP (1977) Staining methods for sectioned material. In: AM Glauert (ed.), *Practical Methods in Electron Microscopy*, Vol. 5, Part I. Elsevier, New York.

Murphy J (1993) Image interpretation. In: Robards AW & Wilson AJ (eds), *Procedures in Electron Microscopy*, Module 17:3, pp. 17:3.1–13 1. Wiley & Sons, Chichester.

Nagarajan P & Bates LS (1981) A rapid poly-L-lysine schedule for SEM studies of acetolyzed pollen grains. *Pollen et Spores* 23: 273–279.

Otten MT (1991) High-angle annular dark field imaging on a TEM/STEM system. *Journal of Electron Microscopy Techniques* 17: 221–230.

Pelttari A & Helminen HJ (1979) The effects of various fixatives on the relative thickness of cellular membranes in the ventral lobe of the rat prostate. *Histochem. J.* 11: 599.

Reynolds ES (1963) The use of lead citrate at high pH as an electron opaque stain in electron microscopy. *Journal of Cell Biology* 17: 208.

Robards AW & Wilson AJ (1993) *Procedures in Electron Microscopy*. Wiley, Chichester.

Shah JS (1990) High pressure scanning electron microscopy. *Laboratory Practice* 39(12): 65–73.

Takahashi G. (1990) Ruthenium tetroxide vapour staining of dry specimens for scanning electron microscopy. *Proceedings of the XIIth Intrnational Congress EM*, p. 714.

Watson ML (1958a) Staining of tissue sections for electron microscopy with heavy metals. *Journal of Biophysics, Biochemical and Cytology* 4, p. 475.

Watson ML (1958b) Staining of tissue sections for electron microscopy with heavy metals II. Application of solutions containing lead and barium. *Journal of Biophysics and Biochemical Cytology* 4: 727.

Weibel ER (1979) *Stereological Methods*, Vol. 1. *Practical Methods for Biological Morphometry*. Academic Press, New York.

Wergin WP (1993) Image recording and interpretation In: Robards AW & Wilson AJ (eds), *Procedures in Electron Microscopy*, Module 14:11, pp. 14:11.1–14:11.7. Wiley, Chichester.

Williams MA (1977a) Quantitative methods in biology. In: AM Glauert (ed.), *Practical Methods in Electron Microscopy*, Vol. 6. Part II. Elsevier, New York.

Williams MA (1977b) Autoradiography and immunocytochemistry. In: Glauert AM (ed.), *Practical Methods in Electron Microscopy*, Vol, 6, Part 1. North-Holland, Amsterdam.

Wilson AJ (1993a) Specimen support grids and films. In: Robards AW & Wilson AJ (eds), *Procedures in Electron Microscopy* Module 4:6, pp. 4:6.9–4:6.12. Wiley, Chichester.

Wilson AJ (1993b) Critical-point drying. In: Robards AW & Wilson AJ (eds), *Procedures in Electron Microscopy*, Module 11:4, pp. 411:4.1–11:4.17. Wiley, Chichester.

Yacobi BG & Holt DB (1990) *Cathodoluminescence Microscopy of Inorganic Solids*. Plenum Press, New York.

3.7 CELL DEATH IN CULTURE SYSTEMS (KINETICS OF CELL DEATH)

Cell death in eukaryotic cells may be divided into two morphologically and biochemically distinct modes, those of apoptosis and necrosis. Necrosis is the classically recognized form of cell death that results from severe cellular insults and involves rapid cell swelling and lysis and is a process over which the cell has little or no control. Apoptosis, on the other hand, is the mode of cell death observed primarily under physiological conditions, and is an organized, preprogrammed response of the cell to changing environmental conditions. This form of cell death is characterized by cell shrinkage, nuclear and DNA fragmentation and breaking up of the cell into membrane-bounded vesicles, termed 'apoptotic bodies', which are subsequently ingested by neighbouring cells or macrophages *in vivo*. Apoptosis is an ATP-dependent process that in some systems also requires RNA and protein synthesis. The activation of a Ca^{2+}/Mg^{2+}-dependent, zinc-inhibitable endonuclease, which cleaves the cell's DNA into fragments of 200 base pairs or multiples thereof, is the main biochemical hallmark of apoptosis.

In cell culture systems, cell death via apoptosis is quite common and can occur under a variety of circumstances. For example, growth-factor-dependent cell lines die by apoptosis following removal of the growth-promoting agent from the culture medium (Nieto & Lopez-Rivas 1989). Terminal differentiation of cells *in vitro* results in apoptosis (Martin *et al.* 1990), as does normal turnover of cells with a limited lifespan, e.g. neutrophils (Savill *et al.* 1989) or freshly isolated thymocytes (McConkey *et al.* 1989). Altered culture medium conditions, e.g. removal of zinc from the medium (Martin *et al.* 1991), or cultures that are allowed to overgrow to produce high cell densities, also induce apoptosis. Finally, cells may also be induced to undergo apoptosis by exposure to a wide range of cytotoxic agents (Lennon *et al.* 1991). Under *in vivo* conditions, apoptotic cells are recognized and rapidly removed by phagocytic cells. However, in cell culture this cannot occur and instead the cells undergo secondary necrosis. Cells at this stage are Trypan blue positive. Up to this point, however, apoptotic cells maintain the ability to exclude vital dyes, and hence this method underestimates the health of a cell culture.

Reagents and solutions

Note: Lysis buffer (without proteinase K) and the TE buffer should be autoclaved to remove contaminating deoxyribonuclease activity.

Cell and Tissue Culture for Medical Research, edited by A. Doyle and J.B. Griffiths.
© 2000 John Wiley & Sons, Ltd

- *Lysis buffer*: 10 mM EDTA, 50 mM Tris (pH 8.0) containing 0.5% *N*-lauroyl-sarcosine and 0.5 mg ml^{-1} proteinase K
- *TE buffer*: 10 mM Tris·HCl (pH 8.0) and 1 mM EDTA
- *Electrophoresis loading buffer*: 10 mM EDTA, 0.25% bromophenol blue and 50% glycerol
- *TBE buffer*: 2 mM EDTA (pH 8.0), 89 mM Tris and 89 mM boric acid

PROCEDURE: MORPHOLOGICAL CHARACTERIZATION OF CELL DEATH

The two modes of cell death may be identified easily on morphological grounds. Apoptosis involves cell shrinkage, nuclear fragmentation and apoptotic bodies budding off, maintaining the ability to exclude vital dyes. Necrosis involves cell swelling, chromatin flocculation and direct cell lysis. Thus, cells undergoing necrosis rapidly lose their ability to exclude vital dyes.

Materials and equipment

- Cell fixative – a haematoxylin and eosin stain (e.g. Rapi-Diff II, Diachem Diagnostic Developments, Southport, UK)
- Trypan blue
- Cytocentrifuge

 Trypan blue is a suspected carcinogen. Gloves should be worn when using all stains.

1. Cytocentrifuge cells onto glass slides. The speed of centrifugation will depend on the cell type; a typical cytocentrifuge setting would be 300 rpm for 2 min.
2. Remove glass slides and place in fixing solution for 20 s ('A' of Rapi-Diff II stain). Remove excess fluid from slide and repeat for the cytoplasmic and nuclear stains ('B' and 'C' of Rapi-Diff II stain, respectively). Rinse with distilled water.
3. Examine under the light microscope.

The key morphological criteria for recognizing apoptotic cells are nuclear condensation and fragmentation (Figure 3.7.1). Apoptotic cells also maintain their ability to exclude vital dyes such as Trypan blue. Hence cultures with a high proportion of cells displaying nuclear condensation and fragmentation, indicative of apoptosis, should also have a relatively high number of cells with the ability to exclude vital dyes.

Necrotic cells may be recognized morphologically by cell swelling and nuclear flocculation (Figure 3.7.2). Cultures with a high proportion of necrotic cells should have a relatively low number of cells with the ability to exclude vital dyes.

PROCEDURE: BIOCHEMICAL CHARACTERIZATION OF CELL DEATH

The biochemical hallmark of apoptosis is cleavage of the nuclear DNA into ~200 base pair multiples (Wyllie 1980). This specific DNA cleavage is thought to result

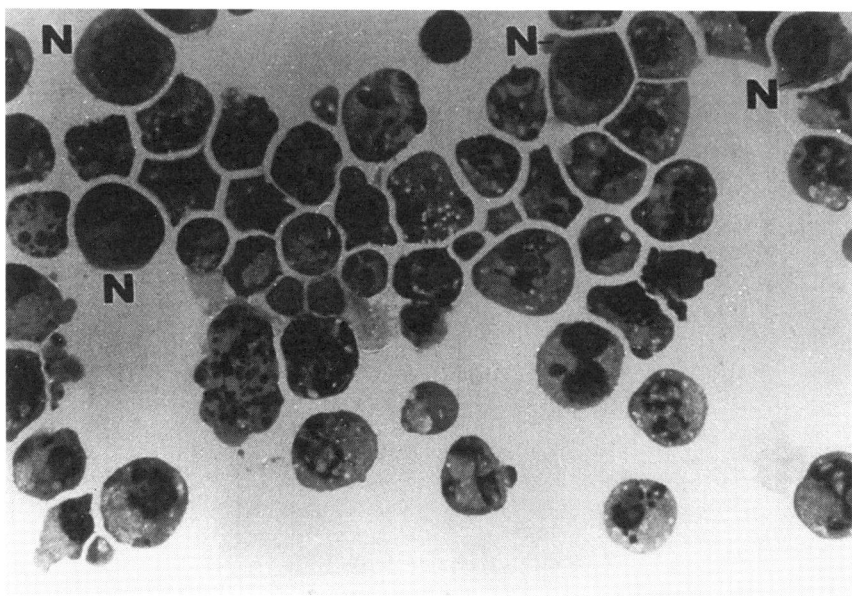

Figure 3.7.1 Morphological characteristics of apoptosis in human promyelocytic leukaemic HL-60 cells, illustrating nuclear fragmentation typical of apoptosis in HL-60 cells. *N* indicates normal cells. All the other cells are undergoing various phases of apoptosis. Original magnification × 500.

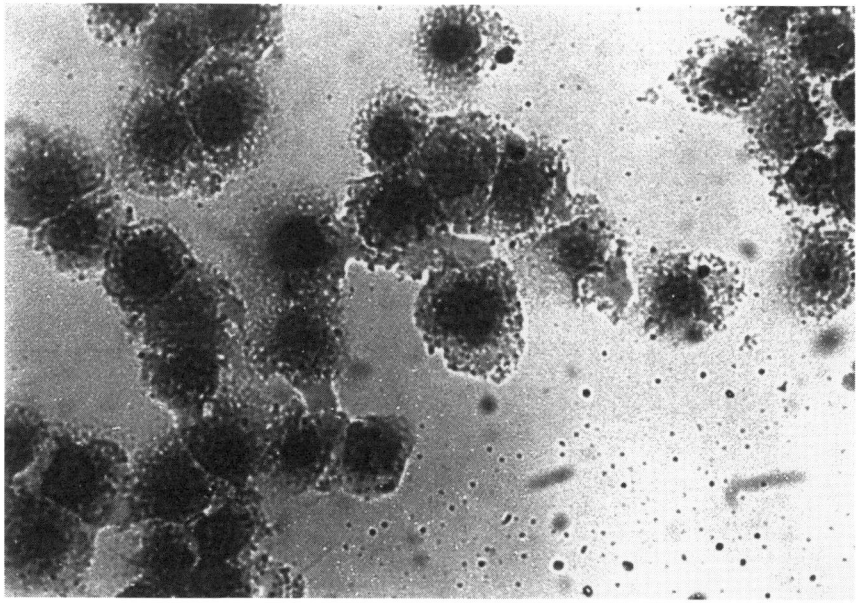

Figure 3.7.2 Morphological characteristics of necrosis in HL-60 cells. Original magnification × 500.

from the activation of an endogenous endonuclease that cleaves at the exposed linker regions between nucleosomes. Necrosis, on the other hand, is not associated with the ordered form of DNA cleavage observed during apoptosis. Hence a biochemical examination of the DNA cells can be used to characterize the mode of cell death occurring in cell cultures.

Isolation of DNA

- Lysis buffer
- Rnase A (previously heat treated to remove contaminating deoxyribonuclease activity)
- Phenol, buffered with 0.1 M Tris·HCl (pH 7.4)
- Chloroform/isoamyl alcohol (24 : 1)

 Gloves should be worn throughout this procedure, both to protect the worker from phenol and chloroform and to prevent contamination of samples with deoxyribonucleases. Pipette tips and tubes should be autoclaved before use.

1. Wash cells twice in phosphate-buffered saline (PBS).
2. Resuspend cell pellets at 2×10^7 ml^{-1} in lysis buffer and incubate in a 50°C water-bath for 1 h.
3. Add Rnase A to a concentration of 0.25 mg ml^{-1} and continue incubation at 50°C for 1 h.
4. Extract the crude DNA preparations twice with an equal volume of phenol (retain the aqueous layer after each extraction).
5. Extract twice with an equal volume of chloroform/isoamyl alcohol (retain the aqueous layer after each extraction).
6. Centrifuge the DNA preparations at 13 000 g for 15 min to separate intact from fragmented chromatin.
7. Place the supernatant (containing the fragmented chromatin) into a separate tube.
8. Precipitate the DNA in two volumes of ice-cold ethanol overnight at −70°C. The DNA may be stored in this condition for long periods.

Electrophoresis of DNA

Materials and equipment

- TE buffer
- Electrophoresis loading buffer
- Agarose
- TBE buffer
- Ethidium bromide
- UV source (Transilluminator UV: 302 nm), UVP Inc., San Gabriel, CA, USA)

 Ethidium bromide is an irritant and known mutagen. Gloves should be worn throughout this procedure.

1. Recover the DNA precipitates by centrifugation at 13 000 g for 15 min. Allow the tube to air-dry at room temperature for 10 min and resuspend in TE buffer. Store at 4°C.

2. Add electrophoresis loading buffer in a 1 : 5 ratio. Place samples into a 65°C water-bath for 10 min and then maintain on ice.
3. Place samples into wells of a 1% agarose gel.
4. Electrophoresis is carried out in TBE buffer at 6 V cm^{-1} of gel.
5. After electrophoresis, soak the gel in TBE buffer containing 1 mg ml^{-1} ethidium bromide. Destain briefly in TBE buffer.

⚠ Visualize the DNA by UV fluorescence. Care should be taken not to expose skin or eyes to UV light.

The DNA isolated from apoptotic cells separates into bands of 200 or multiples of 200 base pairs (Figure 3.7.3, lanes 3 and 4), whereas DNA isolated from control cells shows relatively little degradation (Figure 3.7.3, lanes 1 and 2). The DNA isolated from necrotic cells shows no ordered DNA fragmentation.

Note: The use of DNA molecular size markers enables an estimate of the size of the fragmented DNA.

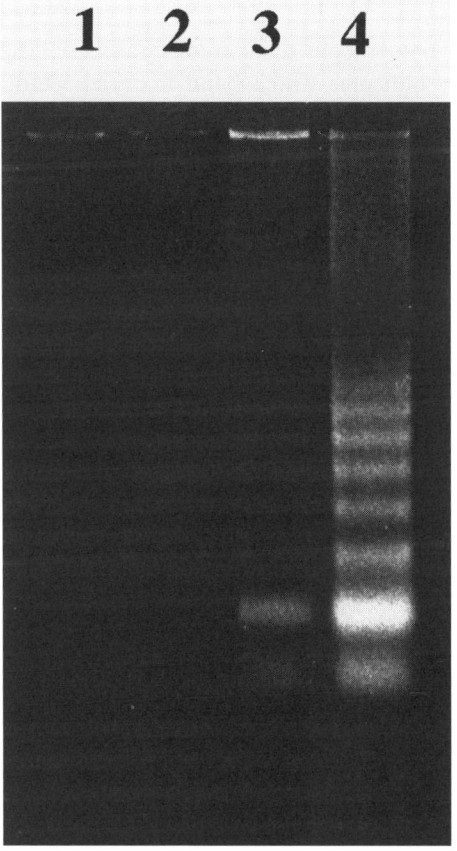

Figure 3.7.3 Cleavage of DNA into 200 or multiples of 200 base pairs during apoptosis in HL-60 cells (lanes 3 and 4). Control cells do not display any DNA degradation (lanes 1 and 2).

SUPPLEMENTARY PROCEDURE: PURIFICATION OF APOPTOTIC CELLS

If apoptosis is only occurring at relatively low levels in cultures, it may be necessary to obtain a purified population of apoptotic cells prior to DNA isolation and electrophoresis. This may be achieved by exploiting the fact that apoptotic cells are more dense than normal cells. Hence it is possible to purify apoptotic cells by isopycnic centrifugation. Percoll can be used to create solutions of different densities. The precise Percoll densities used for isolation of apoptotic cells will depend on the cell type under investigation.

1. Prepare Percoll solutions of various densities. These densities should typically range from 1.05 to 1.08 in the case of most mammation cells.
2. Carefully layer the Percoll solutions sequentially into a test tube. Wash cells in PBS, resuspend at a high concentration (typically 20×10^6 ml^{-1}) in PBS and place on top of the Percoll column.
3. Centrifuge at $400\,g$ for 30 min in a swing-out rotor. Cells distribute to their isopycnic points, creating bands representing normal, dead and apoptotic cells.
4. Elute the different bands from the Percoll gradients. Morphological examination of these bands indicates which band is enriched for apoptotic cells. This provides an estimate for the density of the apoptotic cells.
5. Adjust the Percoll gradients, placing the solution with the estimated density of apoptotic cells at the bottom of the tube.
6. Repeat the above procedure until apoptotic cells pellet to the bottom of the test tube after centrifugation. It is possible to obtain a purified population of apoptotic cells ($> 90\%$) by this method.

Figure 3.7.4 shows a typical Percoll gradient for isolation of apoptotic cells from a human promyelocytic leukaemic cell culture (HL-60).

DISCUSSION

Apoptosis was first described in 1972 (Kerr et al. 1972). Since then it has become apparent that this mode of cell death plays a role in a number of important physiological processes. Thus it is advantageous to be able to recognize when this mode of death is occurring. Investigations into the mechanistic aspects of apoptosis have revealed a regulatory role of different ions in this process. For example, apoptosis occurs in certain human cell lines under conditions of zinc deficiency (Martin et al. 1991). Likewise, apoptosis may be inhibited by the addition of zinc to the culture medium (Martin & Cotter 1991) or by removal of calcium (McConkey et al. 1989). Hence it is apparent that different culture conditions may activate or inhibit this mode of cell death. The above morphological and biochemical procedures should be used to complement each other in identifying the mode of cell death occurring in cells.

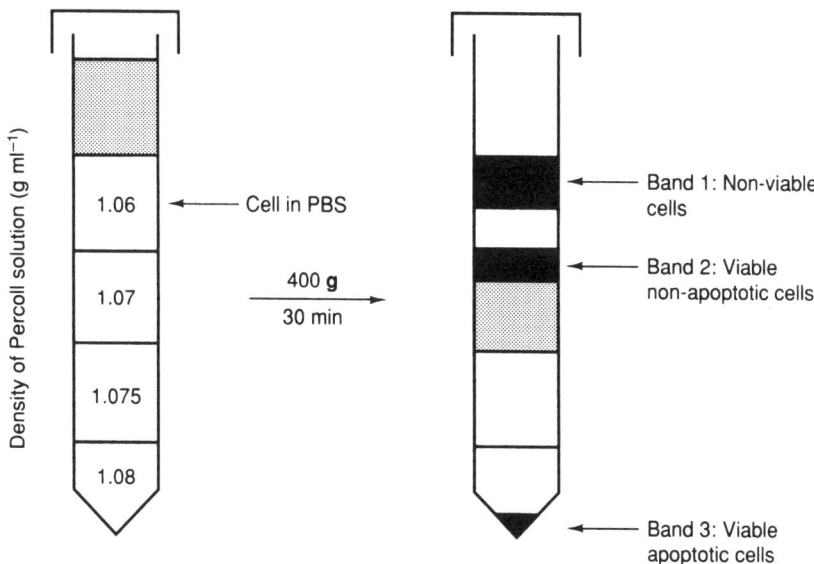

Figure 3.7.4 Percoll gradient for isolation of apoptotic cells from an LH-60 cell culture.

REFERENCES

Kerr JFR, Wyllie AH & Currie AR (1972) Apoptosis: a basic biological phenomenon with wide ranging implications in tissue kinetics. *British Journal of Cancer* 26: 239–257.

Lennon SV, Martin SM & Cotter TG (1991) Dose-dependent inducation of apoptosis in human tumour cell lines by widely diverging stimuli. *Cell Proliferation* 24: 203–214.

Martin SJ & Cotter TG (1991) Ultraviolet B irradiation of human leukaemia HL-60 cells *in vitro* induces apoptosis. *International Journal of Radiation Biology* 59: 1001–1016.

Martin SJ, Bradley JG & Cotter TG (1990) HL-60 cells induced to differentiate towards neutrophils subsequently die via apoptosis. *Clinical Experimental Immunology* 79: 448–453.

Martin SJ, Mazdai C, Strain JJ, Cotter TG & Hannigan BM (1991) Programmed cell death (apoptosis) in lymphoid and myeloid cell lines during zinc deficiency. *Clinical and Experimental Immunology* 83: 338–343.

McConkey DJ, Hartzell P, Amador-Perez JF, Orrenius S & Jondal M (1989) Calcium-dependent killing of immature thymocytes by stimulus via the CD3/T cell receptor complex. *Journal of Immunology* 145: 1801–1806.

Nieto MA & Lopez-Rivas A (1989) IL-2 protects T lymphocytes from glucocorticoid-induced DNA fragmentation and cell death. *Journal of Immunology* 143: 4166–4170.

Savill JS, Wyllie AH, Henson JE, Walport MJ, Henson PE & Haslett C (1989) Macrophage phagocytosis of aging neutrophils in inflammation: programmed cell death in the neutrophil leads to its recognition by macrophages. *Journal of Clinical Investigation* 83: 865–871.

Wyllie AH (1980) Glucocorticoid-induced thymocyte apoptosis is associated with endogenous endonuclease activation. *Nature (London)* 284: 555–556.

CHAPTER 4

CELL LINE DERIVATION METHODS

4.1 CELL LINE DERIVATION – AN OVERVIEW

Cell culture for medical research embraces a wide range of cell types and systems each of which have their particular merits and associated culture requirements. In order to put the following procedures in perspective the definitions and characteristics of *in vitro* cells is summarized in this overview. A broad classification of cells in culture is:

- Organ culture: composed of functional tissue in short-term culture e.g. tissue slices
- Primary cells: derived from tissues (normally enzymatically) and put immediately into short-term culture
- Finite cell lines: cells derived from normal tissue which can replicate and undergo limited subculture *in vitro*; they may be sufficiently well characterized to achieve the status of 'cell strain', e.g. MRC.5
- Continuous cell lines: cells that have the ability to be cultured *in vitro* indefinitely. Often derived from tumour tissue or have undergone an immortalization step *in vitro*, e.g. HeLa, tumour -derived 293 cells, embroyonal kidney transformed with seared human Ad 5 DNA.

This classification is given in more detail in Table 4.1.1, together with information on how these cells types may be initiated in culture.

Finite cell lines have been used very effectively for many years but offer a fairly limited range of cell types (mainly fibroblasts) as well as a limited lifespan. In general, the use of continuous cell lines overcomes these two limitations. The downside of the latter is the transformed nature of the cells with the presence of a transforming element (including oncogenic DNA) and the cell may well not express all the characteristics of differentiated cells. However, culture in three-dimensional (e.g. Histiotypic) systems or the use of specialist surfaces and media may enhance their differentiated characteristics. It must be emphasized that there is a degree of mutual exclusivity between proliferation and the retention of differentiated functions.

Spontaneous cellular transformation *in vitro* can occur, but this is an ill-understood process that is found far more frequently with rodent cells in culture compared with human cells.

The proliferative capacity and relative stability of both finite and continuous cell lines has enabled the establishment of routines to create bulk stocks or cell banks to ensure the research worker can have a reliable, stable and secure stock of cells for the foreseeable future.

The type of cell being cultured dictates to some extent the actual culture system that can be used and this is summarized in Figure 4.1.1.

Cell and Tissue Culture for Medical Research, edited by A. Doyle and J.B. Griffiths.
© 2000 John Wiley & Sons, Ltd

Table 4.1.1 The major categories of cell culture including source of material and derivation

Category	Provenances of material	Transformation/immortalization strategies
Cell culture • Denotes maintenance and growth of cells *in vitro*, including the culture of single cells • May be derived from primary explants or dispersed cell suspension	*In vivo derivations*: e.g. • Whole embryo: disaggregated • Foetal tissue: normal or pathological • Postnatal/adult tissues/organs from e.g.: normal, X-irradiated, radiosensitive, cancer-prone organisms Tissue from e.g. primary/secondary human tumours, chemically, viral- or radiation-induced animal tumours • Bone marrow • Peripheral blood • Body fluids • Tissue scrapings • Transgenic animals: carrying bacterial or viral gene	*In vitro transformation*: • A heritable change, occurring in cells in culture, either intrinsically or from treatment with e.g. – chemical, carcinogens – oncogenic viruses – irradiation – selection in e.g. BUdR; azaguanine – transfection with oncogenes; DNA vectors etc. • Leads to acquisition of altered morphological, antigenic, neoplastic proliferation or other properties. • May not always include the ability of the cells to produce tumours in appropriate hosts.
Primary culture • Started from cells, tissues organs taken directly from organisms: a primary culture may be regarded as such until the first successful subculture • A cell line arises from a primary culture at the time of the first successful subculture • A cell strain is derived either from a primary culture or a cell line by selection or cloning of cells having specific properties or markers	*In vitro derivations*: e.g. • Explant cultures • Cell line/mutant subline	*In vitro neoplastic transformations*: • Acquisition, by cultured cells, of the property to form neoplasms, benign or malignant, when inoculated into animals. – Many transformed cell populations (arising *in vitro* intrinsically or through deliberate manipulation) produce only benign tumours (show no local invasion or metastatasis following animal inoculation). – The terms '*in vitro* malignant neoplastic transformation' or '*in vitro* malignant transformation' can, with supporting *evidence*, be used to indicate that an injected cell line does, indeed invade or metastasize.
Finite cell culture • Denotes a culture which is capable of only limited number of population doublings after which the culture ceases proliferation • *In vitro senescence* denotes the property attributable to finite cell cultures i.e. their inability to be propagable beyond a finite number of population doublings • Immortalization denotes the attainment by a finite cell culture, whether by perturbation or intrinsically, of the attributes of a continuous cell line	*Continuous cell culture* • Denotes a culture which is apparently capable of an unlimited number of population doublings: often referred to as an immortal cell culture. Such cells may or may not express the characteristics of *in vitro* neoplastic or malignant transformation.	

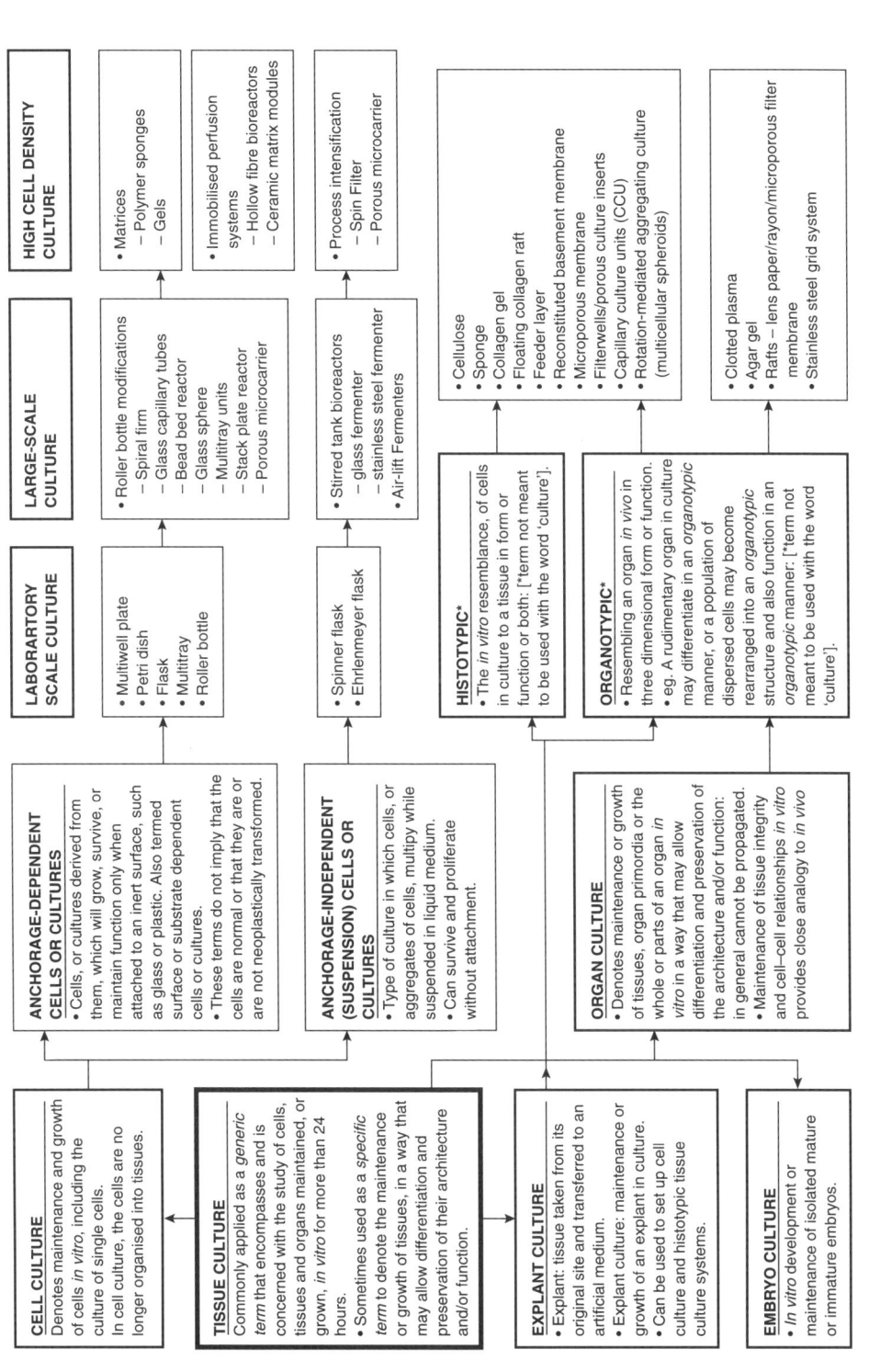

Figure 4.1.1 *In vitro* culture systems.

HUMAN CELL LINES

Human cell lines are utilized for a wide variety of purposes in medical research and the number and range available are increasing, especially with the advent of new cell immortalization technologies. It is necessary, however, to consider all the relevant information regarding the derivation of new material for a particular project or the application of existing cell lines (often generated for a completely different purpose) before embarking on a new cell culture-based project. In particular, safety and ethical issues have to be considered.

RISK ASSESSMENT

In choosing an *in vitro* system with human cells there is the fundamental question of safety. Some of the issues are outlined below:

- Does an appropriate cell line already exist?
- What hazards may the source materials (cells and media) possess?
- If an existing cell line is to be used and is available from a reliable source, has it undergone any quality assessment i.e. authenticity, functionality, microbial contamination and mycoplasma test.
- What level of containment is necessary for hazards either identified or potential?

A risk assessment should be carried out and the protocol should include:

- Selection of patient material in order to reduce the risk of serious infection and coding of specimens to retain anonymity
- Transportation including notification of recipients, mode of transport, containment and storage conditions
- Laboratory reception and recording, including labelling
- Laboratory procedures for manipulation of the tissue, cell culture, routine testing or screening, preservation, storage and disposal.

The criteria used in a decision process for choosing a cell system in the context of safety is given schematically in Figure 4.1.2. The resulting protocol should also be approved by the local biological safety officer or safety committee.

Ethical aspects of the work should receive early attention to identify the requirements of both local and national ethics committees. (NBAC 1999) Questions of patient consent, ownership of any potential products and commercial exploitation need to be clarified. Failure to do so can mar a scientifically successful project on human *in vitro* models. Thus relevant authorities should be approached for approval well in advance of any work. In some countries, notably France, there is great concern over the exploitation of human material, and recipients of human cell lines may need to be registered with a national government body.

In work with human material it is important to find medical partners who are willing to ensure that the right material, in an appropriate state, is obtained reliably and efficiently. This will minimize the number of samples required to achieve the aims of the project. It is also important to have detailed preliminary discussions to define any potential hazards with the material.

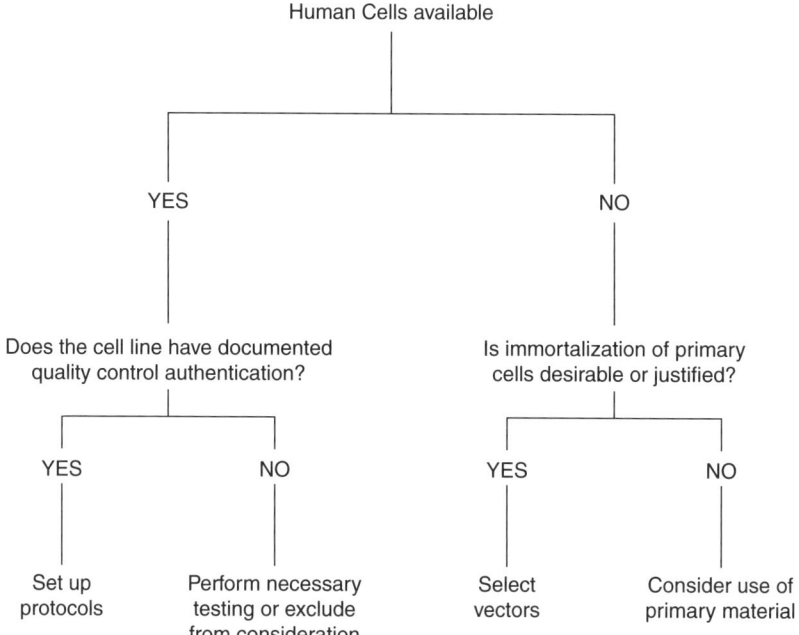

Figure 4.1.2 Schematic for the criteria for an informal choice of system in the context of safety and risk assessment.

REFERENCE

United States National Bioethics Advisory Commission (1999). Research Involving Human Biological Materials: Ethical issues and policy guidance. Report & Recommendations, Rockville, Maryland, USA.

4.2 CELL LINES FOR MEDICAL RESEARCH

INTRODUCTION

From its earliest beginnings in the 1940s cell culture has played a pivotal role in medical science. The success of this technique has stimulated the development of a wide range of cell lines derived mainly but not exclusively from avian and mammalian species. Many areas of medical research have benefited from the use of well-characterized cell lines. One of the earliest applications for such lines was to cultivate and assay viruses. This technique has had a great impact on general medicinal science and contributed significantly to public health. Production of vaccines requires large amounts of pure virus (measles, mumps, rubella and polio) and these are grown in avian or human diploid cell lines. Indeed it was the need for mass cultivation of viruses for vaccine production that stimulated the development of improved culture methods, media and reagents.

The field of animal cell biotechnology itself was born from the need to cultivate such cells on an industrial scale (Spier & Griffiths 1985). The ability to fuse cells from different species to form a hybrid cell led directly to the development of numerous hybridoma cell lines. These lines are capable of secreting monoclonal antibodies (Kohler & Millstein 1975). This technology has itself revolutionized medical diagnostic techniques, as well as proving an invaluable research tool. In a second medical revolution genetically modified cell lines have revealed information about the control of gene transcription, translation and differentiation. Indeed the insertion of appropriate genes into mammalian cells in order to produce a pharmacological product is having a tremendous impact on medical science. Gene therapy is at present being implemented and validated through the science of cell culture. It is this concept more than any other that holds the greatest potential for future advances in medical science in the next century.

The greatest impact cell culture has had on a single medical discipline is in the field of cancer research. The ability of malignant cells to grow indefinitely in culture revolutionized the understanding of this disease. Many types of tumours were successfully cultivated and provided a means to study both the biology of cancer and the means to treat it. It is within the area of cancer research that we see the greatest diversity of cell lines (Hay *et al.* 1994). Tumour cell lines have contributed significantly to the understanding of the genetics and biochemistry of cancer cells. However, research into the drug and radiation sensitivity of cancer cells has had the greatest impact on therapy and patient survival. The following sections deal with a number of cell lines used in medical research with some emphasis on cancer research. The cell lines themselves are merely a small portion

Cell and Tissue Culture for Medical Research, edited by A. Doyle and J.B. Griffiths.
© 2000 John Wiley & Sons, Ltd

of the many hundreds available from accredited cell banks. Indeed many thousands are available from individual laboratories worldwide. It should be stressed however, that cell lines should wherever possible, be obtained from accredited sources and such sources are listed in Table 4.2.1.

VERO

A culture initiated in 1962 from the kidney of an African green monkey. The cell line has a fibroblastic morphology and has been used extensively for virus propagation and mycoplasma detection (Chen *et al.* 1977). The use of these cells has been reported during current investigations into Marburg (MBGV) virus vaccine development (Hevey *et al.* 1998) and Ebola virus cultivation (Bray *et al.* 1998). This cell line is commonly used for investigations involving Toxoplasma, Trichomonas, Trypanosome and Chlamydia infections. It is the sensitivity of this cell line to infection by viruses and protozoan parasites that has ensured its widespread use. In addition the cell line offers a large flat cytoplasm ideal for visualization of microtubules, golgi complex and endoplasmic reticulum (de Melo *et al.* 1992).

A number of variants of this line have been developed principally by transfection.

Media

The cell line has been cultured using a number of different media and DEM + 10% FBS is the most commonly used.

Table 4.2.1 Availability of cell lines from accredited cell banks

Cell line	ECACC	DSMZ	ATCC
A2780	93112519	na	na
A6	89072613	na	CCL 102
C6 glial cells	85040101	na	CCL 107
FDCP-Mix A4	na	ACC 401	na
HIT-T15	na	na	CRL 1777
HUV-EC-C	na	na	CRL 1730
K-562	89121407	na	CCL 243
MCF-7	na	ACC 115	HTB 22
MRC-5	na	na	CCL 171
NCI-H460	na	na	HTB 177
NCI-H596	na	na	HTB 178
pt K2 (NBL-5)	88031601	na	CCL 56
SK-Mel-1	na	ACC 303	HTB 67
WiDr	85111501	na	CCL 218
VERO	88020401	ACC 33	CCL 81

Note: **ECACC**: European Collection of Animal Cell Cultures, Porton Down, Salisbury SP4 0JG, Wiltshire, UK. **DSMZ**: Deutsche Sammlung von Mikroorganismen und Zellkulturen GmbH (DSMZ), Mascheroder Weg 1b, D-38124 Braunschweig. **ATCC**: American Type Tissue Culture Collection, 10801 University Boulevard, Manassas, VA 20110–2209, USA).
na = not available at the time of printing.

A2780

A monolayer culture of an ovarian carcinoma established from a previously untreated patient, forming a cobblestone or polygonal growth pattern (Eva *et al.* 1982). This cell line will form tumours if implanted into athymic mice and has proved useful as an *in vitro* model of ovarian cancer (Louie *et al.* 1985). Many drug-selected A2780 lines have been used extensively in investigations into multidrug resistance, in particular the P-gp overexpressing line A2780ADR and various platinum drug-resistant lines. The line has recently been used to describe complex drug–time effects with clinical anticancer drugs (Levasseur *et al.* 1998). A2780 cells have wild-type p53 (Wetzel & Berberich 1998), and has been shown to express low levels of Bcl-xl (Liu *et al.* 1998). It is oestrogen receptor negative, and has been used in a comparative role to investigate C-kit receptor status in ovarian tissues and primary cultures. (Wrigley *et al.* 1996). This cell line can be grown in suspension and will form spheroids if grown in spinner cultures.

Medium

RPMI 1640 + 10% FBS.

NCI-H460

A non-small cell lung (NSCL) tumour cell line derived in 1982 from the plural effusion of a patient with large cell carcinoma of the lung. The cell line has been used extensively in anti-cancer drug screens and *in vitro* models forming xenografts in athymic mice. The cell line has detectable p53 and expresses a high level of the NAD(P)H quinone oxidoreductase (NQO1 or DT-diaphorase, (Phillips 1996). DT-diaphorase is a two-electron reducing enzyme involved in the detoxification of xenobiotics. This enzyme has been shown to be overexpressed in NSCL tumours (Ross *et al.* 1994) and has been exploited as a bioreductive anti-cancer pro-drug activating enzyme (Beall *et al.* 1995).

Medium

RPMI 1640 + 10% FBS.

NCI- H596

A adenosquamous carcinoma of the lung which characteristically has no functional DT-diaphorase. This line is often used in bireeductive activation studies alongside the NCI-H460 line described above. The lack of functional DT-diaphorase is due to a point mutation at position 609 in the NQO1 gene which confers an C to T transition. This transition translates as a serine to proline substitution at position 187 of the amino acid sequence (Traver *et al.* 1997). This line has also been used in recent studies investigating p53 mutations in human tumour cell lines (Wenz *et al.* 1998).

Medium

RPMI 1640 + 10% FBS.

MRC-5

A normal human diploid cell line showing fibroblastic morphology. Derived from foetal lung tissue in 1966 (Jacob *et al.* 1970). This line has a finite lifespan in culture but can attain 40+ population doubling before degeneration sets in. Used extensively for vaccine production and along with WI-38 cells has been exstensivly used in ageing studies. As described for the Vero cell line above MRC-5 cells are also susceptible parasitic infections. The cell line is often used as example of a normal untransformed human cell line. In such instances studies involving mutagenesis (Wollons *et al.* 1997), DNA damage (Zhuang *et al.* 1996) and cell signalling (Guy *et al.* 1994) have been reported. An immortalized (SV40 transformed) variant of this line designated as MRC-5 V1 exists, but is less commonly used (Rhode & Paradiso 1989).

Medium

MEM + 10% FBS.

MCF-7

A human breast adenocarcinoma obtained from a plural effusion described as retaining many characteristics of a differentiated mammary epithelium. Oestrogen receptor positive, these cells are often used in models of human breast cancer and oestrogen receptor studies. Hormone-resistant sublines have been reported (Jorgensen *et al.* 1997). Interestingly this line has featured in studies on the conversion of androgens to oestrogen's catalysed *in vivo* by aromatse/P450 complexes (Burak *et al.* 1997). Because of its oestrogen receptor status MCF-7 cells have featured prominently in work involving the anti-cancer drug Tamoxifen. Tamoxifen-resistant breast cancer is a serious clinical problem and a number of studies have tried to resolve it by using parental and drug-resistant MCF-7 cell lines (Jiang *et al.* 1992).

Because of their flat cytoplasm MCF-7 cells are often used in various imaging and immunofluorescence studies. The cell line will form Xenografts in nude mice and several drug-resistant variants are available.

Medium

MEM + 10% FBS.

FDCP-MIX CLONE A4

The FDCP-mix is a mouse bone marrow culture established from a long-term bone marrow culture derived from B6D2F1 mice. These cells are highly IL-3-dependent and have all the characteristics of freshly isolated haematopoietic stem cell lines. The clone A4 is a multipotent, haemopoietic progenitor line derived from FDCP-mix and may proliferate as a culture or be induced to differentiate (Spooncer *et al.* 1986). The clone A4 will differentiate into mature neutrophils in response to low levels of IL-3 and GM-CSF and GSF. In contrast low levels of IL-3 and erythro-poietin and haemin promote the development of erythoid cells. This line has also been reported to be able to form osteoclasts *in vitro* (Hagenaars *et al.* 1991). More recently this line has been used in studies concerning the effects of low-frequency magnetic fields on haemopoietic progenitor cells (Reipert *et al.* 1997).

Medium

80% Iscove's MDM + 20% horse serum + 10 ng ml^{-1} murine IL-3. It is possible to use the conditioned medium from the IL-3-producing cell line WEHI-3B as a substitute for IL-3. This is fine for the generic FDCP-mix but A4 cells grow better with purified IL-3.

HIT-T15

A Syrian hamster cell line derived from a primary culture of pancreatic islet cells. These cells were subsequently transformed and immortalized using SV40 virus. The line is capable of secreting insulin in response to glucose stimulation and offers a useful model for diabetes. The line has been used in evaluating novel anti-diabetic agents, (Jones *et al.* 1997) and glyceraldehyde transport, (Davies *et al.* 1994).

Medium

Ham's F12K medium + 2.5% FBS and 10% horse serum.

C6 GLIAL CELLS

Originally cloned from a rat glial tumour, this line produces S-100 protein unique to the brain tissue of vertebrates and which is abundant in the brain tumours of humans and rodents. The line has been used to study the physiological function of glial cells (Brismar 1995). More recently C6 cells have been used to study neuro-toxicity (Mead and Penreath 1998) as well as the effect of antidepressants on nervous tissue (Cookson *et al.* 1998).

Medium

DMEM + 10% FBS.

SK-Mel-1

This line was isolated from a patient with widespread metastatic melanoma (Oettgen *et al.* 1968). The cells are pigmented and grow in suspension unlike other melanoma cell lines which usually grow as monolayers. The line will produce tumours in xenogeneic mice. A series of human melanoma cultures was initiated by different workers during the period 1968–1976, all of which bear the prefix SK-MEL. Most metastatic melanoma cell lines come from patients who have been treated with one or more DNA-damaging agents. Over 1000 melanoma cell lines have been reported (Hay *et al.* 1994). The SK-Mel cell lines have been used extensively in screening potential anti-tumour drugs, and as models of metatstic invasion both *in vivo* and *in vitro*. Recent work has focused on detecting peripheral malignant melanoma cells in blood using PCR technology (Brossart *et al.* 1998). In a different approach Meghdadi and associates have measured technetium-99m uptake using imaging technology (Meghdadi *et al.* 1998).

Medium

Originally MEM + 10% FBS, but more commonly RPMI 1640 + 10% FBS.

WiDr

A colon adenocarcinoma derived as a primary culture from a 78-year-old female (Noguchi *et al.* 1979). The cell line has been used extensively in drug-screening programmes as well as cell tumourigenicity assays. The line has a distinct large metacentric marker chromosome and is positive for carcinoembryonic antigen. This line has the ability to form large (1–2 mm) spheroids when cultured in suspension. Such spheroids have been used as models to study hypoxia, phototoxicity, and DNA damage (Olive *et al.* 1996; Olive & Banath 1997; West & Moore 1992). More recently clones of WiDr as well as A2780-expressing nitroreductase (NR) activity have been used in gene-directed prodrug therapy (Friedlos *et al.* 1998).

Medium

RPMI 1640 + 10% FBS for monolayer culture, BME + 10% FBS for spheroid culture in spinner flasks.

K-562

Erythroleukaemic cells from a patient with chronic myeloid leukaemia in blast crisis (Lozzio & Lozzio 1975). K-562 grows as a suspension culture and represents a highly undifferentiated granulocytic cell line. The line is CD34 positive and capable of spontaneous or induced differentiation into members of the erythrocytic, granulocytic and monocytic series. Used widely in immunology as a target cell to assess natural killer cell activity, numerous drug resistant sublines have been developed. It is used extensively for studies into anti-cancer drug screening, drug resistance, and chemoprotective gene transfer (Ward *et al.* 1995, 1997; Hickson *et al.* 1998). K-562 cells lack detectable *p53* protein, (Ehinger *et al.* 1995) a characteristic which makes the line suitable for apotosis and *p53* mutation studies.

Medium

RPMI 1640 + 10% FBS.

A6

Primary culture of an adult male toad initiated in 1965 by K.A. Rafferty. Amphibian cell lines such as A6 represent useful research tools. The cell line grows in standard culture media modified to reduce the degree of osmolarity. Cultures will grow at room temperature without the need for a CO_2 atmosphere, and as such provide a valuable teaching aid. A6 cells are often used to complement studies using Xenopus oocytes.

Cell lines from haploid embryos are useful for genetic studies since mutant genes are not obscured by normal alleles. A6 cells have been used to study cell motility and the cytoskeleton (Terasaki & Reese 1994). Studies using A6 include the apical distribution of Na+ channels, nuclear ion channel activation, endoplasmic reticulum, microtubules, and surface bead movement. These cells are virus sensitive and can be transfected with exogenous DNA in the same manner as mammalian cells.

Medium

Leibovitz L-15 61%, FBS 10%.

HUV-EC-C

A primary human umbilical vein endothelial cell culture, often regarded as representative of large blood vessel endothelial tissue. These cells have a finite lifetime in culture, usually about 50 population doublings. The culture media is complex often supplemented with heparin, EGF, HGF, hydrocortisone and insulin. The need to supplement with numerous growth factors may be partially circumvented by the use of high levels of foetal calf serum (~20%). As a result growth is often slow and erratic. The term HUV-EC-C is a generic one referring to any umbilical cord culture and many laboratories prefer to isolate fresh cells on a regular basis (Jaffe *et al.* 1973). Pooled samples from several donors are common. These cells have been used extensively to study vascular endothelial cell biology. The most common areas of research involve tumorogenesis, angiogenesis, hypertension and blood brain barrier integrity. HUV-EC-C cells are used in permeability studies as a model of vascular integrity (Watts & Woodcock 1992; Cho *et al.* 1998). The limitations of a finite life in culture of these cells has prompted researchers to develop immortalized variants using papiloma virus (Rhim *et al.* 1998).

Medium

Many specialized media have been formulated for cells of endothelial origin. These are supplemented with low serum (< 5%) and several growth factors. However, a common medium for the propagation of these cells is medium 199 + 15% FBS + 5%, human serum + EGF and heparin.

Pt K2 (NBL-5)

A cell line derived from the kidney of a male potoroo, *Potorous tridacylis*, a marsupial commonly known as the kangaroo rat. This cell line has a large cyto-plasm and has the unusual characteristic of remaining virtually flat while proceeding through mitosis. In consequence this line is favoured for studies on chromosomal replication, organelle organization and cytoskeletal structure. Such studies include the involvement of myosin in postmitotic spreading and time-lapse video imaging of cell margin movement (Cramer & Mitchison 1995, 1997).

Medium

EMEM (EBSS) + 1% non essential amino acids (NEAA) + 10% FBS.

The number and diversity of cell lines now available to the medical researcher is too vast to be covered here. The Table 4.2.2 shows some additonal applications in which cell lines have been used.

Table 4.2.2 Additional applications of cell lines

Cell line	Type	Application
MG63,HOS	Osteoblasts	Hormone research/ bone metabolism
ROS 17/2.8	Osteosarcoma	Hormone research
T-47D, T-471	Breast carcinoma	Hormone research
MFM-223	Human mammary carcinoma	Hormone research Responds to androgens only
Caco-2	Intestinal	Oxidative damage, coeliac disease
BEAS-2B	Human bronchio epithelial line	Pollution, metal exposure
A549	Human lung tumour	Bronchitis, cystic fibrosis
NCI-H-322, NCI-H-358	Human lung cell lines	Pollution
HaCat	Human hyperpoliferative keratinocytes	Psoriasis
BAG2-GN6TH	Rat liver tumourigenic	Metastasis
SP2/0	Myeloma	Hybridoma fusion partner
P3-x63-ag8.653	Mouse myeloma	Hybridoma fusion partner
NAT-30, HO-323	Human fusion partner, IGM production	Human hybridoma fusion partner
U118	Human glioblsatoma cell line	Multicell spheroids
9L	Rat glioblastoma	Multicell spheroids
LLC-PK	Renal epithelial cell line	Multicell spheroids
GaL23	Large cell lung cancer	Multi-spheroids
A4H12	Human T-lymphoma. non-Igg producing.	Human hybridoma fusion partner

REFERENCES

Beall HD, Murphy AM, Siegel D, Hargreaves RH, Butler J & Ross D (1995) Nicotinamide adenine dinucleotide (phosphate): quinone oxidoreductase (DT-diaphorase) as a target for bioreductive antitumor quinones: quinone cytotoxicity and selectivity in human lung and breast cancer cell lines. *Molecular Pharmacology* 48(3): 499–504.

Bray M, Davis K, Geisbert T, Schmaljohn C & Huggins J (1998) A mouse model for evaluation of prophylaxis and therapy of Ebola hemorrhagic fever. *Journal of Infectious Diseases* 178(3): 651–661.

Brismar T (1995) The physiology of transformed glial cells. *Glia* 15(3), 231–243.

Brooks SC, Locke ER & Soule HD (1973) Oestrogen receptor in a human cell line (MCF-7) from breast carcinoma. *Journal of Biological Chemistry* 248(17): 6251–6253.

Brossart P, Grunebach F, Stuhler G, Reichardt VL, Mohle R, Kanz L & Brugger W (1998) Generation of functional human dendritic cells from adherent peripheral blood monocytes by CD40 ligation in the absence of granulocyte-macrophage colony-stimulating factor. *Blood* 92(11): 4238–4247.

Burak WE Jr, Quinn AL, Farrar WB & Brueggemeier RW (1997) Androgens influence oestrogen-induced responses in human breast carcinoma cells through cytochrome P450 aromatase. *Breast Cancer Research and Treatments* 44(1): 57–64.

Chen TR (1977) In situ detection of mycoplasma contamination in cell cultures by fluorescent Hoechst 33258 stain. *Experimental Cell Research* 104(2): 255–262.

Cho MM, Ziats NP, Abdul-Karim FW, Pal D, Goldfarb J, Utian WH & Gorodeski GI (1998) Effects of oestrogen on tight junctional resistance in cultured human umbilical vein endothelial cells. *Journal of the Society of Gynaecological Investigation* 5(5): 260–270.

Cookson MR, Slamon ND & Pentreath VW (1998) Glutathione modifies the toxicity of triethyltin and trimethyltin in C6 glioma cells. *Archives of Toxicology* 72(4): 197–202.

Cramer LP & Mitchison TJ (1995) Myosin is involved in postmitotic spreading. *Journal of Cell Biology* 13(1): 179–189.

Cramer LP & Mitchison TJ (1997) Investigation of the mechanism of retraction of the cell margin and rearward flow of nodules during mitotic cell rounding. *Molecular Biology of Cells* 8(1): 109–119.

Davies J, Tomlinson S, Elliott AC & Best L (1994) A possible role for glyceraldehyde transport in the stimulation of HIT-T15 insulinoma cells. *Biochemistry Journal* 304(1): 295–299.

de Melo EJ, de Carvalho & De Souza W(1992) Penetration of *Toxoplasma gondii* into host cells induces changes in the distribution of the mitochondria and the endoplasmic reticulum. *Cell Structure and Function* 17(5): 311–317.

Ehinger M, Nilsson E, Persson AM, Olsson I & Gullberg U (1995) Involvement of the tumour suppresser gene p53 in tumour necrosis factor-induced differentiation of the leukaemic cell line K562. *Cell Growth and Differentiation* 6(1): 9–17.

Eva A, Robbins KC, Anderson PR, Srinivasan A, Tronick SR, Reddy EP, Ellmore NW, Galen AT, Lautenberger JA, Papas TS, Westin EH, Wong-Staal, Gallo RC & Aaronson SA (1982) Cellular genes analogous to retroviral onc genes transcribed in human tumour cells. *Nature* 295(5845): 115–119.

Friedlos F, Court S, Ford M, Denny WA & Springer C (1998) Gene-directed enzyme prodrug therapy: quantitative bystander cytotoxicity and DNA damage induced by CB1954 in cells expressing bacterial nitroreductase. *Gene Therapy* 5(1): 105–112.

Guy GR, Phillips R & Tan YH, (1994), Analysis of cellular phosphoproteins by two-dimensional gel electrophoresis: applications for cell signalling in normal and cancer cells. *Electrophoresis* 15(3–4): 417–440.

Hagenaars CE, Kawilarang de Haas EW, van der Kraan AA, Spooncer E, Dexter TM & Nijweide PJ (1991) Interleukin-3-dependent hematopoietic stem cell lines capable of osteoclast formation in vitro. *Journal of Bone and Mineral Research* 6(9): 947–954.

Hay RJ, Park JG & Gazdar A (1994) *Atlas of Human Tumour Cell Lines*. Academic Press, London.

Hevey M, Negley D, Pushko P, Smith J & Schmaljohn A (1998) Marburg virus vaccines based upon alpha virus replicons protect guinea pigs and nonhuman primates. *Virology* 251(1): 28–37.

Hickson I, Fairbairn LJ, Chinnasamy N, Lashford LS, Thatcher N, Margison GP, Dexter TM & Rafferty JA (1998) Chemoprotective gene transfer I: transduction of human haemopoietic progenitors with O6-benzylguanine-resistant O6-alkylguanine-DNA alkyl transferase attenuates the toxic effects of O6-alkylating agents *in vitro*. *Gene Therapy* 5(6): 835–841.

Jacobs JP, Jones CM & Baille JP (1970) Characteristics of a human diploid cell designated MRC-5. *Nature* 227(254): 168–170.

Jaffe EA, Nachman RI & Becker CG (1973) Culture of human endothelial cells derived from umbilical veins. Identification morphologic and immunologic criteria. *Journal of Clinical Investigations* 14: 2745–2756.

Jiang SY & Jordan VC (1992) A molecular strategy to control Tamoxifen resistant breast cancer. *Cancer Surveys* 14: 55–70.

Jones RB, Dickinson K, Anthony DM, Marita AR, Kaul Cl & Buckett WR (1997) Evaluation of BTS 67 582, a novel antidiabetic agent, in normal and diabetic rats. *British Journal of Pharmacology* 120(6): 1135–1143.

Jorgensen L, Brunner N, Spang-Thomsen M, James MR, Clarke R, Dombernowsky P & Svenstrup B (1997) Steroid metabolism in the hormone dependant MCF-7 human breast carcinoma cell line and its two hormone resistant subpopulations MCF-7/LCC1 and MCF-7/LCC2. *Journal of Steroid Biochemistry and Molecular Biology* 63(4–6): 275–281.

Kohler G & Milstein C (1975) Continuous cultures of fused cells secreting antibody of predefined specificity. *Nature* 256 (5517): 495–497.

Levasseur LM, Slocum HK, Rustum YM & Greco WR (1998) Modeling of the time-dependency of *in vitro* drug cytotoxicity and resistance. *Cancer Research* 58(24): 5749–5761.

Liu JR, Fletcher B, Page C, Hu C, Nunez G & Baker V (1998) Bcl-xl is expressed in ovarian carcinoma and modulates chemotherapy-induced apoptosis. *Gynaecology and Oncology* 70(3): 398–403.

Louie KG, Behrens BC, Kinsella TJ, Hamilton TC, Grotzinger KR, McKoy WM, Winker MA & Ozols RF (1998) Radiation survival parameters of antineoplastic drug-sensitive and resistant human ovarian cancer cell lines and their modification by buthionine sulfoximine. *Cancer Research* 45(5): 2110–2115.

Lozzio CB & Lozzio BB (1975) Human chronic myelogenous leukaemia cell-line with positive Philadelphia chromosome. *Blood* 45(3): 321–334.

Mead C & Pentreath VW (1998) Evaluation of toxicity indicators in rat primary astrocytes, C6 glioma and human 1321N1 astrocytoma cells: can gliotoxicity be distinguished from cytotoxicity? *Archives of Toxicology* 72(6): 372–380.

Meghdadi S, Karanikas G, Schlagbauer Wadl H, Jansen B, Chehne F, Rodrigues M, Pehamberger H & Sinzinger H (1998) Technetium-99m-tetrofosmin: a new agent for melanoma imaging? *Anticancer Research* 18(4A): 2759–2762.

Noguchi P, Wallace R, Johnson J, Earley EM, O'Brien S, Ferrone S, Pellegrino MA, Milstien J, Needy C, Browne W & Petricciani J (1979) Characterisation of the WiDr: a human colon carcinoma cell line. *In Vitro* 15(6): 401–408.

Oettgen HF, Aoki T, Old LJ, Boyse EA, Harven E de & Mills GM (1968) Suspension culture of a pigment-producing cell line derived from a human malignant melanoma. *Journal of the National Cancer Institute* 41(4): 827–843.

Olive PL & Banath JP (1997) Multicell spheroid response to drugs predicted with the comet assay. *Cancer Research* 57(24): 5528–5533.

Olive PL, Vikse CM & Banath JP (1996) Use of the comet assay to identify cells sensitive to tirapazamine in multicell spheroids and tumours in mice. *Cancer Research* 56(19): 4460–4463.

Phillips RM (1996) Bioreductive activation of a series of analogues of 5-aziridinyl-3-hydroxymethyl-1-methyl-2-[1H-indole-4, 7-dione] prop-beta-en-alpha-ol (EO9) by human DT-diaphorase. *Biochemistry and Pharmacology* 52(11): 1711–1718.

Raffery KA (1969). Mass culture of amphibian cells: Methods and observations concerning stability of cell type. In: Mizell M (ed.) *Biology of Amphibian Tumours.* pp. 52–81. Springer-Verlag, Berlin.

Reipert BM, Allan D, Reipert S & Dexter TM (1997) Apoptosis in haemopoietic progenitor cells exposed to extremely low-frequency magnetic fields. *Life Science* 61(16): 1571–1582.

Rhim JS, Tsai WP, Chen ZQ, Van-Waes C, Burger AM & Lautenberger JA (1998) A human vascular endothelial model to study angiogenesis and tumorigenesis. *Carcinogenesis* 19(4): 673–681.

Rhode SL & Paradiso PR (1989) Parvovirus replication in normal and transformed human cells correlates with the nuclear translocation of the early protein NS1. *Journal of Virology* 63(1): 349–355.

Ross D, Beall H, Traver RD, Siegel D, Phillips RM & Gibson NW (1994). Bioactivation of quinones by DT-Diaphorase, molecular, biochemical and chemical studies. *Oncology Research* 6: 493–500.

Spier RE & Griffiths (1985) *Animal Cell Biotechnology.* Academic Press, London.

Spooncer E, Heyworth CM, Dunn A & Dexter TM (1986) Self-renewal and differentiation of interleukin-3-dependent multipotent stem cells are modulated by stromal cells and serum factors. *Differentiation* 31(2): 111–118.

Terasaki M & Reese TS. (1994) Interactions amongst endoplasmic reticulum, microtubules, and retrograde movements of the cell surface. *Cell Motility and Cytoskeleton* 29(4): 291–300.

Traver RD, Siegel D, Beall HD, Phillips RM, Gibson NW, Franklin WA & Ross D (1997) Characterisation of a polymorphism in NAD(P)H: quinone oxidoreductase (DT-diaphorase). *British Journal of Cancer* 75(1): 69–75.

Ward TH, Butler J, Shahbakhti H & Richards JT (1997) Comet assay studies on the activation of two diaziridinylbenzoquinones in K562 cells. *Biochemistry and Pharmacology* 53(8): 1115–1121.

Ward TH, Haran MS, Whittaker D, Watson AJ, Howard TD & Butler J (1995) Cross-resistance studies on two K562 sublines resistant to diaziridinylbenzoquinones. *Biochemistry and Pharmacology* 50(4): 459–464.

Watts ME & Woodcock M (1992) Flavone acetic acid induced changes in human endothelial permeability: potentiation by tumour-conditioned medium. *European Journal of Cancer* 28A(10): 1628–1632.

Wenz HM, Ramachandra S, O'Connell CD & Atha DH (1998) Identification of known *p53* point mutations by capillary electrophoresis using unique mobility profiles in a blind study. *Mutation Research* 382(3–4): 121–132.

West CM & Moore JV (1992) Mechanisms behind the resistance of spheroids to photodynamic treatment: a flow cytometry study. *Photochemistry Photobiology* 55(3): 425–430.

Wetzel CC & Berberich SJ (1998) DNA binding activities of *p53* protein following cisplatin damage of ovarian cells. *Oncology Research* 10(3): 151–161.

Wollons A, Clingern PH, Price ML, Arlett CF & Green MH (1997) Induction of mutagenic DNA damage in human fibroblasts after exposure to artificial tanning lamps. *British Journal of Dermatololgy* 137(5): 687–692.

Wrigley EC, McGown AT, Ward TH, Ewen C & Crowther (1996) E C-kit receptors in ovarian tumours and the response of ovarian carcinoma cell lines to recombinant human stem cell factor. *International Journal of Gynaecology: Cancer* 6: 273–278.

Zhuang ZX, Shen Y, Shen HM, Ng V & Ong CN (1996) DNA strand breaks and poly(ADP-ribose) polymerase activation induced by crystalline nickel subsulfide in MRC-5 lung fibroblast cells. *Human Experimental Toxicology* 15(11): 891–897.

4.3 EXPLANT CULTURES

Primary organized cultures are prepared from whole intact tissue explants set up as organ cultures or organotypic (histiotypic) cultures, or they are reconstructed from two or more components of which at least one may be a tissue explant. This section describes the method for obtaining, transferring and maintaining suitable tissue explants. In all the related culture methods, tissues are microdissected and transferred in an orientation-selected manner to the chosen culture chamber. Organ cultures consist of explants or multi-explant tissue constructs grown at the medium–gas interface. Organotypic cultures consist of explants grown submerged in medium for at least a part of the culture duration. Growth is usually in medium supplemented with a high percentage (15–40%) of serum, but for some culture types, serum-free media compositions may be possible or even desirable. For general guidance, Table 4.3.1 shows the main primary organized culture formats.

Collectively, explant cultures from normal foetal or adult stages, and pathological tissues from human and animals, provide optimal cytodifferentiation and tissue cell interactions, and in many cases give insights into tissue structure development and pathological progression which are not possible *in situ*. The principal disadvantage of explant culture procedures is the high rate of culture failure. This may result from limitations of the culture environment or from operator inexperience. The solution is to adhere very rigidly to published protocols, and to repeat the preparation a sufficient number of times on a weekly basis to ensure a significant number of cultures of high quality survive over the desired period.

The procedures given below describe only the initial stages of tissue preparation and transfer *in vitro*. A lengthy treatment of the assembly of different explant culture formats is beyond the scope of this book and the reader is recommended to refer to more specialized publications (Enami & Taukada, 1993).

PROCEDURE: PREPARATION OF HUMAN TISSUE EXPLANTS

Tissue explants are typically prepared from the following sources:

- Foetus
- Tissue biopsies (punch)
- Surgically removed tissues
- Naturally derived living tissue decidua (e.g. placenta)

It is essential to establish and maintain a positive working relationship with the surgeon, the midwife and, where appropriate, the histopathologist. When possible,

Cell and Tissue Culture for Medical Research, edited by A. Doyle and J.B. Griffiths.
© 2000 John Wiley & Sons, Ltd

Table 4.3.1 The main primary organized culture formats

Culture format	System	Duration	Principal application[b]	Pilot Reference
Organ	Organ culture chamber	3 days–2 months	d,g,h,t,u,v	Silbermann & Manor (1984)
Organotypic	Rocker chamber/ roller tube	2–9 months	c,d,h,p	Costero & Pomerat (1951)
	Maximow slide assembly/ Cruickshank chamber			Toran-Allerand (1990)
	Gelfoam/collagen matrix			Flynn *et al.* (1982)
	Dermal equivalent			Regnier *et al.* (1981)
Reconstructed tissue cultures	Microwell/cell 'dot'[a]	1–12 weeks	c,d,g,h,u	Solursh (1988)
	Confrontation			Marcel *et al.* (1979)
	Reaggregating[a]			Honegger *et al.* (1979)

[a] These systems are prepared from dissociated tissue cells, although they take on the appearance of organized three dimensional cultures.
[b] Applications code: c – cell interaction; d – differentiation; g – growth kinetics; h – histopathology; m – microanatomy/cytology; p – physiology; t – tumorigenesis; u – ultrastructure; v – virology.

attend the operation in person or arrange with theatre staff for the aseptic collection of the (carefully selected and labelled) samples and their prompt delivery to the culture laboratory. Principal sources of failure are contamination deriving from the tissue of origin (e.g. urogenital or gastrointestinal epithelia) or tissue anoxia caused by excessive delay prior to preparation.

Tissue pieces must be washed completely free of blood clots, plasma and other unwanted biological tissues, dissected to a size small enough to facilitate adequate gas exchange to the majority of the explant and, for explants attached to a substratum, to prevent loss of anchorage during the first day *in vitro*. Necrotic tissue must be removed as completely as possible as its presence, and leukocytes and phagocytic cells within, will be harmful to the growth of otherwise healthy explants.

Materials and equipment

- Tissue fragments collected aseptically in a tightly capped specimen jar
- Hanks' balanced salt solution (HBSS) without phenol red
- Eagle's minimal essential medium (EMEM) with penicillin and streptomycin (500 U ml^{-1} each) without serum
- Four or five Petri plates
- Spemann pipettes with 0.6–1.0 mm orifice
- Operating microscope ($\times$1-3 objectives)[1]

---

[1] For preference, a microscope with a long working distance (> 5 cm for all magnifications) is desirable to reduce operator-derived asepsis. Ensure that all microscope surfaces are swabbed with 10% hibitane or 70% ethanol prior to commencing work.

- Sterilized stainless steel dissecting instruments[2]
- Iridectomy scissors
- Two pairs of watchmaker's forceps, straight and curved
- Two micro-scalpels or two small scalpel handles with pre-sterilized No. 11 or 15 blades
- Two electrolytically sharpened mounted needles (optional)
- Disposable operator face masks

 All work should be carried out in a Class II (minimum operational safety) containment cabinet dedicated to work with human tissue samples, wearing gloves.

1. Cut large (> 0.5 cm) tissue fragments into smaller (2 mm cube) fragments in HBSS (10 ml in Petri plate).
2. Transfer fragments twice to fresh HBSS.
3. Wearing face mask, dissect blood clots away using operating microscope and watchmaker's forceps.
4. Removing connective tissue, transfer portions of tissue required for culture to fresh HBSS.
5. Using fine forceps or sharpened needles, cut tissue fragments, where applicable, into 300–400 1-μm cubes for adherent substratum-based cultures and up to 800 μm cubes for non-adherent explants.
6. To transfer tissue fragments without contact, use minimum-displacement (Spemann) pipettes, transferring the explant in 10 μl HBSS (organ cultures).
7. To transfer tissue fragments with precise positioning in a minimum fluid volume, use curved watchmaker's forceps. Pick up each by the edge and carefully blot away HBSS between the forceps tips with paper tissue, leaving 1–2 μl. Position the explant. Explants may be placed within a preformed collagen gel by this method.
8. Using mounted needles, orientate the explant rapidly before the fluid evaporates with the aid of the operating microscope.

PROCEDURE: PREPARATION OF ANIMAL TISSUE EXPLANTS

The range of tissue sources and techniques for animal explant culture is greater. The condition of the tissue explants is normally superior to that of explants of human origin since delivery from the living organism and macrodissection are under the control of the operator. Terminal anaesthesia, when used, and delivery and killing of all animal foetuses when past the mid-stage of gestation, are now regulated procedures in the UK under the Animals (Scientific Procedures) Act 1986 and appropriate authorization must be obtained in advance.

---

[2] Preferred instruments are manufactured by Roboz, Moria or S & T. Certain foetal tissues have been shown to yield viable organotypic cultures from explants when frozen to –70°C for short periods. Use techniques for cell cryopreservation. If dissection is to commence immediately (< 15 min) the tissue may remain in an enclosed air environment. If a delay is expected, it should be transferred to EMEM pre-gassed with 5% CO_2. Sample tubes should be kept on ice until dissection commences.

The best arrangement for tissue explant dissection is a clean room ventilated at positive pressure with sterile air and equipped with sterile surgical instruments, an operating microscope and, where appropriate, a mechanical tissue chopper. If this is impracticable, a clean, suitably equipped area should be set aside for operating work in the preparation room. The operator should wear clean disposable clothing and a face mask.

Animal dissection in a laminar flow cabinet is not recommended. If necessary a simple cabinet with a large operator access area may be dedicated for operations. Animals should never be operated on in the same cabinet used for culture preparations.

Additional materials and equipment

- Sterilized glass or disposable 'blunt' Pasteur pipettes cut back to the shaft
- Two 2-litre polypropylene beakers
- Glass evaporating basin
- Instruments suitable for macrodissection of chosen organ
- Mcllwain tissue chopper or vibratome (BioRad Laboratories, Bio-Rad House, Maylands Avenue, Hemel Hempstead, Herts, HP2 7TD, UK) with sterile blades

Adult and foetal mammalian tissue

General procedures relating to the condition of isolated tissues apply as in 'Preparation of human tissue explants'.

1. Anaesthetize adult animal intraperitoneally or intramuscularly with anaesthetic (fentanyl/fluanisone) according to the manufacturer's directions. A terminal dose is normally given when the operation is complete. A muscle relaxant may be given if skeletal muscle tissue is to be explanted.
2. Sterilize animal by total or partial immersion in 70% ethanol.
3. Remove dissected tissue to sterile tubes. If pregnant uterus, place segments with live foetuses each into separate sterile Petri plates.
4. Transfer tissue vessels to Class 11 containment cabinet.
5. After dissecting foetus from uterus, transfer to successive Petri plates of fresh saline to rinse free of cord blood.
6. Continue as for 'Preparation of human tissue explants' or first mount and cut tissues in tissue chopper.

Avian foetal tissue

Work with eggs should be conducted in the Class II containment cabinet.

1. In the preparation room, transfer eggs from egg trays to a sterilized 2-litre beaker, swabbing each egg liberally with 70% ethanol.
2. In the containment Class II cabinet, pierce eggs with a mounted needle and rest one by one in the evaporating basin with blunt end uppermost, supported with sterile paper towelling.

3. Cut a disk of shell off the top with sterile medium scissors and remove the embryo. Use 'blunt' Pasteur pipettes for embryos up to 6 days *in ovo* or use long-handled medium forceps with care from 7 days *in ovo* onwards. Use the other hand to cut perivitelline membranes if necessary. Transfer the embryo to fresh HBSS.
4. Wash embryo in successive Petri plates of fresh HBSS to remove yolk platelets.
5. Continue as for 'Preparation of human tissue explants' or first mount and cut tissues in tissue chopper.

CULTURE VESSELS

The successful cultivation of explants in all types of format is critically dependent on the quality and cleanliness of the culture vessels. Very few vessel types are supplied commercially as presterilized disposable products in culture-grade plastics. Examples of some are given in Table 4.3.2.

For these and the remaining applications given in Table 4.3.1 the only alternative vessels are normally made of re-usable glassware and must be rigorously cleaned, sterilized and recycled under the operator's direction.

The principal items for culture chamber/vessel assembly are listed below and are of borosilicate glass unless shown otherwise:

Organ culture chamber
Maximow depression slide
Re-usable Cruickshank chamber (Perspex)
Coverslips (No. 1–2 thickness 16–22 mm round or square)
De Long rotation culture flask

Some of these items are supplied 'precleaned' (e.g. coverslips) but cleanliness to culture quality should never be assumed and steps must be taken to both clean and prepare glassware for use. These steps are not only lengthy and exhaustive, but must be repeated to exactly the same protocol for each batch. It is therefore worth while preparing large batches of glassware at one time. Two alternative procedures for glassware cleaning are given below.

PROCEDURE A

1. Wash with Contra D-70 (5% v/v) (Curtis-Matheson Scientific Inc., Houston) for12 h.

Table 4.3.2 Vessel types (presterilized disposables in culture-grade plastic)

Product code	Supplier	Application
Falcon 3010 or 3037	Becton-Dickinson	Grid-support organ cultures [a]
309	Sterilin	Hanging-drop organotypic cultures

[a] Steel mesh from Expanded Metal Co Ltd, PO Box 14, Stanton Works, Hartlepool, UK.

2. Wash with deionized water (3×10 min).
3. Dry.
4. Sterilize.

PROCEDURE B

1. Wash in 2% Isoclean at 100°C in H_2O (Markson, Phoenix, AZ, USA) for 2×5 min.
2. Wash in hot running tap water for 2 h.
3. Wash in 1% HCl for 12 h or concentrated sulphuric acid: No-chromix oxidant 1.1 (Godax Labs, New York).
4. Wash in cold running tap water for 6 h.
5. Wash in deionized water (3×1 h).
6. Wash in acetone (2×15 min).
7. Dry.
8. Sterilize.

Note: New glassware requires an additional cleaning step in fuming nitric acid.

For reaggregation cultures, glassware should be siliconized with dimethyl-dichlorosilane (DD3879, Sigma Chemical Co.) vapour and then washed with deionized water before heat sterilization by autoclave (25 min at 1 kg cm^{-2}).

DISCUSSION

The tissue culture method was first employed by the developmental biologist Harrison (1907) in the study of growth in explant cultures of the amphibian central nervous system. It was developed further by Carrel (1912) and later by others. The technique has remained a preferred experimental tool. The particular disadvantages are that it is labour-intensive and a high level of expertise is required to optimize growth conditions. However, it proves a very powerful and revealing model of normal development and pathological progression in the hands of an expert when coupled with appropriate equipment for anatomical and physiological investigation. Those who have used this group of techniques will attest that a wealth of information can be derived from organized cultures when studied over a long period, far more than is generally published.

Anchorage of adherent explants

The substratum onto which an explant culture is attached must provide both a suitable biological surface and sufficient mechanical strength to retain structural integrity when the culture chamber is moved. Reconstituted collagen matrices (Enami & Taukada 1993) are commonly used. The source of collagen for this purpose should be animal tendon (e.g. rat tail, bovine achilles; type 1) rather than from placenta, cartilage or basement membrane, as a greater degree of persistent structural crosslinking occurs in the former type. Collagen type 1 is best prepared freshly in the laboratory from rat (Enami & Taukada 1993). Greater mechanical stability can also be obtained by gelling collagen (1.5–2.0 mg/ml) in the presence of 0.02% riboflavin 5'-phosphate in an atmosphere of ammonia.

Growth of explants in a plasma clot has the advantage that it derives a source of nutrition from the clot itself. Furthermore, the formation of the clot is more readily controlled by inserting the tissue explant in a small drop of gelling agent. Thrombin (0.1 U ml^{-1}) or chick embryo extract (50%) both activate the enzymic crosslinking of fibrin if chicken plasma is used.

REFERENCES

Carrel A (1912) On the permanent life of tissues outside the organism. *Journal of Experimental Medicine* 15: 516–528.

Costero I & Pomerat CM (1951) Cultivation of neurons from the adult human cerebral and cerebellar cortex. *American Journal of Anatomy* 89: 405–467.

Enami J & Taukada Y, (1993). Use of collagen gel as three-dimensional matrix for explant culture. In: Doyle A, Griffiths, JB & Newell, D (eds) *Cell & Tissue Culture: Laboratory Procedures*, 3A5:1–8. J Wiley, Chichester.

Flynn D, Yang J & Nandi S (1982) Growth and differentiation of primary cultures of mouse mammary epithelium embedded in collagen. *Differentiation* 22: 191–194.

Harrison RG (1907) Observations on the living developing nerve fibre. *Anatomical Record* 1: 116–118.

Honegger P, Lenoir D & Favrod P (1979) Growth and differentiation of aggregating foetal brain cells in a serum-free defined medium. *Nature* 282: 305–307.

Marcel M, Kint J & Meyvisch C (1979) Methods of study of the invasion of malignant C3H-mouse fibroblasts into embryonic chick heart *in vitro. Virchows Archiv B Cell Pathology* 30: 95–111.

Regnier M, Prunieras M & Woodley D (1981) Growth and differentiation of adult epidermal cells on dermal substrates. *Frontiers in Matrix Biology* 9: 4–35.

Silbermann M & Manor G (1984) Organ and tissue culture of cartilage and bone. In: Dickson GR (ed.) *Methods of Calcified Tissues Preparation, pp* 467–530. Elsevier/North-Holland, Amsterdam.

Solursh M (1988) Environmental regulation of limb chondrogenesis. In: Keuttner K, Schleyerbach R & Hascall VC (eds) *Articular Cartilage Biochemistry,* pp. 145–161. Raven, New York.

Toran-Allerand CD (1990) Long-term organotypic culture of central nervous system in Maximow assemblies. In: Conn PM (ed.) *Methods in Neurosciences*, Vol. 2, *Cell Culture,* pp. 275–296. Academic Press, New York.

4.4 HAEMOPOETIC CELLS IN CULTURE: HUMAN B AND T LYMPHOCYTES

INTRODUCTION

Peripheral blood has been widely used for many years as a readily exploitable source of human cells for a range of biochemical, immunological and genetic studies. The techniques for cell separation and isolation are well established, but they are largely dependent upon success on the quality of the blood sample provided. Consideration has to be given to the time the material is in transit if shipped to a laboratory from the clinical point of sampling (even shipped across continents) – a maximum of 72 hours is recommended although this should be kept to a minimum. Another issue is the type of anti-coagulant using either heparin or acid citrate dextrose (ACD) tubes – there is anecdotal evidence to suggest the latter improve cell survival. Finally, the temperature of storage and transit are important considerations (avoiding extremes), but refrigeration is not recommended as cell survival is best at standard room temperature.

The following procedures outline the techniques required to obtain short- and long-term cultures of both B- and T-lymphocytes for a large number of purposes, if only in the case of EBV-transformed B-lymphoblastoid cell lines, an indefinite source of DNA. The absolute requirement for informed patient consent before obtaining samples for research purposes and possible local ethical committee approval is discussed in 4.1.

Cell and Tissue Culture for Medical Research, edited by A. Doyle and J.B. Griffiths.
© 2000 John Wiley & Sons, Ltd

4.4A EBV Transformation

The generation of human B lymphoblastoid cell lines is a relatively straight-forward procedure with a success rate of over 95%. Two critical steps in the following procedure should be highlighted; the condition of the blood sample prior to separation, and obtaining B95–8 cell line supernatant with a sufficiently high virus titre. Fresh blood, preferably within 24 h of sampling, should be used. However, cell lines can be successfully established with blood which is up to 72 h old, as long as it has been kept at room temperature. It greatly impairs cell survival to store at 4°C. It is possible to freeze separated lymphocytes and store them in liquid nitrogen, prior to immortalization with EBV, without affecting the success rate.

It also possible to generate large-scale long-term T-cell cultures from the same fresh blood sample or cryopreserved separated lymphocytes (see Section 4.4C), allowing T-cells and virally transformed B-cells to be made in parallel.

PROCEDURE: SEPARATION OF MONONUCLEAR CELLS FROM PERIPHERAL BLOOD

The method described below is essentially that of Boyum (1964, 1968) using a separation medium consisting of an aggregating agent sodium metrizoate together with Ficoll to give a density of 1.077 g ml^{-1} and osmolality of 280 mOsm kg^{-1}. The technique is a one-step centrifugation technique and isolates the mononuclear cells from whole blood.

Materials and equipment

- RPMI 1640 containing 10 units/ml preservative-free heparin
- Ficoll-Isopaque, e.g. Lymphoprep (Nycomed), JPREP (Techgen International)
- Accuspin tubes (optional – this depends on separation method)
- Sodium heparin tubes

1. The whole blood sample taken in sodium heparin tubes should be stored protected from light at room temperature prior to use. If the blood is to be sent by post or courier then it should be sufficiently insulated to avoid temperature extremes (hot and cold).
2. Dilute the blood in RPMI containing 10 units ml^{-1} heparin (preservative-free) at a ratio of 1 : 1. Addition of heparin is not necessary if blood is to be separated immediately.
3. Carefully layer blood/RPMI mixture onto the separation medium at a ratio of one volume separation medium to two volumes diluted blood. This is usually achieved by adding 10 ml to 5 ml in a 15 ml centrifuge tube.
4. Centrifuge at 500 *g* for 20 min at room temperature.

5. Carefully remove the mononuclear cell layer by using a wide-bore Pasteur pipette. Cells adhering to the walls of the tube can be dislodged by gentle scraping with the pipette tip. The tubes can be centrifuged for a further 10°20 min if a good separation has not been obtained. This is more often the case with blood samples separated after 48 h.
6. Wash the cells by adding RPMI to a volume of 15 ml and gently mixing.
7. Centrifuge at 400 *g* for 10 min.
8. Check to see that a cell pellet has formed before discarding the supernatant. If a pellet is not visable repeat Step 7. If the blood has a very low lymphocyte count then the pellet may only just be visible by eye.
9. Resuspend the cell pellet in the same volume as in Step 6 and repeat Steps 6, 7 and 8 twice.
10. Finally, if a viable cell count is to be performed dilute the cells in 5 ml RPMI.

If the cells are to be frozen prior to transformation they should be resuspended in freeze media such as RPMI 1640 + 20% fetal bovine serum (FBS) + 10% dimethylsulfoxide (DMSO), or using 90% FBS + 10% DMSO, and cryopreserved by rate-controlled freezing i.e. 1°C per minute. The lymphocytes should then be stored until required, preferably in liquid nitrogen. However, they can be stored temporarily, i.e. up to 1 month, in a –80°C freezer. If the cells are to be transformed immediately proceed to the next section.

Comment

If heparin or heparin-coated tubes are unavailable then EDTA-coated tubes or ACD tubes (Becton Dickinson) may be used without any detrimental effect on EBV transformation. In fact ACD tubes have been reported to preserve the blood for longer than 72 h, which is currently being evaluated in the author's laboratory. If large numbers of blood samples are to be processed then Accuspin tubes (Sigma Diagnostics) are recommended. They have a filter part-way down the tube which prevents mixing of the two layers. This removes the need for careful layering and the blood need not be diluted before use. The separation is as good as the traditional method described above but much quicker.

SUPPLEMENTARY PROCEDURE: PREPARATION OF EBV-CONTAINING SUPERNATANT

The procedure below outlines a temperature-induced method for the production of EBV-containing supernatant from the marmoset cell line B95–8.

Materials and equipment

- B95–8 marmoset cell line (ECACC, Salisbury, UK; ATCC, Manassas, Virginia, USA)
- 0.45 μm filters (syringe or bottle top type)
- –80°C or liquid nitrogen freezer

EBV is an ACDP Category 2 pathogen (Advisory Committee on Dangerous Pathogens 1984).

1. Maintain a culture of the B95–8 cell line in RPMI 1640 + 5% FBS at between 3×10^5 and 9×10^5 cells ml^{-1} by diluting the cells 1 : 2–1 : 4 every 3–4 days.
2. When a sufficient number of cells have been obtained (depending on the amount of supernatant required) dilute the cells to 2×10^5 in RPMI 1640 + 2% FBS.
3. Gas the flasks with 5% CO_2, tightly screw the caps and incubate overnight at 37°C then at 33°C for 2–3 weeks without any changes of medium. By this time the medium should have turned bright yellow.
4. On the day of harvest up-end the flasks to allow cells to settle out.
5. Decant the supernatant into tubes and centrifuge at 400 g for 10 min at room temperature to remove any cell debris.
6. Filter the supernatant through a 0.45 μm filter to remove cell debris and large particles, but not the virus.
   This is not intended as a sterilization procedure as the material remains sterile throughout processing. In the author's experience the use of a 0.2 μm filter at this stage reduces the virus titre.
7. Aliquot the supernatant at required volumes (usually 1 ml) and store at –80°C or below, preferably in vapour-phase nitrogen.
   Freeze-thawing of the virus should be avoided as this will reduce the efficacy of EBV infection.
8. Prior to use, thaw at 37°C and use immediately. If the virus is not to be used immediately, thaw at room temperature and keep on ice.

Comment

Some laboratories recommend the use of chemical inducing agents such as 12-O-tetradecanoylphorbol-13-acetate (final concentration 20 ng ml^{-1}) or sodium-n-butyrate (2 mM) – see Walls & Crawford (1987). However, the titres produced by the above procedure are comparable with those obtained by chemical induction and the above chemical agents may interfere with B lymphocyte activation by the virus. The virus preparation may be concentrated 50–100-fold by ultra-centrifugation at 27 000 g for 2 h at 4°C and then resuspended in culture medium.

Unfortunately, many samples of the B95–8 cell line distributed informally between research laboratories are mycoplasma-contaminated. It is therefore of the utmost importance that the basic starting material for the technique is obtained from a reliable source, such as an established culture collection. It is also worth noting that the B95–8 cell line displays both suspension and fibroblast morphologies.

SUPPLEMENTARY PROCEDURE: TITRATION OF EBV SUPERNATANT USING CORD BLOOD MONONUCLEAR CELLS

Outlined below is a method for evaluating the EBV supernatant produced in the previous section. This step can be omitted but it does have the advantage of providing a way of comparing different batches of virus-containing supernatants.

Materials

- Cord blood (5–10 ml)
- EBV supernatant

1. Separate the mononuclear cells from the cord blood as described above and resuspend in RPMI 1640 + 10% FBS at approximately $2–3 \times 10^6$ cells ml^{-1}.
2. Remove adherent cells by incubation on plastic for 20 min. This step may be omitted if the cells have been cultured overnight.
3. Prepare serial 10-fold dilutions from neat to $1 : 10^6$ of the virus supernatant to be tested in RPMI 1640 + 10% FBS.
   These and all subsequent titrations should be carried out on ice. It is also important to change pipette tips between dilutions.
4. Aliquot the cord mononuclear cells into centrifuge tubes to give 2×10^6 cells per 0.9 ml per tube.
5. Add 100 µl of each virus dilution to an aliquot of the above cells. As a negative control add 100 µl RPMI 1640 + 10% FBS to a tube of cells. A positive control can be set up using a previously prepared supernatant with a known titre.
6. Resuspend cells by tapping the base of each tube gently and incubate at 37°C for 1 h.
7. Add 1 ml RPMI 1640 + 10% FBS to each tube and mix gently.
8. Plate cells out onto a 96-well flat-bottomed microtitre plate (200 µl per well). Use two separate plates for each titration and two different batches of media when feeding.
9. Feed once a week by removing 100 µl medium without disturbing the cells and replacing with fresh medium.
10. Incubate plates at 37°C in 5% CO_2/95% air.

Comment

Successful immortalization of mononuclear cells can be assessed by examining the cultures using an inverted microscope. After 10–14 days, proliferating foci of B lymphocytes can be seen, which after about 28 days are usually visible macroscopically as large clumps of cells. The control cultures and cultures failing to be immortalized should contain only dying cells and debris. The efficiency of immortalization is defined as the negative log to the base 10 of the virus dilution which induced 50% immortalization of the cultures. For routine use, the supernatant should have a titre of at least 10^{-3} and preferably 10^{-4}. If it is not possible to obtain cord blood, the virus can be evaluated by testing its ability to immortalize lymphocytes from several normal donors.

PROCEDURE: ESTABLISHMENT OF HUMAN B LYMPHOBLASTOID CELL LINES FROM PERIPHERAL BLOOD MONONUCLEAR CELLS

The procedure below is derived from the methods of Walls & Crawford (1987) and Doyle (1989) and from experience of generating over 15000 cell lines using EBV in the author's laboratory.

Materials

- EBV supernatant
- RPMI 1640 + 20% FBS
- Polymixin-B-sulfate (100 units ml^{-1}) – Life Technologies Ltd
- Mouse peritoneal macrophages (optional, see below)

The following procedure is for $5–10 \times 10^6$ cells, i.e. lymphocytes from a 10-ml blood sample.

1. Thaw vial (approximately 5×10^6) of lymphocytes (fresh lymphocytes can be used) at 37°C and transfer to a 15 ml centrifuge tube. Add prewarmed RPMI + 10% FBS to a final volume of 15 ml and mix. Centrifuge at 150 g for 5 min.
2. Discard supernatant and resuspend cell pellet in 1 ml EBV supernatant.
3. Incubate at 37°C for 1–1.5 h with gentle agitation halfway through incubation.
4. Centrifuge at 150 g for 10 min. Ensure a pellet forms before discarding the supernatant.
5. Resuspend cells in RPMI + 20% FBS, PHA (1% v/v), penicillin/streptomycin (100 units ml^{-1}), polymyxin-B-sulphate (100 units ml^{-1}) to give $1–2 \times 10^6$ cells ml^{-1}. Pipette 1 ml cells onto each well of a 24-well flat-bottomed culture dish and incubate at 37°C in a 5% CO_2/95% air atmosphere.
6. The cultures should be fed twice weekly by removing half the supernatant and replacing it with fresh medium (no PHA added) without disturbing the cells. Ensure that pipettes or pipette tips are changed between wells to prevent cross-contamination.
7. After 1–2 weeks foci of B cells should be visible under an inverted microscope and the initial proliferation of T cells should regress.
8. The cultures may be expanded into 25 cm^2 tissue culture flasks once established (usually after 2 and 3 weeks). This is often indicated by the media turning yellow overnight and lots of large clumps of cells being present. The flasks should be kept upright and the cultures split 1 : 3 every 2–4 days depending on growth rate.

Comment

The use of 24-well plates prepared with mouse peritoneal macrophages is recommended for low numbers of peripheral blood lymphocytes (PBLs) (less than 1×10^6), PBLs contaminated with red blood cells or PBLs isolated from old blood samples (over 3 days).

Cyclosporin A may be used to suppress T-cell proliferation which is the opposite effect to that of PHA. However, cyclosporin A does not increase the efficiency of EBV transformation, and itself is often difficult to handle. PHA results in the proliferation of T-cells and has a mitogenic effect on B cells. There is no evidence of specific T-cell killing following PHA cell stimulation. In laboratory practice this has proved to be the most efficient technique available: hence, it is reproduced here. (One possible drawback is that mitogenesis leads to differentiation of committed B cells to plasma cells and then the loss of the C3 receptor, and thus the cells are not transformed.)

During routine subculture of established lymphoblastoid cell lines it is important not to overdilute the cells, i.e. split at ratios 1 : 3 or lower, as this can often halt their growth. HEPES buffer may be added to the medium (final concentration 25 mM) if required. If the culture is not growing well, growth supplement can be added at any stage of the establishment of a cell line. 2-Mercaptoethanol (50 mM), non-essential amino acids (Life Technologies Ltd), sodium selenite (2×10^6 M), transferrin (4 mg/ml), sodium pyruvate (50 mM) used at a 1 : 50 dilution is recommended.

REFERENCES

Advisory Committee on Dangerous Pathogens 1995 *Categorization of Pathogens According to Hazard and Categories of Containment*. Fourth Edition HMSO, London.

Boyum A (1964) Separation of white blood cells. *Nature* 204: 793.

Boyum A (1968) Separation of leucocytes from blood and bone marrow. *Scandinavian Journal of Clinical Investigation* 21 (supplement 97): 77–89.

Doyle A (1989) Establishment of lymphoblastoid cell lines. In: Poll JW & Walker JM (eds) *Animal Cell Culture*, pp. 43–47. Humana Press, New Jersey.

Walls EV & Crawford DH (1987) Generation of human B lymphoblastoid cell lines using Epstein-Barr virus. In: Klaus GGB (ed.) *Lymphocytes – a Practical Approach*, pp. 149–162. IRL Press, Oxford.

4.4B Cryopreservation of Human Blood for B-cell Immortalization[1]

Whereas most currently available techniques for B-cell immortalization require viral infection of fresh cells, the immortalization technique outlined in this section is designed to transform lymphocytes from either fresh or frozen blood. This technique extends the usefulness of previous methods by allowing investigators to expand an unlimited number of cells from a limited blood sample and to freeze samples of whole blood or isolated lymphocytes for transformation at a later date. This scaled-down protocol is therefore ideally suited to the collection and storage of lymphocytes from large numbers of subjects. As this approach removes the requirement for immediate transformation, it offers greater flexibility in scheduling and eliminates some of the logistic and financial difficulties associated with large groups of samples.

Alternative ways to process lymphocytes for transformation are shown in Figure 4.4B.1.

1. Blood can be drawn and cryopreserved (without fractionation) for future lymphocyte isolation and transformation (Figure 4.4B.1(a)). This third approach has a somewhat lower success rate (85–90%) but is substantially simpler and involves less immediate time investment and expense than the other two approaches.
2. The lymphocytes can be isolated and transformed immediately (Figure 4.4B.1(b)). This standard approach has a high success rate (of over 95%) but also a relatively high cost in terms of immediate time investment and initial financial outlay.
3. Lymphocytes can be isolated from freshly collected blood and cryopreserved, and then thawed at a later day and transformed (Figure 4.4B.1(c)). This alternative also has a relatively high success rate (90–95%) but a lower initial time investment and initial cost.

These three alternatives make it feasible for investigators to develop a transformation strategy that takes into consideration their specific requirements with regard to cost, transformation efficiency and workflow. One useful strategy is to combine approaches for each blood sample collected: to set aside and freeze an aliquot (e.g. 1 ml) from each freshly drawn blood sample, isolate the lymphocytes from the remainder of the fresh sample, and then freeze or transform an aliquot of the fresh lymphocytes immediately. This combined strategy is costly but ensures nearly 100% efficient transformation and at the same time provides a back-up source should additional cells be needed in the future. However, if facilities for immortalizing cells are not immediately available, samples of the lymphocytes can

[1] From Penno *et al.* (1993).

be isolated and transformed. The success of this method is based on several underlying principles:

• minimizing haemolysis during blood collection (by direct collection with a 21 G
  needle into Vacutainer tubes).

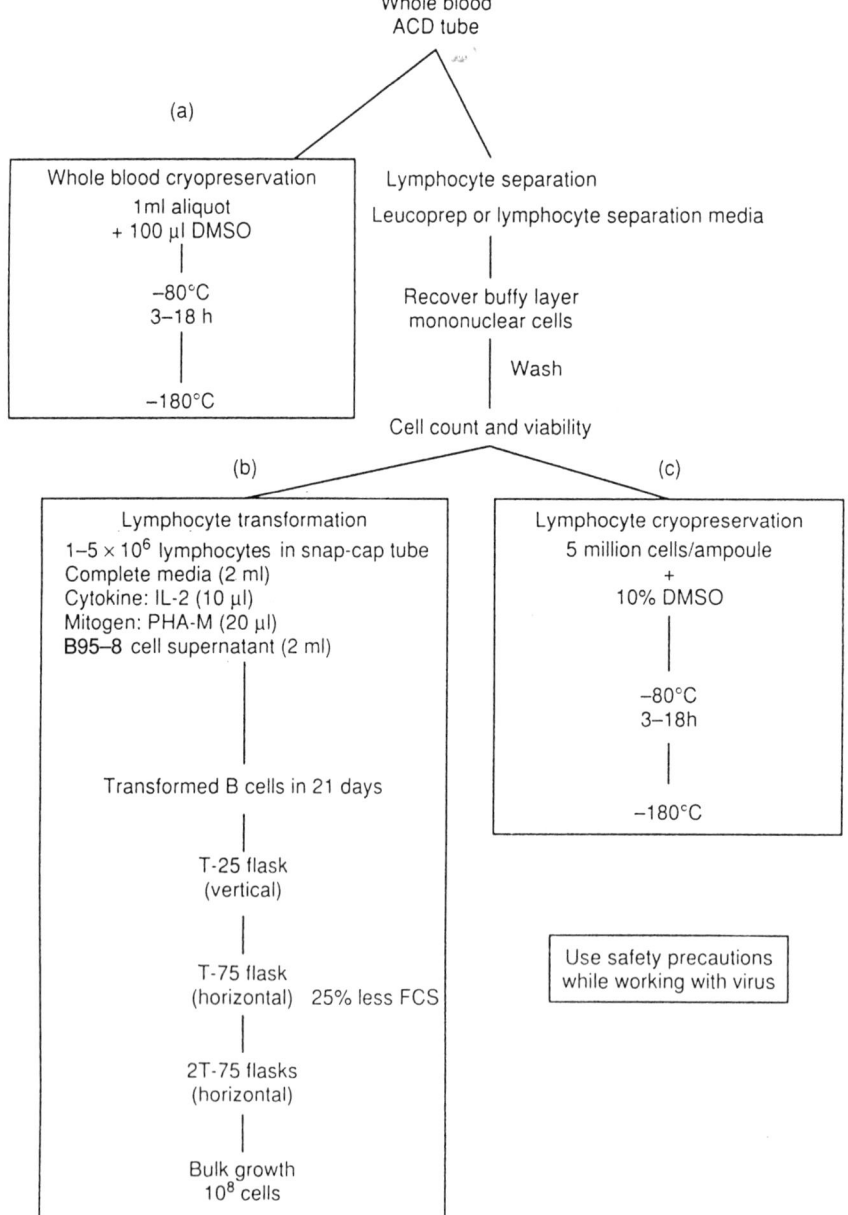

Figure 4.4B.1 The establishment of lymphoblastoid cell lines: options and procedures for
processing blood.

- careful transport of blood to maintain viability (at room temperature in acid citrate dextrose (ACD) for no more than 72 h).
- controlled freezing and thawing of cells to minimize formation of intracellular ice crystals (by slow freezing in preservative at $-1°C$ min^{-1}, storage at $-180°C$, rapid thawing at $37°C$).
- for aliquots of whole blood, removing red cell debris to prevent interference with the transformation process (by panning on treated tissue culture plates).
- stimulating production of B-cell growth factors (by including interleukin-2 (IL-2)).

Advance preparations

Advance preparation is necessary for certain steps in the protocol:

1. EBV-infected marmoset cells must be tested for the absence of *Mycoplasma* and grown for at least a week at the required cell density for supernatants to have high titres of shed virus. Supernatants should be used immediately to infect the waiting lymphocytes, since freezing the virus may impair infectivity.
2. Cells should be transported to and from the freezer on dry ice and should not be thawed until all other reagents are in place.
3. For transformation from whole blood, the pretreated tissue culture plates used for removing red cell debris must be prepared 3–24 h in advance.

PROCEDURE: CRYOPRESERVATION AND IMMORTALIZATION PROCEDURE

Materials and equipment

- B95–8 Marmoset cell line infected with EBV (ECACC, ATCC) (store at $-180°C$ in freezing medium until use)
- Opti-MEM (Gibco) with 2.5% foetal bovine serum (FBS; Gibco) (Opti-MEM medium)
- Dulbecco's phosphate-buffered saline (PBS), Ca^{2+} and Mg^{2+} free, sterile (Gibco) DMSO (Sigma)
- RPMI with 10% FBS, 1% Penn/Strep, and 1% L-glutamine (complete medium)
- Isopropanol
- RPMI with 10% FBS, 10% DMSO, and 1% L-glutamine (freezing medium)
- Recombinant interleukin-2(IL-2), 10 000 U ml^{-1} (Boehringer Mannheim)
- Lymphocyte separation medium (Boehringer Mannheim)
- Phytohemagglutinin M (PHA-M) (Gibco)
- Dry ice
- Biosafety cabinet (SterilGard, Baker)
- Filters, Millipore Millex-GS or Falcon Easy-Flow cellulose acetate, 0.22 μm (Millipore) (low protein binding)
- Blood collection needles, PrecisionGlide, 21 ½ G prepackaged sterile (Becton Dickinson)

- Blood collection tubes, Vacutainer, with ACD, 8.5 ml-draw yellow top (Becton Dickinson)
- Leucoprep separation tubes, 10 ml (Becton Dickinson), or lymphocyte separation medium (such as Ficoll-Hypaque, Boehringer Mannheim)
- Cryogenic storage vials, 2 ml, internal thread (Vangard)
- Cryopreservation container (Mr Frosty, VWR Scientific; Biotech Research Labs, Rockville, MD); or home-made
- Liquid nitrogen storage system (if possible, one that permits storage of vials in the vapour phase)
- Incubation tubes (sterile): 13-ml round-bottomed snap-cap

Blood collection and cryopreservation

1. Draw blood using a Precision-Glide needle (21 G) into Vacutainer tubes containing ACD, maintain tubes at room temperature, and deliver to the laboratory within 72 h (see Figure 4.4B.2).
2. Remove 1 ml of whole blood and combine evenly with $100 \mu l$ DMSO in a cryovial, place immediately in a cryopreservation container at $-80°C$ for 3–18 h, and then transfer on dry ice to the vapour phase of liquid nitrogen. This procedure will freeze cells at approximately $-1°C$ min^{-1}.

INSTRUCTIONS FOR DRAWING AND SHIPPING A BLOOD SAMPLE FOR LYMPHOCYTE STORAGE OR TRANSFORMATION

Thank you for helping to obtain this blood sample. Please consider this a prescription to draw the patient's blood. In order for us to receive the sample in suitable condition for analysis, please follow the guidelines below.

NOTE: The blood draw should take place as early as possible in the day in order to arrange for a morning express pick-up. This will ensure that a priority overnight delivery to the lab can be made.

1. Draw blood into two 10 ml ACD (yellow top) Vacutainer tubes. Fill each tube completely, as the preservative contained within the tube is appropriate for a full volume. Collect blood directly into Vacutainer tubes. Please do not remove the cap of the Vacutainer tube, as this increases the risk of contamination.

2. Use a $21\frac{1}{2}$ G needle. Do not use small-bore needles such as 23 or 25 G, as they increase the chance of cell lysis. If you prefer to use a butterfly needle, please use the 21 G Vacutainer set.

3. Record the following on the tube:

   (Note: Names will be entered into the data base, so write clearly)

   Date of blood draw
   Patient's first name, middle initial, last name, Jr/Sr etc.

4. Blood tubes should be shipped at room temperature (22–24°C) in a styrofoam tube carrier. If this is not available, place all tubes in shockproof wrapping, place in a styrofoam box and seal with several wraps of tape. Enclose the styrofoam box in a cardboard box and seal with tape.

5. Attach the express mail carrier airbill to the outside of the package. Contact your express mail carrier by phone to arrange for a pick-up. In most states this pick-up should occur before noon to ensure overnight delivery.

Thank you again. If you have any questions please call:
Study coordinator:
Phone number:

Figure 4.4B.2 The establishment of lymphoblastoid cell lines: instructions for phlebotomists.

3. Separate the lymphocytes from the remaining whole blood by density gradient centrifugation as described previously.
4. Resuspend the pelleted cells in 5 ml PBS without Ca^{2+} and Mg^{2+}. Obtain a cell count and determine viability by Trypan blue exclusion.

Preparation of lymphocytes from frozen whole blood

1. At least 3–24 h in advance, pretreat a 24-well tissue culture plate by incubating the wells for 2–24 h with sterile 50 mM Tris buffer, pH 9.5 (at 37°C in 5% CO_2). Block the wells with 'complete' RPMI medium for 1 h (37°C in 5% CO_2), and then rinse with PBS without Ca^{2+} and Mg^{2+}.
2. Remove ampoules of whole blood from the liquid nitrogen freezer and place on dry ice. When all reagents are ready, rapidly thaw 1 ml samples at 37°C (take about 2 min) and dilute with 1 ml PBS without Ca^{2+} and Mg^{2+} in the cryovial. Layer diluted cells onto 3 ml of separation medium in a 15 ml conical test tube.
3. Centrifuge at 400 g for 25 min at room temperature. Recover the mononuclear layer, wash in 10 ml PBS without Ca^{2+} and Mg^{2+}, and then resuspend in 1 ml of the same buffer.
4. To remove the red cell debris, pan the lymphocytes as follows: add the cells to the plate and incubate for 1 h at 37°C in 5% CO_2, and then carefully transfer cells to a 15 ml conical centrifuge tube.
5. Wash wells with 1 ml PBS without Ca^{2+} and Mg^{2+}, collect any additional non-adherent cells and add to the 15 ml tube. Cells in the tube are ready for immediate transformation.

Transformation of lymphocytes

1. Pellet lymphocytes (1×10^6 fresh cells, 5×10^6 frozen cells, or isolated lymphocytes from 1 ml of frozen whole blood) by centrifuging at 150 g (1200 rev min^{-1}, $r = 12$) for 5 min at room temperature. Resuspend in 2 ml of complete RPMI and transfer to a 13 ml sterile round-bottomed snap-cap test tube.
2. Add 10 μl IL-2 (10 000 U ml^{-1}), 20 μl PHA, and 2 ml of filtered B95-8 cell supernatant (containing EBV), and grow cells at 37°C in 5% CO_2 in the snap-cap tube (close cap only to first 'snap' position). After 7 days, feed cells by adding 1 ml of complete RPMI. At 14 and 21 days after infection, feed cells by removing half the supernatant and replenishing with an equal volume of fresh complete RPMI (do not disturb the pellet). On day 21 after infection, remove 2 ml from the rube and add 1 ml of fresh complete RPMI, transfer to a 25-cm^2 flask positioned upright and then maintain at 5×10^6 cells ml^{-1}. Feed when medium is acidic (yellowish).
3. Transfer 1×10^7 cells in 10 ml medium to a 75 cm^2 flask with Opti-MEM medium. Position flask horizontally in the incubator and feed by adding enough medium to maintain between 5×10^6 and 1×10^6 cells ml^{-1}. Split after approximately 7 days into two flasks (1 : 2) and grow to a final cell number of 5×10^7 cells per flask (cells can be subdivided indefinitely).
4. Identify signs of B-cell transformation: exponential increases in cell number, with concomitant production of acidic (yellowish) medium; clumping of cells,

with appearance of individual teardrop-shaped cells (Figure 4.4B.3); ability to be subdivided; high post-freeze viability.

DISCUSSION

Previous experience has indicated that cells from whole blood or isolated lymphocytes stored for as long as 4 years in the vapour phase of liquid nitrogen can be efficiently transformed with EBV after thawing. Within this time frame, success rates do not vary appreciably with length of storage. For each approach outlined above, the average time to reach 5×10^6 transformed cells was 28–31 days, well beyond the life expectancy (21 days) of the T-cells present at early stages of the transformation. Cell viability is also excellent for each stage of the immortalization process: previous experiments (Penno *et al.* 1993) indicated a lymphocyte viability of 95–100% upon receipt of whole blood, a post-freeze viability of 90% for lymphocytes isolated from frozen whole blood, a pre-freeze viability of 90–100% for transformed cells (lymphoblasts), and a post-freeze viability of 85–90% for lymphoblasts. Furthermore, flow cytometric analysis of revived lymphocytes from frozen whole blood showed normal representation of B and T lymphocytes (12.2% and 75.7%, respectively). Results obtained with this protocol indicate that lymphocytes from as little as 0.5 ml of frozen or fresh whole blood can be successfully immortalized with EBV.

General considerations

Phlebotomy and blood storage

As careful blood drawing is absolutely critical for successful cryopreservation and immortalization, phlebotomists need to be given specific instructions about blood collection and handling. A one-page protocol (Figure 4.4B2) should be provided

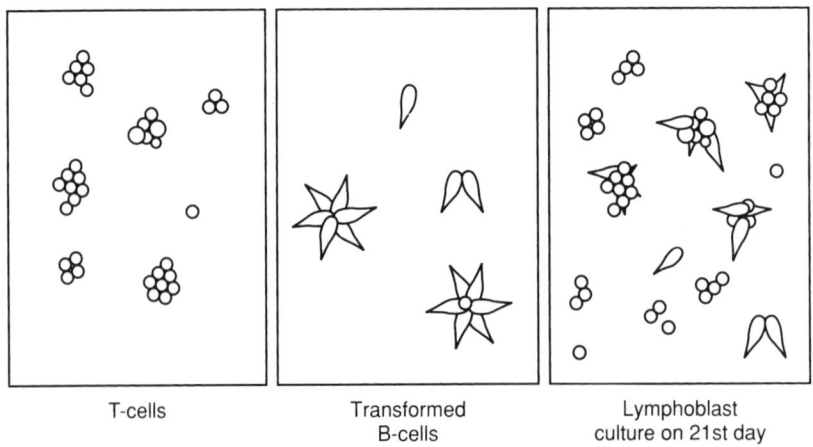

| T-cells | Transformed B-cells | Lymphoblast culture on 21st day |

Figure 4.4B.3 The establishment of lymphoblastoid cell lines: cellular morphology.

to the phlebotomist each time a blood sample is requested. It is also prudent to supply the phlebotomist with the necessary materials (21 G needles, ACD Vacutainer tubes, storage containers, airwaybill) rather than assume that these will be on hand.

Lymphocyte separation media

Lymphocytes can be separated as described in Section 4.4A. The authors use either of two gradients: a lymphocyte separation medium such as Ficoll-Hypaque (the less expensive choice) or Leucoprep (easier to use).

Cryovials and cryopreservation containers

Special attention must be paid to cryopreservation techniques in order to ensure slow, even freezing of isolated lymphocytes and whole blood. Rapid freezing causes formation of intracellular ice crystals, which rupture cell membranes and organelles, resulting in drastically reduced post-freeze viability. In contrast, slow and controlled freezing in the presence of dimethylsulphoxide (DMSO) helps preserve cell integrity and viability. Frozen cells can best be stored in special cryovials that will not crack at ultra-low temperatures. If markers are used to label the cryovials, care must be taken to ensure that marks do not rub off with time. Because of the need for permanence, use of a labelmaker and labels suitable for ultra-low temperatures is strongly suggested. For controlled freezing, cryovials are placed into a commercially available cryopreservation container or a home-made cryopreservation container consisting of a new quart-sized paint can that contains a sponge-like test tube holder submerged in isopropanol or ethanol. The cryo-preservation container (Mr Frosty) is designed to prevent loss of gummed labels during the cryopreservation procedure by shielding samples from direct contact with the alcohol bath. The cells are then slowly cooled at a rate of about $-1°C\ min^{-1}$ by storing the cryovials in the cryopreservation container at $-80°C$ for 3–18 h, and then transferring them on dry ice to the vapour phase of liquid nitrogen ($-180°C$).

Critical parameters

These precautions are critical to obtaining high viability and efficient transfor-mation:

1. Draw blood with a 21 G Vacutainer needle and ACD yellow top Vacutainer tube to minimize hemolysis and preserve viability (see Figure 4.4B.2); fill tubes all the way, since the preservative is calculated for a full volume; to preserve sterility, do not remove the top of the tube; store blood at room temperature for no longer than 72 h; protect samples from drastic changes in temperature (keep insulated from cold and hot weather during transport to the laboratory).
2. Keep all frozen cryovials of cells on dry ice when transporting to the hood for processing. Do not thaw any cells until all other reagents and equipment are ready.
3. If transforming lymphocytes from frozen whole blood, prepare a plate for panning at least 3–24 h before thawing cells.

Safety precautions

Since the transformation protocol requires infectious virus (EBV), all procedures involving EBV must be carried out in a Biosafety level 2 facility. The authors use a free-standing BioGard hood for this purpose. Normal precautions for dealing with human blood must also be taken, and anyone handling blood should be vaccinated against hepatitis B. Gloves should be worn at all times and care should be taken when handling the DMSO, which readily penetrates the skin.

REFERENCE

Penno MB, Pedrotti-Krueger M & Ray T (1993) Cryopreservation of whole blood and isolated lymphocytes for B-cell immortalization. *Journal of Tissue Culture Methods* 15: 43–48.

4.4C T-Cells and Thymocytes

Advances in the understanding of lymphocyte biology have been accompanied by improvements in the techniques of long-term culturing of human T-lymphocytes (Fathman & Fitch 1984; Feldmann *et al.* 1989). Long-term culturing of T-cells can lead to the establishment of stable lines and clones. These cultures respond to and produce a wide range of cytokines. More importantly, T-cell lines and clones can be generated which are antigen-responsive. Many aspects of the T-cell response to antigen have been described using antigen-specific cell lines and clones (Modlin *et al.* 1986; Fisch *et al.* 1990; Lamb *et al.* 1982; Clayberger *et al.* 1987).

Human T-cells can be isolated from whole blood by centrifugation using a commercially available high-density medium. This allows a single-step gradient separation of blood which yields the mononuclear cell fraction. Depending on the frequency of the T-cell of interest in the blood, further purification may not be required for the production of cell lines or even clones.

A culture started with the mononuclear cell fraction would contain several different cell types. Nevertheless, the T-lymphocytes can be cultured specifically and will outgrow the others, eventually resulting in a highly enriched population. To achieve this, the antigen-reactive T-cells must be first stimulated with antigen followed by expansion of cell numbers using the T-cell-specific growth hormone, interleukin 2 (IL-2). It should be noted that the absolute frequency of antigen-response T cells is low, being between 0.01% and 0.001% (van Oers *et al.* 1978), even when blood known to contain the relevant T-cells (i.e. blood from an immune donor) is used. Thus the majority of cells in the starting culture will die and only the T-cells activated by antigenic stimulation will survive. This stimulation not only selects out the antigen-specific T-cells but results in the cellular expression of the receptor for IL-2 (Smith 1988). When an exogenous source of IL-2 is now added to the activated T-cells, they will undergo further rounds of proliferation. This expansion phase of *in vitro* growth is more vigorous and prolonged as compared with the stimulus phase. By first exposing the cells to a stimulus followed by expansion of cell numbers with IL-2, a T-cell line can be established. Since T-cells eventually become refractory to the effects of IL-2, they must be restimulated for receptor reexpression (Gulberg & Smith 1986). A T-cell line can thus be maintained in culture by the alternate stimulation with antigen and expansion with IL-2.

As mentioned above, besides T-cells, other mononuclear cell types are present in the initial culture. These cells have the capacity to activate T-lymphocytes by 'processing' and 'presenting' antigen to the T-cell receptor and provide additional T-cell growth factors (Oppenheim *et al.* 1968; Germain 1986; Kurt-Jones *et al.* 1985). It is only after the induction of IL-2 receptors that T-cells can grow independently of other cell types.

The phenotypes of T-cells isolated from the peripheral blood fall into two main categories: CD4$^+$ and CD8$^+$ cells. In general, T-cells that are primed to exogenous soluble antigens express the CD4 cell surface marker and will proliferate upon re-exposure to that antigen. T-cells primed to endogenous cell surface antigens express the CD8 cell surface marker and upon differentiation are cytotoxic for cells expressing the antigen. Therefore, the type of antigen used can 'preselect' for the T-cell subset that will be expanded from the peripheral blood. Procedures are described which will lead to the generation of T-cell lines and clones of both phenotypes.

PROCEDURE: SOURCES OF T-CELLS

Most T-cell lines and clones are generated from peripheral blood mononuclear cells (PBMCs) obtained by venipuncture and purified by density gradient separation. Thymocytes are easily obtained by disruption of the thymus, and T-cells may also be prepared from a variety of healthy and diseased tissues, although specialized extraction methods may be required.

Peripheral blood mononuclear cell fractions

Materials

- Heparin
- Separation medium (Lymphoprep, Nycomed, Oslo, Norway)
- Hanks' balanced salt solution (HBSS) supplemented with 2% heat-inactivated foetal calf serum (FCS)
- Complete medium – RPMI 1640 supplemented with 10% AB$^+$ or A$^+$ serum, 2 mM L-glutamine, 100 U/ml penicillin/streptomycin, and 25 mM HEPES

1. Dilute blood 1 : 1 with HBSS-FCS supplemented with 10 U ml^{-1} heparin and gently layer over Lymphoprep by pipetting (2–3 vols blood : 1 vol. Lymphoprep).
2. Centrifuge at 800 *g* for 20 min.
3. Aspirate the buffy coat containing the PBMCs and pipette into a 25 ml tube, to no more than half the volume. Discard the remaining Lymphoprep, which contains granulocytes and red blood cells.
4. Wash aspirated PBMCs with HBSS-FCS twice by centrifugation for 10 min at 200 *g*.
5. Resuspend the pellet in 5–10 ml complete medium. Count using a haemocytometer, and then dilute to the required concentration (e.g. 10^6 ml^{-1}).

The mononuclear cell fraction may then be used fresh, or cryopreserved for an indefinite period and recovered, to prepare T-cell clones, lines or mass cultures. For cryopreservation, the cells are suspended at $2-3 \times 10^6$ cells per ml in serum supplemented with 10% dimethyl sulphoxide, then carefully frozen at a controlled rate, i.e. 1°C per minute.

Thymocytes

The thymus is the organ in which T-lymphocytes undergo the little-understood processes of maturation and 'education'. The ability to culture thymocytes has enabled many of the mechanisms involved in these processes to be studied.

Thymocytes are not homogeneous in nature and cells from every stage of T-cell ontogeny can be isolated. The growth requirements of each of these thymocyte populations appear to differ from those of mature T-cells (Gotlieb *et al.* 1991). Thus the addition of different growth factors (IL-2, IL-3, IL-4, IL-6, IL-7 and TNF) or accessory cells (e.g. thymocyte subpopulations required. The thymus is a soft organ which can be easily disrupted and thereby release the thymocytes into a single cell suspension.

Materials and equipment

- HBSS supplemented with 2% heat-inactivated FBS
- 9 cm Petri dish
- Nylon gauze taped to the tp of a 100 ml beaker
- 5 ml syringe plunger

1. Place the thymus in the Petri dish containing 10 ml HBSS-FBS and tease apart with forceps and scissors to release the thymocytes.
2. Remove the clumps of tissue by passing the suspension through the gauze. The remaining clumps can be further disrupted by rubbing them on the gauze with a 5 ml syringe plunger. Flush the gauze with 40 ml HBSS-FBS.
3. Pipette the suspension into a 50 ml conical centrifuge tube and centrifuge at 200 *g* for 10 min.
4. Wash twice in HBSS-FBS by centrifugation and resuspend in 10 ml complete medium. Count using a haemocytometer and adjust to the required concentration.

Generation of an antigen-specific T-cell line

The ease of raising an antigen-specific T-cell line will depend on the nature of the antigen and the frequency of the antigen-responsive T-cells present in the source. For example, it is relatively easy to derive T-cell cultures specific for a foreign protein antigen such as tetanus toxoid, where the peripheral blood from an immune donor contains sufficient numbers of antigen-reactive T-cells for *in vitro* stimulation. Likewise, the generation of cytotoxic T-lymphocytes (CTLs) derived from PBMCs and specific for alloantigens is facilitated by the relatively high frequency of alloreactive T-cells normally present in PBMCs. In comparison, raising autoreactive T-cells from patients specific for self proteins may require quite different cloning strategies (Allegretta *et al.* 1990).

PROCEDURE: PROCEDURES FOR ESTABLISHING T-CELLS REACTIVE WITH EXOGENOUS SOLUBLE ANTIGEN

The following procedures would be appropriate for the generation of T-cell lines and clones to typical foreign antigens such as tetanus toxoid or purified protein derivative (PPD). In addition to describing the establishment of a T-cell line, two supplementary procedures, restimulation with antigen and treatment of antigen-presenting cells, are included. These procedures must be followed to maintain the T-cell line in continuous culture once it has been established.

Establishment of a T-cell line

Materials and equipment

- Complete medium
- IL-2
- Antigen (approximately 1 mg ml^{-1}, dialysed against phosphate-buffered saline (PBS), pH 7.2)
- ^{3}H-thymidine (500 μCi ml^{-1})
- 96-well round-bottomed plate
- Liquid scintillation counter

1. It is advisable to test the T-cells for antigen reactivity before using them for long-term culturing. Wash PBMCs, used as the source of T-cells, and resuspend to a concentration of 1×10^6 ml^{-1} in complete medium. Prepare a dilution series of antigen, made with complete medium, at varying concentrations, for example: 3.125 μg ml^{-1}, 6.25 μg ml^{-1}, 12.5 μg ml^{-1}, 25 μg ml^{-1}, 50 μg ml^{-1}, and 100 μg ml^{-1}.
2. Grow cultures using a 96-well round-bottomed plate. Add to each well: 0.1 ml cell suspension and 0.1 ml antigen dilution. As a negative control, add 0.1 ml complete medium instead of antigen. Prepare triplicate wells for each test culture and the negative control.
3. Incubate cells at 37°C in 5% CO_2/95% air for 6–7 days. During the last 6–8 h of incubation, pulse cultures with 0.5 μCi of ^{3}H-thymidine. At the end of the incubation period, harvest cells onto glass fibre filters and measure the degree of ^{3}H-thymidine incorporation by liquid scintillation counting.
   Data are expressed as mean counts/min of replicate wells. PBMCs are considered antigen-responsive if a concentration-dependent increase in ^{3}H-thymidine incorporation is observed, at least over a portion of the concentration range.
4. Resuspend the PBMCs to 10^6 ml^{-1} in complete medium and add antigen at a concentration known to induce maximal proliferation (see above).
5. Dispense 2 ml PBMC suspension into each well of a 24-well tissue culture plate and incubate the cultures at 37°C for 6 days in a humidified atmosphere containing 5% CO_2/95% air.
6. After 6 days, clusters of dividing cells should be apparent in the wells. Vigorously pass the clusters up and down a pipette to form a single cell suspension and count using a haemocytometer.
7. In 5 ml complete medium, resuspend the cells to a concentration of 2×10^5 ml^{-1} and add IL-2 at a final concentration of 10 ng ml^{-1}.
8. Transfer the cells to the wells of a 6-well plate and incubate as before.

Restimulation with antigen

1. Aspirate the T-cells from the culture plate and wash twice in HBSS-FBS. Resuspend in complete medium and count using a haemocytometer.
2. Adjust the cell concentration to 5×10^5 cells ml^{-1} in complete medium and dispense 1 ml well^{-1} into a 24-well plate.

3. Add 1 ml antigen-presenting cells (APC; treated to inhibit their growth; see below) to each well to give an APC to T-cell ratio of 3 : 1 and add the appropriate concentration of antigen.
4. Incubate the cells at 37°C for 4 days as described previously and expand using IL-2 as described above.

Treatment of antigen-presenting cells

The cell population was derived from PBMCs which contain approximately 60% T-cells while the remainder are B cells and monocytes which act as APCs in the initial culture phase. However, once expanded, the T-cell population does not contain any APCs and therefore these must be provided for restimulation with antigen. These APCs are treated in a manner which will prevent their proliferation but retain their ability to present antigen. This can be achieved in a number of ways which are described below. T-cells require histocompatible APCs to be stimulated by antigen and therefore autologous PBMCs are used in most cases.

Irradiation

Materials and equipment

- HBSS/2% FBS
- Complete medium
- Ionizing-radiation sources

1. Wash and resuspend cells in HBSS-FBS to a concentration of 5×10^6 cells ml^{-1}.
2. Irradiate (40 gy) by exposure to either ^{137}Cs or ^{60}Co or from an X-ray source.
3. Wash the cells once in HBSS-FBS and resuspend to the desired concentration in complete medium.

Mitomycin C treatment

Materials and equipment

- Mitomycin C
- HBSS
- HBSS/5% FBS

 Mitomycin C is very toxic and is a possible carcinogen. Use gloves when handling.

1. Dissolve the mitomycin C in HBSS (serum free) to a concentration of 0.5 mg ml^{-1}, sterilize by membrane filtration and store protected from the light. Discard the solution if a precipitate forms during storage.
2. Dilute cells to a concentration of 1×10^7 cells ml^{-1} in HBSS (serum free).
3. Add 0.05 ml mitomycin C (25 μg ml^{-1}) per ml cell suspension and incubate at 37°C for 30 min protected from the light.
4. Wash four times in HBSS/5% FBS and resuspend in complete medium. It is important to wash the cells thoroughly as any residual mitomycin C carried into the culture will prevent the growth of the T-cells.

Glutaraldehyde fixation

Chemical fixation, unlike irradiation or mitomycin C treatment, crosslinks the cell surface molecules of the APCs. Therefore the cells must be pulsed with antigen prior to fixation if antigen processing is required. The APCs should be incubated with the antigen for 1–2 h before fixation to allow the antigen to be processed and expressed on the cell surface. If peptides are being used to stimulate the T-cells the pulsing step can be eliminated and the peptide added to the APCs after fixation.

Materials and equipment

- 0.05% Glutaraldehyde in PBS, pH 7.2
- 0.2 M Glycine in HBSS
- HBSS/2% FBS
- Complete medium
- 15 ml polypropylene tube

1. Suspend the APCs in complete medium at 1×10^7 ml^{-1} and add the optimal concentration of antigen. Dispense into a polypropylene tube and incubate at 37°C for 1–2 h in a humidified atmosphere of 5% CO_2/95% air.
2. Wash the cells in HBSS-FBS by centrifugation at 200 g for 10 min. Decant the supernatant and resuspend the pellet in the remaining fluid.
3. Add 1 ml 0.05% glutaraldehyde solution and incubate for 1 min. To neutralize the glutaraldehyde, add 1 ml 0.2 M glycine buffer and wash by centrifugation.
4. Wash the cells once more in HBSS-FBS and resuspend in complete medium.

PROCEDURE: ASSAY FOR THE ANTIGEN SPECIFICITY OF T-CELLS REACTIVE WITH EXOGENOUS SOLUBLE ANTIGEN

Proliferative response to antigen

The antigen specificity of T-cells can be assayed by measuring their proliferative response to antigen. A summary of the method is as follows. Individual 0.2 ml cultures of T-cells, PBMCs and antigen are incubated for 3 days. Proliferative capacity of the T-cells is measured as incorporation of the metabolic label ^{3}H-thymidine, during the last 6–8 h of incubation.

1. For the test cultures, make a dilution series of antigen with complete medium at four times the desired final concentration. Four doubling dilutions, starting with the concentration used for initial stimulation of the culture, are usually sufficient to observe concentration-dependent proliferation. For the positive control cultures, supplement a volume of complete medium with IL-2 at a concentration of 80 ng/ml.
2. Prepare the PBMCs by treating with mitomycin C, glutaraldehyde or γ-irradiation to prevent their proliferation. After treatment resuspend the cells at 8×10^5 ml^{-1} in complete medium.

3. Harvest the T-cells to be assayed, at the end of the IL-2 expansion phase, when they are most responsive to antigen. It should be noted that T-cells are refractory to antigenic stimulation if harvested too soon after receiving IL-2. Typically, cells can be used 5–8 days following a fresh exposure to IL-2. Wash twice and resuspend at a concentration of 2×10^5 ml^{-1} in complete medium.
4. Cultures are grown using 96-well round-bottomed plates. To each well add: 0.05 ml PBMC suspension, 0.05 ml antigen dilution, and 0.1 ml T-cell suspension. For the positive control cultures, add 0.05 ml IL-2-containing medium instead of antigen. For the negative control cultures, add 0.05 ml complete medium only. Prepare triplicate wells for each test culture and both sets of controls.
5. Incubate cells at 37°C in a humidified atmosphere containing 5% CO_2/95% air for 3 days. During the last 6–8 h of incubation, pulse the cultures with 0.5 μCi of ^{3}H-thymidine. At the end of the incubation period, harvest the cells onto glass fibre filters and measure radioactivity using liquid scintillation counting.
6. Data are expressed as mean ± SD of replicate wells. SD should be less than 15% of the mean. Positive stimulation has occurred when the stimulation index, defined as the ratio of the mean counts/min of the test cultures and the counts/min of the negative control cultures, is greater than 2 and the values of the test cultures exceed 1000 counts min^{-1}.

Other methods

T-cell antigen specificity can be assessed in a number of ways of which the proliferation assay outlined above is the most commonly used. An alternative method for quantitating T-cell activation is to measure the levels of T-cell-specific cytokines such as IL-2, IL-4 and IFN-γ released into the medium upon stimulation (Palliard *et al.* 1988).

SUPPLEMENTARY PROCEDURE: CLONING OF T-CELLS REACTIVE WITH EXOGENOUS SOLUBLE ANTIGEN

Limiting dilution

It is possible to generate T-cell clones by limiting dilution. If the T-cells are plated at a frequency of one cell for every three wells, then according to Poisson distribution (Lefkovits & Waldmann 1984) there is an 83% chance that the wells will contain one T-cell only. The T-cells are cultured in the presence of PBMCs, antigen and IL-2. Cloning efficiency can range from 10% to 50%.

Materials and equipment

- HBSS/2% FBS
- IL-2
- Antigen
- Complete medium
- Terasaki plates
- 96-well round-bottomed plates

1. Take cells either from a T-cell line or directly from tissue and wash in HBSS-FBS and resuspend in complete medium to a concentration of 3×10^4 cells ml^{-1}.
2. Further dilute the cells to a concentration of 30 cells ml^{-1} in 20 ml.
3. Prepare APCs by γ-irradition or mitomycin C treatment and adjust to 1×10^6 cells ml^{-1} in 20 ml complete medium containing 20 ng/ml IL-2 (10 ng ml^{-1}, final concentration).
4. Add the APCs and T-cells together in a 1 : 1 ratio (20 ml T0-cells to 20 ml APCs), add antigen and mix thoroughly.
5. Dispense 0.02 ml cell suspension per well into Terasaki plates and incubate at 37°C as previously described for 7–14 days. This will give 20 Terasaki plates containing 0.3 T-cells $well^{-1}$. All the plates will contain 10^4 APCs $well^{-1}$, IL-2 and antigen.
6. The plates should be monitored for growth using an inverted microscope and the positive wells marked.
7. After 7 days any wells which have shown significant growth can be transferred into the wells of a 96-well round-bottomed plate. To each well add 0.08 ml complete medium and 0.05 ml containing 2×10^6 APCs ml^{-1} (1×10^5 $well^{-1}$), antigen and 30 ng ml^{-1}, final concentration).
8. Incubate for 7 days at 37°C as previously described. Add 0.05 ml IL-2 at 40 ng ml^{-1} and incubate at 37°C for another 7 days.
9. At this stage the cells should have expanded to the stage where they can be transferred into 24-well plates. Aspirate each clone from the 96-well plate, transfer to one of the wells of a 24-well plate and add complete medium to a final volume of 1 ml. To each well containing a clone, add 10^6 APCs, antigen and IL-2 to bring the volume to 2 ml $well^{-1}$.
10. The clones can then be passaged on a 2-week cycle of antigen stimulation, IL-2 being added and the cells diluted as required.
11. The clones can be tested for antigen proliferation at the end of the IL-2 expansion phase of the cycle.

If the APCs are not in short supply then 96-well round-bottomed plates can be used for cloning in preference to Terasaki plates (use 10^5 APCs $well^{-1}$). This method of cloning reduces both the number of cell manipulations and the time required to expand the clones to the numbers required for the testing of antigen specificity.

SUPPLEMENTARY PROCEDURE: IMPROVED METHOD FOR THE EXPANSION OF HUMAN T-CELL CLONES

Lectin-free expansion using oxidized feeder cells

Assays detecting HPRT-deficient mutants present in human peripheral blood T-lymphocytes have been developed to monitor the population for somatic mutations (Cole *et al.* 1988). In order to investigate the nature of the mutations, colonies must be expanded further to give sufficient numbers (~10^7) for molecular and biochemical characterization, and this has been achieved conventionally by repeated lectin stimulation. However, repeated stimulation with PHA allows only around 50% of T-cell clones to be expanded to this extent so an improved method, using stimulation with oxidized feeder cells rather than lectin, was developed which enabled 100% of clones tested to be expanded (Beare *et al.* 1993). In this method, galactose residues on the surface of B-lymphoblastoid feeder cells are oxidized enzymatically with neuraminidase and galactose oxidase and stimulation occurs by covalent crosslinking between 'reactive' feeder cells and the T-cells. An additional advantage is that, for the investigation of HPRT-deficient clones, a B-lymphobastoid line (RJK853) known to be *hprt* by total deletion of the gene could be used, avoiding potential contamination of the T-cell *hprt*. The protocol below is for the expansion of 6-thioguanine (6TG)-resistant HPRT-deficient clones but can be modified for any clonal type. Depending on the precise application, cells may be cryopreserved when there are sufficient numbers at any point in the expansion procedure.

Materials and equipment

- Culture medium: RPMI 1640 (Dutch modification) buffered with HEPES (RPMI), supplemented with 10% heat-inactivated pooled human AB serum, 200 IU ml^{-1} recombinant IL-2, glutamine (2 mM), sodium pyruvate 0.2 mg ml^{-1}), penicillin (100 U ml^{-1}), streptomycin (100 μg ml^{-1}) and Amphotericin B. (2.5 μg ml^{-1})
- EBV-transformed lymphoblastoid B-cells (e.g. RJK853).
- Medium for feeder cell oxidation: 0.05 units/ml galactose oxidase (Sigma) and 0.02 units ml^{-1} neuraminidase (Sigma) in RPMI
- Wash medium: 0.01 M D(+)galactose in RPMI
- 24-well and 6-well tissue culture plates
- 6-TG stock solution, 334 μg ml^{-1} (2 mM) in 0.5% sodium carbonate.

1. Identify mutant clones in 96-well plates, ensuring that each colony has arisen from a single cell. Resuspend each by pipetting and transfer into 24-well plates, making the final volume up to 0.5 ml with culture medium; determine the cell number by counting with a haemocytometer.
2. Prepare oxidized feeder cells. Lethally irradiate (40 Gy from a gamma source) feeder cells then incubate 2×10^6/ml in medium for oxidation for 90 min at 37°C. Wash the cells by pelleting (400 *g* for 10 min), resuspending in wash medium, pellet and wash again, then finally resuspend the washed cells in culture medium at 5×10^6 cells ml^{-1}.
3. Stimulate the T-cell clones. Add oxidized feeder cells to each clone (in the ratio 6 feeder cells to 1 T-cell) and make up the volume in each well to 1 ml with culture medium. Incubate at 37°C in a humidified 5% CO$_2$/air incubator. After 18–24 h, resuspend the cell aggregates by pipetting and feed each well with 1 ml culture medium.
4. Re-feed the clones. Check the progress of each clone, daily, using an inverted microscope. Every 1–3 days re-feed by removing 1 ml and replacing with 1 ml fresh culture medium. (*Note*: Medium colour is a valuable indicator of cell growth, and T-cells grow far more readily in acidic culture conditions).
5. Expand the clones. When the clone fills its well in a 24-well plate, transfer to a 6-well plate, diluting in fresh culture medium to maintain the cell density at $2–5 \times 10^5$ ml^{-1}. At 7-day intervals, count the cells and re-stimulate with oxidized feeder cells as in (3) above. The optimal feeder cell/T-cell ratio will depend on

how well the clone is growing. At poor growth rates and low cell density the maximum 6 : 1 ratio may be needed, but this may be reduced to 1 : 1 in cultures at high densities growing well.

6. Confirm 6-TG resistance. When a clone has been expanded sufficiently, resuspend the cells and divide into two aliquots. Dilute the TG stock solution 1 : 20 with culture medium and add at 5% v/v (to give final 5 μM) to one. A true 6-TG-resistant HPRT-deficient clone will continue to expand in the presence of 6-TG, although sometimes at a reduced growth rate. A non-resistant clone may undergo one population doubling, but will die out after 1–2 weeks in culture.

7. Harvest an expanded clone. When sufficient cells are available, harvest by centrifugation (300 g for 10 minutes) then resuspend 2–3 × 10^6 ml^{-1} in serum supplemented with 10% DMSO. Dispense 1–2 ml aliquots and cryopreserve by controlled-rate freezing, i.e. 1°C per minute.

PROCEDURE: PROCEDURES FOR ESTABLISHING T-CELLS REACTIVE WITH ENDOGENOUS CELL SURFACE ANTIGEN

Procedures will be described for the raising of CTLs specific for major histocompatibility complex (MHC) class I molecules. These endogenous cell surface molecules are highly antigenic for histoincompatible T-cells (i.e. T-cells of a different MHC genotype from the target cells). A lymphoblastoid cell line (Daudi) will be used as the antigenic stimulus since these cells express class I antigens, are relatively easy to maintain and have been previously shown to act as cellular targets for CTLs specific for class I antigens (Meuer *et al.* 1982). In addition, a procedure for measuring the cytotoxicity of T-cells will be described.

Establishment of a primary cell line

Additional materials

• Daudi or any lymphoblastoid cell line expressing HLA class I antigens that are different from that of the donor PBMCs

1. Isolate PBMCs from the blood and resuspend to 1 × 10^6 cells ml^{-1} in complete medium.
2. Daudi cells are γ-irradiated (16 500 rads) and then prepared at 1 × 10^6 cells ml^{-1} in complete medium.
3. To a well of a 24-well plate, add 1 ml of each cell suspension and incubate for 5 days.
4. After 5 days, observe cell growth.
5. Aspirate the cells, resuspend thoroughly and count.
6. In 5 ml complete medium resuspend the cells to a concentration of 2 × 10^5 ml^{-1} and add IL-2 at a final concentration of 10 ng ml^{-1}.
7. Transfer the cells to the wells of a 6-well plate and incubate as before. The cells will undergo a phase of rapid division and may require diluting after 4 days of culture and fresh IL-2.

The IL-2 stimulation will lead to the formation of large clusters of dividing cells which will disperse towards the end of the IL-2 expansion cycle, leaving single non-dividing cells. The cells are now ready to be tested for their antigen specificity in a cytotoxicity assay and for restimulation with the Daudi cell line.

Restimulation with antigen

1. Aspirate the T-cell line from the culture plate and wash twice in HBSS-FCS. Resuspend the cells in complete medium and count using a haemocytometer.
2. Adjust the cell concentration to 5×10^5 cells ml^{-1} in complete medium and dispense 1 ml well^{-1} into a 24-well plate.
3. Prepare the Daudi cell line as for the establishment of the T-cell line. To the well of a 24-well plate add 1 ml of each cell suspension.
4. Incubate the cells for 4 days at 37°C as described previously and expand using IL-2 as described above.

PROCEDURE: ASSAY FOR THE ANTIGEN SPECIFICITY OF T-CELLS REACTIVE WITH CELL SURFACE ANTIGEN

Cytotoxic response to target cells

The antigen specificity of CTLs can be quantitated using a cytotoxicity assay. In brief, target cells are labelled internally with ^{51}Cr. If these cells and the CTLs which are specific for the surface antigen are now co-cultured, target cell lysis will occur, resulting in the release of ^{51}Cr into the medium. Culture supernatants are collected and the amount of radioisotope present can be used as a measure of cell lysis.

Materials and equipment

- 1% aqueous solution of NP-40
- ^{51}CrO$_4$ (Amersham International, UK)
- 96-well V-bottomed plate
- Centrifuge carriers for 96-well plates
- Gamma counter

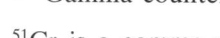 ^{51}Cr is a gamma-ray-emitting isotope. In addition to following the usual guidelines for the handling of radioisotopes, insure that addquate shielding is used when dispensing and storing ^{51}Cr ($T_{1/2} = 28$ days).

1. Resuspend Daudi cells to a concentration of 5×10^6 cells ml^{-1} in complete medium. For labelling, 100 μ Ci^{-1} of ^{51}CrO$_4$ is added to 500 μl of the target cell suspension for 1½ h at 37°C.
2. Wash the cells three times with cold HBSS-FBS and resuspend to 5×10^4 cells ml^{-1} in complete medium.
3. Prepare a doubling dilution series of the T-cell suspension starting from 1×10^6 ml^{-1} to 6.25×10^5 ml^{-1} in complete medium.

4. Dispense 100 μl target cell suspension and 100 μl T-cell suspension from each of the dilutions in triplicate into the wells of a 96-well V-bottomed microtest plate. In this way, the T-cells are titrated against the target cells at effector to target cell ratios starting from 20 : 1 to 1.25 : 1. Two types of control wells are also prepared.

5. To determine the spontaneous background release of ^{51}Cr by target cells, add 100 μl target cell suspension to the wells along with 100 μl complete medium. Total release is determined by adding 100 μl 1% NP-40 solution to 100 μl target cell suspension.

6. Gently pellet the cells by centrifugation (100 *g* for 5 min) and incubate at 37°C in a humidified atmosphere of 5% CO_2 for 4 h.

7. To harvest the culture supernatant, pellet the cells by centrifugation of the 96-well plate for 5 min at 250 *g*. Carefully remove 100 μl supernatant from each well and transfer into individual tubes for counting by γ-spectrometry.

8. Data are expressed as percentage lysis, calculated for each effector/target ratio, using formula: percentage lysis = (test counts/min – background release counts/min)/(total release counts/min – background release) × 100. If a plot of percentage lysis versus effector/target ratio is made, a linear relationship should be observed.

The most common problems with the ^{51}Cr-release assay for cytotoxicity are either low ^{51}Cr uptake by the target cells or high levels of spontaneous release. These problems can be avoided by ensuring that the target cells are in log phase growth when labelled and by taking care to ensure that the cells remain at 37°C both before and during the labelling (i.e. warm medium). The washing of the target cells with cold medium after the labelling procedure will also reduce the degree of spontaneous release by disrupting any fragile cells prior to the assay.

SUPPLEMENTARY PROCEDURE: CLONING OF T-CELLS REACTIVE WITH ENDOGENOUS CELL SURFACE ANTIGEN

Limiting dilution

As for T-cells reactive with soluble antigen, it is possible to generate CTL clones by limiting dilution. As described below, the procedure is very similar to that for CD4$^+$ cells, although it is best to start with a CTL line or at least cells that have been primed previously with the target cells.

Materials and equipment

- HBSS/2% FBS
- IL-2
- Daudi cell line
- Complete medium
- Terasaki plates
- 96-well round-bottomed plates

1. Using cells from a T-cell line, wash in HBSS-FBS and resuspend in complete medium to a concentration of 3×10^4 cells ml^{-1}.
2. Further dilute the cells to a concentration of 30 cells ml^{-1} in 20 ml.
3. Prepare Daudi cells by γ-irradiation or mitomycin C treatment and adjust to 1×10^6 cells ml^{-1} in 20 ml complete medium containing 20 ng ml^{-1} IL-2 (10 ng ml^{-1}, final concentration).
4. Add the Daudi cells and T-cells together in a 1 : 1 ratio (20 ml T-cells to 20 ml Daudi), and mix thoroughly.
5. Dispense 0.02 ml cell suspension per well into Terasaki plates and incubate at 37°C as previously described for 7–14 days. This will give 20 Terasaki plates containing 0.3 T-cells $well^{-1}$. In addition, each well will contain 10^4 Daudi cells and IL-2.
6. Monitor the plates for growth using an inverted microscope and mark the positive wells.
7. After 7 days any wells which have shown significant growth can be transferred into the wells of a 96-well round-bottomed plate. To each well add 0.08 ml complete medium and 0.05 ml containing 2×10^6 Daudi cells ml^{-1} (1×10^5 $well^{-1}$) and 30 ng ml^{-1} IL-2 (10 ng ml^{-1}, final concentration).
8. Incubate for 7 days at 37°C as previously described. Add 0.05 ml IL-2 at 40 ng ml^{-1} and incubate at 37°C for another 7 days.
9. At this stage the cells should have expanded to the stage where they can be transferred into 24-well plates. Aspirate each clone from the 96-well plate, transfer to one of the wells of a 24-well plate and add complete medium to a final volume of 1 ml. To each well containing a clone, add 10^6 Daudi cells, and IL-2 to bring the volume to 2 ml $well^{-1}$.
10 The clones can then be passaged on a 2-week cycle of Daudi cell stimulation, IL-2 being added and the cells diluted as required.
11. Test the clones for cytotoxicity at the end of the IL-2 expansion phase of the cycle.

As an alternative to Terasaki plates, 96-well round-bottomed plates can be used for cloning (use 10^5 Daudi cells $well^{-1}$). This method of cloning reduces both the number of cell manipulations and the time required to expand the clones to the numbers required for the testing of antigen specificity.

PROCEDURE: MASS LONG-TERM CULTURES OF HUMAN T-CELLS

Expansion of particular genotypes without viral transformation

A method was developed to generate mass cultures of human T-lymphocytes, originally intended for use in genotoxicity testing (O'Donovan et al. 1995). The protocol was simplified as much as possible, utilizing foetal bovine serum, recombinant IL-2 and freeze-killed B-lymphoblastoid feeder cells. Freeze-killing was used in preference to lethal irradiation to prepare the feeder cells to allow the method to be used in laboratories without access to a convenient radiation source, and feeder cell killing with mitomycin C was not considered in order to avoid possible mutagenic contamination in cells intended for subsequent use in genotoxicity experiments. In conjunction with PHA, freeze-killed feeder cells provide sufficient accessory antigenic stimulus for mass cultures where, unlike growth at clonal density, provision of cell bulk and medium conditioning by the feeder cells seems not to be important. Finally, it should be noted that this method allows mass cultures to be generated from cryopreserved, unstimulated peripheral blood mononuclear cell fractions just as easily as from freshly prepared samples.

Starting with a 10 ml sample of human peripheral blood, this protocol can easily generate 2×10^8 cells after 8 days' growth, allowing at least 50 aliquots of 2–3×10^6 cells to be cryopreserved. Each of these aliquots can then be restimulated on thawing and expected to give 4–8×10^7 cells within a further 7 days in culture. After this time, the cells are karyotypically normal and are all T-cells (expressing CD3), comprising both CD4$^+$ and CD8$^+$ populations. It was then realized that this method is ideally suited to producing very large amounts of material from humans with genetic disease or polymorphisms without infection with viral DNA, and it has been used to expand cultures xeroderma pigmentosum, ataxia-telangiectasia and trichothiodystrophy patients. While there may be applications where the absence of viral DNA makes these T-cell cultures more appropriate than B-lymphoblastoid lines, it should be noted that it is possible to establish both from a 10–20 ml peripheral blood sample.

Finally, although the basic method generates T-cell cultures comprising both CD4$^+$ and CD8$^+$ populations, it can be modified very simply to provide either purified subset. This may be desirable for a number of reasons, including the fact that there are continuing interactions between the two cell types, with direct contact resulting in *de novo* synthesis of the CD8 antigen by some of CD4$^+$ cells (O'Donovan *et al.* 1999)

Materials and equipment

- Basic medium (RPMI): RPMI 1640 (Dutch Modification) buffered with HEPES and supplemented with sodium pyruvate (1 mM), glutamine (2 mM), penicillin (100 U ml^{-1}) and streptomycin (100 μg ml^{-1})
- Growth medium: RPMI supplemented with 10% heat-inactivated foetal bovine serum and 250 IU ml^{-1} recombinant IL-2 (Chiron, Proleukin)
- Freeze-killed feeder cells: Harvest B-lymphoblastoid feeder cells (RJK 853 or GM1899A) by centrifugation, resuspend in RPMI (NB *no* supplements for cryopreservation) at 3×10^6 per ml, dispense 1 ml aliquots into ampoules and freeze rapidly to $-80°$C or below. To use, thaw an ampoule and incorporate the contents directly into stimulating medium.
- Stimulating medium: growth medium supplemented with 0.4 μg ml^{-1} PHA (Murex purified HA16).
- 2-Mercaptoethanol. Can be added to stimulating and growth medium to give 50 μM; it is not essential, but improves growth rates.

Growth and cryopreservation of mass cultures

1. Collect 10–15 ml blood into lithium heparin tubes and separate the mononuclear cell fraction; yield should be at least 10^6 cells per ml whole blood.

2. Initiate cultures by seeding $1-2 \times 10^5$ mononuclear cells per ml in stimulating medium supplemented with $1-2 \times 10^5$ freeze-killed feeder cells per ml (NB For prolonged growth, feeder cells are essential at this stage) use 5 ml per well in 6-well dishes, 5 ml per 25 cm^2 flask or 20–30 ml per 75 cm^2 flask. Incubate at 37°C in a humidified 5% CO_2/air incubator for 4 days.

3. Day 4, and daily thereafter. Disaggregate cell clumps by pipetting and count. When cell density reaches 10^6 ml^{-1} (around day 6) dilute with growth medium. (*Note*: T-cells grow far more readily in acidic culture conditions; do not reduce the cell density below 2×10^5 ml^{-1}).

4. Day 7/8. When there are sufficient cells, harvest by centrifugation (300 g for 10 min), resuspend at 3×10^6 ml^{-1} in serum supplemented with 10% DMSO. Dispense 1–2 ml aliquots and cryopreserve by controlled-rate freezing, i.e. 1°C per min. These may then be used for biomolecular characterization, or thawed and grown to allow further expansion of cell numbers.

Recovery of cryopreserved mass cultures

1. Rapidly thaw an ampoule of cryopreserved T-cells by immersing in water at 37°C. Wash by centrifugation (300 g for 10 min) in three changes of RPMI + 10% serum, then resuspend at 2×10^5 ml^{-1} in stimulating medium (*Note*: feeder cells are *not* necessary at this stage) and seed 5 ml per well in 6-well dishes, or 20–30 ml in 75 cm^2 flasks. Incubate at 37°C in a humidified 5% CO_2/air incubator for 3 days.

2. Day 3, and daily thereafter, disaggregate cell clumps by pipetting and count. Maintain cell density between 2 and $10-15 \times 10^5$ per ml by dilution with growth medium.

SUPPLEMENTARY PROCEDURE: MASS CULTURES OF CD4⁺ AND CD8⁺ CELLS

Either subset can easily be prepared from the mass cultures above by the the use of antigen-coated magnetic beads. The procedure here for both CD4⁺ and CD8⁺ cells uses negative selection, i.e. beads are used to remove the unwanted subset, then discarded, and results in preparations which are > 99% pure (O'Donovan *et al.* 1999)

Materials and equipment

- M450 Dynabeads (Dynal UK Ltd), CD4 (product number 111.05/6) and CD8 (product number 111.07/8)
- Dynal MPC1 magnetic separator
- Hanks' balanced salts solution +2% heat-inactivated foetal bovine serum (HBS-2)

1. Initiate mass cultures as above and grow to day 6.
2. Prepare the Dynabeads by washing twice in HBS-2, Resuspend in 2 ml HBS-2, then dispense 1 ml into each of two tubes (= 7×10^7 beads per tube).

3. To prepare purified CD4⁺ cells. Centrifuge $10–12 \times 10^7$ cells ($300\,g$ for 10 min) and resuspend in 6 ml HBS-2. Add the first 1 ml aliquot of CD8 Dynabeads. Incubate on a blood roller at room temperature for 25 min. Remove the Dynabeads using a magnetic separator and discard. Add the second aliquot of Dynabeads and repeat. Centrifuge the remaining cells and resuspend the pellet in growth medium at 5×10^5 ml⁻¹.

4. To prepare purified CD8⁺ cells. The method is exactly for CD4⁺ cells, but CD4 Dynabeads are used.

5. Grow the purified subset(s) for two further days, then cryopreserve as for unseparated mass cultures.

DISCUSSION

Preservation of T-cell cultures

T-cell cultures can survive cryopreservation and storage in a liquid nitrogen refrigerator and upon recovery be shown to retain biological activity. This feature of T-cells is important as the biological activity of T-cell lines and clones in continuous cultures can vary over time and cryopreservation allows the 'stockpiling' of valuable cultures. Once thawed and washed thoroughly, the cells can be used immediately for functional studies, cultured with IL-2 or restimulated with antigen.

IL-2 source

Human recombinant IL-2 (rIL-2) can be purchased from a number of biotechnology companies (e.g. Amgen Biologicals, Thousand Oaks, USA) and used in the procedures described in this section. If IL-2 cannot be obtained then conditioned medium, either purchased commercially (Lymphocult T, Biotest Folex, Frankfurt, Germany), or made from mitogen-stimulated PBMCs, can be used (Gillis *et al.* 1978). Mitogen-conditioned medium contains IL-2 as well as a large number of other cytokines which aid T-cell growth.

Antigen stimulation

Mitogens or antibodies can be used instead of the antigen to stimulate T-cells from the primary source (e.g. PBMCs) or used in conjunction with APCs to maintain the growth of established T-cell lines or clones (Londei *et al.* 1988). The use of other stimuli may be useful, especially if antigen is in short supply. T-cells can be activated by mitogenic lectins such as phytohaemagglutinin (PHA) or concanavalin A (ConA), or by agonistic antibodies specific for the T-cell receptor-CD3 complex which mimic antigen-mediated T cell activation. Concentration of lectin, particularly PHA is critical and can show batch-to batch variation. If the source of supply is changed, the optimal stimulating concentration should always be determined.

B-Lymphoblastoid cells as accessory antigenic stimulus

A major limiting factor in the long-term culture of T-cell lines and clones is their requirement for autologous APCs for antigen stimulation. If PBMCs were continually used as the APC source then the individual from whom the line was derived would need to be bled at regular intervals. This would be impractical in most cases, especially if the T-cell line was derived from a patient. A renewable autologous source can be created by the transformation of B-cells by infection with Epstein–Barr virus (EBV) (Neitzel 1986). EBV transformed B-cells can be expanded in culture and will act as a constant source of autologous antigenic stimulation.

There are at least three methods for preparing B-lymphoblastoid cells use in T-cell cultures, all with slightly different applications.

- Conventional killing by irradiation or mitomycin C treatment. This gives cells which are metabolically active but are no longer capable of cell division. This type of feeder cell has been used for a range of applications in the growth of T-cell clones, lines and mass cultures.
- Oxidized feeder cells. These are perpared by treating conventionally killed cells with neuraminidase and galactose oxidase. These do not require supplementary stimulation with lectins and are better for expansion of T-cell clones, where repeated stimulation with lectins appears to give suboptimal results.
- Freeze-killed feeders. These can be used in conjunction with lectins to generate mass cultures where the role of the feeder cells is only to provide accessory stimulus.

Finally, it should be noted that conventionally killed and oxidized feeder cells can be prepared in bulk, then cryopreserved using standard procedures to provide a convenient source of material to be used at a later date. Freeze-killed feeders are also obviously available for use when required.

Monitoring the growth of cultures by direct observation

Although the growth of the T-cell lines and clones is, to a large extent, synchronized by IL-2, some cultures will grow faster than others. It is only by examining the cells on a routine basis with the inverted microscope that the fast-growing cultures can be identified and cultured accordingly. In addition, direct observation is the only way of verifying that the cells are actually growing, on a day-to-day basis.

REFERENCES

Allegretta M, Nicklas J, Sriram S & Albertini R (1990) T cells responsive to myelin basic protein in patients with multiple sclerosis. *Science* 247: 718–721.

Beare DM, Aldridge KE, O'Donovan MR & Cole J (1993) An improved procedure for the in vitro expansion of human T-lymphocyte clones for mutant analysis. *Mutation Research* 291: 207–212.

Clayberger C, Parham P, Rothbard J, Ludwig D, Schoolnik G & Krensky A (1987) HLA-A2 peptides can regulate cytolysis by human allogeneic T lymphocytes. *Nature* 330: 763–765.

Cole J, Gren MHL, James SE, Henderson L & Cole H (1988) A further assessment of factors influencing measurements of thioguanine-resistant mutant frequency in circulating T-lymphocytes. *Mutation Research* 204: 493–507.

Fathman C & Fitch F (1984) Long term culture of immunocompetent cells. In: Paul WE (ed.) *Fundamental Immunology*, pp. 781–795. Raven Press, New York.

Feldmann M, Lamb J & Owen M (eds) (1989) *T Cells*. Wiley, Chichester.

Fisch P, Malkovsky M, Kovats S, Sturm E, Braakman E, Klein B, Voss S, Morrissey L, DeMars R, Welch W, Bolhuis R & Sondel P (1990) Recognition by human $V_\gamma 9$. $V_\delta 2$ T cells of a Gro EL homolog on Daudi Burkitt's lymphoma cells. *Science* 250: 1269–1273.

Germain R (1986) The ins and outs of antigen processing and presentation. *Nature* 322: 687–689.

Gillis S, Ferm MM, Ou W & Smith KA (1978) T cell growth factor: parameters of production and a quantitative microassay for activity. *Journal of Immunology* 120: 2027–2032.

Gotlieb WH, Durum SK, Gregorio TA & Mathieson BJ (1991) Selective stimulation of thymocyte precursors mediated by specific cytokines. Different CD3+ subsets are generated by IL-1 versus IL-2. *Journal of Immunology* 146: 2262–2271.

Gullberg M & Smith K (1986) Regulation of T cell autocrine growth: T4+ cells become refractory to interleukin 2. *Journal of Experimental Medicine* 163: 270–284.

Kurt-Jones E, Beller D, Mizel S & Unanue E (1985) Identification of a membrane associated interleukin 1 in macrophages. *Proceedings of the National Academy of Sciences of the USA* 82: 1204–1208.

Lamb J, Eckels D, Lake P, Johnson A, Hartzman R & Woody J (1982) Antigen specific human T lymphocyte clones: induction, antigen specificity, and MHC restriction of influenza virus-immune clones. *Journal of Immunology* 128: 233–238.

Lefkovits I & Waldmann H (1984) Limiting dilution analysis of the cells of the immune system. I. The clonal basis of the immune response. *Immunology Today* 5: 265–268.

Londei M, Grubeck-Loebenstein B, de Berardinis P, Greenall C & Feldmann M (1988) Efficient propagation and cloning

of human T cells in the absence of antigen using OKT3, IL-2 and antigen presenting cells. *Scandinavian Journal of Immunology* 27: 35–46.

Meuer SC, Schlossman SF & Rheinherz EL (1982) Clonal analysis of human cytotoxic T lymphocytes: T4+ and T8+ effector T cells recognise products of different major histocompatibility complex regions. *Proceedings of the National Academy of Sciences of the USA* 79: 4395–4401.

Modlin R, Kato H, Mehra V, Nelson E, Xue-dong F, Rea T, Pattengale P & Bloom B (1986) Genetically restricted suppressor T cell clones derived from lepromatous leprosy lesions. *Nature* 322: 459–461.

Neitzel H (1986) A routine method for the establishment of permanent growing lymphoblastoid cell lines. *Human Genetics* 73: 320–326.

O'Donovan MR, Freemantle MR, Hull G, Bell DA, Arlett CF & Cole J (1995) Extended-term cultures of human T-lymphocytes: a practical alternative to primary human lymphocytes for use in genotoxicity testing. *Mutagenesis* 10: 189–201.

O'Donovan MR, Jones DRE, Robins RA, Li KF, Shim HK, Zheng Z, Arlett CF, Capulas E & Cole J (1999). Co-cultivation of CD4+ and CD8+ human T-cells leads to the appearance of CD4 cells expressing CD8 through de novo synthesis of the CD8 α-subunit. *Human Immunology* (in press).

Oppenheim J, Leventhal B & Hersch E (1968) The transformation of column-purified lymphocytes with non-specific and specific anigenic stimuli. *Journal of Immunology* 101: 262–270.

Palliard X, de Waal Malefijt R, Yssel H, Blanchard D, Chretien I, Abrams J, de Vries J & Spits H (1988) Simultaneous production of IL-2, IL-4, and IFN-γ by activated human CD4+ AND CD8+ T cell clones. *Journal of Immunology* 141: 849–855.

Smith K (1988) Interleukin-2: inception, impact and implications. *Science* 240: 1169–1176.

van Oers M, Pinkster J & Zeijlemaker W (1978) Quantification of antigen-reactive cells among human T lymphocytes. *European Journal of Immunology* 8: 477–484.

4.5 ESTABLISHMENT OF HUMAN LEUKAEMIA CELL LINES

ESTABLISHMENT OF HUMAN LEUKAEMIA CELL LINES: BASIC PROTOCOL

Introduction

Continuous human leukaemia cell lines have become invaluable tools for haematological diagnosis and research. Over the last 35 years several hundred cell lines spanning almost the whole spectrum of haematopoietic cell lineages have been described. The cardinal features of leukaemia cell lines are their monoclonal origin, arrest of differentiation, genetic alterations, and unlimited proliferation; the major advantages of cell lines are the unlimited supply of cell material and the infinite storability and recoverability at will of the cells (Table 4.5.1).

It is still extremely difficult to establish new leukaemia cell lines, and the majority of attempts fail. In the following we will describe some of the more promising techniques to establish new leukaemia cell lines.

Seeding of human leukaemia cells into suspension cultures is a common procedure in attempts to establish leukaemia cell lines. The incidence of success in the establishment of continuous leukaemia cell lines is low and unpredictable and the reasons for this failure usually remain unclear. According to general experience, the major causes appear to be culture deterioration with cessation of multiplication of leukaemia cells and the development of lymphoblastoid, fibroblast or macrophage cell lines. Despite the fact that the proliferation of most human leukaemia cells *in vivo* seems to be independent of the normal regulatory mechanisms, they usually fail to proliferate autonomously *in vitro* even for short periods of time. *In vivo*, at least initially, leukaemia cells seem to require one or probably several haematopoietic growth factors for their proliferation. The addition of regulatory proteins, e.g. so-called haematopoietic growth factors such as erythropoietin (EPO), granulocyte-macrophage colony-stimulating factor (GM-CSF), granulocyte-CSF (G-CSF), interleukin-2 (IL-2), IL-3 or IL-6, or stem cell factor (SCF), mitogens such as phytohemagglutinin (PHA), or conditioned medium (CM) secreted by certain tumour cell lines (often containing several factors), is a culturing method that increases the frequency of success by overcoming the 'crisis' period in which the neoplastic cells cease to proliferate. These molecules enable the leukaemia cells from the majority of patients to multiply for about 2–4 weeks. Out of these short-term cultures a few continuous leukaemia cell lines derived from the malignant cells can be established.

Cell and Tissue Culture for Medical Research, edited by A. Doyle and J.B. Griffiths.
© 2000 John Wiley & Sons, Ltd

Table 4.5.1 Common features of leukaemia cell lines

Major advantages:
• Unlimited supply of cell material
• Infinite storability and recoverability
Inherent cellular characteristics
• Monoclonal origin
• Differentiation arrest at a discrete maturation stage
• Genetic alterations
• Sustained proliferation in culture

Modified after Drexler *et al.* (1998) and Drexler & Matsuo (1999).

Materials

• Heparinized peripheral blood or bone marrow aspirate from patients with leukaemia
• Ficoll-Hypaque solution (density 1.077 g litre^{-1})
• Standard culture medium (e.g. RPMI 1640, Iscove's Modified Dulbecco's Medium, α-MEM, or McCoy's 5A) and foetal bovine serum (FBS)
• Neubauer haematocytometer, Trypan blue solution and 96-well microplate
• Centrifuge
• Pasteur pipettes and 2 ml, 5 ml and 10 ml pipettes
• 15 ml or 30 ml centrifuge tubes
• 50 ml flasks, 24-well plates or 96-well microplates
• Conditioned medium (CM) from tumour cell line cultures or purified/recombinant growth factors (e.g. CM from the human bladder carcinoma cell line 5637)

All solutions and utensils must be sterilized prior to use.

Procedure

1. Dilute the heparinized fresh peripheral blood or bone marrow sample with culture medium at a ratio of 1 : 2. When isolating cells from a leukapheresis patient, dilute the blood with culture medium at 1 : 4. Leukaemia cell lines can also be established from cryo-preserved specimens.
2. Pipette Ficoll-Hypaque solution into a 15 or 30 ml conical centrifuge tube. Slowly layer the mixture of medium and sample over the Ficoll-Hypaque solution. Use 5 ml Ficoll-Hypaque solution per 5 ml sample mixture. Do not disturb the Ficoll-Hypaque/sample interface. It is helpful to hold the centrifuge tube at a 45° angle.
3. Centrifuge for 20–30 min at $450 \times g$ at room temperature (with the brake of the centrifuge switched off). A layer of mononuclear cells should be visible on top of the Ficoll-Hypaque phase as they have a lower density than the Ficoll-Hypaque solution. Red blood cells and granulocytes will be concentrated below the Ficoll-Hypaque layer.
4. Using a sterile pipette carefully transfer the interface layer containing the mononuclear cells to a centrifuge tube.

5. Wash the mononuclear cells by adding culture medium plus 2% FBS (add about five times the volume of the mononuclear cell solution) and centrifuge for 10 min at $200 \times g$ at room temperature.

6. Discard the supernatant, resuspend the cells again in culture medium plus 2% FBS, wash the cells by repeating step 5. The washing steps described above are performed to remove the acidic heparin which may harm the cells, to remove the patient's serum which may inhibit the cell growth, and to remove the Ficoll-Hypaque which is hypertonic for the cells. Two washes generally suffice. However, the second wash must include at least 2% FBS in order to prevent cells from adhering to one another. FBS should be inactivated prior to use in a 56°C waterbath for 30–45 min.

7. Gently remove the supernatant, resuspend the washed mononuclear cells in culture medium with 20% FBS plus additional 10% CM or with an appropriate concentration of purified or recombinant growth factors. Count the cells and determine the viability of the cells by Trypan blue vital staining. Adjust the cells to a concentration of 2×10^6 ml^{-1}. In general, more than 90% of the mononuclear cells are recovered by this process. Cell yields depend on the number of leukaemia cells in the specimen and are highly variable from patient to patient. On average, each ml of leukaemia bone marrow yields $15–30 \times 10^6$ mononuclear cells and each ml of peripheral blood yields $1–2 \times 10^6$ mononuclear cells, the latter depending obviously on the WBC.

8. Place 5 ml of the cell suspension in a 50 ml culture flask at a concentration of 2×10^6 cells ml^{-1}. If 24-well plates are used, add 1 ml cell suspension containing 2×10^6 cells into each well. Add 100 μl of cell suspension containing 2×10^5 cells into wells of 96-well microplates. As many of the leukaemia cells as possible should be used in attempting to establish a cell line. The number of flasks or wells used depends on the number of leukaemia cells available. In theory, a cell line starts from one single cell. Thus, the more attempts, the higher the chances. It is absolutely mandatory to freeze aliquots of the original cells and to store them in appropriate locations for later documentation and comparisons.

9. Place the cells in a humidified incubator at 37°C and 5% CO_2 in air. Alternatively, incubate the cells in a humidified 37°C incubator with 6% CO_2, 5% O_2 and 89% N_2. Some research data indicate that the leukaemia cells may prefer growing in a low-oxygen environment with nutrient media. Oxygen concentrations of 1% to 10% have been shown to enhance the *in vitro* growth of both normal and malignant haematopoietic cells.

10. Expand the cells by exchanging half of the spent culture volume with culture medium plus 20% FBS plus 10% CM (or with appropriate concentrations of growth factors) once a week. After 4 h, some cells become adherent. These adherent cells are the source of colony-stimulating factors for both normal and leukaemia cells. During the first 2 weeks, it is not necessary to remove the adherent cells from the culture unless there is a specific reason to do so, for example the addition of a purified growth factor to the medium in order to obtain a unique type of leukaemia cell line. After 2 weeks, if the suspension cells grow very rapidly, the adherent cells can be removed simply by passing the suspension cells into new flasks or plates in order to reduce the

potential for overgrowth of fibroblasts and normal lymphoblastoid cells. During the first 2 weeks, the leukaemia cells might appear to proliferate actively. If the medium becomes acidic quickly (yellow in the case of RPMI 1640 medium), it is necessary to change half of the volume of medium at 2- or 3-day intervals. If the number of the cells increases rapidly, resuspend the cells weekly at a concentration of at least 1×10^6 ml^{-1} in fresh complete medium by dilution or subdividing the cells to new flasks. The blasts from the majority of the patients with leukaemia undergo as many as four doublings in 2 weeks, but after 2 weeks most of the cells stop proliferating. Following a lag time of 2–4 weeks, a small part of the population may still proliferate actively and may continue to grow, forming a cell line.

11. If the blasts continue to proliferate for more than 2 months, there is a high possibility of generating a leukaemia cell line. In these cases, the work of characterizing the proliferating cells should be done as soon as possible. Prior to the characterization of the cells, it is wise to freeze at least 2 ampoules of the proliferating cells containing a minimum of 3×10^6 cells ampoule^{-1} in liquid nitrogen in order to avoid loss of the cells due to occasional contaminations or other accidents.

12. Limiting dilution leads to the generation of monoclonal leukaemia cell lines. After prolonged culture *in vitro*, the cell line will become oligoclonal or monoclonal due to the outgrowth of selected cell clones. In most cases, it is not absolutely necessary to subclone the cell line by limiting dilution. In some types of leukaemia cell lines, e.g. immature T- and precursor B-cell lines, it might be very difficult or even impossible to 'clone' the cells.

13. There are at least three reasons to characterize established leukaemia cell lines: (i) to determine that the cell line is derived from the original leukaemia cells; (ii) to investigate whether the established cell line was transformed by viruses such as Epstein–Barr virus (EBV) or human T-cell leukaemia virus (HTLV)-1 or -2; (iii) to characterize the basic biological features of the cultured cells.

Characterization

1. The newly established cell line needs to be categorized (Table 4.5.2).

2. There are six cardinal requirements for the description and publication of new leukemia cell lines (Table 4.5.3).

3. Since leukaemia cell lines grow as single or clustered cells in suspension or only loosely adherent to the flask, single-cell populations can be easily prepared and the cells can thus be characterized and classified. Table 4.5.4 lists a variety of parameters useful for the description of the cells and a panel of possible tests applicable to the phenotypic and functional characterization of most cell lines. This necessary multiparameter examination of the cellular phenotype provides important information on the likely cell of origin, the variable stringency of maturation arrest, and any discrepancies in the pattern of normal gene expression. The list is not intended to cover comprehensively all possible informative parameters as with new techniques becoming available and research areas extending to new avenues, other or entirely new features might be of interest

Table 4.5.2 Categories of malignant haematopoietic cell lines

Main type	Physiological haematopoietic cell lineage	Subtype of neoplastic cell line	Prototype of cell line [a]
Lymphoid	B-cell	B-cell precursor cell line	REH
		B-cell line	U-698-M
		Plasma cell line	RPMI-8226
	T-cell	Immature T-cell line	CCRF-CEM
		Mature T-cell line	SKW-3
	NK cell	NK cell line	YT
Myeloid	Myelocytic	Myelocytic cell line	HL-60
		Promyelocytic cell line	NB4
		Eosinophilic cell line	EoL-1
		Basophilic cell line	KU-812
	Monocytic	Monocytic cell line	U-937
	Erythrocytic	Erythrocytic cell line [b]	K-562
	Megakaryocytic	Megakaryocytic cell line [b]	MEG-01
Hodgkin/ALCL	Lymphoid? Other?	Hodgkin cell line	L-428
	Lymphoid? Other?	ALCL cell line	SU-DHL-1
Dendritic	Lymphoid? Mono-myeloid?	Dendritic cell line [c]	—— [c]

Modified after Drexler *et al.* (1998) and Drexler & Matsuo (1999).
[a] The best known cell lines for each of these categories are often the 'oldest' cell lines (nearly all are available from major cell line banks).
[b] It is often difficult to assign cell lines to either the erythrocytic or megakaryocytic cell lineage as most of these cell lines express features of both lineages, e.g. (haemo)globin, specific transcription factors, surface antigens, differentiation potential, etc. Thus, it is preferable to use the term 'erythrocytic-megakaryocytic cell line'.
[c] At present, no continuous human dendritic cell line has been published.

Table 4.5.3 Cardinal requirements for new leukaemia cell lines

- Immortality
- Verification of neoplasticity
- Authentication
- Scientific significance
- Characterization
- Availability

Modified after Drexler *et al.* (1998) and Drexler & Matsuo (1999).

to scientists. Thus, only some of the features of the phenotypic profiles of cell lines which are most often studied are highlighted. It is also important to indicate when in the life of a cell line particular data were generated and also whether alterations in the phenotypic features of the cells might occur during prolonged culture.

5. While the scope and extent of the analytical characterization of leukaemia cell lines is certainly facultative, a core data set is obligatory and essential for the identification, description and culture of a cell line; these data include the clinical and cell culture description of the cell line (an example for the presumably most often used human leukaemia cell line HL-60 is given in Table 4.5.5).

6. Subclones of any given cell line and sister cell lines must be distinguished and properly described (Table 4.5.6).

Table 4.5.4 Analytical characterization of leukaemia cell lines

Parameter	Details and examples
Most important data	
Clinical data:	• Patient's data (see Table 4.5.5)
In vitro culture:	• Growth kinetics, proliferative characteristics (see Table 4.5.5)
Immunophenotyping:	• Surface marker antigens (fluorescence microscopy, flow cytometry)
	• Intracytoplasmic and nuclear antigens (immunoenzymatic staining)
Cytogenetics:	• Structural and numerical abnormalities
	• Specific chromosomal rearrangements
Further characterization	
Morphology	• *in-situ* (flask, plate) under inverted microscope
	• Light microscopy (May–Grünwald–Giemsa staining)
	• Electron microscopy (transmission and scanning)
Cytochemistry	• Acid phosphatase, α-naphthyl acetate esterase, others
Genotyping	• Southern blot analysis of T-cell receptor (TCR) and immunoglobulin (Ig) heavy and light chain gene rearrangements
	• Northern analysis of expression of TCR and Ig transcripts
Cytokines	• Production of cytokines
	• Expression of cytokine receptors
	• Response to cytokines, dependency on cytokines
Functional aspects/ specific features	• Phagocytosis
	• Antigen presentation
	• Immunoglobulin production/secretion
	• (Haemo)globin synthesis
	• Capacity for (spontaneous or induced) differentiation
	• Positivity for EBV or HTLV-I or other viruses
	• Heterotransplantability into mice or other animals
	• Colony formation in agar/methylcelluose – clonogenicity
	• Production/secretion of specific proteins
	• Natural killer cell activity
	• Oncogene expression
	• Transcription factor expression
	• Unique point mutations
Date of analysis	• Age of cell line at time of analysis
	• Possible changes in the specific marker profile during prolonged culture

Modified after Drexler *et al.* (1994, 1998) and Drexler & Matsuo (1999).

MAINTENANCE OF HUMAN LEUKAEMIA CELL LINES: BASIC PROTOCOL

Introduction

The protocol described below refers to leukaemia cell lines growing in standard culture media supplemented with FBS. Other cell lines which need special media and growth supplements can be maintained similarly with necessary changes of the culture system.

Table 4.5.5 Clinical and cell culture data for leukaemia cell lines

Parameter	Example
Cell lines	
Name of cell line	HL-60
Cell phenotype	Myelocytic cell
Clinical data	
Original disease of patient	Initially AML M3, later corrected to AML M2
Disease status	At diagnosis
Patient data (age, race, sex)	35-year-old Caucasian woman
Source of material	Peripheral blood
Year of establishment	1976
Cell culture data	
Culture medium	90% RPMI 1640 + 10% FBS
Subcultivation routine	Maintain at $0.1–0.5 \times 10^6$ cells ml^{-1}, split ratio 1 : 5 to 1 : 10 every 2–3 days
Minimum cell density	$0.5–1.0 \times 10^5$ cells ml^{-1}
Maximum cell density	$1.0–1.5 \times 10^6$ cells ml^{-1}
Doubling time	24–36 hours
Cell storage conditions	70% RPMI 1640 + 20% FBS + 10% DMSO
In situ morphology	Round, single cells in suspension
Mycoplasma contamination	None – checked with PCR
EBV status	Negative – checked by EBNA staining

Modified after Drexler *et al.* (1994, 1998) and Drexler & Matsuo (1999).

Table 4.5.6 Definition of subclones and sister cell lines

Subclone	Derived from an original (parental) cell line harbouring divergent and unique features
Sister cell line	
Simultaneous	Established from the same patient at the same time, but possibly from different sites or the primary sample was split into several aliquots prior to culture
Serial/longitudinal	Established from the same patient, but at different time points, e.g. at diagnosis and at relapse

Modified after Drexler *et al.* (1998) and Drexler & Matsuo (1999).

Materials

- Leukaemia cell line
- Standard culture medium (e.g. RPMI 1640, Iscove's Modified Dulbecco's Medium, α-MEM, or McCoy's 5A) and foetal bovine serum (FBS)
- 15 ml or 30 ml centrifuge tubes
- 50 ml or 260 ml flasks, 24-well plates or 96-well microplates
- Neubauer haematocytometer and Trypan blue solution
- 1 ml, 2 ml, 5 ml and 10 ml pipettes
- Centrifuge
- Waterbath at 37°C

All solutions and utensils coming into contact with cells must be properly sterilized prior to use and sterile techniques must be used throughout the procedure.

Procedure

1. The leukaemia cell lines might be obtained either as live cells or as frozen cells. Frozen cells must be thawed carefully in order to minimize cell loss.
2. Remove the frozen ampoule from liquid nitrogen, thaw the cells rapidly in a 37°C waterbath by gently shaking the ampoule in the water. It is important that the frozen cell solution be thawed in about one minute. Rapid warming is necessary so that the frozen cells pass quickly through the temperature zone between –50°C to 0°C where most cell damage is believed to occur. Slow thawing will harm the cells by formation of ice crystals in the cells causing hypertonicity and breakage of cellular organelles.
3. Wipe the ampoule with a tissue pre-wetted with 70% ethanol before the vial is opened. Transfer the cell suspension into a centrifuge tube using a pipette.
4. Dilute the cell suspension slowly by adding 10 ml culture medium plus 10% FBS to the tube, shake the tube gently: 1 ml per minute for 4 min, then twice 3 ml per minute. Cells frozen with DMSO are usually dehydrated. During the washing steps with medium, water will diffuse into the cells. Diluting the suspension slowly is supposed to reduce the loss of electrolytes, to counteract extreme pH changes and to prevent denaturation of cellular proteins.
5. Centrifuge the cells at $200 \times g$ for 10 min. Discard the supernatant.
6. Wash the cells again using another 10 ml culture medium plus 10% FBS by repeating steps 4 and 5. During the centrifugation determine the total cell number and the percentage of viable cells in the Neubauer haematocytometer with Trypan blue vital staining.
7. Adjust the cell concentration to $0.3–0.6 \times 10^6$ ml^{-1} with complete medium. Most leukaemia cell lines grow better at a higher concentration than at a lower one. The most often used concentration is about $0.3–0.6 \times 10^6$ ml^{-1}. Occasionally, some cell lines (e.g. some precursor B-cell lines) prefer a concentration higher than 1.0×10^6 ml^{-1}. Usually, the optimal concentration of a cell line for expansion will be recommended by the presenter of the cell line. If the cells do not grow well with many dead cells after 2–3 days of culture, try to culture the cells at a higher cell density by concentrating the cells. It is recommended that the cells be resuspended first in medium containing 20% FBS; should the cells start to multiply and resume their expected growth activity, the percentage of FBS can be decreased stepwise, e.g. 20% $\rightarrow$ 15% $\rightarrow$ 10% $\rightarrow$ 5% over a week or two.
8. Distribute the cell suspension to flasks or 24-well plates. Usually, add 5 ml suspension into a 50 ml flask, 20–30 ml suspension into a 260 ml flask, or 1 ml suspension into each well of a 24-well plate. When dealing with 'difficult' cell lines it may be advantagous to suspend some cells in a flask and another aliquot in a 24-well plate (or even a 96-well microplate). There are distinct differences between flask and plate regarding exposure to CO_2, accessability to microscopic observation and possibilities of manipulation.

9. Incubate the cells in a humidified 37°C incubator with 5% CO_2 in air. Loosen the top of the flask slightly to allow for free gaseous exchange into and out of the flask.

10. Feed the cells by exchanging half of the culture volume with culture medium plus FBS at 2–3-day intervals. Before changing the medium, set the flasks upright in the incubator for at least 30 min in order to let the cells sink to the bottom of the flask. Remove gently half of the spent medium from the flask using a 10 ml sterile pipette and then add the same volume of new complete medium into the flask. If the cells proliferate actively, the culture medium will soon change colour due to a pH change caused by cellular metabolism. In this case, it is necessary to change the medium more frequently. Should the cells have doubled, subdivide the cells from the original flask into a second flask by diluting the suspension 1 : 2 with new medium. When changing the medium, it is important to calculate the total cell number by determining the density as well as the viability of the cells using Trypan blue dye exclusion. A careful documentation of all manipulations, macroscopic and microscopic observations, intentional and accidental changes in the cellular conditions, and data on cell density, viability and total cell number at different time points is mandatory.

11. The cell lines can be maintained as long as required. The cells can be harvested at any time for different uses. If the cells proliferate more quickly than needed, keep the cell growth at a slower pace by decreasing the FBS to a lower percentage in the medium, changing the medium at longer intervals, or discarding a certain amount of the cells (up to 75%) during the exchange of medium. There is a fundamental difference between 'expansion' and 'maintenance' of a leukaemia cell line. Some cell lines will deteriorate over long-time culture in a maintenance schedule; then it might be of more advantage to freeze and rethaw the cells when needed. As culture flasks after longer usage will often contain a lot of unused ingredients of the medium, metabolized molecules and cell detritus, it is recommendable to change the plastic flask once every 1–2 months.

STORAGE OF HUMAN LEUKAEMIA CELL LINES: BASIC PROTOCOL

Introduction

It is generally assumed that leukaemia cell lines can be stored at −196°C in liquid nitrogen for more than 10 years without any significant changes in their biological features. The viable cell lines can be recovered at any time when needed. The traditional method for storing the cell lines is freezing the cells in liquid nitrogen with appropriate medium containing 20% FBS and 10% dimethylsulphoxide (DMSO) which can lower the freezing point in order to protect the frozen cells from damage caused by ice crystals. It should be mentioned that no single suspending medium and procedure will be ideal for processing and cryogenic storage of all cell cultures. However, the procedures recommended here are suitable for most of the human leukaemia cell lines. When the procedure is carried out properly, it seems to be compatible with prolonged preservation of viability and other characteristics of the cell lines.

Materials

- Leukaemia cell line
- RPMI 1640 medium and heat-inactivated foetal bovine serum (FBS)
- Dimethylsulphoxide (DMSO)
- 30 ml or 50 ml centrifuge tubes
- Freezing ampoules (plastic cryo vials)
- Neubauer haematocytometer and Trypan blue solution
- 1 ml, 2 ml and 5 ml pipettes
- Centrifuge
- Controlled rate freezer (cryo freezing system)
- Cryo Freezing Container (plastic cryo box)

Procedure

1. Harvest the suspension cells by transfer of the flask contents to 30 ml or 50 ml centrifuge tubes. It is important that the cells be harvested in the logarithmic growth phase. It must be remembered that freezing and storing the cells will not improve the status and quality of the cell culture prior to the freezing; at best, the status quo will be preserved, but most often will be diminished to various degrees.

2. Remove 50 μl cell suspension to a well of a 96-well plate, add 50 μl Trypan blue solution, mix well. Use about 8 μl of the mixed suspension to fill the haematocytometer counting chamber. Determine the total number of viable cells by counting the cell density as well as the viability of the cells under a light microscope.

3. Centrifuge the cells at $200 \times g$ for 10 min, discard the supernatant.

4. The freezing medium consists of RPMI 1640 medium containing 20% FBS and 10% DMSO. Adjust the cells to a concentration of 5×10^6 ml^{-1} using freshly prepared freezing medium. The 'freezing medium' should be added quickly to the cells. Long-time exposure to DMSO at room temperature can trigger significant cellular changes such as activation and so-called 'induced differentiation'. Keeping cells in DMSO-containing media on ice could minimize the effect of DMSO on the cells. It is not necessary to sterilize the DMSO solution as pure DMSO is lethal to bacteria.

5. Distribute the cells into freezing ampoules (plastic cryo vials) with 1 ml per ampoule, thus containing at least 5×10^6 cells. Seal the ampoules which have been properly labelled with the name of the cell line and the date of freezing; keep a written record. More cells per ampoule can be frozen if needed: depending on the cell type up to 50×10^6 ml ampoule^{-1}.

6. Place the sealed and labelled ampoules in a computer-controlled cooling apparatus (cryo freezing system). Set the controls to achieve a cooling rate of 1°C per minute. When the temperature reaches –25°C, the cooling rate can be

increased to 5–10°C per minute. When the temperature of the specimen reaches –100°C, the ampoules can be transferred quickly to a liquid nitrogen container for storage. Permanent storage should be in the liquid phase of the liquid nitrogen. Alternatively, if only a few ampoules are frozen, the sealed ampoules can be placed in a 'Nalgene™ Cryo 1°C Freezing Container' (Cat. No. 5100–0001) that contains 250 ml isopropanol and then put into a –70°C refrigerator for at least 4 h. The ampoules should later be transferred quickly to liquid nitrogen. With this 'Cryo 1°C Freezing Container', a 1°C per minute cooling rate can be achieved.

COMMENTARY

Background information

Establishment of leukaemia cell lines provides cells for a variety of practical applications: (i) to study the basic biology of leukaemia and lymphoma cells; (ii) to obtain insights into the origin, proliferation and differentiation of leukaemia cells; (iii) to better define the molecules that regulate the proliferation and differentiation of leukaemia cells; (iv) to develop culture assays on leukaemia cells for numerous purposes; and (v) to use leukaemia cell lines as *in vitro* model systems for therapeutical studies. In recent years, several hundred human leukaemia cell lines have been established (an educated guess would be in the range of 500–800 verified and authenticated cell lines, not counting EBV+ Burkitt lymphoma and HLTV-1+ T-leukaemia cell lines). These cell lines have contributed a significant wealth of information to the studies on leukaemia.

Although *in vivo* the leukaemia cells enjoy a selective growth advantage over normal haematopoietic cells, *in vitro* the leukaemia cells are so difficult to grow and to maintain that attempts to establish cell lines meet much more often with failure than with success. However, despite the fact that currently there is not a single cell culture system to consistently establish leukaemia cell lines, several methods for immortalizing leukaemia cells have been developed. The technique of seeding leukaemia cells in suspension cultures as described in this section is certainly the most often used. Other methods recommended by several researchers have their advantages and might meet with success in some attempts.

Growth of leukaemia cells in soft agar or methylcellulose offers the advantage that the colonies formed are well fixed and can be easily removed from the supporting medium using a Pasteur pipette for further culture in other environments. Thus, a cell line might be established by easily passaging single colonies.

Some lymphocytic leukaemia cells can be immortalized using transforming viruses. EBV can promote growth of malignant B-cell lines from some patients with chronic B-lymphocytic leukaemia (B-CLL). But the EBV can also transform normal B-lymphocytes. HTLV-1 and HTLV-2 allow the growth of malignant T-cells by inducing the IL-2 receptor. In consideration of the fact that EBV and HTLV can also transform normal cells, it is necessary to ascertain the leukaemic origin of the established cell lines by means of karyotype and molecular genetic analysis and, possibly, the demonstration of the absence of the virus genome.

The growth of leukaemia cells and normal haematopoietic cells *in vitro* and *in vivo* is the result of complex interactions between growth factors and their respective receptors. The addition of some factors into the culture medium can support the proliferation of the leukaemia cells and induce the formation of cell lines. The most often used molecules are EPO, G-CSF, GM-CSF, SCF, thrombopoietin (TPO), and several interleukins (IL-2, IL-3, IL-5, IL-6). Another protocol used insulin-like growth factor 1 (IGF-1). The IGF-1-induced cell proliferation appeared to be restricted to a low-oxygen environment and was blocked at high-oxygen concentrations. Malignant T-cell lines could be established frequently when the patients' samples were cultured in an environment of 6% CO_2, 5% O_2 and 89% N_2 with nutrient media supplemented with IGF-1.

As purified or recombinant factors are expensive, the CM of some malignant human cell lines can be used alternatively. Such cell lines are 5637 (an adherent cell line from a patient with bladder carcinoma), Mo-T (HTLV-2 transformed T-leukaemia cell line), WEHI-3 (mouse monocytic cell line) and HDLM-3 (Hodgkin's disease). These cell lines generate several growth factors (e.g. G-CSF, GM-CSF, IL-1, IL-3, IL-6, IL-9, SCF and others) that are secreted into the culture supernatant. The supernatant from cultures of these tumour cell lines can be stored at $-20°C$ for several months prior to use. These CM should be used at a final concentration of 10–20% (vol). CM from PHA-stimulated lymphocytes is also an ideal and inexpensive source of these biomodulators.

A number of completely synthetic media such as RPMI 1640, Minimum Essential Medium (Eagle's MEM or α-MEM), Dulbecco's Modified Eagle's Medium (DMEM), Iscove's Modified Dulbecco's Medium (IMDM), Ham's F-10 and F-12, L-15, McCoy's 5A, and others, including several media designed specifically for unique types of leukaemia cells, have been used by researchers for establishment and maintenance of leukaemia cell lines in suspension cultures. It appears that no single medium is well suited for the growth of all types of leukaemia cells. Although the most widely used of these media is RPMI 1640 medium which is usually employed together with 10–20% FBS, IMDM, McCoy's 5A, DMEM and α-MEM are also widely used. Should one medium fail to support the leukaemia cell growth, it might become necessary to try another kind of medium.

FBS is the standard supplement in the suspension culture system of leukaemia cell lines. It is commonly used in concentrations of 5–20%. Prior to usage, it is recommended that batches of serum be pretested for their ability to support the growth of the leukaemia cells and for viral (in particular for bovine viral diarrhoaea), mycoplasmal and other bacterial contamination. If possible, a large supply of the FBS from a pretested batch that supports cell growth well and has no contamination should be purchased and stored at $-20°C$ for future use. Alternatives are newborn calf serum (NCS, usually at only 25% of the price of the expensive FBS) or serum-free media. However, not all leukaemia cell lines will grow in NCS as well as in FBS. Although serum-free media do not provide monetary advantages, they do provide experimental advantages in comparison with FBS as researchers are able to control all the substances to which the cells are exposed; there are many unidentified ingredients at variable concentrations in bovine serum.

A minimal amount of oxygen is essential for the growth of most types of leukaemia cells in suspension cultures. Most leukaemia cells grow well when they are incubated in a humidified 37°C incubator with 5% CO_2 in air. However, the partial pressure of oxygen (pO_2) in the normal body fluids is significantly less than that of air. Studies have shown that the pO_2 in human bone marrow is 2–5%. This is considerably lower than the pO_2 (15–20%) existing in the typical cell culture incubator maintained at 5% CO_2 in air. It has been reported that growth of cultured leukaemia cells could be improved by reducing the percentage of oxygen in the gaseous phase to between 1% and 10%. Growing leukaemia cells under low-oxygen conditions of 6% CO_2, 5% O_2 and 89% N_2 may be a useful method for establishment of leukaemia cell lines.

Safety considerations

When working with human blood, fresh leukaemia cells, established leukaemia cell lines and pathogenic and infectious agents, the biosafety practices must be followed rigorously.

Troubleshooting

In the attempts to establish leukaemia cell lines, the overgrowth of fibroblasts and normal lymphoblastoid cells is the most common problem. Should the nutrients in the medium become exhausted too quickly, the adherent cells should be removed by passaging the suspension cells into new flasks containing fresh medium. The overgrowth of B-lymphocytes with EBV genome can become visible as early as 2 weeks after seeding of the new culture. The EBV+ B-lymphocytes look small and have irregular contours with some short villi. They prefer to proliferate in large floating clusters or colonies. The colonies can be picked out with a Pasteur pipette and the EBV genome should be detected as soon as possible.

It is always necessary to freeze aliquots (at least two ampoules) of the fresh leukaemia cells in liquid nitrogen before culture. If a cell line should subsequently become established, the original leukaemia cells can be used as control for characterization of the established leukaemia cell line. In case of failure to immortalize a cell line, the fresh leukaemia cells can be used for another attempt.

Maintenance of leukaemia cell lines requires careful work. Every leukaemia cell line has its suitable growth environment such as culture medium, cell density, nutrition supplements and pH. If cell growth becomes suboptimal or cells inexplicably die during culture, some of the following problems should be considered: suitability of the culture medium or the growth supplements for this particular cell line; proper functioning of the incubator at the appropriate temperature, humidity and CO_2 levels; and selection of the appropriate cell density. Some leukaemia cell lines clearly grow better in 24-well plates than in culture flasks.

Contamination with mycoplasma, other bacteria, fungus, viruses and other 'foreign' cells is the most common problem encountered in the maintenance of leukaemia cell lines. Therefore, analyses at regular intervals must be undertaken to ensure a contamination-free environment for cell growth. All solutions and utensils coming into contact with cells must be sterilized prior to use; sterile techniques and good laboratory practices must be followed strictly. Although antibiotics can be added to the culture medium to prevent bacterial infection, they do not usually inhibit virus, fungus or mycoplasma infection. Therefore, antibiotics such as penicillin and streptomycin are not absolutely necessary if care is taken regarding cell culture techniques. Because contamination can cause the loss of valuable cell lines, it is important to cryopreserve a sufficient amount of cells from each cell line for future use. In general, 5×10^6 cells ampoule^{-1} is adequate. In order to prevent cellular contamination and misidentification, it is mandatory to use a separate bottle of medium for each cell line. Furthermore, cell culturists should not deal with and feed more than one cell line at the same time.

Time considerations

Generation of a leukaemia cell line takes 2–4 months. However, a cell culture should not be considered a continuous cell line until the cells had been passaged and expanded for at least half a year – better, one year. Expansion of the cell line takes 2–3 weeks. Depending on the parameters analysed, 2–4 months might be needed for a thorough characterization of the established cell line.

BACKGROUND READING

Baserga R (ed.) (1989) *Cell Growth and Division: A Practical Approach*. IRL Press, Oxford.

Drexler HG, Gignac SM & Minowada J (1994) Hematopoietic cell lines. In: Hay RJ, Gazdar A & Park JG (eds) *Atlas of Human Tumor Cell Lines*, pp. 213–250. Academic Press, San Diego.

Drexler HG & Minowada J (1998) History and classification of human leukemia-lymphoma cell lines. *Leukemia Lymphoma* 31: 305–316.

Drexler HG, Matsuo Y & Minowada J (1998) Proposals for the characterization and description of new human leukemia-lymphoma cell lines. *Human Cell* 11: 51–60.

Drexler HG & Minowada J (in press) Human leukemia-lymphoma cell lines: Historical perspective, state of the art and future prospects. In: Masters JRW & Palsson BO (eds) *Cancer Continuous Cell Lines: Leukaemias and Lymphomas*. Kluwer Academic, Dordrecht.

Drexler HG & Uphoff CC (in press) Mycoplasma contamination of cell cultures. In: Spier E *et al.* (eds) *The Encyclopedia of Cell Technology*. Wiley, New York.

Drexler HG & Matsuo Y (1999) Guidelines for the characterization and publication of human malignant hematopoietic cell lines. *Leukemia* 13: 835–842.

Jones GE (ed.) (1996) *Human Cell Culture*. Humana Press, Totowa, New Jersey.

Lange B, Valtieri M, Santoli D, Caracciolo D, Mavilio F, Gemperlein I, Griffin C, Emanuel B, Finan J, Nowell P & Rovera G (1987) Growth factor requirements of childhood acute leukemia: Establishment of GM-CSF-dependent cell lines. *Blood* 70: 192–199.

Matsuo Y & Drexler HG (1998) Establishment and characterization of human B cell precursor-leukemia cell lines. *Leukemia Research* 22: 567–579.

Smith SD, McFall P, Morgan R, Link M, Hecht F, Leary M & Sklar J (1989) Long time growth of malignant thymocytes in vitro. *Blood* 73: 2182–2187.

Stong RC, Korsmeyer ST, Parkin JL, Arthur DC & Kersey JH (1985) Human acute leukemia cell line with the t(4;11) chromosomal rearrangement exhibits B lineage and monocytic characteristics. *Blood* 65: 21–31.

4.6 DERVIATION AND SUBCULTURING OF HUMAN DIPLOID CELL STRAINS

The cultivation of human diploid cell strains from primary tissue does not differ from the traditional techniques used to initiate any monolayer culture from intact tissue. Contrary to popular misconception, there are no 'tricks' necessary to maintain normal human cells in the diploid state providing, of course, that the tissue has been obtained from a normal donor. Indeed, the 'trick' is how to transform cells *in vitro* to cells having, among other properties, a heteroploid karyology. These concepts have been discussed extensively elsewhere (Hayflick & Moorhead 1961; Hayflick 1965, 1980, 1984, 1987, 1989).

PROCEDURE: DERIVATION AND SUBCULTURING

Materials

- Eagle's basal medium

Note: The medium should be prewarmed to 37°C before use. The final pH should be 7.2 before serum addition. The pH of the medium after equilibration of the culture at 37°C must be less than 7.4.

- 10% foetal or adult bovine serum
- Trypsin – 0.25% in phosphate-buffered saline (PBS) or any balanced salt solution. Final pH of the trypsin solution **must** be at least 7.4.
- Human Tissue (see 4.1 for ethical issues)

Initiation of cultures

1. Mince the human foetal tissue with paired scissors or scalpels, under aseptic conditions, in a Petri dish containing a minimum amount of growth medium. Appropriate antibiotics may be used.
2. Transfer the resulting macerated tissue to an Erlenmeyer flask containing sufficient trypsin to freely suspend the minced tissue mass.
3. Incubate the suspension at 37°C and stir constantly using a magnetic mixer. At approximately 1 h intervals, stop the mixing, allow large fragments to settle and decant the supernatant. Repeat this procedure until the entire tissue mass is 'digested'. Immediately after each collection, centrifuge the supernatant trypsin fluid at low speed (1000–5000 g), discard the trypsin and resuspend the

Cell and Tissue Culture for Medical Research, edited by A. Doyle and J.B. Griffiths.
© 2000 John Wiley & Sons, Ltd

cell pellet in culture medium in an appropriate vessel. An inoculation density of the order of 10^5 viable cells/ml is optimum.

4. Incubate the cultures at 37°C and feed at 3–4-day intervals until a confluent monolayer is formed at which time subcultivations can be made.

Subcultivations

1. Decant all the growth medium. The few drops remaining on the culture will not interfere with the procedure.
2. Add enough prewarmed (37°C) trypsin to cover the sheet of a culture just removed from the incubator. (The trypsin action is fastest in the alkaline range and at 37°C.) Allow the culture to stand at room temperature for 1 min.
3. Decant all the trypsin or, if many cultures are being processed, invert the culture vessels so that the trypsin is on the 'floor' and the monolayer on the 'ceiling'. Decant the trypsin from the inverted vessels and return to their original position (monolayer down).
4. Allow the culture to stand for 5–20 min longer at room temperature or at 37°C with the residual trypsin solution remaining on the cell monolayer. Tilt the vessel occasionally to ensure complete trypsin coverage of the monolayer.

   The trypsinization process is completed when the cell sheet is loosened from the glass surface. This can be seen macroscopically by holding the bottle up to the light in a vertical position and observing the cell sheet sloughing off the glass surface. If the cell sheet is still adherent to the vessel surface, dissociate by slapping your hand against a hard surface while holding the vessel. The sloughing process may be monitored microscopically.
5. After the cells have sloughed, add a small amount of growth medium and gently splash it over the cell sheet. All the cells should be removed. Using a pipette, vigorously aspirate the medium plus the cells back and forth onto the vessel surface to remove all attached cells. It is essential that this aspiration be done as completely as possible with a small-bore 5 or 10 ml pipette so as to obtain mostly single dispersed cells. This is one of the most crucial steps, for if the cells are not dispersed completely, the new culture will contain numerous microcolonies or 'explants' that will ultimately lead to an apparently premature senescence of the culture.
6. Add sufficient fresh medium to the aspirated suspension so that the total volume will cover the surface of the two daughter vessels; each having the same surface area as the mother vessel (or use a single vessel having twice the floor area of the original vessel). This is called a 1 : 2 split. The mother vessel may be reused without washing.
7. Incubate at 37°C. No intervening culture feedings are necessary. The cell type that predominates after a few subcultivations is the fibroblast.
8. Subcultivation of 1 : 2 split cultures should be done on a rigid 3- or 4-day schedule. By the end of each culture period confluent cell sheets should have been obtained. Surplus cells can be preserved cryogenically.

   Note: The pH of the medium is of extreme importance. The final pH of the medium must not exceed 7.4 after equilibration of the culture at 37°C. A higher pH may result if: too few cells are contained in the culture; the original pH of the medium is too high; or, and most importantly, the gas phase of the culture vessel is too large. If the last is unavoidable, it is essential that the Eagle's medium be prepared in Hanks' balanced salt solution or in 28 μM HEPES buffer. Alternatively, use a CO_2 incubator which will obviate the need to control the pH by adjusting medium conditions.

9. Mark the population doubling level on the vessel. It is critical to note that the passage level of the culture is equivalent to the population doubling level only when the split ratio is 1 : 2.

A passage is the physical removal of cells from one vessel to another. The term has no quantitative value. Population doubling level (PDL) is the only accurate and scientifically valid designation for the point at which the cell population is located as it traverses its fixed lifespan. The PDL is increased by one for each 1 : 2 split. If other split ratios are used, e.g. 1 : 4 once per week, the PDL increases by two for each weekly passage. For production of large quantities of cells for various purposes, make repeated 1 : 2 splits twice a week. The number of culture vessels will increase geometrically (1, 2, 4, 8, 16, 32, 64 etc.) in a short period of time.

A human diploid embryonic cell strain has a population-doubling potential, or limited *in vitro* lifetime of about 50 PDs if a 1 : 2 subcultivation ratio is used. If 1 : 4 split ratios are used then 25 passages or subcultivations will occur but the result will still be a maximum of 50 population doublings. At this time the cells will stop dividing and eventually die, a phenomenon that was named the phase III phenomenon (Hayflick & Moorhead 1961).

Cell populations derived from normal adult tissues have a population doubling limit less than that found for human embryos. The population doubling limit is found to be reduced and proportional to the age of the donor (Hayflick 1965, 1987, 1989).

Although the diploid cell strain will be lost as a continuously passaged population, it will not be lost for use since frozen ampoules can be preserved at almost every population doubling. Thus the strain can be restored to continuous culture again, up to a cumulative total of 50 PDs. By repeating these procedures, the numbers of cells that can be obtained are virtually unlimited for all practical purposes.

Using split ratios higher than 1 : 2 results in the advantage of minimizing the number of manipulations necessary to obtain a specific cell density or the number of culture vessels. Human diploid cell strains pass through a finite number of population doublings *in vitro*. It is therefore necessary to keep a record of the number of population doublings that have elapsed. With a 1 : 2 split ratio this is achieved by simply adding '1' to each split since this ratio yields one population doubling although larger splits ratios can be used. For example, a split ratio of 1 : 4 would yield two population doublings per 1 : 4 split and a 1 : 10 split ratio would yield 3.25 population doublings per 1 : 10 split. In order to have knowledge of the approach of phase III it is essential to keep records of the number of elapsed population doublings.

The frequent use of the term 'cell doublings' in lieu of population doublings is erroneous. The number of individual cell doublings that occur during a population doubling is almost impossible to determine. For example, if a culture vessel holds at confluency one million cells, and if a $1:2$ split is made, then 5×10^5 cells will be introduced into each of two daughter vessels. At confluency each of the two daughter vessels now contains one million cells. The population has doubled once in each vessel. Nothing can be said about the number of 'cell' doublings, which can vary between two extremes. First, it is possible that only one of the cells in the 5×10^5 introduced into the vessel divided rapidly to produce a clone of 5×10^5 cells, while the reamining 5×10^5 cells do not divide. The second extreme possibility is that all the 5×10^5 cells introduced into the vessel divide once. In both extreme cases the yield is 1×10^6 cells. The actual events that transpire clearly fall somewhere between these two extremes, making it virtually impossible to measure 'cell doublings'.

THEORY OF POPULATION INCREASE BY SUBCULTIVATION

Since human diploid cells multiply by fission, the increase in population numbers may be expressed as follows:

1		2		4		8		16 ... number of cells
0		1		2		3		4 ... n (number of generations)

one   one   one   one
generation generation generation generation

Expressed exponentially, the population after n generations is 2^n per cell in the inoculum, or total population N is the initial population, X_0, multiplied by 2^n or:

$$N = X_0 2^n \tag{4.6.1}$$

The data needed to determine the number of generations, n, will be the number of cells per unit volume in the inoculum, X_0, at time t_0, and the final population, N, at time $= t_2$. The number of generations, n, can be most readily evaluated by expressing equation 4.6.1 in logarithmic form. Using logarithms to the base 10, this equation becomes:

$$\log N - \log X_0 + n \log 2 \tag{4.6.2}$$

or, rearranging:

$$n = (\log N - \log X_0)/\log 2 \tag{4.6.3}$$

$$\text{Since } \log_e 2 = 0.301: n = 3.32 (\log N - \log X_0) \tag{4.6.4}$$

Logarithms to the base 2 should be used for biological systems because an increase of one logarithmic unit corresponds to one doubling or one generation. If this is done, the log 2 drops out of the denominator of equation 4.6.3. Natural logarithms (base e, written ln) may also be used. Regardless of the base of the logarithms used, the equation will take the same form, and conversion from one form to another can be made by multiplying by a constant, for example:

$$\log 2 = 0.301 \text{ or } 1/\log 2 = 3.32, \text{ so } \log_2 = 3.32 \log_{10}$$

The multiplication rate, r, or number of generations per unit time can be obtained or for equation 4.6.4 by dividing by the time interval between inoculation, t_0, and the time at which the final population, N, was taken, i.e. t_2. Therefore, the multiplication rate, r, is

$$r = N/(t_2 - t_0) \qquad (4.6.5)$$

or

$$r = 3.32 \; (\log N - \log X_0)/t_2 - t_0 \qquad (4.6.6)$$

Units used in cell culture systems must be specified: both r, the base of the logarithms used, and units of time, usually in units of 24 hours, i.e. doublings in population per 24 hours.

To write equations 4.6.2 and 4.6.5 in general form, the number of generations or multiplication rate over any interval in which the initial count is X_1 at any selected time t_1, and the final count X_2, at time t_2, may be determined thus:

$$n = (\log X_2 - \log X_1)/\log 2 = 3.32(\log X_2 - \log X_1) \qquad (4.6.7)$$

or

$$n = 3.32 \log X_2/X_1 \quad \text{or} \quad n = (\log X_2/\log X_1) \qquad (4.6.8)$$

and

$$r = (3.32 \log X_2/X_1)/t_2 - t_1 \qquad (4.6.9)$$

Since the generation time, g, is the time for the population to double, it is the reciprocal of the doubling, per unit time:

$$g = 1/r \qquad (4.6.10)$$

generation time = time elapsed per doubling in number.

REFERENCES

Hayflick L & Moorhead PS (1961) The serial cultivation of human diploid cell strains. *Experimental Cell Research* 25: 585–621.

Hayflick L (1965) The limited *in vitro* lifetime of human diploid cell strains. *Experimental Cell Research* 37: 614–636.

Hayflick L (1980) The cell biology of human aging. *Scientific American* 242: 58–66.

Hayflick L (1984) The coming of age of WI-38. In: Maramorosch K (ed.) *Advances in Cell Culture*, vol. 3, pp. 303–316. Academic Press, New York.

Hayflick L (1987) Aging in cultured human cells. In: Butler R & Kent B (eds) *Human Aging Research, Concepts and Techniques*, vol. 34, pp. 133–148. Raven Press, New York.

Hayflick L (1989) Antecedents of cell aging research. *Experimental Gerontology* 24: 355–367.

4.7 RAPID INITIATION OF ADHERENT MONOLAYER CELL CULTURES: THE 'COLD COCKTAIL' METHOD

INTRODUCTION

Detachment of firmly adherent cells during routine subcultivation of cell lines is normally accomplished enzymatically by short-term incubation in buffer-containing trypsin; purely mechanical cell detachment being usually reserved for certain trypsin-resistant cell lines or for terminal cell harvests where speed may be of the essence. In contrast, for the initiation of cultures from solid tumours, even though many of these contain fibrous stroma and therefore resist mechanical disruption (Freshney 1987), mechanical means of disruption are used more often than incubation with enzymes, usually either collagenase or trypsin, used singly. The use of enzyme-combinations (cocktails) for disruption of solid tissues or tumours is but infrequently cited in publications describing the initiation of cell cultures.

This seeming lack of interest in enzymatic methods for tissue disruption probably reflects gaps in the relevant literature giving rise to uncertainty regarding the net advantages of procedures which might risk toxicity – by damage of the cell membrane, or selectivity, by favouring growth of unwanted stromal cells at the expense of desirable tumour or differentiated cells. Furthermore, the requirement for continuously available supplies of sterile enzymes and buffers, given that human diagnostic samples, in particular, tend to arrive at the culture laboratory with little or no prior warning, might be regarded a significant drawback. In addition, the overwhelming variety of possible enzyme cocktails tailored to each tissue / tumour/ species combination has discouraged the collection of such data in anything appoaching a systematic or complete way, fostering reasonable doubts regarding the applicability of analogous protocols developed using other tissues or species.

Given the lack of controlled experimentation with enzyme cocktails, it is scarcely surprising that evidence for the systematic selection of specific types of cell by diverse enzyme treatments is patchy and conflicting. The success of standardized enzyme protocols suited to certain types of cell, e.g. collagenase perfusion widely used for establishing liver hepatocyte cultures (Seglen 1975) illustrates how the possible selective effects of enzyme treatments to bring otherwise reluctant types of cell into growth in culture may be exploited.

The ability to offset the selective effects of different enzymes present in cocktails against one another is a further area which has received undeservedly little attention. It is sometimes claimed that disruption by enzymes, when used singly,

Cell and Tissue Culture for Medical Research, edited by A. Doyle and J.B. Griffiths.
© 2000 John Wiley & Sons, Ltd

may lead to unwanted selection of particular cell types in mixed biopsies. Thus collagenase, while effective against both human skin fibroblasts and internal soft tissue, is unsuited to disaggregating many epithelial cells; while trypsin, while reasonably effective with some normal tissue, has been reported to exhibit selective toxicity against tumour cells. Thus, disaggregation by trypsin alone may lead to reduced cell viabilities when compared to that by cocktails which include both trypsin and collagenase (Ensley *et al.* 1987; Cerra *et al.* 1990), as recommended in the present protocol.

Improved release and preservation of differentiated cells has been demonstrated for a long-term enzymic disruption method using trypsin, previously described for use with mouse embryos (Cole and Paul 1966; Freshney 1987). The present protocol has adopted the key feature of this method which involves soaking minced tissue in enzyme overnight at 4°C. This allows thorough tissue penetration in the absence of enzymic activity – unlike conventional short-term enzyme dissociations at 37°C. Next day the cocktail containing still intact pieces tissue is briefly incubated at 37°C, thereby activating the enzymes which have by then deeply infiltrated the tissue – thus effecting a faster and more productive disaggregation. The ensuing mass release of cells more than offsets problems occasioned by the rather low plating efficiencies characteristic of primary cells. Hence, the main potential advantages of enzyme-based over purely mechanical methods of tissue disruption are simultaneous gains in efficiency and rapidity of initiation and in throughput. Traditional methods which rely on the outgrowth of explanted tissue fragments are relatively slow and unpredictable which, for those wishing to obtain larger cell numbers as quickly as possible (e.g. for diagnostic procedures) is a significant drawback.

This *cold cocktail* method has been used to disrupt human diagnostic tumour biopsies, and both human and rodent skin as well as human embryonic trophoblast tissue, where it has been used as an aid to prenatal genetic diagnosis. In all these systems it led to higher yields of viable cells and more productive cell cultures when compared to both standard explant or orthodox *warm* enzyme disruption techniques. Awkward or time-consuming procedures and non-standard apparatus have been avoided, rendering the technique equally suitable for routine or occasional use. This and similar cocktails have been shown to be effective in dissociating human solid tumours (Ensley *et al.* 1987; Cerra *et al.* 1990), so the cold cocktail method should be able to accommodate a wider range of species or tissues. (Some other cocktail recipes are suggested in the section on alternative protocols.)

MATERIALS

Required for initial cutting (day 1)

- Two scalpels with supply of large curved blades
- Petri dishes (plastic)
- Glass beaker containing sterilized distilled water (for cooling blades after flaming)
- Centrifuge tubes 10–15 ml conical (gamma irradiated with screwtops)

Required additionally for setting up cultures (day 2)

- Cell-counting chamber, Neubauer or equivalent
- Trypan blue (Sigma) 0.4%
- Micropipette 200 μl size and tips
- Multiwell plates, 96 wells (for mixing cells and trypan blue)
- Tissue culture flasks 25 or 75 cm^2

Additional items

- Multiwell plates, 24 wells (for troubleshooting)
- Dissection microscope (optional)

REAGENTS, SOLUTIONS AND MEDIA

All solutions and dissecting instruments should be autoclaved before use. It is important that all non-autoclavable solutions such as enzymes be sterilized using 0.2 μM pore filters. All concentrations refer to those of the final solutions unless indicated otherwise.

Disruption buffer

- Earle's Balanced Salt Solution (EBSS) (w/o phenol red, Ca^{2+}, Mg^{2+} or $NaHCO_3$). Either purchase commercially from any reputable supplier or make up to 963 ml in milli-Q deionized water the following, and autoclave.
- $CaCl_2.2H_2O$ (0.147 g)
- $MgSO_4$ anhydrous (0.0977 g)
- KCl (0.400 g)
- NaCl (6.800 g)
- NaH_2PO_4 anhydrous (0.122 g)
- D-Glucose (1.000 g)

Add 20 ml 1M sterile HEPES buffer solution (Flow ICN) (20 mM) and 17 ml of standard 7.5% $NaHCO_3$ (15 mM). Store for up to 6 months refrigerated in 100 ml aliquots. (Aliquot(s) in use may be kept at room temperature to facilitate detection of microbial contamination.)

Transport medium (if required for clinical material)

- Disruption buffer (or any suitable cell culture medium containing HEPES buffer)
- Supplemented with double-strength penicillin (5000 U ml^{-1}) and streptomycin (10 mg ml^{-1}) (Sigma) 20 ml l^{-1}.
- Store refrigerated for up to 2 months 15 ml aliquots in dated, plastic *universal* containers labelled 'pathological specimen' with instructions for refrigerated storage and immediate despatch.
- Specimen bag with separate pocket for paperwork

Stock enzyme solutions (times concentrated)

- Trypsin: (Difco 1 : 250) 0.5% (2 ×)
- Collagenase: (Sigma IV) 2000 U ml⁻¹ (2 ×)
- Make up each enzyme separately in disruption buffer, filter-sterilize and aliquot × 1 ml.
- DNase I (Roche) 0.1% (50 ×)
- Dissolve the lyophilized powder gently (without vortexing) in a storage buffer comprising:
- Glycerol 50% w/v
- MgCl₂ 1 mM, dissolved in milli-Q water containing
- Tris-Cl 20 mM, pH 7.5
- Filter-sterilize and aliquot × 100 μl.

All enzymes may be stored at (–20°C) up to 6 months.

Enzyme cocktail 'A' (required on day 1)

- 1 ml stock trypsin solution
- 1 ml stock collagenase solution

Freshly thaw frozen aliquots on crushed ice and keep chilled.

Enzyme cocktail 'B' (required on day 2)

- 2.5 ml stock trypsin solution
- 2.5 ml stock collagenase solution
- 100 μl stock DNase solution

Culture medium

Choose either a 'rich' medium – e.g. McCoy's 5A/Ham's F12 – or a specialized medium – e.g. MCDB 110 in the case of human fibroblasts. Supplement with 10–20% foetal calf serum, obtained from a reputable supplier which allows batch testing prior to bulk purchase.

PROCEDURE (SEE FIGURE 4.7.1)

Sterile techniques should be employed throughout: thus all steps during which the tissue/cells are exposed should be performed in a laminar flow cabinet (S2 or equivalent).

Safety

It may be literally vital that investigators wishing to establish cultures should consider the likelihood that human biopsy material may harbour infectious pathogens – particularly retroviruses. There is always the suspicion that

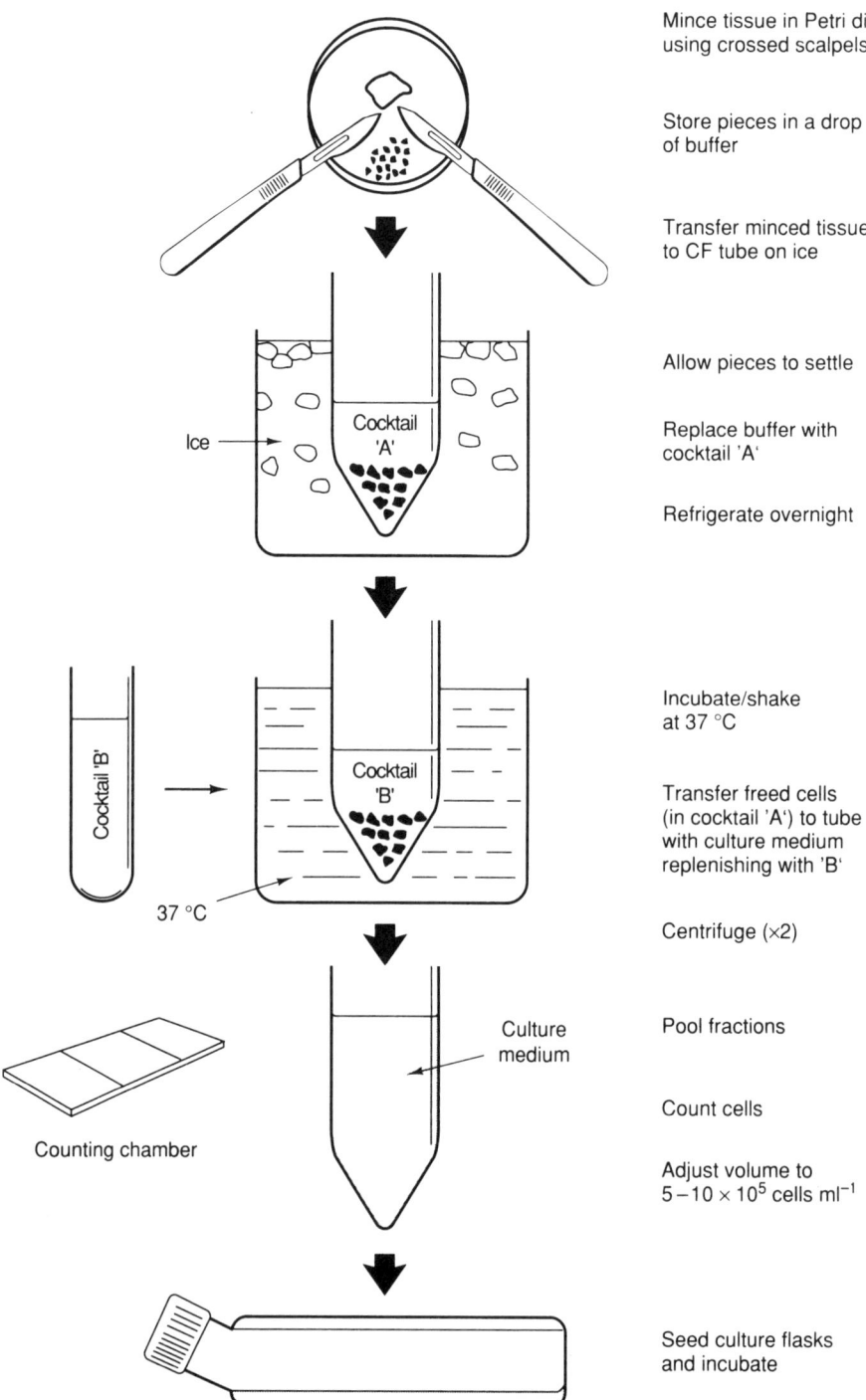

Mince tissue in Petri dish
using crossed scalpels

Store pieces in a drop
of buffer

Transfer minced tissue
to CF tube on ice

Allow pieces to settle

Replace buffer with
cocktail 'A'

Refrigerate overnight

Incubate/shake
at 37 °C

Transfer freed cells
(in cocktail 'A') to tube
with culture medium
replenishing with 'B'

Centrifuge (×2)

Pool fractions

Count cells

Adjust volume to
$5-10 \times 10^5$ cells ml^{-1}

Seed culture flasks
and incubate

Figure 4.7.1 Enzymic disruption procedure.

post-mortem samples (which form a disproportionately large number of human cell culture initiations) may be hazardous. Staff handling such material should be made aware of the risks involved and trained accordingly. A strategically placed reminder of the possible dangers (e.g. attached to a flow cabinet) may be a wise precaution.

Ethics and legality

A second important consideration regards the ethics or legality of any procedures involving human or animal material; and the necessity for obtaining appropriate certification beforehand. Visiting investigators should acquaint themselves with any local rules which may apply.

Preparation (day 1)

1. First obtain your sample! External surfaces coming in contact with or forming part of the biopsy should be swabbed with 70% ethanol.
2. Pre-wash the biopsy in disruption buffer ($\times 3$ with agitation).
3. (Optional) If contamination is suspected, the biopsy may be pre-incubated in disruption buffer containing quadruple strength pen/strep for ~ 1 h.
4. Meanwhile prepare/thaw enzyme cocktail 'A' and hold on ice.
5. Transfer sample into a plastic Petri dish lid with a few drops of disruption buffer and briefly examine under a dissecting microscope (if available) noting its size and general condition. (If desired the specimen may be weighed at this stage to facilitate determination of relative cell yield.)

Cutting and mincing (day 1)

6. Separate biopsy from medium on the Petri dish lid as 'dry' cutting is easier enabling the lid to be used as a 'chopping board' at the rather acute angles necessary within the flow cabinet.
7. Using a pair of scalpels (or a scalpel and a dissecting needle) as a 'knife and fork' mince biopsy cleanly into small (~ 1–2 mm) pieces. Dead or fatty material should be discarded. Instruments should be flamed periodically and cooled in the beaker containing sterile distilled water.
8. During cutting, small pieces may be transferred to and held in the drop of disruption buffer while the uncut larger piece(s) are dissected.

Cold enzyme treatment (day 1)

9. On completion, transfer minced tissue in disruption buffer to a conical centrifuge tube on ice and allow to settle.
10. After 5 min chilling remove disruption buffer and replace with 1–2 ml ice-cold enzyme cocktail 'A' (according to the size of the biopsy) keeping tube on ice the whole time. Centrifugation is not usually needed as the tissue should quickly settle.
11. Refrigerate overnight on ice (i.e. ~ 15–20 h).

Warm enzyme treatment (day 2)

12. Next day prepare:
    - ~ 5 ml enzyme cocktail 'B'
    - ~ 100 ml complete culture medium with serum (*exact volume according to number of living cells liberated/cultures desired*)
    - shaking water bath set at 37°C
13. Transfer tube with minced tissue to water bath and shake vigorously at 37°C.

Washing to remove enzyme (day 2)

14. After a few minutes the enzyme cocktail will become opaque, indicating release of cells. Transfer this supernatant to a tube of culture medium, and replenish the tube containing tissue with cocktail 'B' before returning to waterbath.
15. Centrifuge the supernatant containing the dispersed cells in culture medium (5 min, $200 \times g$) twice, inverting tube each time to remove traces of enzyme.
16. Resuspend washed cells in tube containing culture medium and hold at room temperature.
17. Meanwhile repeat steps 12 to 14 above, pooling all cells after final wash, until further incubation fails to release cells in quantity. (*Experience shows that little benefit is gained in total warm enzyme incubation times exceeding 40 min.*)
18. Thus any remaining small pieces of resistant tissue may be collected and washed as above; then pooled with the final sample.

Setting up cultures (day 2)

19. Using a micropipette, withdraw 25–50 μl suspension, mix with an equal amount of Trypan blue solution in the multiwell plate and count both the total number of cells and the percentage viability.
20. Adjust cell concentration according to the expected plating efficiency of the released cells if known. When this is unknown a safe seeding density of $5–10 \times 10^5$ cells per ml (approx $1–2 \times 10^5$ cells per cm^2) is recommended.
21. Set up cultures in small or larger flasks according to the number of living cells released and transfer to a 37°C incubator, which should be humidified and supplied with CO_2 when bicarbonate buffered mediums are used.
22. At the first medium change any non-adherent cells may be conveniently removed; (or sooner if present in large numbers).

Mycoplasma testing

It is advised that all primary cultures are checked regularly for possible mycoplasma contamination using DAPI and agar culture detection methods and, if the required facilities are available, by using PCR or DNA-RNA hybridization. While it is clearly preferable that cultures be carried out in antibiotic-free medium, antibiotics may be necessary due to contamination of the specimen prior to receipt. In such cases testing for mycoplasma may yield false negative results due to suppression of the infection.

TROUBLESHOOTING

Problems tend to fall into one of the three classes listed in Table 4.7.1. Effective problem solving, which is contingent upon correct diagnosis, may be assisted by microscopic observation of test disruptions performed in multi-well plates, e.g. when trying novel cocktails.

ALTERNATIVE PROTOCOLS

Should the foregoing protocol fail to achieve effective disruption while maintaining an adequate level of cell viability/growth, it recommended that the modifications listed below be tested.

Alternative cocktails

While the foregoing protocol was developed to carry out genetic/biochemical studies in primary human cell cultures it has been also tested with murine epidermal fibroblasts; and it should accommodate other cells/species by the simple expedient of altering the strength or composition of the enzyme cocktail. Some alternative cocktails are listed in Table 4.7.2 together with species/tissues which have

Table 4.7.1 Diagnosis and problem solving

Potential Problem	Possible causes	Suggested solutions
Failure to achieve disruption	Weak cocktail	Increase strength of trypsin and/or collagenase
	Cocktail unsuitable	Try different cocktail [a]
	Presence of serum/ inhibitory ions	Wash biopsy before chopping
	Biopsy contains epithelia	Include ionic chelator [a]
	Cells trapped in clump of DNA	Increase amount of DNase
		Add cocktail 'B' immediately after overnight cold step
Low viability of cells after disruption	Cocktail too potent	Reduce amount of trypsin and/or collagenase
	Unsuitable cocktail	Try different cocktail [a]
	Collagenase contaminated with clostripain	Obtain new batch
		Inhibit clostripain with TLCM [a]
	Incubation time(s) too long	Reduce overnight incubation to 6 h and/or warm incubation to 15 min
Failure of viable cells to proliferate	Cells obtained from terminally differentiated / senescent tissue	No straightforward solution
	Cells are growth factor dependent	Supplement with growth factor(s) Supply feeder layer
	Cells fail to attach	Coat flask with attachment factor [a]
	Unsuitable culture medium	Try alternative medium

[a] See section on alternative protocols for details.

Table 4.7.2 Some alternative cocktails

Enzymes	Concentration	Buffers	Species/tissues
Collagenase II DNase I	• 1200 U ml^{-1} • 6 U ml^{-1}	PBS/EBSS (w/o Ca^{2+}, Mg^{2+})	Human solid tumours
Collagenase I Hyaluronidase I	• 1500–3000 U ml^{-1} • 0.1–0.2%	EBSS (w/o Ca^{2+}, Mg^{2+})	Rat liver
Pancreatin	• 0.1–0.2%	EBSS +EDTA 2mM (w/o Ca^{2+}, Mg^{2+})	Mammalian epithelia
Trypsin	• 0.1–0.3%		
Collagenase Pronase	• 0.025% • 0.05%	PBS w/o Ca^{2+}/Mg^{2+}	Human tumour Xenografts
DNaseI Dispase and collagenase	• 0.04% – 0.1%	PBS (w/o Ca^{2+}, Mg^{2+})	See product leaflet for details

responded well to given combinations. Constraints of space have severely restricted the number of permissible citations. (*Note that trypsin-free cocktails may permit inclusion of FBS in the disruption buffer.*)

Inhibition of clostripain

Even some batches of purified collagenase may contain clostripain the presence of which may lead to loss of viability. To inhibit clostripain present in collagenase, add TLCM (tosyl-L-lysyl-chlormethane, 50 nM) (Calbiochem, San Diego CA) to the collagenase prior to aliquotting (Hefley *et al.* 1983; Sitar *et al.* 1987). (*Note*: TLCM tends to inhibit trypsin).

Improved disruption of epithelial tissue

Addition of cationic chelators (e.g. EDTA or EGTA) undoubtedly assists in the disruption of intractable epithelia. To promote selective disruption of epithelial cells cocktail 'B' should be supplemented with 1 mM EGTA. This action may be further increased by addition of EDTA to buffer 'A' – though this may interfere with the adherent capacity of the cells. (Note that significant quantities of EDTA (> 0.1 μM) will inhibit the action of DNase which requires Mg^{2+}.)

Should toxicity be a persistent problem it may be found that the viability of epithelial cells is improved after disruption by cocktails which omit collagenase.

Attachment factors

The failure of apparently viable cells to attach and yield active cultures (see Table 4.7.1) may be counteracted by the use of attachment factors. Some such as poly-lysine (D- or L- forms), vitronectin, gelatin or collagen type I may be used for a wide variety of cell types; whereas others may favour attachment of particular types, including: laminin (epithelial, endothelial, tumour hepatocytes); fibronectin (epithelial, mesenchymal, neuronal fibroblasts); collagen type II (chondrocytes); and collagen type IV (epithelial, endothelial, muscle, nerve).

Other technical modifications

In some samples the optimal duration of the cold enzyme treatment may be shortened to fit that of a working day or ~6 h say. For larger samples the warm treatment may be further shortened by performing it in an Erlenmeyer flask with a stir bar. Specially designed flasks indented near the base to induce maximum turbulence are available for this purpose (Bellco, NJ.)

Simplified procedure

In many instances excess tissue will be available. Maximizing cellular viability will therefore be unnecessary. Such cases may warrant the omission of the steps requiring DNase containing cocktail 'B'. The entire disruption may then be carried out using cocktail 'A'.

BACKGROUND

The foregoing 'cold-cocktail' (CC) technique has proved effective in growing cell cultures to be harvested for either DNA extraction or cytogenetic analysis from human tissue biopsied from both chorionic villi or skin.

Expected cell yield and viability

Cell yields approximated $3-10 \times 10^7$ cells per gram of tissue. Mean viability of released cells (as assayed by Trypan blue exclusion) varied rather widely for the villus specimens obtained by pregnancy termination, the condition of which is often poor: average = 74% (range 52–91%), compared to respective averages of 81% (72–88%) and 82% (67–95%) for diagnostic villi and human skin specimens.

It is neither necessary, nor perhaps desirable, to aim for complete tissue disruption prior to culture, unless selective loss of viability among cells near the surface of the tissue is desired: e.g. with villus culture it is desirable preferentially to culture cells from the mesenchyme core at the expense of epithelial cells. Small cell clumps surviving disruption may in some samples yield foci of actively dividing cells which tend to promote culture initiation and proliferation.

Expected time required for cultures to grow

From skin punch biopsies (50–100 mm³) obtained to investigate the cause of death in stillborn babies or those who died neonatally we found that four 75 cm² flasks could be grown to confluence in a median three days (range 2 to 4 days) using the CC protocol. This was a significant improvement over both conventional single-enzyme disrupted tissue and explant cultures which respectively took a median 10 and 9 days longer. The explant cultures evidenced a higher rate of failure. The median time to harvest villus cultures was 8 days compared to 20 days for the explant cultures; and, as a bonus, CC cultures exhibited a more representative array of cell morphologies, probably due to exposure of the mesenchyme core cells.

ACKNOWLEDGEMENT

The author would like to thank Ms M. Kaufmann for expert technical advice and assistance.

REFERENCES

Allalunis-Turner MJ & Siemann DW (1986) Recovery of cell subpopulations from human tumour xenografts following dissociation with different enzymes. *British Journal of Cancer* 54: 615–622.

Bijman JT, Wagener DJT, van Rennes H, Wessels JMC & van den Broek P (1985) Flow cytometry evaluation of cell dispersion from human head and neck tumours. *Cytometry* 6: 334–341.

Cerra R, Zarbo RJ & Crissman JD (1990) Dissociation of cells from solid tumours. In: Darzynkiewicz Z and Crissman H A, (eds) *Methods in Cell Biology* 33: Academic Press, New York.

Cole RJ & Paul J (1966) The effects of erythropoetin on haem synthesis in mouse yolk sac and cultured fetal liver cells. *Journal of Embryology Experimental Morphology* 15: 245–260.

Ensley JF, Maciorowski Z, Pietraskiewicz H, Klemic G, KuKuraga M, Sapareto S, Corbett T & Crissman J (1987) Solid tumour preparation for flow cytometry using a standard murine model. *Cytometry* 8: 479–487.

Freshney RI (1987) Disaggregation of the tissue and primary culture. In: *Culture of Animal Cells* (2nd edn.), Wiley-Liss, New York.

Hefley TL, Stern PH & Brand JS (1983) Enzymatic isolation of cells from neonatal calvaria using two purified enzymes from clostridium histolyticum. *Experimental Cell Research* 149: 227–236.

Iype PT (1971) Cultures from adult rat liver cells: I Establishment of monolayer cell-cultures from normal liver. *Journal of Cellular Physiology* 78: 281–288.

Nguyen DH, Beuerman RW, Halbert CL, Ma Q & Sun G (1999) Characterization of immortalized rabbit lachrimal gland epithelial cells. *In Vitro Cellular and Developmental Biology* 35A: 198–204.

Seglen PO (1975) Preparation of isolated rat liver cells. *Methods in Cell Biology* 13: 29–83.

Sitar G, Brusamolino E, Scivetti P & Borroni R (1987) The disaggregation of Hodgkin's lymph node: A preparative technique using a new dissociation chamber. *Haematologica* 72: 23–28.

Slocum HK, Pavelic ZP & Rustum YM (1981) An enzymatic method for the disaggregation of human solid tumours for studies of clonogenicity and biochemical determinants of drug action. In: Salmon S (ed.) *Cloning of Human Tumour Stem Cells* Alan Liss and Co., New York.

4.8 CULTURES OF HUMAN KERATINOCYTES

Different methods exist for the culture of keratinocytes. One is the growth of explants either as epidermal or as organ cultures. In these cultures, epidermal cells are grown in the presence of remaining connective tissue. Different from the explant cultures is the use of dispersed keratinocytes plated at either high or low density. In these cultures, dermal constituents are discarded and keratinocytes represent 90% of the cell population; melanocytes and Langerhans cells are also present. To enhance the growth of keratinocytes seeded at low density, feeder cells are used, and the culture medium is supplemented (or not) with bovine serum and various growth factors. Recently a third method has been proposed. It consists of reconstructing an epidermis on substrates composed of macromolecules present in the papillary dermis, which may be colonized with dermal cells. Since these substrates are firm and easy to handle, the reconstructed epidermis can be maintained either submerged or exposed at the air–liquid interface.

Note: All cultures should be kept at 37°C and 98% relative humidity. When the culture medium is buffered with sodium bicarbonate, air enriched with 7% CO_2 should be used.

PROCEDURE: SKIN BIOPSIES

Reagents and solutions

Maintenance media

- *Solution A* Eagle's minimum essential medium (contains Earle's salts, L-glutamine and 25 mM HEPES) with:
  4% antibiotic–antimycotic solution × 100 (contains 100 000 U penicillin, 10 mg streptomycin and 25 μg amphotericin B per ml in 0.9% sodium chloride)
  40 mg gentamicin solution (stock concentration 50 mg ml^{-1}) for 100 ml of medium
- *Solution B* As solution A, but containing 2% antibiotic–antimycotic solution and 20 mg gentamicin per 100 ml medium.
- *Solution C* As solution A, but without antibiotics.

Materials and equipment

- Maintenance medium A, B and C
- Cork board
- Electrical dermatome

Cell and Tissue Culture for Medical Research, edited by A. Doyle and J.B. Griffiths.
© 2000 John Wiley & Sons, Ltd

Human skin is generally obtained after plastic surgery.

1. Store samples at 4°C, for 2 days in maintenance medium solution A.
2. Remove fat, pin the skin on a cork board, dermis down, and cut it into strips.
3. Dermatomize each strip at 0.4 mm thickness with an electrical dermatome.
4. Rinse flaps of skin for 5 min in maintenance medium solutions A, B and C.

PROCEDURE: EXPLANT CULTURE (KARASEK 1966; FLAXMAN & HARPER 1975)

Reagents and solutions

Explant culture medium

Dulbecco's modified Eagle's medium containing 10% newborn bovine serum, 1% non-essential amino acids solution ($\times$ 100), 2 mM L-glutamine, 1 mM sodium pyruvate and 2.8 g l^{-1} sodium bicarbonate.

Epidermal cultures

1. Cut dermatomized skin into 2–5 mm pieces with a scalpel.
2. Place dermal side down on substrates like plastic, glass or collagen.
3. Dry at room temperature for 15–30 min in a tissue culture cabinet.
4. Gently add medium.

Observations

Epidermal cells migrate from the explant, forming a crown of cells. The explant becomes disorganized after some time. Growth of epithelial cells is horizontal and can be measured by planimetry (Hammar & Halprin 1981).

After 10 days of culture, fibroblasts from the explant can completely invade the epidermal culture. They can be identified by specific immunolabelling with an antivimentin antibody.

Organ cultures

1. Place large split-thickness dermatomized skin fragments of 2×4 cm on a bacteriological-grade plastic dish.
2. Cover the fragment with culture medium.

Commentary

The substrate does not allow outgrowth of epidermal cells; the explant retains its original organization (Pruniéras *et al.* 1987).

DNA synthesis is estimated by counting radiolabelled cells after incorporation of tritiated thymidine on autoradiographs of histological sections.

PROCEDURE: DISPERSED EPIDERMAL CELL CULTURE (MEDAWAR 1941; CRUICKSHANK *ET AL.* 1960; REGNIER *ET AL.* 1973)

Reagents and solutions

Culture medium for high-density culture

Mix 3 parts Dulbecco's modified Eagle's medium (DMEM) containing 4.5 mg l^{-1} glucose and 1 part Ham's F-12 medium. Add 2.8 g l^{-1} sodium bicarbonate, 2 mM L-glutamine, 1 mM sodium pyruvate, 1% non-essential amino acids ($\times$ 100, 2 mM L-glutamine, 1 mM sodium pyruvate and 0.24% sodium bicarbonate (7.5%).

Growth medium

DMEM/F-12 (3 : 1) containing 2.8 g l^{-1} sodium bicarbonate; 2 mM L-glutamine; 1 mM sodium pyruvate; 1% non-essential amino acids; 10% newborn bovine serum; 10 ng ml^{-1} epidermal growth factor (EGF); 0.4 μg ml^{-1} hydrocortisone; 10^{-6} M isoproterenol; 5 μg ml^{-1} transferrin; 2×10^{-9} M triiodothyronine; 1.8×10^{-4} M adenine and 5 μg ml^{-1} insulin.

Materials and equipment

- Bovine serum
- Culture medium for high-density culture
- Culture medium for low density culture
- Growth medium
- Percoll
- Collagen type I or IV

Dissociation of human skin

1. Float split-thickness dermatomized skin (dermal side down) on a solution containing trypsin (0.25% trypsin in phosphate-buffered solution (PBS) without Ca^{2+} and Mg^{2+}) at 4°C overnight.
2. Separate epidermis from dermis in a solution of 0.05% trypsin/0.02% EDTA in PBS.
3. Discard the dermis.
4. Dissociate the epidermis with gentle agitation on a vortex mixer for 1 min.
5. Filter the epidermal cell suspension on a beaker covered with gauze to retain cornified cells and upper granular cells.
6. Stop the effect of trypsin by the addition of an equal volume of bovine serum.
7. Centrifuge the cell suspension at 1000 *g* for 10 min.
8. Resuspend the pellet in culture medium.
9. Dilute an aliquot of the suspension in Trypan blue solution (0.05% in sodium chloride 0.9%). The number of viable cells (i.e. those not stained) is counted in a haemocytometer chamber.

Purification of keratinocytes

The epidermis contains keratinocytes at different stages of differentiation. Three methods to enrich basal cells are described.

Percoll gradient (Fusenig & Worst 1975; Brysk et al. 1981; Gross et al. 1987)

1. Mix keratinocyte cell suspension with Percoll at 20% (density: 1.03 g/ml).
2. Place it on a step gradient (8 ml of 70%, density 1.085 g ml^{-1}; 12 ml of 60% density 1.075 g ml^{-1}; and 12 ml of 40%, density 1.05 g ml^{-1}).
3. Centrifuge at 1200 *g* for 25 min. Basal cells are concentrated in fractions 3 and 4 (1.06 and 1.08 g ml^{-1}), and squamous and spinous cells are in fractions 1 and 2 (1.02 and 1.04 g ml^{-1}).

   Cells can be identified by immunological methods with specific antibodies (see 'Characterization of Human Epidermal Cells', below).

Size and refringence (Liu & Karasek 1978; Barrandon & Green 1985)

Suspended basal cells are relatively small (less than 13 μm) and very bright under the microscope. It is possible to extract them with a narrow Pasteur pipette and seed them to obtain clonal cultures of keratinocytes (see 'Low-density Cultures of Keratinocytes', below).

Attachment to the substrate (Skerrow & Skerrow 1983)

Basal cells attach more rapidly to collagen than suprabasal cells.

1. Seed an epidermal keratinocyte suspension onto a dish coated with a thin film of collagen type I or IV.
2. After 1 h at 37°C, 90–95% of the attached cells are basal cells.

Characterization of human epidermal cells (Régnier *et al.* 1985)

Indirect immunofluorescence labelling allows the rapid identification of keratinocytes.

1. Basal cells express the bullous pemphigoid antigen at their basal pole.
2. Suprabasal cells react with monoclonal antibodies specific for keratin 1 and 10.
3. Melanocytes, Langerhans cells and fibroblasts are stained with an antivimentin antibody.

High-density cultures of keratinocytes (Marcelo *et al.* 1978; Pruniéras *at al.* 1980)

The concentration of cells per cm^2 should exceed 5×10^5 cells. These cultures are confluent after 1–4 days. For example, a culture seeded with 5×10^5 viable epidermal cells per cm^2 forms a coherent sheet of epidermis after 24 h.

Observations

Cell proliferation is evaluated after incorporation of tritiated thymidine into DNA. A first peak of DNA synthesis is generally observed during the first few days of culture thereafter cells begin to differentiate. Differentiation can be estimated either by measuring keratin synthesis (after incorporation of ^{35}S-labelled methionine) or by counting the detached cells in the culture medium. Subculture is difficult; three to four passages are possible at maximum.

Low-density cultures of keratinocytes

Puck & Marcus (1955) showed that a monolayer culture in which growth is stopped after X-irradiation still synthesizes proteins which can serve as growth factors for other cells co-cultivated with these irradiated cells.

Rheinwald & Green (1975) co-cultivated irradiated conjunctival mouse (3T3) and human epidermal cells seeded at low density. The 3T3 cells served as a feeder layer for the keratinocytes. If the culture medium is enriched with certain growth factors, large amounts of human cells are obtained.

Maintenance of 3T3 cells (ATCC: CCL92)

1. Detach confluent 3T3 cells from the flask with 4 volumes PBS (containing 0.02% EDTA and 0.1% glucose) and 1 volume trypsin 0.05%/EDTA 0.02% at 37°C.
2. Add medium and centrifuge at 1000 g for 10 min.
3. Seed the cells at a concentration of 1.5×10^3 per cm². Cells may be subcultured once or twice a week.

Preparation of the feeder layer

1. For growth arrest, expose 3T3 cells to X-irradiation at a dose of 6000 rads (Rheinwald & Green 1975) or treat 3T3 cells with mitomycin C (Kubilus *et al.* 1981). For mitomycin C treatment, rinse with PBS and add DMEM containing 0.001% of mitomycin C at 37°C for 2 h.
2. Rinse the cells with PBS without Ca^{2+} and Mg^{2+}.
3. Detach as described above.
4. Centrifuge at 1000 g and resuspend the pellet in growth medium.
5. Seed the cells at 1.5×10^4 cells per cm².

Cultures of keratinocytes on a 3T3 feeder layer

1. Seed keratinocytes (4×10^3 cells per cm²) directly onto the mitomycin-treated or irradiated 3T3 cells, prepared 24 h earlier. The medium is not changed. With time, the keratinocytes push the 3T3 cells away.
2. Before keratinocytes reach confluency, detach 3T3 cells by rinsing the culture with PBS-glucose-trypsin-EDTA (see 'Maintenance of 3T3 Cells', above). Detachment is very rapid (2 min) and must be controlled under the microscope.

3. Trypsinize the keratinocytes with 0.05% trypsin containing 0.02% EDTA for 5–10 min at 37°C.
4. Count the cells.
5. For freezing, suspend keratinocytes in culture medium containing 10% bovine serum and 10% glycerol but no growth factors. Cell concentration in the cryotubes should be about 4×10^6 cells per ml.

Observations

It is relatively easy to subpassage epidermal cells cultured under these conditions. After 3 weeks of culture, the amount of epidermal cells harvested can be 300–400 times greater than the initial seeding number of cells. Epidermal sheets, suitable for grafting onto burns patients, are obtained after detachment of the epidermal cell layers from the substrate by dispase treatment (2.5 mg per ml of DMEM medium, 37°C, 1 h). Clonal cultures can be obtained (Barrandon & Green 1987).

- *Defined medium* For a detailed discussion of serum-free media to grow keratinocytes, see 'Cultures of Human Keratinocytes'.
- *Disadvantages* Cells obtained under these culture conditions are highly proliferative and are far from their normal physiological state.

PROCEDURE: RECONSTRUCTED EPIDERMIS

Various approaches have been used to provide better physiological culture conditions for keratinocytes *in vitro*.

Basal membrane equivalents

Endothelial cells from bovine cornea cultivated on a filter synthesize a film of macromolecules equivalent to basal membrane constituents. After detachment of endothelial cells, keratinocytes are seeded on this basal membrane (Tseng *et al.* 1981; Arruti & Courtois 1982).

Frozen pig skin

Explants of human skin are cultivated on the dermal side of thin strips of pig skin (Freeman *et al.* 1976; Régnier *et al.* 1981). These cultures can be lifted on a grid and exposed to air.

Collagen gels

Acellular collagen gels (type I collagen) can be used as substrates for keratinocytes (Lillie *et al.* 1980). They can be made of a mixture of type I and III collagens covered with a type IV collagen film to improve the attachment of keratinocytes and facilitate the formation of hemidesmosomes (Tinois *et al.* 1991).

Viable human fibroblasts mixed with a collagen solution (Bell *et al.* 1979; Asselineau & Pruniéras 1984; Coulomb *et al.* 1984; Boyce & Hansbrough 1988) form a dermal equivalent after retraction of the gel. Size and thickness of the substrate can be controlled and depend on the concentrations of cells and collagen. Fibroblasts arc still viable after retraction of the gel. Keratinocytes are seeded on these dermal equivalents. They are first cultivated by immersion for a few days and are then air-exposed. Different cell types can be introduced into the collagen gel which allow the study of cellular interactions. Growth and differentiation of keratinocytes are assessed using histological techniques. It is possible to evaluate cytotoxicity of drugs and to measure percutaneous absorption of formulations.

Microporous membranes (Bernstam *et al.* 1986)

Marrow-Tech, Inc., San Diego, CA (Triglia *et al.* 1991) has developed a three-dimensional human skin model in which fibroblasts and keratinocytes are grown on a nylon mesh.

De-epidermized dermis (Régnier *et al.* 1981)

Preparation (Pruniéras et al. 1979)

1. Immerse split-thickness human skin (0.7 mm) in Ca^{2+}- and Mg^{2+}-free PBS for 7–10 days at 37°C.
2. Peel off the epidermis with forceps. The dermis will still be covered with its organized lamina densa containing collagen IV, laminin and proteoglycans.
3. Inactivate the remaining dermal cells by successive freezing (–20°C) and thawing (37°C). Repeat 10 times.
4. Store de-epidermized dermis below –70°C until use.
5. Living dermis obtained after enzymic separation of epidermis can also be used as substrate (Mackensie & Fusenig 1983).

Keratinocyte culture on de-epidermized dermis

- *Differentiation medium* DMEM/F-12 containing 10% newborn bovine serum, 1% non-essential amino acis × 100, 2 mM L-glutamine, 1 mM sodium pyruvate and 2.8 g l^{-1} sodium bicarbonate.

1. Spread the dermis on the bottom of a dish with the basal membrane upwards.
2. Rinse with growth medium.
3. One day later, remove the medium and place a steel ring on the basal membrane to delineate the area of seeding.
4. Seed either suspended keratinocytes obtained from 3T3 co-cultures or fresh keratinocytes dissociated from the skin into the ring (5×10^5 cm^{-2}). Add growth medium. Incubate at 37°C.
5. After 1–3 days remove the ring and lift the culture onto a grid.
6. Add the differentiation medium underneath the grid. The epidermal cells are exposed to the atmosphere while differentiation medium feeds the cells through the dermis. Medium is changed every 2 days.

Observations

Epidermal cells are attached to the basal membrane covering the dermis. They form a stratified epidermis. When maintained emerged, the cultured epidermis has the aspect of normal epidermis. Several layers of keratinocytes are present with granular cells covered by a stratum corneum. This reconstructed epidermis is treated as a skin specimen. Histological sections are necessary to know the state of the culture. Keratinocyte proliferation is studied by autoradiography after incorporation of tritiated thymidine. Epidermal differentiation is observed using frozen sections after immunological reaction with various antibodies (Régnier & Darmon 1989). Effects of environmental agents such as humidity, medium constituents and other cells can be studied. Since the epidermis is exposed to the atmosphere, topical applications of drugs or formulations is possible (Régnier *et al.* 1990). The model can also be used for percutaneous absorption studies (Régnier *et al.* 1989; Ponec *et al.* 1990). Protein synthesis is evaluated after incorporation of labelled amino acids (Régnier *et al.* 1986). Lipid composition of the reconstructed epidermis (horny layer) is determined after extraction and separation by thin layer chromatography (Ponec *et al.* 1988). As with skin sections, *in situ* hybridization can be performed. Dissociated keratinocytes can be replaced by skin explants (Basset-Séguin *et al.* 1990). A particular disadvantage is that the cultured cells are not visible under the microscope and sections must be prepared.

Since the substrate is of human origin, it is difficult to provide enough material for large-scale studies, particularly for the routine screening of new molecules for dermatopharmacologic purposes.

REFERENCES

Arruti C & Courtois Y (1982) Monolayer organization by serially cultured bovine corneal endothelial cells: effects of a retinal derived growth promoting activity. *Experimental Eye Research* 34: 735–747.

Asselineau D & Pruniéras M (1984) Reconstruction of 'simplified' skin: control of fabrication. *British Journal of Dermatology* 111 (Suppl. 27): 219–222.

Barrandon Y & Green H (1985) Cell size as a determinant of the clone forming ability of human keratinocytes. *Proceedings of the National Academy of Sciences of the USA* 82: 5390–5394.

Barrandon Y & Green H (1987) Three clonal types of keratinocyte with different capacities for multiplication. *Proceedings of the National Academy of Sciences of the USA* 84: 2302–2306.

Basset-Séguin N, Culard J-F, Kerai C, Bernard F, Watrin A, Demaille J & Guilhou JJ (1990) Reconstituted skin in culture: a simple method with optimal differentiation. *Differentiation* 44: 232–238.

Bell E, Ivarsson B & Merrill C (1979) Production of a tissue-like structure by contraction of collagen lattices by human fibroblasts of different proliferative potential *in vitro*. *Proceedings of the National Academy of Sciences of the USA* 76: 1274–1278.

Bernstam LI, Vaughan FL & Bernstein IA (1986) Keratinocytes grown at the air-liquid interface. *In Vitro Cellular and Developmental Biology* 22: 695–705.

Boyce ST & Hansbrough JF (1988) Biologic attachment, growth, and differentiation of cultured human epidermal keratinocytes on a graftable collagen and chondroitin-6-sulfate substrate. *Surgery* 103: 421–431.

Brysk MM, Snider JM & Smith EB (1981) Separation of newborn rat epidermal cells on discontinuous isokinetic gradients of percoll. *Journal of Investigative Dermatology* 77: 205–209.

Coulomb B, Dubertret L, Merrill C, Touraine R & Bell E (1984) The collagen lattice: a model for studying the physiology, biosynthetic function and pharmacology of the skin. *British Journal of Dermatology* 111 (Suppl. 27): 83–87.

Cruickshank CND, Cooper JR & Hooper C (1960) The cultivation of cells from adult epidermis. *Journal of Investigative Dermatology* 34: 339–342.

Flaxman BA & Harper RA (1975) Primary cell culture from biochemical studies of human keratinocytes. *British Journal of Dermatology* 92: 305–309.

Freeman AE, Igel HJ, Herrman BJ & Kleinfeld KL (1976) Growth and characterization of human skin epithelial cell cultures. *In Vitro Cellular and Developmental Biology* 12: 352–362.

Fusenig N & Worst PKM (1975) Mouse epidermal cell cultures. *Experimental Cell Research* 93: 443–457.

Gross M, Fürstenberger G & Marks F (1987) Isolation, characterization, and *in vitro* cultivation of keratinocyte subfractions from adult NMRI mouse epidermis: epidermal target cells for phorbol esters. *Experimental Cell Research* 171: 460–474.

Hammar H & Halprin K (1981) Epidermal cell growth. In: Marks R & Christophers E (eds) *The Epidermis in Disease*, pp. 242–271. MTP Press, Lancaster.

Karasek M (1966) *In vitro* cultures of human skin epithelial cells. *Journal of Investigative Dermatology* 47: 533–540.

Kubilus J, Rand R & Baden HP (1981) Effects of retinoic acid and other retinoids on the growth and differentiation of 3T3 supported human keratinocytes. *In Vitro Cellular and Developmental Biology* 17: 786–795.

Lillie JH, MacCallum DK & Jepsen A (1980) Fine structure of subcultivated stratified squamous epithelium grown on collagen rafts. *Experimental Cell Research* 125: 153–165.

Liu S & Karasek M (1978) Isolation and serial cultivation of rabbit skin epithelial cells. *Journal of Investigative Dermatology* 70: 288–293.

Mackensie IC & Fusenig NE (1983) Regeneration of organized epithelial structure. *Journal of Investigative Dermatology* 81: 189s–194s.

Marcelo CL, Kim YG, Kaine JL & Voorhees JJ (1978) Stratification, specialization and proliferation of primary keratinocyte cultures. *Journal of Cell Biology* 79: 356–370.

Medawar PB (1941) Sheets of pure epidermal epithelium from human skin. *Nature* 148: 783.

Ponec M, Weerheim A, Kempenaar J, Mommaas AM & Nugteren DH (1988) Lipid composition of cultured human keratinocytes in relation to their differentiation. *Journal of Lipid Research* 29: 949–961.

Ponec M, Wauben-Penris PJJ, Burger A, Kempenaar J & Boddé HE (1990) Nitroglycerin and sucrose permeability as quality markers for reconstructed human epidermis. *Skin Pharmacology* 3: 126–135.

Pruniéras M, Régnier M & Schlotterer M (1979) Nouveau procédé de culture des cellules épidermiques humaines sur derme homologue ou hétérologue: préparation de greffons recombinés. *Annales de Chirurgie Plastique* 24: 357–362.

Pruniéras M, Delescluse C & Régnier M (1980) A cell culture model for the study of epidermal (chalone) homeostasis. *Pharmacology and Therapeutics* 9: 271–295.

Pruniéras M, Régnier M & Démarchez M (1987) The regeneration of adult human skin *in vitro* and *in vivo*. *Giornale Italiano di Chirurgi Dermatologia* ED Oncolog 2, 216–219.

Puck TT & Marcus PI (1955) A rapid method for viable cell titration and clone production with HeLa cells in tissue culture: the use of X-irradiated cells to supply conditioning factors. *Proceedings of the National Academy of Sciences of the USA* 41: 432–437.

Régnier M & Darmon M (1989) Human epidermis reconstructed *in vitro*: a model to study keratinocyte differentiation and its modulation by retinoic acid. *In Vitro Cellular and Developmental Biology* 25: 1000–1008.

Régnier M, Delescluse C & Pruniéras M (1973) Studies on guinea-pig cell cultures. I. Separate culture of keratinocytes and dermal fibroblasts. *Acta Dermato Venereologica* 53: 241–247.

Régnier M, Pruniéras M & Woodley D (1981) Growth and differentiation of adult human epidermal cells on dermal substrates. *Frontiers of Matrix Biology* 9: 4–35.

Régnier M, Vaigot P, Michel S & Pruniéras M (1985) Localization of Bullous Pemphigoid antigen (BPA) in isolated human keratinocytes. *Journal of Investigative Dermatology* 85: 187–190.

Régnier M, Schweizer J, Michel S, Bailly C & Pruniéras M (1986) Expression of high molecular weight (67kD) keratin in isolated human keratinocytes cultured on dead de-epidermized dermis. *Experimental Cell Research* 165: 63–72.

Régnier M, Darmon M & Schaefer H (1989) Human skin reconstructed *in vitro* as a model for studying epidermal permeability. Effect of retinoic acid. In: von Tscharner C & Halliwell REW (eds) *Advances in Veterinary Dermatology*, Vol. I, pp. 278–289. Baillière Tindall, London.

Régnier M, Asselineau D & Lenoir MC (1990) Human epidermis reconstructed on dermal substrates *in vitro*: an alternative to animals in skin pharmacology. *Skin Pharmacology* 3: 70–85.

Rheinwald JG & Green H (1975) Serial cultivation of strains of human keratinocytes: the formation of keratinizing colonies from single cells. *Cell* 6: 331–343.

Skerrow D & Skerrow CJ (1983) Tonofilament differentiation in human epidermis, isolation and polypeptide chain composition of keratinocyte subpopulations. *Experimental Cell Research* 143: 27–35.

Tinois E, Tiollier J, Gaucherand M, Dumas H, Tardy M & Thivolet J (1991) *In vitro* and post-transplantation differentiation of human keratinocytes grown on the human type IV collagen film. *Experimental Cell Research* 193: 310–319.

Triglia D, Braa SS, Yonan C & Naughton GK (1991) *In vitro* toxicity of various classes of test agents using the neutral red assay on a human three-dimensional physiologic skin model. *In Vitro Cellular and Developmental Biology* 27A: 239–244.

Tseng SCG, Savion N, Gospodarowicz D & Stern N (1981) Characterization of collagens synthesized by cultured bovine corneal endothelial cells. *Journal of Biological Chemistry* 256: 3361–3365.

Wille JJ, Pittelkow MR, Shipley GD & Scott RE (1984) Integrated control growth and differentiation of normal human prokeratinocytes cultured in serum-free medium: clonal analyses, growth kinetics, and cell cycle studies. *Journal of Cellular Physiology* 121: 31–44.

BACKGROUND READING

Boyce ST & Ham RG (1985) Cultivation frozen storage and growth of normal human epidermal keratinocytes in serum. *Journal of Tissue Culture Method* 9: 83–93.

Tsao MC, Walthall BJ & Ham RG (1982) Clonal growth of normal human epidermal keratinocytes in a defined medium. *Journal of Cellular Physiology* 110: 219–229.

4.9 THE ISOLATION AND MAINTENANCE OF HUMAN GASTRIC EPITHELIAL CELLS IN PRIMARY CULTURE

Remarkable advances in endoscopy techniques applicable to the alimentary tract canal, widely used nowadays in diagnosis and monitoring both the course of disease and the outcome of treatment, have opened new avenues for the study of gastric epithelial cell biology in health and disease. Gastric epithelial cells have become a unique source of information not only in areas of histopathology but also in eukaryotic cell biology and molecular genetics.

There are a variety of well-established cell lines originating from human alimentary tract epithelium. Although they provide information on morphology and function of the surface epithelium, the major drawback inherent in their biology is neoplastic transformation. Since transformed cells exhibit numerous phenotypes, some foreign to normal epithelium, both morphology and function could be significantly aberrant.

Therefore, the search for an optimal culture model of the normal human gastric epithelium, feasible to establish in most tissue culture laboratories, continues. This is justified by the enormous demand for human gastric epithelial cells as an *in vitro* target for the study of cell metabolism and for the evaluation of the effects of a variety of pharmacological agents. Such agents include those developed for strengthening the gastric mucosal barrier with potential use in the treatment of gastric ulcer, as well as formulations targeted at other body organs, but possessing side effects that compromise the gastric epithelium.

Although most drug testing has been performed with *in vivo* animal models, there are two major drawbacks. One is the high cost of animal experimentation, and the other is the variation in morphology and function of the gastric mucosa among species. This makes the extrapolation of animal results a significant challenge.

PRELIMINARY PROCEDURE: GASTRIC EPITHELIAL CELL ADHESION REQUIREMENTS

Human gastric epithelium, like most normal epithelial cells, requires attachment to a basement membrane for growth and differentiation. Adhesion is mediated by a family of cell surface proteins, so-called integrins or adhesion receptors. These adhesion receptors bind to corresponding structures in the extracellular matrix, also called matrix ligands, such as fibronectin, laminin, collagens, entactin, tenascin,

Cell and Tissue Culture for Medical Research, edited by A. Doyle and J.B. Griffiths.
© 2000 John Wiley & Sons, Ltd

thrombospondin, von Willebrand factor and vitronectin (Ruoslahti 1990; Hemler 1990; Ruoslahti & Pierschbacher 1987).

There are two ways to initiate attachment *in vitro*:

1. Modification of the surface of a tissue culture dish to imitate some natural ligand-like structures. A range of positively charged chemicals have been described such as polyornithine, polyarginine, polyhistidine, DEAE-dextran, protamine or polylysine that modify the surface or charge of tissue culture wells (McKeehan 1981). The most widely used is D-polylysine. Firm attachment of the human gastric epithelium to this modified artificial attachment factor has not been achieved; further attempts have focused on natural basement membrane structures.

2. Covering the tissue culture dish with molecules which are natural ligands or components of human basement membranes. An ideal cell attachment environment should include all the molecules present *in vivo* in human basement membranes of the lamina propria. As detailed reconstruction of such conditions is tedious, expensive and impossible to achieve in a routine tissue culture laboratory, a variety of single molecules, major components of lamina propria, have been used as a practical compromise. Recently (Antonescu *et al.* 1990; Sarosiek *et al.* 1991) collagen, both type I and IV (Sigma Chemical Co.), laminin and fibronectin have been shown to enhance attachment of human gastric epithelial cells to tissue culture wells. However, the attachment achieved is fragile and cells easily detach during routine medium changes.

Recently, it has been found that both human antral and corporal epithelium exhibit very high affinity for Matrigel (Collaborative Research Incorp.) in culture (See Figure 4.9.1). This strong attachment with Matrigel enables not only proliferation but also migration of newly divided cells with formation of monolayers. By maintaining the cells with unchanged medium for 6 days it is possible to identify a mucus layer covering the mucous cell monolayer, indicating that cultured gastric epithelial cells are differentiated enough to retain an ability to secrete (see Figure 4.9.2). This monolayer forms very tight junctions between neighbouring cells, similar to observations, *in vivo*.

Matrigel is the first complete multi-component preparation of natural solubilized tissue basement membranes. Its major component is laminin, followed by type IV collagen, heparan sulphate proteoglycans, entactin (nidogen), along with transforming growth factor β (TGF-β) and other growth factors which occur naturally in the EHS (Engelbreth-Holm-Swarm) mouse sarcoma tumour; although Matrigel permits satisfactory attachment and growth of human gastric epithelial cells.

The thickness of the Matrigel layer is crucial and should be selected depending on the final purpose of the cell culture. For example, a Matrigel thickness of 0.5 mm is excessive for the study of *Helicobacter pylori* adhesion to the surface of the human gastric epithelium since most of the epithelial cells are embedded or buried within the Matrigel, thereby blocking the access of inoculated bacteria to the surface of gastric epithelium (unpublished data). The optimal thickness of the supportive medium should therefore be defined experimentally in each case.

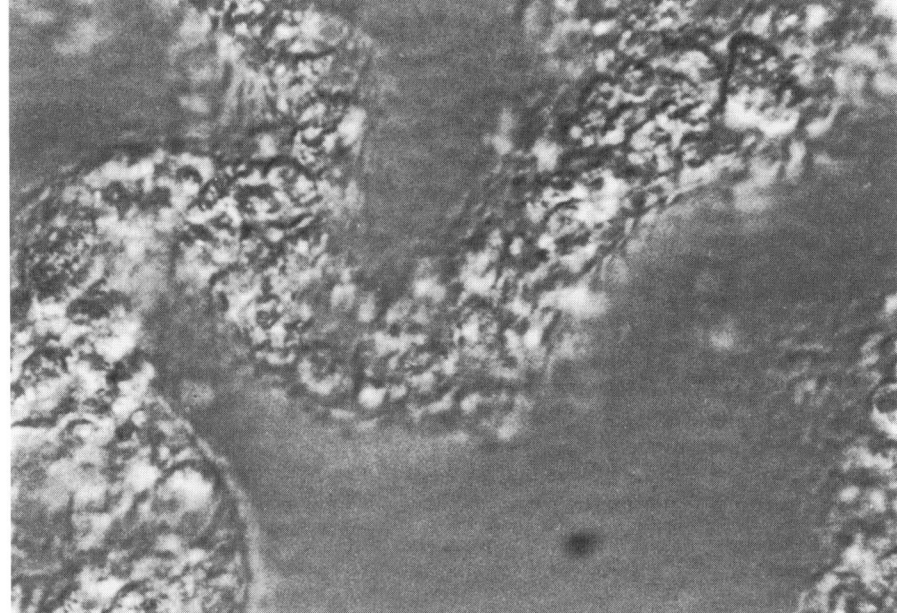

Figure 4.9.1 Human gastric epithelial cells attached to Lab-Tek wells coated with a thin layer (50 μm) of Matrigel. Phase contrast inverted microscope (× 300) with Hoffmann modulation optics. Gastric epithelial cells attach, proliferate and migrate only on the surface area covered with a satisfactory thickness of the Matrigel.

Figure 4.9.2 Monolayer of human gastric epithelial cells maintained on a Matrigel layer with areas of a high confluency and formation of tight type of cell-cell adhesion. (Phase contrast inverted microscope with Hoffmann modulation optics; magnification × 300).

PROCEDURE: PROCESSING A HUMAN GASTRIC BIOPSY SAMPLE

To ensure the best yield of freshly isolated human gastric epithelium, the following standard procedure has been established:

Materials and equipment

- Medium F-12 (Life Technologies, Gaithersburg, MD)
- Penicillin and streptomycin (Life Technologies)
- Lab-Tek 2 (Nunc Inc., Naperville, IL)

1. Incubate F-12 medium (Life Technologies) with EDTA (0.02%), penicillin (100 U ml^{-1}) and streptomycin (100 μg ml^{-1}) in a 5% CO_2/95% air atmosphere at 37°C for 30 min in order to equilibrate pCO_2.
2. Place the equilibrated F-12 medium in Lab-Tek 2 and transport to the endoscopy suite immediately before the biopsy sampling procedure.
3. Immerse the fresh biopsy specimen in medium and transport immediately to the tissue culture laboratory.
4. Place the biopsy specimen, maintained in F-12 medium, in a 5% CO_2/95% air incubator for 30 min to allow for equilibration with CO_2.

SUPPLEMENTARY PROCEDURE: DISSOCIATION OF HUMAN GASTRIC EPITHELIAL CELLS

Additional materials and equipment

- Matrigel (Collaborative Research Incorp., Bedford, MA)
- Collagenase IV (Sigma Chemical Co.)
- D-valine (Life Technologies)
- EDTA (Sigma Chemical Co.)
- Foetal bovine serum (FBS) (Life Technologies)
- Epidermal growth factor (human recombinant – Amgen, Thousand Oaks, CA)
- Penicillin and Streptomycin (Life Technologies)
- Lab-Tek 4 (Nunc Inc., Naperville, IL)

1. In a sterile cabinet identify the luminal surface of the biopsy specimen (no vessels bleeding) and maintain this surface face-up during further processing. Gastric epithelial cells may be isolated with or without collagenase digestion.
2. Mechanically, gently remove the mucus layer from the biopsy tissue using fine forceps and a surgical scalpel blade in order to facilitate penetration of collagenase and the subsequent release of dissociated cells into the medium.
3. Place the biopsy sample, designated for isolation of gastric epithelium without collagenase, in a 37°C incubator, mixing three times by aspiration with a disposable transfer pipette every 1 h during the first 6 h and then leave overnight.
4. Place the biopsy sample with collagenase in similar medium but without EDTA, since collagenase requires calcium ions for optimum activity. The recommended concentration of collagenase IV (Sigma Chemical Co.) is 100 μg ml^{-1}. Place the

biopsy in a 37°C incubator and mix three times every 1 h during the first 6 h and then leave overnight.

5. The following day mix both biopsy samples three times. Aspirate tissue fragments into a transfer pipette and transfer to a separate tube with fresh F-12 medium with either collagenase or EDTA.

6. Centrifuge remaining cells, dissociated from the matrix of the biopsy specimen, at 400 g for 5 min at 4°C. Gently aspirate the supernatant using a 1 ml syringe and 19 G needle, leaving a 3 mm layer of medium covering the sediment at the bottom of the centrifuge tube.

7. Mix the cell pellet with the medium and transfer to Lab-Tek 4 wells covered with Matrigel and filled with F-12 medium (Life Technologies) containing L-valine substituted by D-valine to inhibit the growth of fibroblasts (Gilbert & Migeon 1975) with 1% penicillin/streptomycin, 10% FBS and 3 ng/ml epidermal growth factor (EGF) human recombinant form (Amgen).

8. Continue the collagenase or EDTA treatment of the biopsy specimens over the next 24–48 h. The cell yield, however, is best during the first 24 h of the isolation procedure.

If the gastric epithelial cell isolation procedure does not yield sufficient cells during the EDTA or collagen digestion procedures, a neutral protease isolated from *Bacillus polymyxa*, dispase (Boehringer-Mannhein, Germany), may be used. This enzyme possesses an activity similar to not only collagenase IV but also fibronectinase. Thus it can serve as a powerful tool for dissociation of epithelial–mesenchymal interaction (Stenn *et al.* 1989). Although the authors are not aware of any publication utilizing this enzyme in cleaving the basement membrane zone in human gastric mucosa, the successful use of this enzyme has been confirmed by isolation of intact epidermis from the dermis of newborn foreskin (Stenn *et al.* 1989).

SUPPLEMENTARY PROCEDURE: MAINTENANCE OF HUMAN GASTRIC EPITHELIUM IN LONG-TERM CULTURE

1. Remove the medium every third day, always gently aspirating from the same corner of the Lab-Tek well, leaving a 1 mm layer of old medium. Transfer the aspirated medium to a centrifuge tube.
2. Fill up 80% of the well space with fresh medium.
3. Centrifuge the aspirated medium at 400 g for 5 min.
4. Aspirate the supernatant, leaving a 3 mm meniscus over the cell sediment.
5. Inoculate the cell pellets (sediment) into a fresh Lab-Tek well or replace into the aspirated well to facilitate formation of a confluent monolayer.

SUPPLEMENTARY PROCEDURE: PROPAGATION OF CELLS

Materials and equipment

- Trypsin 0.25% (Life Technologies)
- Matrigel (Collaborative Research Inc., Bedford, MA)
- Lab-Tek

1. Near confluency, treat the cell monolayer with trypsin (Life Technologies) for 5–10 min.
2. Split in a 1 : 2 ratio by diluting cell suspension with fresh medium.
3. Passage to a new Lab-Tek well coated with Matrigel.

DISCUSSION

Freshly isolated human gastric epithelial cells from biopsy specimens exhibit a slow but persistent rate of proliferative activity. They require 2–3 weeks to achieve a monolayer of satisfactory confluency. Up to 3–5 passages can be achieved with a total culture time of over 90 days. Growth-promoting factors such as insulin, cortisone or higher doses of EGF might potentiate the proliferative activity of epithelial cells and therefore may extend the total lifespan of the epithelial cell line isolated from the single biopsy tissue.

ACKNOWLEDGEMENTS

Biomedical Research Support Grants, 5-SO7-RR05431-28 and 5-S07-RR05431-29 and Jeffress Research Grant J-240.

REFERENCES

Antonescu C, Marshall BJ, Anderson J, Sarosiek J, Hamlin J & McCallum RW (1990) Optimization of media for the long-term culture of gastric epithelial cells from a patient with Menetrier's disease. *Nineth Ann. Res. Day. Int. Med.* pp. 2–3. University of Virginia, Department of Internal Medicine.

Gilbert SF & Migeon BR (1975) D-Valine as a selective agent for normal human and rodent epithelial cells in culture. *Cell* 5: 11–17.

Hemler ME (1990) VLA proteins in the integrin family: structures, function and their role on leukocytes. *Annual Review of Immunology* 8: 365–400.

McKeehan W (1981) The use of low-temperature subculturing and culture surface coated with basic polymers to reduce the requirment for serum macromolecules. In: Waymouth C, Ham RG & Chapple RD (eds) *Requirements of Vertebrate Cells in Vitro*, pp. 118–130. Cambridge University Press, Cambridge.

McKeehan WL, Barnes D, Reid L, Stanbridge E, Nurakami H & Sato GH (1990) Frontiers in mammalian cell culture. *In Vitro Cellular and Developmental Biology* 26: 9–23.

Ruoslahti E (1990) Integrins. *Journal of Clinical Investigation* 87: 1–5.

Ruoslahti E & Pierschbacher MD (1987) New perspectives in cell adhesion: RGD and integrins. *Science* (*Washington DC*) 238: 491–497.

Sarosiek J, Marshall BJ, Hoffman S, Barret L, Guerrant R & McCallum RW (1991) The attachment of *Helicobacter pylori* to human gastric epithelium *in vitro*: a model for the study of the pathomechanism of colonization. *Gastroenterology* 100: A155.

Stenn KS, Link R, Moellmann G, Madri J & Kuklinska E (1989) Dispase, a neutral protease from *Bacillus polymyxa*, is a powerful fibronectinase and type IV collagenase. *Journal of Investigative Dermatology* 93: 287–290.

4.10 REPLICATIVE AND FUNCTIONAL CULTURES OF NORMAL HUMAN HEPATOCYTES

INTRODUCTION

The liver is a critical organ responsible for many metabolic and storage functions, and in the activation and detoxification of endogenous and exogenous chemicals. While the liver is usually a quiescent organ in adult mammals including humans, it is able to regenerate itself and replace lost tissue following the surgical removal of two-thirds of its mass, after transplantation, and upon the loss of cells following chemical and viral injury (Bucher 1995; Columbano & Shinozuka 1996; Grisham & Thorgeirsson 1997; Michalopoulos & De Frances 1997; Pistoi & Morello 1996; Steer 1995; Thorgeirsson 1996). Almost immediately following cell loss, the remaining cells undergo compensatory hyperplasia, first with the hepatocytes, followed by other parenchymal and non-parenchymal cell types. Regeneration *in vivo* can be repeated many times, indicating the unique regenerative capacity of this organ in response to injury. While the liver contains many epithelial cell types in various stages of cellular differentiation and regenerative capacity, the quickness of the regenerative response suggests that the hepatocytes themselves have the capacity to replicate (Gibson-D'Ambrosio *et al.* 1998). The replicative capacity of rodent hepatocytes was eloquently demonstrated by the repopulation of diseased mouse liver by transplanted wild-type hepatocytes (Overturf *et al.* 1996, 1997), transfer of adult mouse and rat hepatocytes into transgenic mice, (Pisto & Morello 1996; Rhim, *et al.* 1994, 1995) in transgenic mice expressing β-galactosidase gene (Kennedy *et al.* 1995), and in the retroviral transfer of the β-galactosidase gene into rat hepatocytes (Bralet *et al.* 1994, 1996). In humans, the rapid replacement of hepatocytes following liver injury suggests that a population of replicatively active hepatocytes, or precursor hepatocytes exist in the normal adult liver, that divide and repopulate the liver (Grisham 1995; Grisham & Thorgeirsson 1997; Michalopoulos & De Frances 1997; Steer 1995; Thorgeirsson 1996). This also suggests that functioning human hepatocytes, under appropriate conditions, should be able to proliferate in cell culture (Gibson-D'Ambrosio *et al.* 1998). In this section we describe procedures for the establishment of primary human hepatocytes in culture, and the conditions that permit long-term culturing of replicatively active normal adult human hepatocytes (Gibson-D'Ambrosio *et al.* 1993, 1995, 1998; Gibson-D'Ambrosio & D'Ambrosio 1996). The cultured hepatocytes maintain many of the functions characteristic of normal human hepatocytes *in vivo*. The procedures presented here provide a model system in

Cell and Tissue Culture for Medical Research, edited by A. Doyle and J.B. Griffiths.
© 2000 John Wiley & Sons, Ltd

which biochemical and cellular functions of normal human adult liver hepato-cytes can be evaluated for drug metabolism, toxicity, pharmacokinetics, and carcinogenesis. The culture conditions may also provide a system for banking and subsequent transplantation of normal human hepatocytes to restore liver func-tion lost to disease or injury.

EQUIPMENT AND MATERIAL

Sterile

100 ml stainless steel beaker with gauze in bottom filled part-way with 95% alcohol.
Standard dissecting instruments
Standard glass beakers to hold large and small Stomacher® bag
One Petri dish 100 × 20 mm
One Petri dish 100 × 15 mm with a glass cutting block 5.0 × 7.5 cm such that the block fits into the 100 × 15 mm Petri dish, chemically etched on one side. Etched side up
Stainless steel sieve with 80–100 mesh 190–140 μm.
Stomacher® bag – large and small
4 × 4 surgical gauze.
VacuCap Filter 0.2 μ (Gelman).

Non-sterile

Tekmar Stomacher® lab blender (Tekmar, Cincinnati, OH, USA)
Voltage regulator, to reduce Stomacher® strokes to 150 per min.
Slide warmer 14 × 14 in. (Clinical Scientific Equip. Co.) set at 37°C.

Solutions and reagents

All solutions are made from Milli-Q 5 bowl fresh water and filter-sterilized unless otherwise indicated.
Alkaline phosphate buffer (AP-9.5) 100 mM Tris HCl, 100 mM NaCl, 5.0 mM $MgCl_2$, pH 9.5, 5-bromo-4-chloro-3-indolyl phosphate disodium salt (BCIP) (50 mg ml^{-1}) in 100% dimethylformamide (DMSF)
Collagenase II (Worthington) > 125 μ mgP^{-1}
Copper sulphate ($CuSO_4$), 100 mM
L-Glutamic acid γ-(4-methoxy-β-naphthylamide) 4.5 mg in 100 μl DMSO and 100 μl 1N NaOH Glycine–glycine (15 mg) and Fast Granet GBC (15 mg) dissolved in 15 ml 100 mM Tris HCl pH 7.5 then add 42 ml distilled water
Primary Antibodies
Secondary antibodies – alkaline phosphatase conjugates
Nitro blue tetrazolium chloride (NBT) (75 mg ml^{-1}) in fresh 70% DMSF
Trypsin 2 × lyophilized (Worthington) > 180 μ mgP^{-1}
Blocking solution in D-PBS-free containing 0.5% BSA and 10% normal goat serum. Heat inactivated
Ice cold Hanks Balanced Salt Solution (H-BSS) pH 7.2 containing 5.5 mM D-glu-cose, 5% BCS, 10 U penicillin, 10 ng ml^{-1} streptomycin and 500 ng ml^{-1} fungizone.

Ca^{2+} and Mg^{2+} free Dulbecco's Phosphate Buffered Salt Solution (D-PBS-free) Hepes Buffered Hanks (H-BSS) pH 7.2 containing 1.78 mM $NaHCO_3$, 5.5 mM D-glucose, 0.001% phenol red, 10 mM Hepes, 0.1% methylcellulose (low viscosity cps). Dissolve methylcellulose in H-BSS stirring at 4°C for 48 h. After dissolving warm to room temperature and autoclave for three hours (slow exhaust). Cool to room temperature and store at 4°C. The methylcellulose will again take a couple of days to dissolve at 4°C.

Ethidium bromide (1 mg ml⁻¹)/acridine orange (0.3 mg ml⁻¹). Dissolve 100 mg ethidium bromide and 30 mg acridine orange in 2 ml absolute alcohol. Add to 98 ml fresh distilled water, dispense and store at –20°C (Liddel & Cryer 1991). This solution does not need to be sterilized.

Trypsin solution (0.01%) is made using D-PBS-free containing 10 mM HEPES, 0.001% phenol red, 0.1% methylcellulose (low viscosity: 15 cps). Handle methylcellulose as above for H-BSS solution. The pH is adjusted to 7.8 and the solution is filtered through a Gelman VacuCap 0.2 μ and stored at –20°C. Sterile EGTA is added at time of use to make a final concentration of 0.1 mM solution.

Ethyleneglycol -N, N, N', N',-tetraacetic acid (EGTA 10 mM). Slowly dissolve 1.9 gm EGTA into 300 ml PBS. Keep pH of solution at 7.0 with 10 N NaOH while slowly adding the EGTA. Bring final volume up to 500 ml. Filter-sterilize, aliquot in small vials, and store at room temperature.

Media

Modified ALPHA Minimal Essential Medium (Gibco Formula No. 96–5017EF) is dissolved in 9 litres of distilled water. Add to this medium the following: 48 mg l⁻¹ L-valine, 155 mg l⁻¹ NaCl, 350 mg l⁻¹ $NaHCO_3$ and 3600 mg l⁻¹ Hepes. pH to 7.2 with 10 N NaOH. Do not use HCl to adjust the pH. The osmolarity should be 320 ± 5 mOsm. Filter-sterilize (0.2 μ) the medium, and fill 1-litre bottles to 900 ml. Store at 4°C for up to 6 months. To one of the stock bottles contaiing 900 ml of medium add the following: 5 mg insulin (use a 10 mg ml⁻¹ aqueous stock solution, Collaborative Res #40305), 2 mg human transferrin (use a 4 mg ml⁻¹ aqueous stock solution, Collaborative Res. #40304), 500 ng hydrocortisone (use a 2 mg ml⁻¹ stock in 70% ethanol, Sigma H0888), and 100 ml foetal bovine serum (FBS), prescreened for optimal growth of the liver cells. After the medium is added to the cells, 25 ng ml⁻¹ epidermal growth factor (10 μg ml⁻¹ stock in D-PBS, Collaborative Res. #40001) is added to the flask or dish. The flasks and dishes are placed in a 37°C incubator.

BASIC PROCEDURE

Isolation of hepatocytes (Figure 4.10.1)

As with all human cells, the following safety precautions should be taken: (1) work in a Class II, Safety Cabinet, (2) wear gloves and lab coat, (3) disinfect material and work area with 20% bleach.

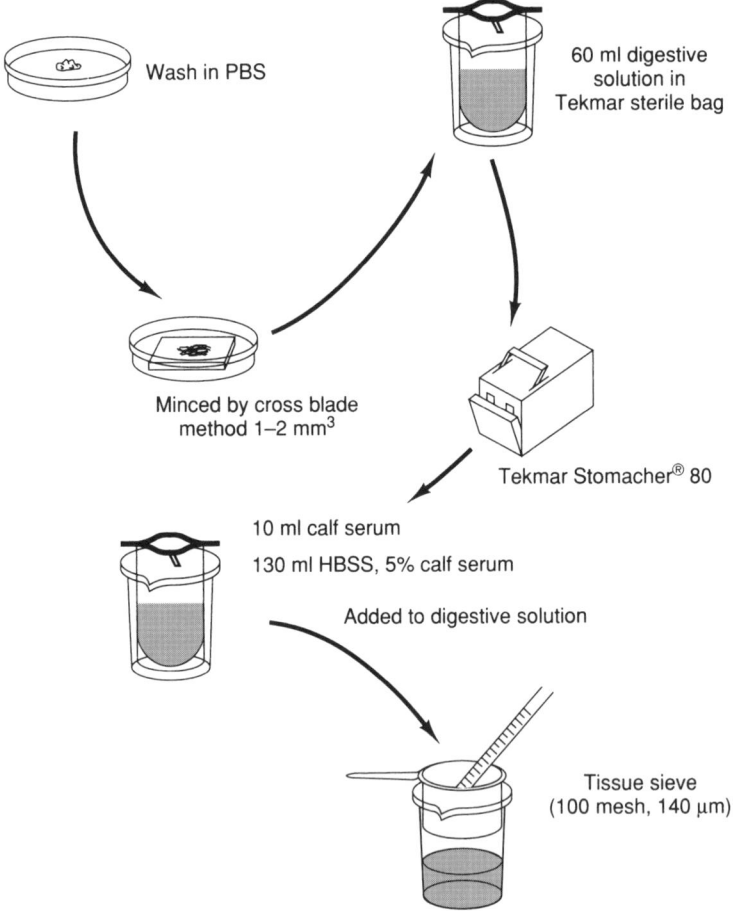

Figure 4.10.1 Experimental flow chart for establishing normal human cells in culture.

1. All liver specimens are shipped overnight on wet ice and received within 24 h after surgery. Shipping solution in H-BSS containing 10 U penicillin, 10 ng ml^{-1} streptomycin, 500 ng ml^{-1} fungizone and 5% calf serum in a 50 ml centrifuge tube. A tissue piece approximately 2×6 cm^3 per shipping tube is cut into small pieces using surgical scissors prior to shipping.

2. After the tissue has been received in the laboratory, prepare the disaggregation solution by mixing Collagenase II (0.05% – 500 ng ml^{-1}) and H-BSS with Hepes Buffer containing methylcellulose. Sterilize this solution by filtering through a Gelman VacuCap and keep at 4°C or on ice until needed. Due to the viscosity of this solution you may need two to three filters per litre of solution.

3. The centrifuge tubes containing liver tissue are wiped with 70% alcohol and the lid passed through a flame to sterilize the mouth of tubes prior to opening in the tissue culture hood.

4. Carefully decant the H-BSS from the tube containing the liver tissue pieces into a waste beaker located in the hood. Aseptically transfer the tissue pieces from the tube into a 100×20 mm Petri dish with fresh, cold D-PBS-free containing 10 U penicillin, 10 ng ml^{-1} streptomycin and 500 ng ml^{-1} fungizone.

5. Tissue dissecting is done at this time, i.e. removal of blood clots, separation of tissue segments, etc. The tissue to be used for primary culture is placed into a new, dry 50 ml centrifuge tube and weighed. After weighing, the tissue is transferred into a fresh Petri dish containing D-PBS-free containing the above antibiotics. The wet, empty centrifuged tube is now weighed and the tissue weight determined.

6. A Petri dish containing a glass cutting block is then wetted with 0.5 ml of the Collagenase (step 2 above) disaggregation solution. Care is taken not to allow fluid to spill over sides of cutting block into dish. At this time the collagenase solution is placed into the sterile Stomacher® bag located in a sterile holding beaker. See Table 4.10.1.

7. The tissue to be used is then placed onto the wet cutting block (~0.5 g per block). Two scalpels are used to mince the tissue into 1–2 mm^3 pieces. Be careful not to tear the tissue. Angle the blades of the scalpels such that the cutting edges are in contact, then pull out both blades smoothly to cut the tissue. If both blades are held at a 90° angle to the glass plate the tissue will be turned rather than cut, and this will reduce viability significantly. This procedure takes no longer than 4 min.

8. Using the scalpel blades as a scoop, carefully transfer the tissue pieces from the cutting block into a Stomacher® bag containing the collagenase solution. Be careful not to touch the inside of the bag with gloved fingers or the system might be contaminated. Once the tissue is in the bag, remove as much air as possible (use solution to displace air). Close the bag by rolling the top down a couple of times and bind the tags.

9. Place the bag containing the tissue pieces into the Stomacher® and incubate at room temperature. Adult liver takes approximately 8–12 min to digest at 150 strokes per minute. Do *not* overdigest. This can be monitored by observing the appearance of single cells within the bag using an inverted microscope. Tissue fragments disappear into 'feather-like' structures. While the tissue is being digested, clear the tissue culture cabinet of used items.

10. At the end of the Stomacher® incubation, the solution bag is again placed into the holding beaker in the culture cabinet. Wipe the lip of bag with a 4×4 sterile gauze wetted with 70% alcohol before opening. Carefully open the bag, and add 10 ml of bovine calf serum (BCS) and 120 ml of ice cold H-BSS containing 5% BCS (Table 4.10.1).

Table 4.10.1 Stomacher® procedural summary

Tissue weight (g)	Stomacher® model	Bag size: sterile	Collagenase solution (ml)	Hanks BSS 5% BCS (ml)
≤ 0.5	STO-80	Small	60	120
≥ 0.5 ≤ 1.0	STO-4CD	Large	300	180

11. Hold in place the sterile stainless steel sieve on the edge of a sterile beaker and gently pipette the cell suspension from the Stomacher® bag through the sieve. Dispense the cell suspension into 50 ml or 150 ml plastic polypropylene test tubes and gently centrifuge at 50 g for 5 min at 4°C to isolate hepatocytes. Gently decant and discard the supernatant from cell pellet leaving approximately 1–2 ml on top of the cell pellet. Longer (6–8 min) centrifuge times might help to stabilize a soft cell pellet but can also increase non-hepatic cell contamination and/or cell loss due to clumping. Repeated centrifuging of the cell supernatant frequently increases the number of haepatic cells isolated, but will also increase the amount of non-viable cells.

12. Gently resuspend each cell pellet into 10 ml growth media (mALPHA MEM with 10% FBS and growth factors). Combine all cell pellets and keep cell suspension on ice until ready for counting. If total cell suspension volume is more than 30 to 50 ml, centrifuge a second time at 150 g for 5 min at 4°C. It should be noted, however, that a second centrifuge may cause cell lose due to clumping. Combine all cell pellets after the second centrifuge into no more than 30 to 50 ml for counting.

13. Cell viability is determined by removing 1.0 ml of freshly resuspended cell suspension to a small glass test tube and 2.5 μl of the ethidium bromide/acridine orange solution is added and gently mixed. Load cells as usual for a haemocytometer count. Immediately view under a fluorescence microscope using the same wavelength settings as for FITC (λ_{ex} 500 nm, λ_{em} 595 nm). Ethidium bromide is excluded from viable cells and hence, non-viable cells appear orange/red. Viable cells take up the acridine orange and hence, appear green (Liddell & Cryer 1991).

14. Cells are then seeded at 2×10^4 viable cells per cm². Total cells and amount of media depends on flask size. (Table 4.10.2) Seeding viable cells at a higher density appears to decrease the growth potential. Likewise when cells are seeded at a lower density, growth response is reduced and time in primary culture can be prolonged.

Primary cell culture

1. Primary flasks with caps tightened are placed inside a 37°C *NON-CO$_2$* incubator.

Table 4.10.2 Cell maintenance number and volume summary

Corning tissue culture flask/dish	Trypsin (ml)	Media used to inactivate trypsin (ml)	Number of cells seeded per vessel	Media used to seed cells (ml)
25 cm² flask	0.5	5	$1 \times 10^5 \leq 2 \times 10^5$	5
75 cm² flask	1.0	10	$3 \times 10^5 \leq 5 \times 10^5$	10
150 cm² flask	1.5	15–20	$7.5 \times 10^5 \leq 1 \times 10^6$	15–20
12-well multi-well plate (1.9 cm²)	n/a	n/a	$3.6 \times 10^4 \leq 5 \times 10^4$	2
35 mm dish (8 cm²)	n/a	n/a	$3.6 \times 10^4 \leq 5 \times 10^4$	2
60 mm dish (21 cm²)	n/a	n/a	$8.4 \times 10^4 \leq 5 \times 10^5$	2

2. On day one (Figure 4.10.2a), flasks are checked for contamination, but moved as little as possible. Be careful not to disturb primary flasks every day as this will disturb the loosely attached cells and inhibit primary growth. On day three (Figure 4.10.2b) all flasks are gently washed with room temperature D-PBS-free and fresh prewarmed growth media is added. On day six (Figure 4.10.2c), flasks are fed by adding half the amount contained within the flasks with fresh prewarmed growth media. This pattern is followed until primary cultures are subcultured.

3. Flasks are than observed every three days. Figure 4.10.2 shows single and clusters of cells that appear at various times after the initial seeding. Primary growth may take 14 to 28 days depending on age of donor, time between surgery and culturing, and shipping conditions.

4. Washing and feeding the primary cultures should be done on the days when the flasks are observed. Alternating between washing cells with D-PBS-free and adding fresh media and merely small amounts of fresh growth media can quicken growth. Gently washing cells with room temperature D-PBS-free will stimulate growth, but is not needed every 3 days.

5. When cultures grow to 80–90% confluency, subpassage the primary. Allowing the cells to grow to > 90% confluent or in large tightly packed cell clusters will greatly decrease population growth potential. We have found that subculturing early is better than waiting until cultures are close to 100% confluent (Figure 4.10.3).

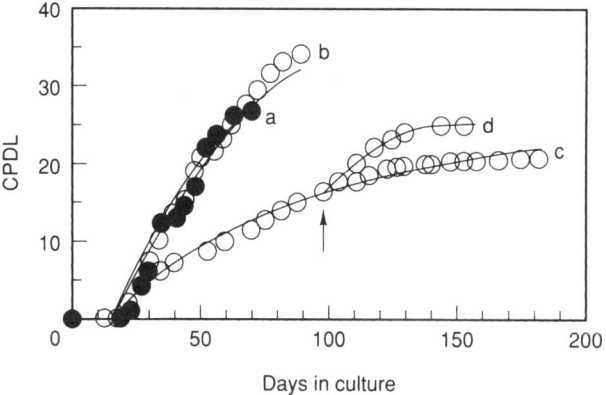

Figure 4.10.2 Cellular growth kinetics of normal adult liver hepatocytes. Cell line 1007 **(b)** and 1014 **(b,c,d)**. Primary cell cultures were established at a density of 2×10^4 cells per cm^2 and subpassaged at either: 4.6×10^3 **(a,b)** or 1.3×10^4 **(c)** cells per cm^2 (Figure1, Gibson D'Ambrosio *et al.* 1993).

Figure 4.10.3 Photomicrographs of normal human cells in primary culture. Immediately after seeding into flask on day 0. **(a)** Notice the single cells, as well as cellular debris floating in the flask. **(b)** Day 3, after first washing and feeding of the cultures, shows single cells (arrows). **(c)** Day 9 shows the formation of a cell colony.

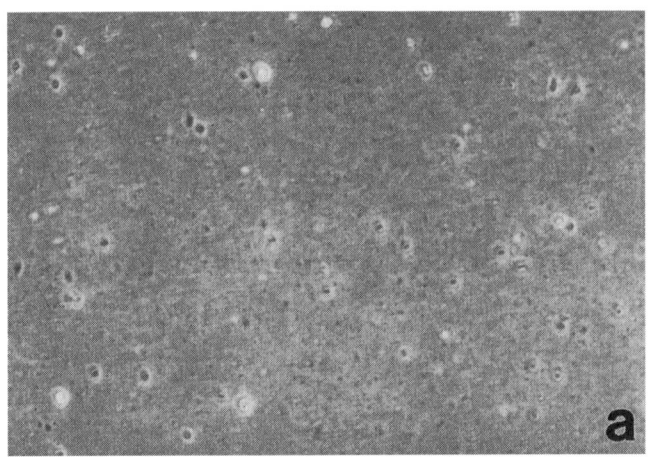

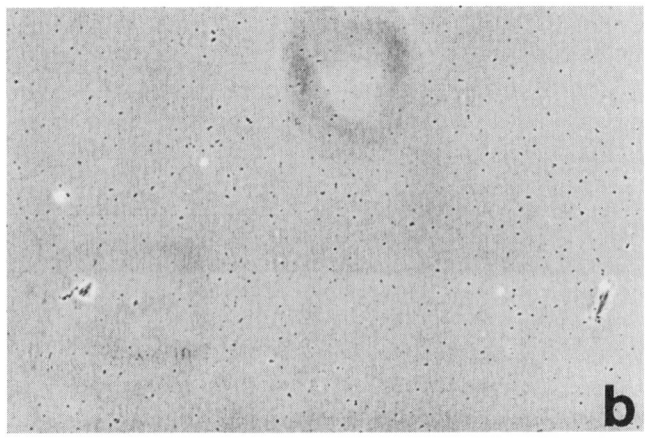

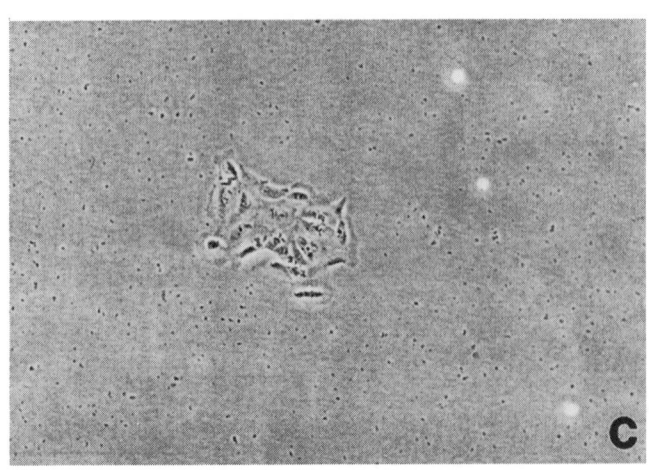

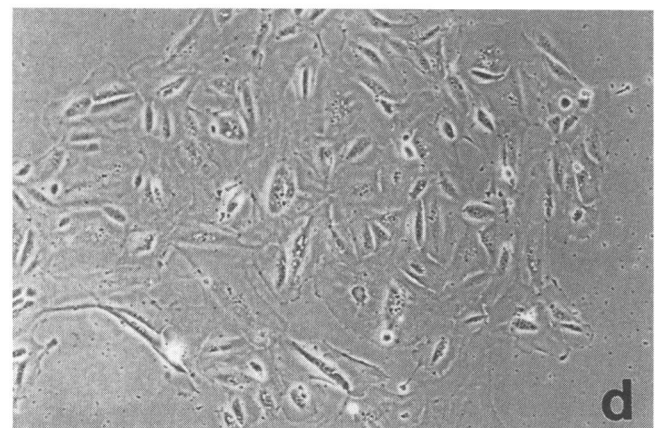

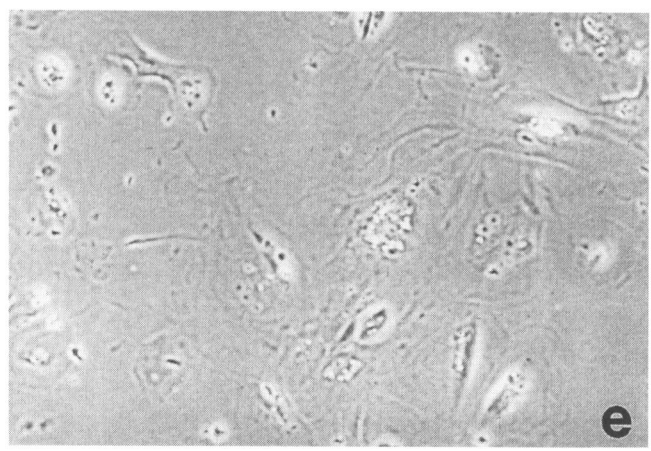

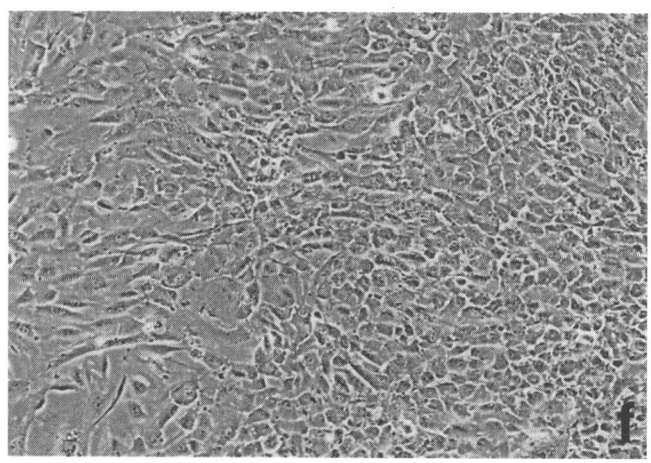

Subculturing

1. Gently wash the flasks twice with room temperature D-PBS-free. Set flask on end and remove as much D-PBS-free as possible.
2. Add an appropriate amount of 0.01% Trypsin/EGTA (Table 4.10.2) and place on a 37°C slide warmer for 5 min. Primary cultures flask may have to be left on the slide warmer for up to 10 min if tight cell clusters are present. If flasks of cells are left longer, the number of viable cells will decrease. Give the flask a couple quick raps against the back of old chair cushion. This will dislodge cells from the plastic surface and break up cell clusters into a single cell suspension.
3. Use an appropriate amount of (Table 4.10.2) fresh growth media that has been warmed to 37°C to wash down the inside of the flasks. This will dilute the trypsin concentration adequately to inactivate it. Centrifuging cell suspensions may create clumps that are hard to dissociate. A useful rule is to seed cells at approximately 4×10^3 viable cells per cm^2 into proliferating cultures. This number is different from the viable cell number used at primary culture.
4. Subpassaging the cultures at regular intervals (5 to 7 days), before cultures achieve 100% confluency, leads to a constant rate of growth (Figure 4.10.3) and more stable culture characteristics.

CHARACTERIZATION OF HEPATIC FUNCTION

A summary of the characteristics of human hepatocytes in culture is described in Table 4.10.3. The expression of hepatic phenotypes in culture are compared to expression of the same phenotypes *in vivo* in human liver tissue sections (Figure 4.10.4).

Immunohistochemistry

1. Seed cells into 12-well tissue culture cluster plates (Corning).
2. Culture the cells for 5 to 7 days until approximately 80% confluent.
3. Rinse the cells twice with D-PBS-free and allow to air dry.
4. Fixed the cells with buffered formalin for 15 min. Wash with PBS and air dry.
5. Endogenous alkaline phosphatase is inactivated by incubating the cells at 37°C overnight in a D-PBS-free solution containing 10 mM EGTA.
6. Block non-specific antigenic sites with D-PBS-free containing 0.5% BSA and 10% normal goat serum for 15 min prior to incubation with the primary antibodies.
7. Incubate cells for 1 h at 37°C with commercially available primary antibodies directed specifically toward human albumin, α-fetoprotein, factor-VIII and

Figure 4.10.3 (cont) (d) By day 12 the colonies have greatly expanded to occupy the field of view. **(e, f)** A representative culture on day 19 showing two different areas of the same flask. **(e)** shows cells at approximately 80% confluency. **(f)** shows an area of increasing cell density from left to right. These are the conditions at which time the primary culture must be subpassaged.

Table 4.10.3 Characterization of normal human liver haepatocytes in culture

Marker	Function	Results[a]
Albumin	Hepatic specific secretory protein (Caron 1990; Herbst & Babiss 1990; Tilghman 1985)	Strongly positive in areas of densely packed cells
Cytokeratin 18	Expressed by normal adult hepatocytes and bile ductal cells (Desmet et al. 1990; Ishii et al. 1989; Lai et al. 1989; Shiojiri et al. 1991)	Strongly positive in > 95% of cells in culture
Cytokeratin 19	Induced in response to retinoids (Crowe et al. 1991)	Weakly positive in cells growing in media containing retinoid
p450	Responsible for metabolizing polyaromatic hydrocarbons, characteristic of hepatocytes (Wrighton & Stevens. 1992)	Positive, but individual levels vary
alpha-fetoprotein	Expressed in liver ductal oval cells or in fetal hepatocytes (Abelev 1989; Sell & Ilic 1997)	Negative
Factor VIII	Expressed in endothelial cell (Ruiter 1989)	Negative
gamma-Glutamyl transpeptidase	Thought to be expressed in oval cells and not hepatocytes (Sell, 1990)	Negative
Glycogen	Metabolic function unique to hepatocytes. Used to store glucose, in the form of glycogen, in the liver.	Positive

[a]See Figure 4.10.4, and previously published data (Gibson-D'Ambrosio et al. 1993, 1995, 1998).

various cytokeratins. These antibodies should first be titred using positive and negative control cell lines to determine the proper working dilutions and conditions of incubation (Harlow & Lane 1988; Speel et al. 1995). The following cell lines can be used as positive control: albumin (human HBG$_2$ (ATCC HB8065) hepatoma cell line); α-fetoprotein (human foetal liver cell lines); keratin (human keratinocytes); and Factor VIII (HUVEG endothelial cell line from Clonetics). Negative control cell lines included for albumin and α-fetoprotein (HUVEG endothelial cell line); keratin and Factor VIII (human skin fibroblasts).

8. Wash the cells with D-PBS-free twice.

9. An alkaline phosphatase conjugated goat-anti-rabbit-IgG or mouse-IgG (Schleicher & Schuell) is incubated with the cells for 1 h at 37°C.

10. Wash twice and incubate the cells with a AP-9.5 containing levamisolc (0.5 mM), pH 9.5 solution and 338 μg ml^{-1} NBT and 175 μg/ml BCIP for 30 min at 37°C.

11. Wash the cells with distilled water, and a round coverslip are mounted with an aqueous mounting agent. The cells are viewed under a microscope. The appearance of a purple/blue colour in the cytoplasm of the cells, as visualized under brightfield microscopy, indicate the presence of antigen.

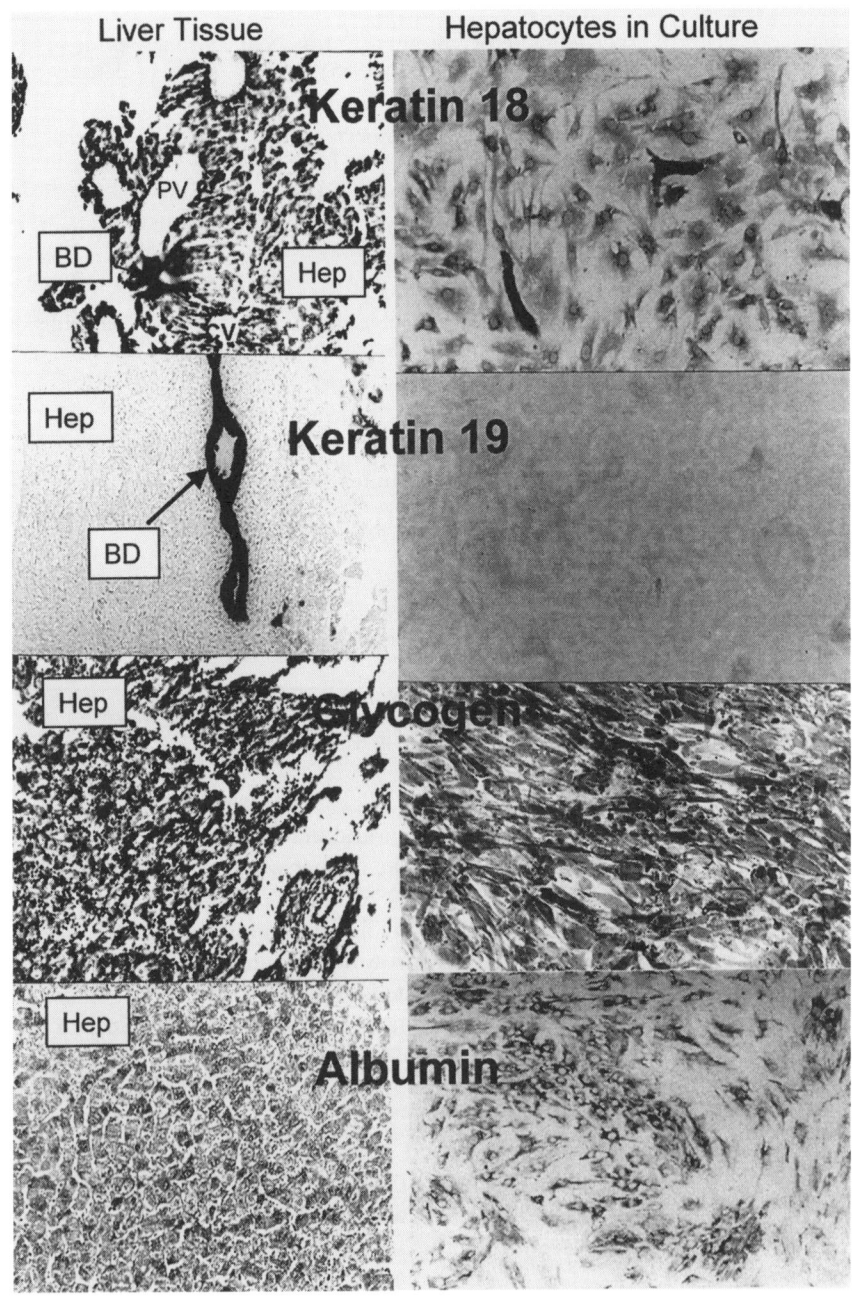

Gamma-glutamyl transpeptidase (GGT) activity

1. GGT activity is determined histochemically (Gibson-D'Ambrosio *et al.* 1987) directly in cells fixed with 95% ethanol.
2. Gamma-*L*-glutamyl-4'-methoxy-β-naphthylamide, glycineglycine and Fast Garnet GBC are dissolved, and incubated with the cells for 30 min at 37°C.
3. Rinse the cells with distilled water and incubate the cells with 100 mM $CuSO_4$ for 1 min. Rinse with distilled water.
4. Cells are viewed under microscope as above. The appearance of red precipitate in the cytoplasm is characteristic of liver ductal oval cells. Hepatocytes should be negative.
5. Use kidney proximal tubule cells as a positive control. A good negative control is the human endothelial cell line, HUVEG from Clonetics, Inc.

p450 Activity

1. The activity of *p450* cytochrome in the liver cells is determined by monitoring the conversion of radiolabelled benzo[a]pyrene (BP) to water soluble metabolites as described by Gibson *et al.* (1993).
2. Seed cells into Corning T25 flasks and allowed cultures to reach approximately 80% confluency.
3. Replace medium with fresh medium containing 50 μl (0.85 nmol) of the ^3H labelled BP (final concentration of 0.17 μM). All procedures are carried out under yellow light.
4. At various times, remove medium and extract the water-soluble metabolites by mixing first with one volume each of acetone and ethylacetate, then with ethylacetate. The amount of radiolabelled material in the growth medium, final aqueous and first organic layers are then determined by scintillation counting. The percentage of BP converted to water-soluble metabolites is calculated from total radioactivity after subtracting background radioactivity.

Glycogen

1. Cells are fixed with a formalin-ethanol (5 : 95) solution.
2. Glycogen is determined cytochemically using Periodic Acid-Schiff Kit (Sigma Cat. 395-B) essentially as described by the manufacturer.

Figure 4.10.4 Human hepatocytes in culture exhibit hepatic specific functions. In this figure we compared haepatocyte specific markers in normal liver tissue sections with cells cultured from liver tissue. In tissue sections, notice that both haepatocytes (HEP) and bile duct (BD) cell immunoreact with monoclonal antibody to keratin 18. In cell culture there is also a strong staining for keratin 18. In tissue, only the BD cells stain for keratin 19. Notice the lack of staining for keratin 19 in the cell culture. This positive staining for keratin 18 and negative staining for 19 indicate the haepatic nature of our cultured cells. In tissue and cell culture, haepatocytes actively synthesize glycogen. In both tissue and cultured cells, the hepatocytes actively synthesize human albumin.

3. The cells are counter-stained with Haematoxylin solution (Gill No.3).
4. Cells are viewed under a brightfield microscope for the appearance of intense red cytoplasmic stain.

DISCUSSION

Despite the high capacity of hepatocytes to proliferate *in vivo*, typically hepatocytes in primary culture exhibit only a few population doublings, deteriorate and die (Block *et al.* 1996; Butterworth *et al.* 1989; Coleman & Presnell 1996; Grisham *et al.* 1993; Grisham & Thorgeirsson 1997; McQueen 1989; Mitaka *et al.* 1995; Shimbara *et al.* 1996; Tateno & Yoshizato 1996; Mitaka *et al.* 1995). While the clonal expansion of rat hepatocytes with three to four population doublings in primary culture has been reported (Block *et al.* 1996), these cells appeared to function as, or be the source of, bipotential facultative hepatic stem cells. The rat cells in culture also exhibited a great deal of plasticity both in morphology and

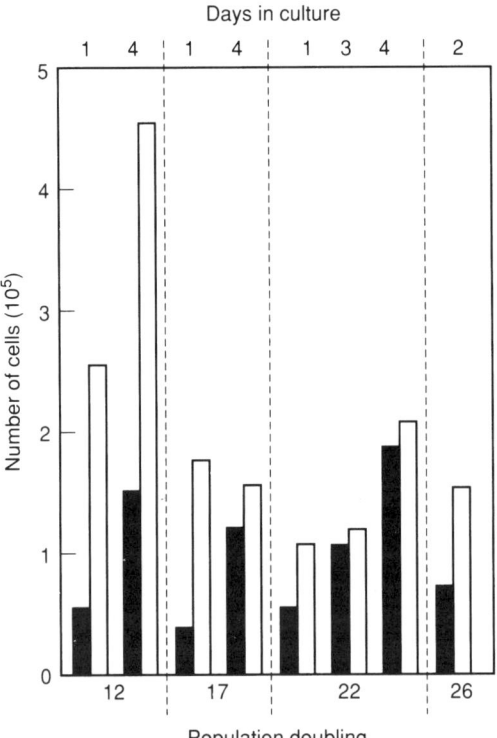

Figure 4.10.5 Expansion of cells producing human albumin with time in culture. Cells from LI1007 were assayed immunohistochemically for albumin at the indicated PDL and post-seeding days in culture. Cultures were viewed under brightfield and phase contrast microscopy and 300–500 cells were counted to determine albumin positive (closed bar) and total number of cells (open bar), respectively (Figure 4, Gibson-D'Ambrosio *et al.* 1993).

in the expression of hepatic (albumin), and non-hepatic cell markers (cytokeratin 19, γ-glutamyl transpeptidase, or α-fetoprotein) expressed by bile duct or oval epithelial cells (Block *et al.* 1996). The human hepatocytes established and cultured as described in this section maintain the expression of haepatic specific markers, i.e. albumin and glycogen, while never acquiring the non-hepatic markers, i.e. cytokeratin 19, γ-glutamyl transpeptidase, or α-fetoprotein (Table 4.10.3, Figures 4.10.4 and 4.10.5).

Human hepatocytes in culture exhibit morphological and functional plasticity which appears to be dependent upon cell density, surface substratum, and medium components. For example, human hepatocytes in sparsely populated areas of the culture dish appear elongated morphologically, while hepatocytes in dense colonies appear polygonal in shape with well-defined edges (Figure 4.10.4). The expression of albumin also appears to be plastic, highly dependent upon cell density, with the greatest level of expression appearing in the tightly packed polygonal shaped cells (Figure 4.10.4) (Gibson-D'Ambrosio *et al.* 1998). Human hepatocytes seeded onto Matrigel®, collagen or poly-D-lysine strongly express albumin and cease to replicate. Low amounts of calcium (< 1.0 mM) encourage polygonal morphology, but the cells differentiate terminally. The mALPHA medium contains a higher level (1.8 mM) of calcium which appears to encourage cellular proliferation. Trace elements, the described levels of amino acids, and prescreened FBS, which we have found to be important for continual growth of many epithelial cell types (Gibson-D'Ambrosio *et al.* 1986), are also essential for the growth of human hepatocytes. This suggest that exogenous signal(s) from the medium and/or surface substratum are important for determining whether the hepatocyte should proliferate or terminally differentiate. This process is analogous to the *in vivo* state, where hepatocytes are quiescent until a signal, i.e. an injury, is given for the hepatocyte to divide and replace lost cells. Upon reaching a critical tissue density, another signal is given for hepatocytes to stop dividing. Thus, the human hepatocyte cell culture system described here provides an important system for investigating mechanisms of hepatocyte growth and function, and potentially in the use of human hepatocytes in gene therapy for treatment of metabolic disorders and for xenograft replacement of hepatocytes lost during liver disease.

Troubleshooting

Care should be taken during the shipping of tissue. This time period is very important to the success of the culture. The solutions we use maintain cells during shipping with high viability and hepatic function for up to 36 h. We do not use University of Wisconsin (UW) solution during shipment because we, like others (Guillouzo *et al.* 1993), have found loss of viability and cell function. During shipping and dissecting, the tissue that remains red and soft will yield the most viable cells. Tissue that appears brown and/or is difficult to cut is less likely to yield high viability. Do not spend longer than 6 min mincing the tissue, or you will decrease total cell yield.

When collecting cells for counting, if total cell yield (both viable and non-viable cells) is high, it is better to resuspend cells into 30–50 ml of media. A high density of suspended cells will result in cell clumping, and increased cell density during the initial phases of growth, greatly hampering establishment and subsequent subpassaging of primary cultures.

Since hepatic cells are generally very fragile, the more centrifuging that occurs, the lower the viability and total cell yield. This is not only true in primary culture but also during the entire lifespan of the cell line. At early passages it is important to characterize the cell line as well as to cryopreserve cells for later use. It is important to monitor the hepatic function continuously during culture. Sufficient time (5–7 days in passage) should pass before monitoring to allow expression of these markers in culture.

REFERENCES

Abelev GI (1989) Alfa-fetoprotein: 25 years of study. *Tumor Biology* 10: 63–74.

Block GD, Locker J, Bowen WC, Petersen BE, Katyal S, Strom SC, Riley T, Howard TA & Michalopoulos GK (1996) Population expansion, clonal growth, and specific differentiation patterns in primary cultures of hepatocytes induced by HGF/SF, EGF and TGFα in a chemically defined (HGM) medium. *Journal of Cellular Biology* 132: 1133–1149.

Bralet MP, Branchereau S, Brechot C & Ferry N (1994) Cell lineage study in the liver using retroviral mediated gene transfer evidence against the streaming of hepatocytes in normal liver. *American Journal of Pathology* 144: 896–905.

Bralet MP, Calise D, Brechot C & Ferry N (1996) In vivo cell lineage analysis during chemical hepatocarcinogenesis using retroviral-mediated gene transfer. *Laboratory Investigation* 74: 781–881.

Bucher NLR (1995) Liver regeneration then and now. In: Jirtle RL (ed.) *Liver Regeneration and Carcinogenesis: Molecular and cellular mechanisms* pp. 1–25. Academic Press, San Diego.

Butterworth BE, Smith-Oliver T, Earle L, Loury DJ, White RD, Doolittle DJ, Working PK, Cattley RC, Jirtle R, Michalopoulos G & Strom S (1989) Use of primary cultures of human hepatocytes in toxicology studies. *Cancer Research* 49: 1075–1084.

Caron JM (1990) Induction of albumin gene transcription in hepatocytes by extracellular matrix protein. *Molecular Cell Biology* 10: 1239–1243.

Coleman WB & Presnell SC (1996) Plasticity of the hepatocyte phenotype in vitro: Complex phenotypic transitions in proliferating hepatocyte cultures suggest bipotent differentiation capacity of mature hepatocytes – Comments. *Hepatology* 24: 1542–1546.

Columbano A & Shinozuka H (1996) Liver regeneration 8. Liver regeneration versus direct hyperplasia. *FASEB Journal* 10: 1118–1128.

Crowe DL, Hu L, Gudas LJ & Rheinwald JG (1991) Variable expression of retinoic acid receptor (RARb) mRNA in human oral and epidermal keratinocytes: Relation to keratin 19 expression and keratinzation potential. *Differentiation* 48: 199–208.

Desmet VJ, Van Eyken P & Sciot R (1990) Cytokeratins for probing cell lineage relationships in developing liver. *Hepatology* 12: 1249–1251.

Gibson-D'Ambrosio RE, Brady T & D'Ambriosio SM (1995). Identifying and isolation of human epithelial cell colonies that express specific gene products. *Biotechniques* 19: 784–790.

Gibson-D'Ambrosio RE, Brady T & D'Ambrosio SM (1998). The regenerative capacity of normal human adult hepatocytes: clonal expression of cells producing albumin in culture. *Hepatology Research* 11: 188–200.

Gibson-D'Ambrosio RE, Crowe DL, Shuler CF & D'Ambrosio SM (1993) The establishment of continuous subculturing of normal human adult hepatocytes: Expression of differentiated liver functions. *Cell Biology and Toxicology* 9: 385–403.

Gibson-D'Ambrosio RE & D'Ambrosio SM (1996) Replicative and functional cultures of normal human hepatocytes. In: Doyal A, Griffiths JB & Newell DG (eds) *Cell & Tissue Culture: Laboratory procedures*, pp. 12B: 17.1–12B: 17.5. Wiley, Chichester.

Gibson-D'Ambrosio RE, Samuel M, Chang CC, Trosko JE & D'Ambrosio SM (1987). Characteristics of long-term human epithelial cell cultures derived from normal human fetal kidney. *In Vitro Cellular and Developmental Biology*, 23: 279–287.

Gibson-D'Ambrosio RE, Samuel M & D'Ambrosio SM (1986) A method for isolating large numbers of viable disaggregated cells from various human tissues for cell culture establishment. *In Vitro Cellular and Developmental Biology* 22: 529–534.

Grisham JW (1995) Hepatic epithelial stem-like cells. *Verh.Dtsch.Ges.path.* 79: 47–54.

Grisham JW, Coleman WB & Smith GJ (1993) Isolation, culture, and transplantation of rat hepatic precursor (stem-like) cells. *Proceedings of the Society of Experimental Biology and Medicine* 204: 270–279.

Grisham JW & Thorgeirsson SS (1997) Liver stem cells. In: Potten CS (ed.) *Stem Cells. Cell differentiation* pp. 233–282. Academic Press, San Diego.

Guillouzo A, Morel F, Fardel O & Meunier B (1993) Use of human hepatocyte cultures for drug metabolism studies. *Toxicology* 82: 209–219.

Harlow E & Lane D (1988) *Antibodies, a Laboratory Manual*. Cold Spring Harbor Laboratory, New York.

Herbst RS & Babiss LE (1990) Regulation of liver gene expression during development and regeneration. In: Fisher PB (ed.) *Mechanisms of Differentiation: Modulation of differentiation by exogenous agents*, pp. 15–48. CRC Press, Baco Raton, FL.

Ishii M, Vroman B & LaRusso N (1989) Isolation and morphological characterization of bile duct epithelial cells from normal rat liver. *Gastroenterology* 97: 1236–1247.

Kennedy S, Rettinger S, Flye MW & Ponder KP (1995) Experiments in transgenic mice show that hepatocytes are the source for postnatal liver growth and do not stream. *Hepatology* 22: 160–168.

Lai Y-S, Thung SN, Gerber MA, Chen ML, & Schaffner F (1989) Expression of cytokeratins in normal and diseased livers and in primary liver carcinomas. *Archives of Pathology and Laboratory Medicine* 113: 134–138.

Liddell JE & Cryer A (1991) *A Practical Guide to Monoclonal Antibodies*. Wiley, Chichester.

McQueen CA (1989) Hepatocytes in monolayer culture: An *in vitro* model for toxicity studies. In: McQueen CA (ed.) *In Vitro Toxicology: Model systems and methods*. Telford Press Inc., Caldwell NJ.

Michalopoulos GK & De Frances MC (1997) Liver regeneration. *Science* 276: 60–66.

Mitaka T, Kojima T, Mizuguchi T & Mochizuki Y (1995) Growth and maturation of small hepatocytes isolated from adult rat liver. *Biochemical and Biophysical Research Communications* 214: 310–317.

Overturf K, Al-Dhalimy M, Ou CN, Finegold M & Grompe M (1997) Serial transplantation reveals the stem-cell-like regenerative potential of adult mouse hepatocytes. *American Journal of Pathology* 151: 1273–1280.

Overturf K, Al-Dhalimy M, Tanguay R, Brantly M, Ou CN, Finegold M & Grompe M (1996) Hepatocytes corrected by gene therapy are selected in vivo in a murine model of hereditary tyrosinaemia type I. *Nature Genetics* 12: 266–273.

Pistoi S & Morello D (1996) Prometheus' myth revisted: transgenic mice as a powerful tool to study liver regeneration. *FASEB Journal* 10: 819–828.

Rhim JA, Sandgren EP, Degen JL, Palmiter RD & Brinster RL (1994) Replacement of of diseased mouse liver by hepatic cell transplantation. *Science* 263: 1149–1152.

Rhim JA, Sandgren EP, Palmiter RD & Brinster RL (1995) Complete reconstitution of the mouse liver with xenogeneic hepatocytes. *Proceedings of the National Academy of Science of the USA* 92: 4942–4946.

Ruiter D (1989) Monoclonal antibody-defined human endothelial antigens as vascular markers. *Journal of Investigative Dermatology* 93: 25S–32S.

Sell S (1990) Cancer markers of the 1990s. Comparison of the new generation of markers defined by monoclonal antibodies and oncogene probes to prototypic markers. *Clin.Lab.Med.* 10: 1–37.

Sell S & Ilic Z (1997) *Liver Stem Cells*. Landes, Austin TX.

Shimbara N, Atawa R, Takashina M, Tanaka K & Ichihara A (1996). Long-term culture of functional hepatocytes on chemically modified collagen gels. *Cytotechnology* 21: 31–43.

Shiojiri N, Lemire JM & Fausto N (1991). Cell lineages and oval cell progenitors in rat liver development. *Cancer Research* 51: 2611–2620.

Speel EJM, Ramaekers FCS & Hopman AHN (1995) Cytochemical detection systems for *in situ* hybridization, and the combination with immunocytochemistry. 'Who is still afraid of Red, Green and Blue?'. *Histochemical Journal* 27: 833–858.

Steer CJ 91995) Liver regeneration. *FASEB Journal* 9: 1396–1400.

Tateno C & Yoshizato K (1996) Long-term cultivation of adult rat hepatocytes that undergo multiple cell divisions and express normal parenchymal phenotypes. *American Journal of Pathology* 148: 383–392.

Thorgeirsson SS (1996) Hepatic stem cells in liver regeneration. *FASEB Journal* 10: 1249–1256.

Tilghman SM (1985) The structure and regulation of the α-fetoprotein and albumin genes. In: Maclean N (ed.) *Oxford Surveys on Eukaryotic Genes* pp. 160–206. Oxford University Press, Oxford.

Wrighton SA & Stevens JC (1992) The human hepatic cytochromes P450 involved in drug metabolism. *Critical Reviews in Toxicology* 22: 1–21.

4.11 EMBRYONIC KIDNEY IN ORGAN CULTURE

The mammalian kidney, the metanephros, is a powerful model system for studying cytodifferentiation and organogenesis, including their guiding mechanisms. The formation of the organ comprises regular branching of an epithelial component, the ureter bud, and transformation of the nephric mesenchyme into epithelial secretory tubules. The tubules segregate into segments, each expressing specific molecules. Finally vascular and neuronal elements participate in this development. The mechanisms by which the different cell lineages interact during their spatially and temporally regulated development are only superficially known. The *in vitro* techniques presented here are designed for studies on these interactive events. The techniques include the transfilter method, designed for studying the kinetics of development.

It is important to use kidneys of strictly defined developmental stages for any quantitative determinations and for starting organotypic cultures. Since rapid, dramatic changes occur, especially during the early stages of nephrogenesis, differences of a few hours may profoundly affect the results. Hence it is important to define the normal development.

In the mouse, the first detectable morphogenetic events leading to the formation of the metanephric kidney occur early on embryonic day 11 (in the rat, these events occur approximately two days later). There are, however, strain-dependent differences in this timing. Thus the developmental stage should always be ultimately checked by morphological criteria. To avoid unnecessary dissection work, screening of some gross features of embryos facilitates choosing those to be dissected and examined in detail.

Many of the experiments based on this model system use the 11-day kidney rudiment as the starting material or the source of the main cell lineages. At this stage an epithelial bud derived from the Wolffian duct has bulged into the nephric mesenchyme where it will soon branch to the T-shaped stage (Figure 4.11.1). In the first screening of the material, these stages can be reasonably well predicted from the shape of the hind limb bud of the embryo as illustrated in Figure 4.11.1. It must be stressed that mesenchyme cells beyond Stage II in Figure 4.11.1 should not be used for transfilter experiments as described below.

Cell and Tissue Culture for Medical Research, edited by A. Doyle and J.B. Griffiths.
© 2000 John Wiley & Sons, Ltd

PRELIMINARY PROCEDURE: THE PREPARATION OF NUCLEPORE FILTERS AND PANCREATIN–TRYPSIN

Reagents and solutions

Serum-free medium

In transfilter cultures, FBS in the culture medium can be substituted by selected horse or human serum. For serum-free culture, the critical growth factor is transferrin. The following serum-free medium has been successfully used:

Richter's medium (Gibco; available on special order) supplemented with glutamine (final concentration 2 mM) or,
1.2 g l^{-1} bovine serum albumin (BSA) and
30 mg l^{-1} transferrin (Boehringer Mannheim Gmbh, Mannheim, Germany).

Routine culture medium

Eagle's minimum essential medium (MEM) with Earle's salts, supplemented with 10% foetal bovine serum (FBS; batch selected for non-toxicity) and penicillin (final concentration 100 IU ml^{-1}) and streptomycin (final concentration 100 μg ml^{-1}).

2% agarose (Sigma) in MEM

Dissolve by boiling, autoclave in glass tubes in 1 ml aliquots, store at 4°C.

Materials and equipment

- Dulbecco's phosphate-buffered saline (PBS), pH 7.2–7.4 (with Ca^{2+} and Mg^{2+})
- Routine culture medium
- 2% agarose (Sigma) in MEM
- 2.25% pancreatin/0.75% trypsin in Ca^{2+}/Mg^{2+}-free Tyrode's solution, pH 7.4, supplemented with penicillin (final concentration 100 IU/ml) and streptomycin (final concentration 100 μg ml^{-1})
- Nuclepore filters (Nuclepore Corp., Pleasanton, CA, USA) – for pretreatment of filters, see 'Pretreatment of Nuclepore filters'
- Culture grids
- Siliconized Pasteur pipettes
- Microburner
- Mouthpiece and tubing for mouth pipette

 Mouth pipetting is not usually permissible for safety reasons. Special care must be taken to avoid any likelihood of infection of the operator. However, the risk of such infection in this particular method is low and we are given to understand from the author that this procedure is only possible using this method.

All instruments should be of highest quality stainless steel. Micropipettes of desired bore sizes are pulled from the siliconized Pasteur pipettes by using the microburner.

PRELIMINARY PROCEDURE: PRE-TREATMENT OF NUCLEPORE FILTERS

1. Soak Nuclepore filters, pore size 0.05–8.0 μm, in 0.5% Pyroneg (Nuclepore Corp., Pleasanton, CA) solution overnight.
2. Rinse in 10 changes of tap water, followed by 10 changes in double-distilled water; avoid air bubbles during the rinsing.
3. Store in 70% ethanol. Use within 2 months.

Before use, rinse in three changes of double-distilled water or Dulbecco's PBS. In the final rinse, cut rectangles of 2–3 mm × 4–5 mm from the filters. The Nuclepore filters have a matt and a glossy side. For convenient identification of the sides during further steps, cut a corner of the rectangle: remove, for example, the upper-right corner from each filter piece with the matt side up.

Preparation of pancreatin–trypsin

1. Dissolve pancreatin and trypsin in Ca^{2+}/Mg^{2+}-free Tyrode's solution by using a magnetic stirrer at minimum speed, at 4°C, overnight.
2. Add penicillin (final concentration 100 IU ml^{-1}) and streptomycin (final concentration 100 μg ml^{-1}).
3. Adjust pH to 7.4.
4. Sterilize by filtering (pore size 0.8 μm).
5. Dispense in 1 ml aliquots.
6. Store at –20°C.

PROCEDURE: SETTING UP THE CULTURES

The stage of the kidneys is routinely expressed as embryonic days. Convention designates the day of the appearance of the vaginal copulation plug as day 0. For the transfilter kidney induction experiments 11-day mouse embryos are used. The animals are killed in the morning of day 11; this corresponds to about 11.5 days post coitum. The staging of the embryos must, however, always be based on morphology (see Figure 4.11.1).

For dissection of the early kidney anlagen, a block is cut by two transverse sections on both sides of the hind limb and divided into halves by a sagittal cut. The anlagen will be found in the locations indicated provided that the transillumination of the preparation microscope is properly adjusted to obtain optimal light conditions.

Organotypic culture of kidney rudiments

1. Dissect the embryos in Dulbecco's PBS (with Ca^{2+} and Mg^{2+}) at room temperature.
2. For Trowell-type tissue culture (Trowell 1959), place the kidneys on the matt side of a Nuclepore filter lying on a screen made of stainless steel wire.

Day	Limb bud	Metanephros

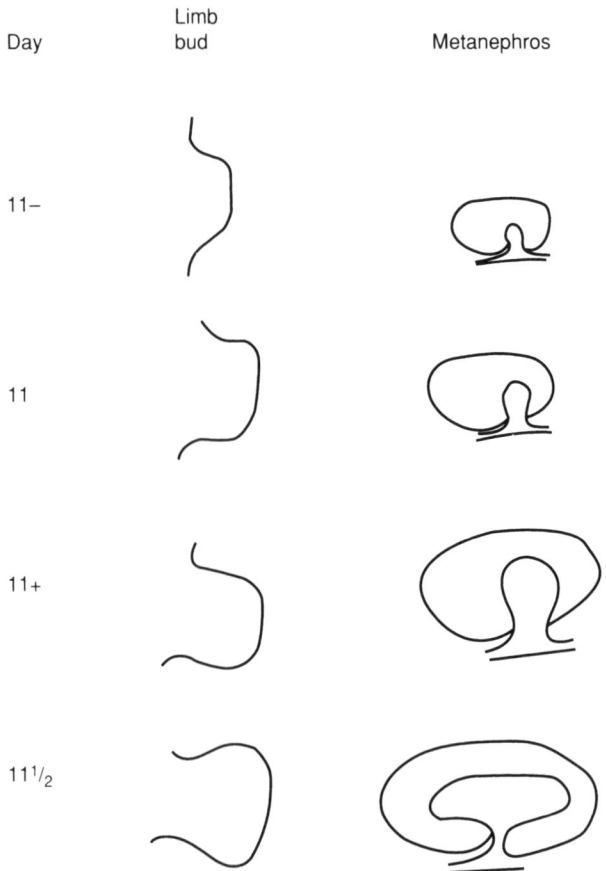

Figure 4.11.1 Scheme for estimating the stage of development of the metanephric kidney around day 11 of mouse embryo development. Left: hind limb bud. Right: metanephros consisting of the invading ureter bud and the surrounding mesenchymal blastema.

3. Culture in medium–gas interface. For this, add medium almost to the level of the upper surface of the supporting grid so that the medium creeps on the filter assembly. Make sure that there are no air bubbles under the filter.

Transfilter cultures

1. For transfilter induction experiments (see Figure 4.11.2), use kidney rudiments and fragments of spinal cord from 11-day mouse embryos.
2. Dissect the embryos in Dulbecco's PBS (with Ca^{2+} and Mg^{2+}) at room temperature. Collect dorsal spinal cord and kidney rudiments.
3. Separate each mesenchyme from the Wolffian duct and the ureter bud. This is done mechanically by using disposable syringe needles, after treatment with EDTA, appropriate enzymes, or without pretreatment.

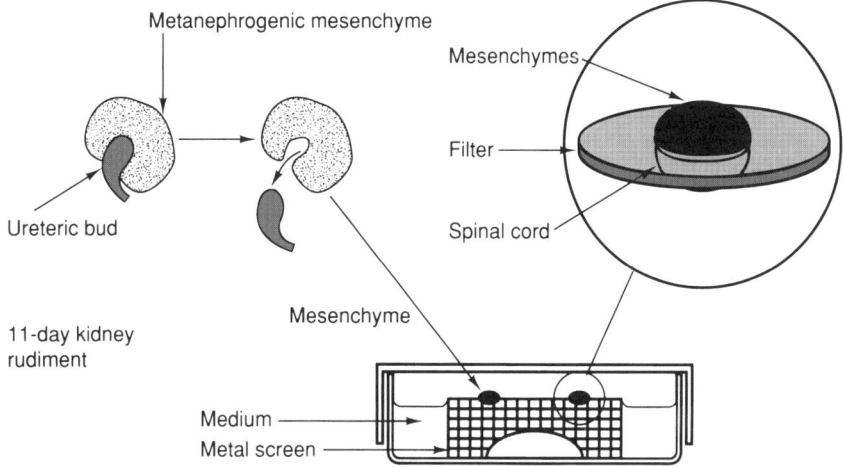

Figure 4.11.2 Scheme of the transfilter technique.

Procedure for EDTA treatment

a. Place the kidney rudiments in 0.02% EDTA in NaCl-P buffer (140 mM NaCl, 10 mM sodium phosphate, pH 7.2) onto a glass Petri dish at room temperature.
b. Separate the mesenchymes by manipulating the rudiments with disposable needles.
c. Transfer the tissues to culture medium in plastic dishes.

Procedure for enzyme treatment

a. Incubate the kidney rudiments in 2.25% pancreatin/0.75% trypsin in Ca^{2+}/Mg^{2+}-free Tyrode's solution, pH 7.4, on ice for 80 s.
b. Incubate he rudiments in serum-containing culture medium at 20°C for 15 min.
c. Separate the mesenchymes from the ureter buds by manipulating with disposable needles in the culture medium.

4. Place a filter piece on the screen and pipette a fragment of spinal cord onto the matt side of the filter piece.
5. Pipette dissolved agarose solution on the spinal cord by using a thin mouth pipette drawn from a Pasteur pipette.
   Note: Take care not to use too much agarose so that it runs onto the lower side of the filter and blocks the pores.
   When melting the agarose on the microburner flame take the following precautions so as not to burn yourself. Use good-quality glass tubes and a non-burnable holder, and agitate a little while melting. Start melting from the surface to guard against the agarose shooting up the tube.
6. Turn the filter upside down and place so that the spinal cord, now hanging below the filter, is visible through a hole punched in the screen.
7. Pipette 1–6 mesenchymes on the filter opposite the spinal cord. Culture in the incubator (Figure 4.11.2).

DISCUSSION

The transfilter technique was originally devised to explore the transmission characteristics of signals emitting morphogenetic interactions (Grobstein 1956). The method has subsequently proved useful in various other approaches in developmental biology. The inducing tissue can be removed at any stage of the transfilter contact, and hence the kinetics of the induction process can be investigated (Saxén & Lehtonen 1978).

Another advantage of the technique is the possibility of analysing the differentiating target tissue separately from the inductor component. Hence, the technique allows analysis based on biochemistry, immunochemistry and molecular biology (Jansson *et al.* 1997). Many of these analyses are, however, hampered by the very small amount of tissue harvested after laborious experiments; an 11-day embryonic mouse nephric mesenchyme contains some 20 000 cells, 100 ng DNA and 100 ng total RNA.

Caution: When the inducing tissue is scraped off mechanically, great care and visual control must be exerted to ensure the elimination of every tissue fragment, as an unremoved fragment can still act as an inducer!

REFERENCES

Grobstein C (1956) Trans-filter induction of tubules in mouse metanephric mesenchyme. *Experimental Cell Research* 10: 424–440.

Jansson S, Olkkonen V, Martin-Parras L, Chavrier P, Stapleton M, Zerial M & Lehtonen E (1997) Mouse metanephric kidney as a model system for identifying developmentally regulated genes. *Journal of Cellular Physiology* 173: 147–151.

Lehtonen E (1976) Transmission of signals in embryonic induction. *Medical Biology* 54: 108–128.

Saxén L (1987) *Organogenesis of the Kidney.* Cambridge University Press, Cambridge.

Saxén L & Lehtonen E (1978) Transfilter induction of kidney tubules as a function of the extent and duration of intercellular contacts. *Journal of Embryology and Experimental Morphology* 47: 97–109.

Saxén L & Lehtonen E (1987) Embryonic kidney in organ culture. *Differentiation* 36: 2–11.

Trowell OA (1959) The culture of mature organs in a synthetic medium. *Experimental Cell Research* 16: 118–147.

4.12 THE CULTURE OF AMNIOCYTES, CHORIONIC VILLI, AND HUMAN FOETAL AND PLACENTAL TISSUE

Reagents, solutions and media

Basal medium

Chang's: received as 90 ml of medium and separate vials of lyophilized supplement. Store at 4°C.

Supplements

- *HEPES buffer* Received as 1 M. Use at 0.5% v/v in culture medium.
- *L-Glutamine* Received as a 200 mM solution. Store in 1 ml aliquots at –20°C. Use at 1% v/v.
- *Mycostatin (Nystatin) fungicide* Optional or only when required to control fungal contamination. Received as 10 000 IU ml^{-1}. Use at 0.5% or 1.0% v/v in culture medium, depending on the extent of the contamination.
- *Penicillin/streptomycin* Received as 10 000 IU ml^{-1} penicillin, 10 000 μg ml^{-1} streptomycin. Store in small aliquots at –20°C. Use at 1% v/v in culture medium.
- *Sodium bicarbonate* This is sometimes supplied as a constituent of the medium. Otherwise, it is supplied as a 7.5% solution. Store at 4°C and use at 1% v/v.

Complete medium for setting up and changing cultures

100 ml Ham's F-10 with sodium bicarbonate buffer
1.2 ml penicillin/streptomycin
1.2 ml L-glutamine
20 ml foetal bovine serum (or 10 ml foetal bovine serum, 10 ml serum substitutes)
0.6 ml HEPES buffer if required

Store frozen for up to 6 months. Thaw out immediately before use.

Collection medium for foetal and placental tissues

This maintains the viability of cells. It contains double-strength antibiotics and antimycotics to control infection.

Cell and Tissue Culture for Medical Research, edited by A. Doyle and J.B. Griffiths.
© 2000 John Wiley & Sons, Ltd

100 ml Ham's F-10 with sodium bicarbonate buffer
2 ml penicillin/streptomycin (final concentration 20 000 IU ml^{-1} penicillin, 20 000
    μg ml^{-1} streptomycin)
1 ml mycostatin (final concentration 20 000 IU ml^{-1})

Store frozen for up to 6 months. Thaw out immediately before use.

Collection medium for chorionic villus preparation

RPMI 1640 tissue culture medium with HEPES buffer and the following supplements (final concentrations):

20% foetal bovine serum (FBS)
1% penicillin (10 000 IU ml^{-1})/streptomycin (10 000 μg ml^{-1})
1% L-glutamine (200 mM)
1% heparin (1 : 100 v/v)

Release medium

Add 2 ml deoxycytidine to complete medium.
Store frozen for up to 6 months. Thaw out immediately before use.

Chang medium

Received as 90 ml of medium (store at 4°C) and separate vials of lyophilized
supplement. Dissolve the supplement in 10 ml sterile distilled water and store in
1 ml aliquots at –20°C. Add 1 ml of the supplement to 9 ml of the medium as
required. Add antibiotics and L-glutamine. This complete Chang medium is stored
at 4°C.

Colchicine

Stock solution 10 mg ml^{-1} in distilled water. Filter-sterilize. Store at –20°C.
Working solution 500 μg ml^{-1} in distilled water. Use at a final concentration of
25 μg ml^{-1}.

Deoxycytidine

This is used in the release medium following exposure of cultures to thymidine.
Stock solution 1 mg ml^{-1} in distilled water. Working solution 10 μg ml^{-1} in distilled
water. Store stock and working solutions at –20°C. Use at a final concentration
of 0.2 μg ml^{-1}.

Dispase

Stock solution: dissolve 400 mg dispase in 10 ml distilled water using a magnetic
stirrer and a hotplate. Filter-sterilize the solution and store in 1 ml aliquots at
–20°C. Use at a final concentration of 4 mg ml^{-1}.

Fixative

The usual fixative is 3 : 1 methanol (Analar)/acetic acid (Analar). Make fresh. Store for short periods at –20°C if required, or at room temperature.

Hypotonic solutions

Various solutions are used as hypotonic agents.

(a) Pure distilled water.
(b) 0.075 M KCl. This is made by adding 5.6 g of KCl to 1 litre of distilled water. The solution is then either filter-sterilized or autoclaved and stored at room temperature or at 4°C.
(c) 0.056 M KCl. This is used if a more dilute hypotonic solution is required. Dissolve 4.19 g of KCl in 1 litre of distilled water. Prepare and store as for 0.075 M KCL.
(d) 1% trisodium citrate: 1 g of trisodium citrate is dissolved in 100 ml of distilled water. The solution is filter-sterilized or autoclaved and stored at room temperature or 4°C.

Thymidine solution

This is used for synchronizing cell cultures. Stock solution: dissolve 1 g of thymidine powder in 100 ml distilled water. Filter-sterilize and store in small aliquots at –20°C for up to 1 month. Use at final concentration of 250 μg ml^{-1}.

PROCEDURE: CULTURE OF AMNIOCYTES

Amniocyte culture relies on the separation of cells from the supernatant amniotic fluid, transfer of the amniocytes into a culture vessel and the addition of suitable tissue culture medium, with appropriate supplements. The cultures are incubated and individual cells settle and grow into colonies. When sufficient growth has taken place, the cells are harvested and chromosome preparations are made. There are several methods available for culturing and harvesting cells from amniotic fluid, and the method of choice usually depends on personal experience.

There are several basic factors which influence the success of amniotic fluid cell culture (Milunsky 1979). These are as follows:

• *Sterility* Good aseptic technique must be employed. All culture processes are carried out in Class II safety cabinets.
• *Constant temperature* 37°C ± 0.5°C
• *Stable pH: 7.2 to 7.4* This is achieved by adding HEPES buffer (for closed systems – see below) and/or by maintaining the cultures in 5% CO_2/95% air with sodium bicarbonate as a buffering agent.
• *Choice of culture medium and use of supplements* Several suitable basal culture media are available. Supplements are added to the basal medium, or some of these may be present in the medium as supplied. These include foetal bovine serum with or without serum substitutes (available commercially), L-glutamine,

antibiotics, antimycotics (optional or as required) and buffers. Basal medium plus supplements is usually referred to as 'complete medium'.

- *Choice of culture system* There are two main types of culture system, 'open' and 'closed'.

Open systems: cells are grown in vessels which are not individually sealed, such as Petri dishes, or in Leighton tubes, flasks or flaskettes which have partially closed lids. The principle of the system is to allow an interchange of 5% CO_2 in air between the vessel and the surrounding atmosphere. The vessels are placed either in a CO_2 incubator or inside sandwich boxes which are gassed with 5% CO_2 in air prior to having the lids sealed. 'Triple' gas (80% nitrogen, 15% oxygen and 5% CO_2) is sometimes used: this seems to encourage more rapid cell growth.

Closed systems: the lids are tightly closed on the culture vessels, which are usually Leighton tubes, flasks or flaskettes. The pH is maintained by gassing the culture with 5% CO_2 in air prior to tightening the lids, using bicarbonate or HEPES-buffered medium. HEPES-buffered medium does not necessarily require gassing but this is a wise precaution to help maintain the correct pH.

- *Choice of culture method* There are two main types of culture method:
1. The cells are grown directly on the surface of the culture vessel, which has been rendered hydrophilic by the manufacturing process. The cells have to be removed from the cell surface for harvesting in suspension.
2. The cells are grown on carefully cleaned glass coverslips. The cells are harvested *in situ* on the coverslips.

- *Choice of harvesting procedure* This may involve the use of thymidine solution (final concentration 250 µg ml^{-1}) and deoxycytidine release (final concentration 0.2 µg ml^{-1}) to obtain extended chromosomes.

The method described is for cultures set up in Leighton tubes in a closed system, with suspension harvest.

Setting up cultures

Materials and equipment

- Complete medium
- Leighton tubes

1. Sample requirements: 10–20 ml of amniotic fluid in a 20 ml Universal container with a conical base.
2. Label a minimum of two Leighton tubes with the laboratory number and the patient's name.
3. Agitate the sample to release any cells attached to the sides of the Universal container.
4. Centrifuge at 200 *g* for 10 min.
5. Using a Pasteur pipette, remove the supernatant fluid, leaving 1–1.5 ml of fluid. Take care not to disturb the cell pellet.
6. Using a 2 ml pipette, resuspend the cells and transfer approximately 0.5 ml cell suspension to each Leighton tube.

7. Add 1 ml complete medium to each Leighton tube. At least two different batches of medium should be used in separate tubes, with the tubes labelled accordingly. Gas the tubes with 5% CO_2/95% air. Tighten the lids of the tubes.
8. Incubate the cultures at 37°C for 5 days.

Inspection of cultures

Materials

- Complete medium

1. After incubation for 5 days, inspect the cultures for cell attachment and growth, using an inverted microscope.
2. Pour off the medium gently.
3. Feed with 1.5 ml fresh complete medium. Gas the tubes with 5% CO_2/95% air and tighten the lids.
4. Replace the culture tubes in the incubator.
5. Change the medium twice a week until there is adequate cell growth for harvesting (usually by 7–10 days).

Harvesting of cultures, using thymidine

Materials

- Complete medium
- Thymidine: (10 mg ml^{-1} working solution)
- Release medium containing 10 μg ml^{-1} deoxycytidine
- Colchicine (500 μg ml^{-1} working solution)
- Hypotonic agent: distilled water
- Fixative: 3 parts methanol to 1 part glacial acetic acid –20°C

1. On the day prior to harvesting, change the medium in the morning, adding 2 ml fresh complete medium.
2. At approximately 4.00 pm add 0.05 ml thymidine working solution (final concentration 250 μg ml^{-1}).
3. After 16–17 h (i.e. at approximately 9.00 am the following morning), check the tubes for rounded-up cells, which are an indicator of mitotic activity.
4. Pour off the medium and gently rinse with 1ml release medium containing deoxycytidine, prewarmed to 37°C.
5. Change the culture with 2 ml prewarmed, fresh medium and incubate at 37°C for 5–6 h.
6. Monitor the mitotic index of the cultures. If there are few rounded-up cells, indicating a low level of mitotic activity, reincubate for 30 min and re-examine. Repeat as necessary.
7. When there is sufficient mitotic activity, or when there is no obvious increase in the numbers of rounded-up cells, add 0.1 ml colchicine to each culture (final concentration 25 μg ml^{-1}). Incubate the cultures for a further 30 min.
8. Pour off the medium carefully.

9. Add 0.5 ml prewarmed trypsin-EDTA solution, wash the culture and pour off gently. Add 1 ml trypsin-EDTA and incubate at 37°C for 2 min.
10. Examine the culture. The cells should be coming off the surface of the tube. Tap the tube gently to get cells into suspension if necessary.
11. Add 1.5 ml prewarmed distilled water to each tube as a hypotonic agent. Leave for 2 min.
12. Centrifuge at 200 *g* for 7 min. Recentrifuge for 2–3 min if the pellet is not solid.
13. Pour off the supernatant and agitate the cell pellet by flicking the base of each tube.
14. Add very cold fixative (–20°C), drop by drop using a Pasteur pipette, to the cell suspension, agitating the cells by flicking the tube. Make up to 2 ml. Tighten the cap. Store at –20°C overnight or until ready to make the slide.

Slide making

Materials and equipment

- Fixative
- Oven at 60°C (optional)

1. Centrifuge the tubes of fixed cells at 200 *g* for 10 min. Pour off the fixative.
2. Add a few drops of fresh fixative.
3. Using pre-cleaned slides, wipe with a tissue to remove dust, and label with the case number, patient's name and the culture identity.
4. Using a Pasteur pipette, drop one drop of cell suspension onto the clean slide from a height of about 1 inch. Allow the drop to spread. Add one drop of fresh fixative as Newton's rings appear but before the slide dries.
5. Allow the slide to dry at room temperature.
6. Examine the slide under a phase contrast microscope. In the event of under-spreading of the chromosomes, adjust the slide-making process by gently blowing on the slide before it dries, or by adding further drops of fixative, or both. If they have overspread, do not add the second drop of fixative and/or place the drop of cell suspension on the slide gently, without dropping.
7. Prior to banding, the slide can be 'aged' by air-drying overnight in an oven at 60°C.

PROCEDURE: CULTURE OF HUMAN FOETAL AND PLACENTAL TISSUES FOR CHROMOSOME ANALYSIS

The basic requirements for successful culture of foetal material are that the tissue is viable and clear of bacterial or fungal contamination. Macerated or dried tissues, or ones fixed in formalin, are not suitable.

Selection of tissues

- Samples from early gestation.

The foetus and placenta are not usually well differentiated in early products of conception from spontaneous abortion or elective termination of pregnancy and foetal tissues are often not recognizable.

The entire sample should be put into in a dry, plastic container. The samples can be stored at 4°C over the weekend if necessary before transporting to the laboratory. Experienced staff will then select the tissues suitable for cell culture. Chorionic villi should be dissected out if there is no evidence of foetal tissues. This helps to reduce the risk of contamination with maternal tissue. Material for culture should be placed into a clean container with collection medium (see below).

- Samples from mid-trimester spontaneous abortion or terminations of pregnancy and from stillbirths and neonatal deaths.

Permission should be gained from the parents for the biopsy procedure and any consequent research studies.

A biopsy specimen can be taken at the site of delivery of the foetus, or the whole foetus can be sent to the laboratory. The foetus may be transported in a heat-sealed strong plastic bag, or in a leakproof plastic container. One of the best foetal tissues for fibroblast cell culture is fascia lata. This is the tissue covering the thigh. Tissue biopsies should be put into collection medium for transportation to the laboratory and all specimens should be delivered to the laboratory as soon as possible, usually within a maximum of 24 h.

If the foetus shows signs of maceration (foul smell, hydropic, jelly-like, leather-like, peeling skin or discoloured) do not attempt to collect or culture the tissue, as it will almost certainly be non-viable and/or contaminated. Select chorionic villi as described above.

Note: foetal cardiac blood is a valuable alternative source of material for chromosome analysis and for the establishment of permanent cell lines.

Setting up cultures

Materials and equipment

- Complete medium
- Collection medium
- Leighton tubes

1. Label a minimum of two Leighton tubes with the laboratory number and the patient's name.
2. Place the tissue in a Petri dish with a few drops of collection medium. Using a scalpel, or forceps and a pair of small curved dissecting scissors, cut away any fat from just below the skin surface, as this will inhibit cell growth. Chop the remaining tissue into small fragments.
3. Using a fine-tipped Pasteur pipette, transfer the fragments to the Leighton tubes and spread over the flat culture surface of the tube.
4. Using forceps, pick up a clean, sterile coverslip and place gently over the fragments. Do not tilt the tube too much, as the fragments will move to the sides of the tube and escape entrapment. Make sure that the coverslip is placed flat down on the explants.

5. Using a 1 ml pipette, add 1 ml complete medium to each tube. At least two different batches of medium should be used in separate tubes, with the tubes labelled accordingly. Gas the tubes with 5% CO_2/95% air. Ensure that the lids of the tubes are tightened.
6. Incubate the cultures at 37°C for 3 days.

Inspection of cultures

1. After incubation for 3 days, check the tubes for signs of infection but take care not to disturb the coverslips or explants. If the samples are infected, do not open the tubes. If necessary, one tube can be sent for identification of the contaminant. The other tubes should be disposed of and an appropriate report sent out.
2. If cultures are free of contamination, the medium should be changed after a further 3–4 days or less if there is adequate cell growth.

Changing of medium

Materials

• Complete medium

1. Pour off the medium.
2. Feed with 1.5 ml fresh complete medium and tighten the lids.
3. Gas the tubes with 5% CO_2/95% air.
4. Replace the tubes in the incubator.
5. Inspect the feed the cultures and change the medium twice a week until adequate growth has occurred (usually by 10–14 days).

Subculturing, dispersal of cells and harvesting

If there is active cell growth, the primary cultures can be harvested directly. Because the cells are growing on the underside of the coverslip, the coverslip should be inverted and transferred to a new Leighton tube for harvesting.

If there is excess growth, primary cultures are subcultured to reduce the cell density prior to harvesting. Similarly, the cells may be dispersed in the original tube if growth is quite dense and patchy.

If there is adequate growth on both the coverslip and the surrounding area of the flat culture surface of the Leighton tube, the coverslip can be removed and transferred to a new Leighton tube, labelled appropriately, to provide an extra source of cells. Fresh medium is added and the new tube is put in the incubator. The cells in the original tube can be harvested immediately, subcultured or dispersed as required.

Materials and equipment

• Complete medium
• Colchicine (500 μg ml^{-1} working solution)
• Fixative: 3 parts methanol/1 part glacial acetic acid at –20°C
• Leighton tubes

Subculturing and dispersal

1. Pour off the medium, wash the culture with 0.5 ml prewarmed trypsin-EDTA and incubate the culture for 1–2 min. Observe the culture under an inverted microscope to determine when the cells round up and start to detach from the surface of the culture vessel. Add 1.5 ml complete medium to stop the action of the trypsin. Allow the cells to disperse in the original tube (shake gently if necessary) or, if the cell density is sufficient, transfer a proportion of the cells to a new Leighton tube, appropriately labelled as a subculture.
2. Inspect the culture for cell attachment, growth and activity. Change the medium.

Harvesting

If longer chromosomes are required, a thymidine block and deoxycytidine release can be used as for cultured aminiocytes.

1. When there is adequate growth of active cells, the medium should be changed on the day prior to harvesting. Use 2 ml complete medium.
2. On the morning when the cells are to be harvested, add 0.1 ml colchicine (25 μg ml^{-1} final concentration).
3. After 2–4 h, examine the culture under an inverted microscope and check for mitotic activity. If there are few rounded-up cells, which would indicate that the colchicine has not effectively arrested the cells, leave for a further 1–2 h and re-examine.
4. Pour off the medium carefully and wash the culture with 0.5 ml prewarmed trypsin-EDTA.
5. Add 1 ml prewarmed trypsin-EDTA for approximately 5 min. Check that the cells have detached. If the coverslip is still in the tube it may be necessary to loosen it by shaking the tube vigorously.
6. Add 1.5 ml distilled water, prewarmed to 37°C, as a hypotonic treatment.
7. Centrifuge at 200 *g* for 5 min.
8. Pour off the supernatant completely, taking care not to disturb the cell pellet.
9. Agitate the cells by gently flicking the base of the tube and gently add 1 ml of cold fixative (–20°C).
10. Place at –20°C for 1 h.
11. Centrifuge at 200 *g* for 5 min.
12. Pour off the fixative and add a few drops of fresh fixative. Store at –20°C until read to make the slides.
13. Prepare slides as for amniocytes.

PROCEDURE: CULTURE OF CHORIONIC VILLI FOR PRENATAL CHROMOSOME ANALYSIS

Investigation of chorionic villi can be undertaken by either direct processing or long-term culture. The direct method involves collecting metaphases which are occurring naturally within the mitotically active cytotrophoblast. It can be applied to the specimen immediately after receipt or after overnight incubation. Although maternal decidual tissue is present, maternal cell contamination is not a problem with the direct method because the decidua has a very low mitotic activity. One problem which can occur with this method, however, is that of confined placental mosaicism, i.e. when the foetus has a normal karyotype but some cells from the chorionic villi of the placenta are abnormal. The long-term culture method involves growth of cells from the mesenchymal core of the villi. This method provides better-quality metaphases than the direct method, allowing more detailed analysis of the karyotype, but care has to be taken to dissect away decidua prior to setting up the cell cultures to avoid problems of maternal cell contamination.

Ideally, the most accurate foetal karyotype is obtained by using both the direct and the long-term culture method on each sample. This avoids false-positive or false-negative results which can arise because of confined placental mosaicism or growth of maternal cells (Miny *et al.* 1989).

Selection of tissue

The clinical procedure of chorionic villus biopsy, together with the nature of the sample, results in possible contamination with maternal tissue. It is important to inspect the sample prior to long-term cell culture or prior to DNA extraction for molecular genetic analysis and to dissect away any maternal decidua (decidua comprises amorphous material, lacking vascularization, whereas the chorionic villi show main villi, which are vascularized, and show regular division into fronds).

Chorionic villus preparation

Materials and equipment

- Collection medium

1. If the sample can be inspected on an inverted microscope where the chorionic villus sampling takes place, the villi can be placed directly into a plastic Petri dish, with medium. The quantity and quality of villi is assessed. If the sample is not suitable, a further sample can be requested immediately. After satisfactory inspection, use a Pasteur pipette to transfer the sample to a Universal container with fresh collection medium, leaving as much of the blood-stained fluid as possible in the Petri dish.
   If the sample is collected without inspection, place it directly into the Universal container with fresh collection medium. Transport the sample to the laboratory as soon as possible and inspect as described above.
2. Wash the specimen by carefully removing the collection medium with a Pasteur pipette and replacing with fresh medium until the medium is no longer blood-stained.
3. Transfer the washed villi to a Petri dish. Using a scalpel and forceps, examine the specimen under a dissecting microscope and remove the decidua.
4. Place selected villi in a new Petri dish. Add a few drops of medium and, using two scalpel, or very finecurved dissecting scissors, chop the villi to a fine pulp.

Culture of chorionic villus

The pulped villi can either be harvested directly to obtain chromosome preparations from spontaneously dividing cells ('direct' method) (Simoni *et al.* 1983) or they can be set up in long-term culture. It is usual to carry out both a direct and a long-term culture on each sample.

Materials and equipment

- Complete RPMI medium
- Complete Chang medium (Chang *et al.* 1982)
- Dispase (1 : 10 v/v)
- 1% Trisodium citrate (hypotonic agent for direct method)
- 3 : 1 Methanol/acetic fixative ($-20°C$)
- Leighton tubes

Direct culture method

1. Place the villi pulp and 2 ml complete RPMI medium in a Leighton tube labelled with the laboratory number and the patient's name.
2. Add 1 : 10 v/v dispase and 0.1 ml colchicine (final concentration 1 µg ml^{-1}). Gas the tube with 5% CO_2 in air. Incubate for 1 h.
3. Shake the tube quite vigorously to break up the tissue.
4. Centrifuge at 200 *g* for 5 min.
5. Using a Pasteur pipette, remove the supernatant.
6. Add 3 ml trisodium citrate. Incubate the tube at 37°C for 10 min.
7. Centrifuge at 200 *g* for 5 min.
8. Using a Pasteur pipette, remove the supernatant.
9. Agitate the cell pellet by gently flicking the base of the tube. Add 2 ml fixative ($-20°C$), continuing to gently tap the tube to mix the fixative with the cells.
10. Repeat steps 7–9 twice.
11. Prepare the slides as for amniocytes.

Note: there will be some undigested tissue at the end of the harvest procedure, because dispase acts on the cytotrophoblast layer and leaves the core of the villi intact.

Long-term culture method

1. Using a Pasteur pipette, transfer the villi pulp to a minimum of two Leighton tubes, labelled with the laboratory number and the patient's name. Spread the tissue over the flat surface of the tubes. Tighten the lids.
2. Invert the tubes and allow the cells to stick to the surface (1–2 min).
3. While the tubes are still inverted, very gently add 1 ml complete Chang medium. Allow the medium to run down the curved sides of the tubes. Gas the tubes with 5% CO_2 in air. Tighten the lid.
4. Place the tubes, still inverted, in an incubator at 37°C.
5. After 1 h, turn the tubes over so that the medium covers the tissue. Reincubate.

Inspection of cultures, subculturing, dispersion and harvesting

Follow the same procedures as for foetal and placental tissues, except that, at harvesting, the colchicine exposure time is only 1 h.

ALTERNATIVE PROCEDURE: *IN SITU* AMNIOCYTE CELL CULTURE AND HARVESTING

This method involves growing cell cultures on glass coverslips, or subculturing cells onto the coverslips, and harvesting the cultures *in situ* on the coverslips for chromosome analysis (Schmid 1975). Amniocytes are often harvested as primary cell cultures, without subculture or dispersion. Cell cultures grown from explants, such as foetal tissue biopsies and long-term chorionic villi cultures, are usually subcultured or dispersed prior to harvesting.

Cleaning of coverslips

Materials and equipment

- Decon (or other suitable cleaning agent)
- Methanol (Analar)

1. Boil 1.5 litres of distilled water (use a 2-litre beaker).
2. Turn off heat, allow the water to come off the boil, and then add 75 ml Decon.
3. Add 22 × 22 mm coverslips and leave to soak overnight.
4. On the next morning, pour off the Decon solution, and then half-fill the beaker with cold water, swirl the water, and pour off. Repeat this three times. This will help to remove the Decon which may have accumulated at the bottom.
5. Rinse the beaker of coverslips under running cold tap water overnight.
6. Next day, pour off the water and wash the coverslips in several changes of distilled water (at least 10 changes over a period of 2 days).
7. Remove as much of the water as possible and replace with methanol.
8. Change the methanol two or three times, and then cover the beaker with foil until ready to use.

Preparation of coverslips for amniocyte cell culture

1. Collect together in a tissue culture cabinet:

Pot of coverslips
Rack to hold coverslips
Trough to hold coverslip rack
Diamond pencil
Plastic boxes

2. Using forceps, pick out individual coverslips and place in a rack.
3. Rinse methanol in a trough.

4. Leave coverslips to air-dry in the tissue culture cabinet.
5. Transfer one dry coverslip to each Petri dish.
6. Prepare four coverslips and Petri dishes for each sample. Using a diamond pencil, write the laboratory number and A/B/C or D on the coverslips. Write the corresponding number and letter on the Petri dish lid. Place the coverslip number side down in the Petri dish.
7. Place Petri dishes in a plastic box.
8. Store boxes at the back of the tissue culture cabinet.

Setting up of amniocyte cultures

Materials

- Complete medium

1. Resuspend the amniotic fluid cells *or* suspension of cells *or* suspension of trypsinized fibroblasts in 6 ml of complete medium.
2. Using a 10 ml pipette, transfer 1.5 ml of this suspension into each of the four culture dishes containing appropriately labelled coverslips.
3. Divide the cultures between two different plastic boxes to reduce risks of total loss of the sample by contamination.
4. Check that the box lid fits well, gas with 5% CO_2 in air for about 30 s, and then seal the box with insulation tape. (Alternatively: place in a CO_2 incubator and ensure seals are closed.)
5. Incubate for 7 days at 37°C.

Harvesting of cells on coverslips

Materials

- Complete medium
- Colchicine (500 μg ml^{-1} working solution)
- Hypotonic solution: 0.056 M KCl
- Fixative: 3 : 1 methanol/acetic acid

1. When the cultures are ready to harvest, add 0.1 ml colchicine to each dish, gas the dishes with 5% CO_2/95% air and leave in the incubator for about 2 h.
2. Warm a bottle of 0.056 M KCl to 37°C.
3. Remove the medium using a pipette and replace with 1.5 ml prewarmed KCl. Leave for 20 min at room temperature.
4. Prepare fresh fixative at room temperature. Each dish processed requires about 6 ml.
5. Add two drops of fixative to each dish (try to avoid placing it directly on the cells) and wait for 2 min.
6. Add six drops of fixative and wait for 2 min.
7. Add 12 drops of fixative and wait for 2 min.
8. Add 2 ml of fixative and wait for 2 min.

9. Remove the KCl/fixative solution and replace with 2 ml of fresh fixative. Perform this operation one dish at a time to avoid the coverslip drying off. Repeat once.

10. Place the dishes in a box and store at 4°C for at least 2 h. Overnight is best.

11. Remove fixative and replace with 2 ml of fresh fixative. Leave for 2 min.

12. The coverslips may now be dried off. The spreading and morphology of the chromosomes is dependent on humidity and temperature. One way to dry off is to remove the coverslip using forceps, remove excess fixative by touching one edge of the coverslip onto a dry paper towel, and then place onto a damp towel and leave to dry. This method can be altered by a process of trial and error until optimum preparations are achieved for the usual ambient conditions in the laboratory.

DISCUSSION

Troubleshooting

The same general principles apply for all the above tissue types when attempting to determine what has gone wrong if the required results are not obtained.

Microbial contamination

All potential sources of contamination should be examined, including the possibility that the incoming sample was infected. This is usually apparent if all other samples set up at the same time are clear. Always initiate investigations by checking cultures which have some process or ingredients in common with the contaminated culture(s). Occasionally, even a batch of articles which are purchased as 'sterile', such as plastic disposables, are the cause of the problem. Never overlook the 'impossible'. Check the operator's sterile technique, the temperature of the sterilizing oven, and the cleaning procedures for the culture areas. It is important to trace the source of infection so that it cannot spread to other cultures. If contamination is an ongoing problem, it is important to identify the type of contaminant and to apply appropriate antibiotics or antimycotics to control and eliminate it.

Failure of cells to attach and initiate growth or failure to grow once attached

Check to see how widespread the problem is. If it is just one specimen, it is not usually possible to identify the cause. If it is affecting several samples, check if a new batch of culture vessels is being used. They may have missed the manufacturer's process to provide a hydrophilic growing surface.

Check the source(s) of the failed samples. Occasionally, the syringes used for the amniocentesis procedure have been found to contain toxic substances in the rubber of the plunger. Check the syringe type and batch numbers to see if these correlate with culture failure. If so, carry out toxicity tests by rinsing samples of the syringes with culture medium and using this for a controlled experiment to check the ability of the medium to support cell growth.

If using coverslips, check that they have been cleaned properly.

Make sure that the serum in the medium has been tested for its ability to support good cell growth, that L-glutamine has been added to the medium and that the correct concentration of antibiotics is being used. If the medium is diluted to single strength from $\times 10$ concentrate, the distilled water may be at fault. Check the temperature of the incubator, make sure that the gas being used really is 5% CO_2 in air (it is not unknown for 5% CO_2 in nitrogen to be delivered by mistake!), and check that the cultures retain their pH, particularly in the first few days before they establish active growth.

Failure of cells to respond to trypsin-EDTA for harvesting, dispersing or subculturing, or failure to settle after trypsinization

True fibroblasts and fibroblast-like epithelial cells (which are commonly grown in amniotic fluid cell cultures) rarely give this problem. If they do, check the strength of the trypsin-EDTA solution. Other cell types may not be so readily responsive (e.g. large epithelial cells, which are often vacuolated, and endothelial cells). Adjust the time of exposure to trypsin-EDTA. Very occasionally, really stubborn cells have to be scraped off the surface – however, these rarely produce successful chromosome preparations. Failure to re-attach may be due to any of the causes given above, or it may be that the cells have been adversely affected by trypsinization. Glial cells (which are occasionally grown in amniotic fluid cell cultures if the foetus has a neural tube defect) do not re-adhere.

Chromosome preparations are suboptimal

When the slides are made, or the coverslips harvested *in situ*, the chromosomes should disperse adequately to result in minimum numbers of overlaps but without causing the loss of individual chromosomes from the metaphase spread. Overspreading, or 'bursting', may be caused by excess hypotonic treatment (too dilute or over-exposure) or by mechanical forces resulting from the cell suspension being dropped onto the slide from too great a height. Overspreading can often be alleviated by not using a second drop of fixative as the cells are settling and by drying the slides off more quickly. Environmental factors – temperature and humidity – also have an effect but it is often different to control them. To increase humidity, try placing the slides on wet paper towels to dry off.

Underspreading may be caused by the converse of any of the reasons for bursting and the conditions should be adjusted accordingly. One other cause may be greasy slides.

Low mitotic index

Some cultures give a low mitotic index because of insufficient cell growth or because the cell type was inappropriate for successful harvesting (this can apply to small round endothelial-type cells or to large, vacuolated epithelial-type cells). If there are consistently inadequate numbers of metaphase spreads, check that the cell growth is not being adversely affected by the quality of the culture medium (is it fresh, has it got all the correct supplements, has the serum been batch-tested for support of cell growth?). Check that the colchicine has been made up to the correct concentration and/or that the hypotonic agent is also at the correct concentration (hypotonic agent which is too dilute can burst the cells so that the chromosomes are lost). Bursting of cells during the slide-making process can also result in a low yield of metaphase spreads.

REFERENCES

Chang HC, Jones OW & Masui H (1982) Human amniotic fluid cells grown in a hormone-supplemented medium. Suitability for prenatal diagnosis. *Proceedings of the National Academy of Sciences of the USA* 79: 4795–4799.

Milunsky A (1979) Amniotic fluid cell culture. In: Milunsky A (ed.) *Genetic Disorders and the Fetus*, 1st edn. Plenum Press, New York.

Miny P, Basaran S, Pawlowitzki I-H, Horst J, Westerdorp A, Niedner W & Holzgreve W (1989) Validity of cytogenetic analyses from trophoblast tissue throughout gestation. *American Journal of Medical Genetics* 33: 136–141.

Schmid W (1975) A technique for in situ karyotyping of primary amniotic fluid cell cultures. *Humangenetik* 30: 325.

Simoni G, Brambati S, Danesino C, Rosella F, Terzoli GL, Ferrari M & Fraccaro M (1983) Efficient direct chromosome analysis and enzyme determination from chorionic villi samples in the first trimester of pregnancy. *Human Genetics* 63: 349–357.

4.13 ENDOTHELIAL CELLS (FROM THE HUMAN UMBILICAL CORD VEIN)

Endothelial cells (ECs) make a continuous monolayer called endothelium, lining the lumens of heart cavities and blood and lymphatic vessels. These cells are very active in a variety of metabolic processes. In particular, they can synthesize both a substance and its antagonist, modulating in a very subtle fashion the vascular metabolic processes in which they are involved.

Their major property, haemocompatibility, results from a balance between their secretions of antithrombotic substances – prostacyclin (PGI_2), thrombomodulin, and plasminogen activator (tPA) – and prothrombotic substances – von Willebrand factor, tissue factor, factor V, and plasminogen activator inhibitor (PAI_1) (Vane *et al.* 1990).

They participate in the vascular tone regulation by secretion of vasodilators – PGI_2, and endothelium-derived relaxing factors (EDRFs), identified as nitric oxide (NO) – and vasoconstrictors – platelet-derived growth factor (PDGF), and endothelin 1, active on the smooth muscle cells (Vane *et al.* 1990; Takuwa *et al.* 1989). The presence, on their luminal side, of the angiotensin-converting enzyme (ACE), which transforms inactive angiotensin I into active angiotensin II, and their ability to degrade bradykinin, reflect their usefulness in the regulation of blood pressure (Vane *et al.* 1990; Børsum 1991).

Under the stimulation of cytokines they can also synthesize interleukin 1 (IL-1), tumour necrosis factor (TNF_α) and transforming growth factor (TGF_β). They secrete specific membrane glycoproteins – endothelial leukocyte adhesion molecule (ELAM-1) and intercellular adhesion molecule (ICAM-1) – which facilitate the adhesion of leukocytes to their surface. They modulate platelet activity with the secretion of anti-aggregants – PGI_2 and EDRF – and a pro-aggregant, platelet-activating factor (PAF), that remains associated with the cell membrane (Vane *et al.* 1990).

ECs synthesize the components of their extracellular matrix and underlying basement membrane: type III and IV collagens, fibronectin, elastin, thrombospondin and heparan sulfates (Kramer *et al.* 1985).

ECs synthesize growth factors: endothelial-derived growth factor (EDGF), insulin-like growth factor (IGF-1), epidermal growth factor (EGF), smooth muscle cell mitogens and acid and basic fibroblast growth factors (FGF_a, FGF_b) stored in the extracellular matrix (Boes *et al.* 1991). The EC proliferation is stimulated by these growth factors and by the endothelial cell growth factor (ECGF) of neural origin and is inhibited by TGF_β and TNF_α (Schwartz *et al.* 1981). In response to

Cell and Tissue Culture for Medical Research, edited by A. Doyle and J.B. Griffiths.
© 2000 John Wiley & Sons, Ltd

viral aggression, ECs secrete interferons α and β. Macroangiopathies (thrombosis, atherosclerosis), micro-angiopathies and high blood pressure are principally due to alterations in the structure and functions of ECs; this presently is a very fruitful research area. However, the localization of ECs makes them very difficult to study *in vivo*, and consequently investigators show a great interest in obtaining EC cultures.

The extraction of these cells from the human umbilical cord vein, a readily available material, permits work on human cells. The extraction technique designed by Jaffe in 1973 (Jaffe *et al.* 1973) and 1980 (Jaffe 1980) is easy, quick, cheap to perform and yields exclusively human ECs.

It is necessary to have a good knowledge of this basic technique before using more complex models such as ECs extracted from different networks of capillaries.

Reagents and solutions

Phosphate buffer ×10 concentrate solution

1 litre of sterile water with: NaCl 0.14 M; KCl 0.004 M; $Na_2HPO_4.12H_2O$ 0.001 M; glucose 0.011 M; pH 7.4. To be filtered with a 0.22 μm filter and stored at 4°C in 100 ml volumes. Use ×1 concentrate.

Collagenase solution (Lyophilized collagenase. Type CLS 1 4196, Worthington Bichemicals Corporation, Freehold, New Jersey)

Dissolve 400 mg collagenase in 200 ml PBS at 4°C. Filter-sterilize. Immediately freeze in 20 ml volumes at –80°C.

EDTA-collagenase solution

This solution must be prepared just before use. Mix 1 volume of collagenase solution with 1 volume of EDTA to obtain a mixture with 0.2% glucose and 0.25% bovine serum albumin (BSA). Filter-sterilize. Although collagenase requires calcium for activity, this mixture is very efficient in releasing ECs from culture vessels.

Culture medium

For 100 ml medium mix: 10 ml M199 × 10 concentrate, 1 ml amphotericin B (250 μg ml⁻¹), 1 ml penicillin (10 00 UI ml⁻¹) + streptomycin (100 000 μg ml⁻¹), 1 ml 200 mM L-glutamine, 1.5 ml 1 M HEPES buffer, 1.65 ml 7.5% sodium bicarbonate and pyrogen-free sterile water to make 100 ml. The pH is adjusted to 7.2–7.3.

1. For seeding, use this medium completed with 20% foetal calf serum, during 24 h.
2. The next day and thereafter, use this medium completed with 10% human serum – non-heat activated.

Human serum

Human serum is obtained from volunteer donors. Blood is taken into dry sterile tubes and kept at room temperature in order to clot, and then centrifuged at room temperature (200 *g* for 15 min). The serum is sterile filtered (0.22 μm filter), pooled and stored at –20°C.

Human umbilical cords

Sterile containers, each containing 50 ml of phosphate-buffered saline (PBS) and colimycin-penicillin (300 000 units), are kept at 4°C in the obstetrics ward. The cords should be fairly long and must come from normal deliveries, without any trauma. If possible, sterile techniques are used for all cord manipulations. The cord is severed from the placenta, placed in the container and kept at 4°C until required. The cords can be stored for 3 days before use.

Materials and equipment

- Phosphate buffer
- EDTA-collagenase solution
- Culture medium
- Collagenase solution
- Foetal calf serum
- Human serum
- Umbilical cords
- 18 mm-diameter steel cannulas
- Syringe: 60 ml, 20 ml, 1 ml
- 0.13 mm-thick glass coverslips
- Aluminium foil
- Rubber tubing

All procedures are carried out in a verticle laminar flow hood.

PROCEDURE: ENDOTHELIAL CELL EXTRACTION

Rinsing the umbilical vein

1. Spread the cord over sterile gauze.
2. Slide a thread under one end of the cord, put a steel cannula (cannula 1) into the vein and secure it by tying the thread over the cord with a double knot.
3. Put a syringe containing 60 ml PBS (at room temperature) on the cannula.
4. Inject the PBS into the vein.
5. The other end of the umbilical vein is cannulated in the same way (cannula 2).
6. Rinse again with 60 ml PBS.
7. Keep the syringe on the cannula 1.

Perfusion with collagenase solution

1. Fill a syringe with 20 ml warmed collagenase solution (37°C) and replace the rinsing syringe with it.
2. Inject the collagenase solution slowly and, as the first drops of the collagenase solution appear at the opening of cannula 2, close with a syringe (1 ml) put on this cannula.
3. Continue the injection until the vein is full and protrudes on the cord surface.
4. Keep the syringe in place. Wrap sterile aluminium foil around each end of the cord.

Incubation

1. Submerge the cord in an isotonic saline bath (at 37°C) with both ends protruding from the bath. The incubation time varies from 10 to 15 min, depending on the collagenase activity level.
2. Prepare two conical tubes with 10 ml culture medium in each.

Cell extraction

1. Remove the cord from the saline bath, spread over sterile gauze and manually knead the umbilical vein in order to increase cell yield.
2. Raise the end of the cord bearing the syringe of collagenase in order to avoid leakage of the effluent under pressure. Carefully remove the syringe and immediately replace with another containing 40 ml PBS at 4°C.
3. Carefully remove the syringe (1 ml) and put the cannula 2 into a conical tube.
4. Raise the other end of the cord with the syringe attached, putting the cord at an oblique angle.
5. Rapidly flush the vein with the PBS. Collect the effluent containing the endothelial cells in each conical tube.

Cell washing

1. Centrifuge at room temperature at $180\,g$ for 15 min. A small cell pellet will be visible, particularly in the first tube.
2. Remove the supernatant and suspend the cell pellet in 5 ml culture medium, with 20% foetal calf serum, taking care not to dissociate the cells completely and thus obtaining 3–5 cell aggregates.
3. Count the cells with a haemocytometer. The cell yield is approximately $1–1.5 \times 10^6$ cells per 10 cm of cord, 90% of which should be viable (as determined by the Trypan blue exclusion test).

Endothelial cells obtained from different human cords can be pooled.

CELL CULTURES

1. Seed the cells in 35 mm diameter non-coated Petri dishes at a density of $3.0–4.0 \times 10^4$ cells cm^{-2} with 3 ml of culture medium completed with 20% foetal calf serum, per dish, during 24 h.

2. Put the culture dishes into a double compartment (three 35 mm diameter culture dishes in the top of a 100 mm diameter Petri dish – cover with the bottom of the dish).
3. Incubate at 37°C under 5% CO_2.

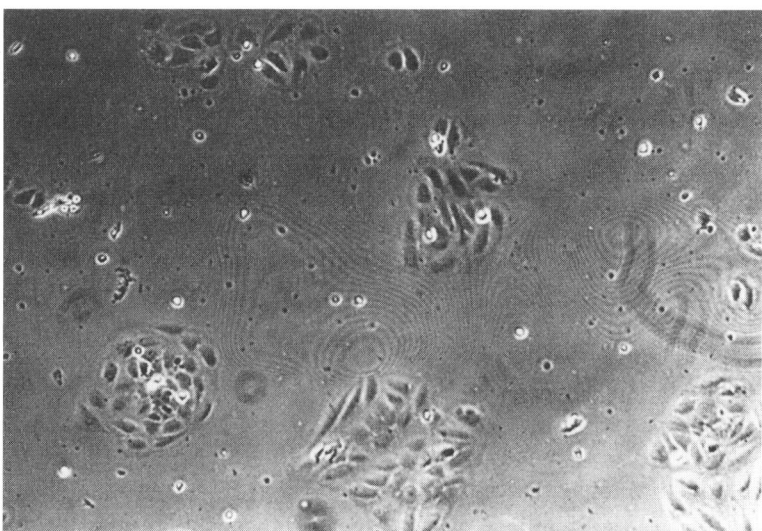

Figure 4.13.1 The endothelial cell cultures (after 24 h); small epithelioid cell clusters adhere at the bottom of the dishes.

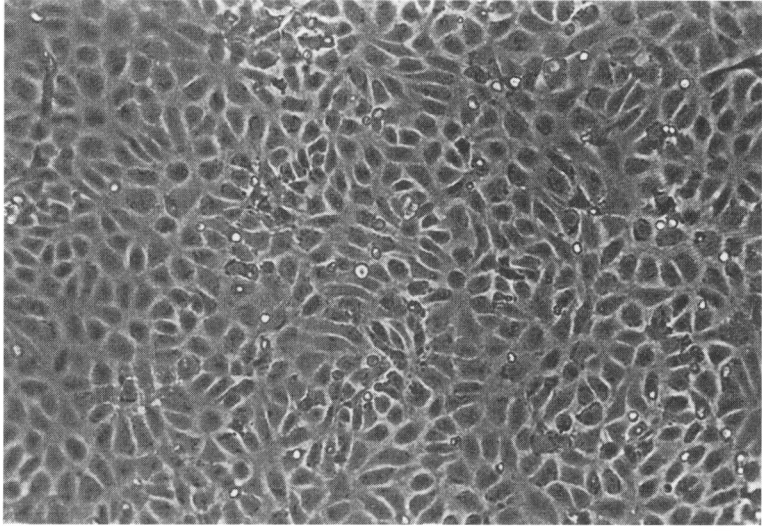

Figure 4.13.2 Seven-day-old endothelial cell cultures: monolayer of confluent polygonal cells.

Cell culture development

The following day, examination by phase contrast microscopy shows that small epithelioid cell clusters adhere at the bottom of each culture dish (Figure 4.13.1).

Gently rinse the cells once or twice with culture medium completed with 10% human serum, prewarmed to 37°C. Finally put 3 ml of this culture medium into each dish and reincubate. The culture medium must be changed every other day.

On about the sixth or seventh day, the bottom of each culture dish will be covered with a complete confluent homogeneous monolayer of polygonal cells with hazy cell limits and ovoid nucleolated nuclei (Figure 4.13.2). Growth is stopped by contact inhibition. Then, use culture medium with only 5% human serum.

Subculture

1. Release endothelial cells from culture dishes when the cell monolayer becomes confluent.
2. Rinse the cell monolayer twice with PBS at 37°C.
3. Add 3 ml EDTA-collagenase solution per dish and incubate at 37°C for 10 min.
4. Transfer all the cell suspensions into a conical tube containing 10 ml culture medium.
5. Rinse the bottom of the culture dishes twice with culture medium in order to recover most of the cells.

Thereafter proceed as for the effluent from the umbilical vein. Cultures remain healthy until the fifth passage.

Complementary Techniques

SUPPLEMENTARY PROCEDURE: CULTURE DISH COATING

Gelatin

Additional materials

- Gelatin, Difco, Detroit, Michigan, USA

1. Dissolve 200 mg gelatin in pyrogen-free sterile water at 80°C for 10 min. Allow to cool. Filter the gelatin solution. Store at 4°C. Fill 3–4 culture dishes with this gelatin solution and leave at 37°C for at least 4 h before use.
2. Just before use, draw up the excess solution. Sterilize by a 10 min ultraviolet exposure. Rinse with culture medium three times.

Fibronectin

Additional materials

- Crystallized fibronectin

- *Solution A* Dissolve 5 mg crystallized fibronectin in 5 ml pyrogen-free sterile water at 37°C and divide into 1 ml volumes. Store at 4°C.
- *Solution B* Put 1 ml Solution A in 20 ml pyrogen-free sterile water at 37°C. Fill 3–4 culture dishes and leave at room temperature for 2 h. Just before use, draw up the excess solution and sterilize by a 10 min ultraviolet exposure. Rinse three times with culture medium.

After desiccation in a dry incubator at 40°C for 24 h, the coated Petri dishes can be stored at 4°C for 2 months.

Fibronectin is also available as a frozen solution (Institut Jacques Boy SA, Reims, Fraince).

Collagen

Additional materials

- Sterile lyophilized rat tail collagen, supplied in 30 mg vials (Boehringer).

Dissolve the collagen in 0.1 M acetic acid (0.3 ml acetic acid in 50 ml pyrogen-free sterile water) in order to obtain 1 μg cm^{-2} in culture dishes. Filter. Fill 3–4 culture dishes with this solution and keep at 37°C for 24 h.

Just before use draw up the excess solution. Sterilize by a 10 min ultraviolet exposure. Rinse three times with culture medium.

SUPPLEMENTARY PROCEDURE: VON WILLEBRAND FACTOR CHARACTERIZATION

Additional materials

- Rabbit anti-human vWF antibody (Immunotech, Marseilles, France)
- Normal rabbit serum (Pasteur Institute, Lyon, France)
- Fluorescein-conjugated goat anti-rabbit IgG (Sigma)
- 5% Evans blue solution
- 50% glycerol/50% PBS solution

1. Seed endothelial cells on gelatin-coated coverslips or in Petri dishes.
2. When confluent, rinse the cell monolayer three times with PBS at 37°C.
3. Fix the cells by immersion of the coverslips in aceton for 10 min.
4. Wash twice with PBS.
5. Separate the control and test coverslips. Cover the test coverslips with one drop of rabbit anti-human HvWF antibody. Cover the control coverslips with one drop of normal rabbit serum.
6. Incubate all coverslips at 37°C for 30 min.
7. Rinse for 30 min in three changes of PBS.
8. Cover all coverslips with one drop of fluorescein-conjugated goat anti-rabbit IgG (1/20 dilution).
9. Incubate at 37°C for 30 min.
10. Rinse for 20 min in two changes of PBS. Put in 5% Evans blue solution for 5 min.

11. Mount the coverslips with a drop of 50% glycerol/50% PBS and examine by immunofluorescence microscopy.

Not all endothelial cells synthesize the von Willebrand factor. This synthesis varies according to the origin of the cells: particularly species and localization within the body.

Coating the Petri dishes with fibronectin, collagen or gelatin allows plating at a lower cell density: $1.5–2.0 \times 10^4$ cell cm^{-2}. Coating is also a requirement for endothelial cells obtained from other vessels than umbilical vein.

Human serum is essential. The cell-serum species specificity seems to be particularly important for human endothelial cells. Partial substitution with growth factors (the endothelial cell growth factor especially) allows the amount of human serum used to be reduced.

The typical morphology of the cell monolayer and von Willebrand factor are sufficient for identification. To date, for endothelial cells of other vessels, the best markers seem to be endothelin (radioimmunoassay: RPA.555 kit, Amersham) and the angiotensin-converting enzyme activity (Ryan US 1987).

CONCLUSION

The culture of endothelial cells extracted from the human umbilical cord vein has led to a better knowledge of the varied activities of these cells, although they were traditionally considered as a passive barrier between blood and internal compartment. These cultures, which are easily conducted, represent a practical and reliable material, frequently used for fundamental, clinical or pharmacologic studies.

REFERENCES

Boes M, Drake B & Bar R (1991) Interactions of cultured endothelial cells with TGF$_\beta$, bFGF, PDGF and IGF$_1$. *Life Sciences* 48: 811–821.

Børsum T (1991) Biochemical properties of vascular endothelial cells. *Virchows Archiv B Cell Pathology* 60: 279–286.

Jaffe EA (1980) Culture of human endothelial cells. *Transplantation Proceedings* XII(3) (suppl 1): 49–53.

Jaffe EA, Nachman RL, Becker CG & Minick CR (1973) Culture of human endothelial cells derived from umbilical veins. Identification by morphologic and immunologic criteria. *Journal of Clinical Investigation* 52: 2745–2756.

Kramer RH, Fuh GH, Bensch KG & Karasek MA (1985) Synthesis of extracellular matrix glycoproteins by cultured microvascular endothelial cells isolated from the dermis of neonatal and adult skin. *Journal of Cellular Physiology* 123: 1–9.

Ryan JW, Chung, A, Martin LC & Ryan US (1978) New substrates for the radioassay of angiotensin converting-enzyme of endothelial cells in culture. *Tissue and Cell* 10: 555–562.

Ryan US (1987) *Endothelial Cells*, vol. 2. CRC Press, Wolf Medical Publications Ltd, London.

Schwartz S, Gajdusek C & Selden S (1981) Vascular wall growth control: the role of the endothelium. *Atherosclerosis* 1: 107–126.

Takuwa Y, Yanagisawa M, Takuwa N & Masaki T (1989) Endothelium, its diverse biological activities and mechanism of action. *Progress in Growth Factor Research* 1: 195–206.

Vane J, Anggtard E & Botting R (1990) Regulatory functions of the vascular endothelium. *New England Journal of Medicine* 323(1): 27–36.

4.14 DISSOCIATED NERVE CELLS

Nerve cells differ from most other cell types in that their proliferation potential *in vitro* is quite limited. This causes two major problems: (a) cell death is not compensated by ongoing proliferation, which means that a relatively high initial plating density may be necessary; (b) nerve cells are generally mixed together with non-neuronal cells such as glio- or fibroblasts, and these cells may proliferate and in fact outgrow the neurons in long-term cultures. The latter problem can be alleviated by using anti-mitotics such as cytosine arabinoside (Dichter 1978) or photo-induced killing of dividing cells (Shine & Sidman 1984). The use of serum-free media also helps to control to some extent the proliferation of non-neuronal cells (Pettman *et al.* 1979; Bottenstein 1985). Serum-free medium (review: Romijn 1988) is, anyway, a good choice whenever the strict control of the culture conditions is essential, or when serum has significant inhibitory or even toxic effects. However, some neuronal populations grow better, or can be maintained for longer *in vitro* periods, in media containing at least small amounts of serum. This should be tested for each cell type and donor age.

Another important feature is the substrate on which the dissociated cells are to be plated. Porous surfaces or three-dimensional matrices generally yield better results than smooth glass surfaces; however, the latter may be preferable for studies where high optical quality is needed, e.g. for computer-assisted morphometry (König *et al.* 1987), or densitometry of immunocytochemical staining (Becquet *et al.* 1991). For biochemical studies where the cultured neurons have to be harvested (e.g. Cerruti *et al.* 1991), plastic dishes or wells may be more appropriate. The attachment properties of glass or plastic surfaces can be modulated to some extent by coating with substances such as collagen or poly-lysine.

The heterogeneity of diverse neuronal populations (or a given population taken at different stages of development) with respect to their medium and substrate requirements cannot be overemphasized. Preliminary studies should be carried out to determine the optimal conditions for each tissue type.

PROCEDURE: CULTURE OF DISSOCIATED NERVE CELLS PLATED ON POLY-LYSINE-COATED GLASS COVERSLIPS OR SMOOTH PLASTIC SURFACES

Media

All recipes are based upon Eagle's minimal essential medium (EMEM) containing Earle's salts, L-glutamine and 2.2 g l^{-1} sodium bicarbonate (Gibco).

All sera are used after heat inactivation (56°C for 30 min). Pool at least two different lots of serum to increase reproducibility.

Cell and Tissue Culture for Medical Research, edited by A. Doyle and J.B. Griffiths.
© 2000 John Wiley & Sons, Ltd

EMEM supplemented with horse and/or bovine serum

EMEM 87%
Horse serum (Sigma) 5%
Foetal bovine serum (Sigma) 5%
Glucose (20% stock solution) 3%

or

EMEM 92%
Foetal bovine serum 5%
Glucose (20% stock solution) 3%

EMEM supplemented with Nu-Serum

EMEM 92%
Nu-Serum (Collaborative Research) 5%
Glucose (20% stock solution) 3%

Serum-free medium

All components are from Sigma. EMEM supplemented with:

Ascorbic acid 17.6 mg l^{-1}
Glucose 6.0 g l^{-1}
Insulin 5.0 mg l^{-1}
Progesterone 6.3 μg l^{-1}
Putrescine 16.1 mg l^{-1}
Pyruvate 110.0 mg l^{-1}
Bovine transferrin 100.0 mg l^{-1}
Sodium selenite 5.2 μg l^{-1}

This is a slight modification (ascorbic acid supplement) of the N2.1 medium described by Mattson & Kater (1988). N2.1 derives from the original N2 medium developed by Bottenstein & Sato (1979).

Materials and equipment

- Milli-Q (Millipore) filtered water
- Poly-D-lysine, low MW (Sigma)
- Trypan blue 0.4% (Gibco)
- Standard tissue culture solutions
- Inverted microscope with phase contrast or Hoffman modulation contrast optics (Nomarski differential interference contrast optics are suitable for glass, but not for plasticware or glass coverslips in plasticware)
- Tissue culture cabinet
- Low-speed centrifuge
- Incubator with temperature, humidity, and CO_2 control
- Haemocytometer

- Standard dissection and tissue culture equipment (all sterile), including ethanol-cleaned glass coverslips, 12 or 30 mm diameter

Culturing of nerve cells is feasible without employing any antibiotics. However, this requires strictly sterile working conditions for the following steps.

Preparation of the substrates

Glass coverslips

1. The day preceding plating, place 12 mm ethanol-cleaned glass coverslips in the wells of 24-well tissue culture boxes, or 30 mm coveslips in 35 mm Petri dishes.
2. Add an aqueous solution of poly-D-lysine to yield 10–20 µg cm^{-2}. For some types of biochemical assays, it may be necessary to coat all surfaces in contact with the liquids.
3. Keep in the incubator overnight.
4. Give two short rinses with EMEM just before plating.

Smooth plastic surfaces

1. Incubate the wells or Petri dishes for at least 30 min with an aqueous solution of poly-D-lysine to yield 5 µg cm^{-2}.
2. Give two short rinses with EMEM before plating.

Dissection

1. Euthanize the tissue donor with ether and an intracardiac injection of pento-barbital. If embryonic or foetal tissue is used, euthanize the pregnant female, wet the abdominal wall with alcohol, open it with sterile scissors, and collect the uterine horns in a sterile Petri dish.
2. Dissect the pieces containing the wanted cell populations (the proper localization may be facilitated by immunocytochemical wholemount studies (König *et al.* 1988)), in a Petri dish filled with glucose-supplemented Hanks' balanced salt solution (HBSS, Gibco). Use blunt dissection as far as possible. If a scalpel is necessary, use sliding rather than pressing motions to minimize tissue damage due to compression.

Dissociation

1. To prepare the mechanical dissociation, transfer the tissue blocks into a centrifugation tube containing glucose-supplemented HBSS without calcium and magnesium. The optimal exposure time to this solution is 10–30 min, depending upon the size and compactness of the blocks. For very compact pieces (e.g. relatively mature structures of the nervous system), the following procedure may be preferable: exposure to trypsin-EDTA for 5 min at 35°C, and then stop the enzymic action by adding 10% serum; wash with glucose-supplemented HBSS without calcium and magnesium.

2. Progressively dissociate the tissue by trituration with a 5 ml glass pipette or a blue (1 ml) micropipette tip. Gradually reduce the cleft between the pipette opening and the bottom of the centrifugation tube until no visible tissue chunks are left over.

Centrifugation and resuspension

1. Gently agitate the tube to avoid premature sedimentation.
2. Centrifuge for 10 min at 70–100 g (determine the best value for each tissue type).
3. Pipette or pour off the supernatant.
4. Break the pellet by vigorous fingernail tapping at the bottom of the tube, and resuspend with a small quantity of medium.
5. Take 45 µl of cell suspension and mix with 5 µl of Trypan blue staining solution.
6. After 10 min, fill the counting chamber of a haemocytometer with a fraction of the mixture and immediately count the cells that are not stained with a microscope set for phase contrast or modulation contrast observation. The number of unstained cells roughly indicates the number of viable cells per unit volume.
7. Dilute the cell suspension with medium to obtain the desired final plating density.

For cultures starting from very small amounts of tissue, where the cell loss due to the centrifugation procedure would be prohibitory, the centrifugation step may be omitted.

Plating and incubation

1. Fill the wells of 24-well plates with 400 µl of cell suspension using a micropipette. For 35 mm Petri dishes, the optimal volume is 1200 µl. Take care to redisperse the cells, which tend to sediment and aggregate, before picking up the cells to be plated.

   For special applications (e.g. video-microscopy or visualization of intracellular calcium concentrations with an inverted microscope (König *et al.* 1994) or image analysis of immunostained cells (Bardoul *et al.* 1997), Petri dishes or large coverslips may be necessary; however, there may be no need to have cells on the whole surface, which would require very large numbers of cells for plating. A way to produce limited spots covered by cells is the following: fill the Petri dish with medium; carefully deposit a drop of cell suspension on the center (during ejection of the cell suspension with the micropipette, the tip should be below the air–liquid interface); let the cells sediment and attach for 10 min; and then transfer to the incubator.
2. Incubate the cultures at 35°C, 99% humidity, and 5% CO_2. Replace one third of the medium with freshly prepared medium every 3 or 4 days. Frequent (if possible, daily) inspection using an inverted microscope helps to optimize the culture conditions for subsequent experiments.

ALTERNATIVE PROCEDURE: CULTURE OF DISSOCIATED NERVE CELLS ON UNCOATED MICROPOROUS MEMBRANES

Materials and equipment

As for basic procedure, but Millicell HA 0.45 μm culture plate inserts from Millipore (12 mm, cellulose ester membrane) instead of the glass coverslips are required.

The steps in this procedure are the same as in the basic procedure, except for the following.

Preparation of the substrate

Fill the wells of the 24-well plate with 300 μl of medium. At least 15 min before plating, transfer the inserts carefully into the wells (no air bubbles should be trapped) using sterile forceps.

Plating and incubation

As in basic procedure, but the optimal quantity of cell suspension to be added to the medium already in the inserts is 100 μl.

Cellulose ester membranes are translucid, but not transparent, in an aqueous environment. This impedes microscopic inspection during the culture period. However, the membranes become transparent after fixation and dehydration of the cultures. For dehydration and mounting between a glass slide and a coverslip, separate the membrane from the cylindrical polystyrene holder using a small surgical blade.

DISCUSSION

Both procedures and all described media work well for short-term cultures (2 h to 8 days) of many types of dissociated nerve cells. We principally have cultured the following: embryonic rat rhombencephalon (König *et al.* 1989, 1994; Becquet *et al.* 1991), and mesencephalon (Cerruti *et al.* 1991); and foetal rat cerebellum (unpublished), hippocampus and neocortex (Rondouin *et al.* 1988; Drian *et al.* 1991). For longer culture periods, the choice of suitable conditions becomes more restricted. For instance, neocortical cells need horse serum for good survival up to, and beyond, 3 weeks. Long-term survival is also influenced by the substrate: plastic surfaces, and particularly porous membranes, generally support prolonged survival better than poly-lysine-coated glass. Confluent cell layers instead of synthetic substrates may also be useful for long-term cultures. In addition, growing dissociated cells on such substrates or on cryostat (Tuttle & Matthew 1991; Zuo *et al.* 1998) or unfrozen tissue sections (Giménez y Ribotta & König 1999) provides unique opportunities to study cell-specific interactions.

Different nerve cell populations often have unequal culture condition requirements (for instance, serotonin-expressing cell populations taken from diverse micro-regions of the embryonic rhombencephalon have very different substrate adhesion properties) (König *et al.* 1989). Therefore, the optimal compromise should be determined for each cell type, taking into account: (1) the maximal culture period needed: (2) the requirement for serum-free conditions; (3) the need for optical quality; and (4) the constraints linked to particular methods such as video-microscopy or biochemical assays.

REFERENCES

Bardoul M, Drian MJ & König N (1997) AMPA/kainate receptors modulate the survival *in vitro* of embryonic brainstem cells. *International Journal of Developmental Neuroscience* 15: 695–701.

Becquet D, Héry F, Héry M, Drian MJ, Faudon M & König N (1991) Population-specific modulation of 5-HT expression in cultures of embryonic rat rhombencephalon. *Journal of Neuroscience Research* 29: 42–50.

Bottenstein JE (1985) Growth and differentiation of neural cells in defined media. In: Bottenstein JE & Sato GH (eds) *Cell Culture in the Neurosciences*, pp. 3–43. Plenum, New York.

Bottenstein JE & Sato GH (1979) Growth of rat neuroblastoma cell line in serum-free supplemented medium. *Proceedings of the National Academy of Sciences of the USA* 76: 514–517.

Cerruti C, Drian MJ, Kamenka JM & Privat A (1991) Localization of dopamine carriers by BTCP, a dopamine uptake inhibitor, on nigral cells cultured *in vitro*. *Brain Research* 555: 51–57.

Dichter MA (1978) Rat cortical neurons in cell culture: culture methods, cell morphology, electrophysiology, and synapse formation. *Brain Research* 149: 279–293.

Drian MJ, Kamenka JM, Pirat JL & Privat A (1991) Non competitive agonists of NMDA prevent spontaneous neuronal death in primary cultures of embryonic rat cortex. *Journal of Neuroscience Research* 29: 133–138.

Giménez y Ribotta M & König N (1999) Embryonic raphe neurons grafted *in vitro* on fetal spinal cord slices recognize specific target areas. *Journal of Neuroscience Research* (in press).

König N, Han VKM, Lieth E & Lauder J (1987) Effects of coculture on the morphology of identified raphe and substantia nigra neurones from the embryonic rat brain. *Journal of Neuroscience Research* 17: 349–360.

König N, Wilkie MB & Lauder J (1988) Tyrosine hydroxylase and serotonin containing cells in embryonic rat rhombencephalon: a whole-mount immunocytochemical study. *Journal of Neuroscience Research* 20: 212–223.

König N, Rajaofetra N, Drian MJ, Favier F, Sandillon F, Fuentes C & Privat A (1989) Serotonin-expressing cells from different microregions of the embryonic rat rhombencephalon: behaviour in cell culture and in transplants to the adult spinal cord. In: Gage F, Privat A & Christen Y (eds) *Neuronal Grafting and Alzheimer's Disease*, pp. 150–164. Springer, Berlin.

König N, Serrano P & Drian MJ (1994) AMPA elicits long-lasting, partly hypothermia-sensitive calcium responses in acutely dissociated or cultured embryonic brainstem cells. *Neurochemistry International* 3: 738–740.

Mattson MP & Kater SB (1988) Isolated hippocampal neurons in cryopreserved long-term cultures: development of neuroarchitecture and sensitivity to NMDA. *International Journal of Developmental Neuroscience* 6: 439–452.

Pettmann B, Louis JC & Sensenbrenner M (1979) Morphological and biochemical maturation of neurones cultured in the absence of glial cells. *Nature* 281: 378–380.

Romijn HJ (1988) Development and advantages of serum-free, chemically defined nutrient media for culturing of nerve tissue. *Biology of the Cell* 63: 263–268.

Rondouin G, Drian MJ, Chicheportiche R, Kamenka JM & Privat A (1988) Non-competitive antagonists of N-methyl-D-aspartate receptors protect cortical and hippocampal cell cultures against glutamate neurotoxicity. *Neuroscience Letters* 91: 199–203.

Shine HD & Sidman RL (1984) Immunore-active myelin basic proteins are not detected when Shiverer mutant Schwann cells and fibroblasts are co-cultured with normal neurons. *Journal of Cell Biology* 98: 1291–1295.

Tuttle R & Matthew WD (1991) An *in vitro* bioassay for neurite growth using cryostat sections of nervous tissue as a substratum. *Journal of Neuroscience Methods* 39: 193–201.

Zuo J, Neubauer D, Dyess K, Ferguson TA & Muir D (1998) Degradation of chondroitin sulfate proteoglycan enhances the neurite-promoting potential of spinal cord tissue. *Experimental Neurology* 154: 654–662.

4.15 NEURONAL AND GLIAL TUMOURS *IN VITRO* – AN OVERVIEW

INTRODUCTION

The adult nervous system displays an apparently bewildering array of cell types organized into a complex set of structures. However, these cells are of only two main lineages, neuronal or glial. In the adult nervous system there are approximately 10^{11} neurones and in excess of 10^{12} glial cells. However, both cell types are derived, embryologically, from a simple tubular ectodermal structure, which during the third week of development thickens forming the neural plate and which subsequently begins to fold inwards generating the neural groove. On either edge of this structure are the neural folds, which as the groove deepens, approach each other and fuse forming the nerual tube. During this process, islands of cells are pinched off the neural fold and become the neural crest. These cells will go on in due course to form the sensory ganglia of the spinal and cranial nerves, the post-ganglionic neurones of the autonomic nervous system and the Schwann cells and satellite cells of the peripheral nervous system. The neural tube will form all the elements of the central nervous system with the cavity forming the ventricular system of the brain. This cellular complexity is reflected in the wide variety of tumours that occur within the nervous system. While there are no tumours which arise from mature neurones, presumably because in the post-natal nervous system, these cells are terminally differentiated and post-mitotic, a significant number of important types of tumour occur in the nervous system. Most malignant tumours that occur in the nervous system are gliomas that are derived from one or other class of glial cells. As there is considerable turnover of these cells throughout life these tumours may occur at any age. The second important class of tumours are those embryonal tumours that are thought to be derived from the primitive, undifferentiated cells that arise early in embryonic development and are consequently almost always restricted to children.

The purpose of this brief review is to describe the major types of malignant tumour which develop in the nervous system and to review the availability of well-characterized established cell lines which are can be used for biological or therapeutic studies of these tumours.

Cell and Tissue Culture for Medical Research, edited by A. Doyle and J.B. Griffiths.
© 2000 John Wiley & Sons, Ltd

NEURONAL NEOPLASMS

Tumours with a neuronal component occur in both the CNS and the peripheral nervous system. These tumours are usually histologically benign and clinically indolent. They are very often composed of a mixed population of glial and neuronal cells and it has been argued that the neuronal component represent a haematomatous rather than neoplastic element of these tumours. In any event, cell lines derived from these tumours in humans have not been reported in the literature. A number of studies have characterized short-term cell lines from central neurocytomas that appear to display both astrocytic and neuronal phenotypes *in vitro*. Whether this occurs because this tumour is composed of undifferentiated bipotential precursor cells (Valdueza *et al.* 1996) or the outgrowth of a mixed population of cells of either neuronal or astrocytic nature (Ishiuchi *et al.* 1998; Ischiuchi & Tamura 1997) is unknown.

GLIAL NEOPLASMS

By far the most common malignant tumours of the nervous system are derived from glial cells. These tumours (Table 4.15.1), the so-called gliomas, are derived from three cell types, astrocytes, oligodendrocytes or ependymal cells and comprise an extremely heterogeneous group of neoplasms with marked differences in clinical and biological behaviour. Astrocytic gliomas comprise about half of all primary brain tumours and range from pilocytic astrocytomas in children which are clinically indolent to highly aggressive neoplasms in adults like glioblastoma

Table 4.15.1 Simplified classification of glial tumours

Glial cell type	Tumour	Subtypes	WHO grade
Astrocyte	Diffuse astrocytomas	Low-grade astrocytoma	II
		Anaplastic astrocytoma	III
		Glioblastoma multiforme	IV
	Pilocytic astrocytoma		I
	Pleomorphic xanthoastrocytoma		I
	Desmoplastic cerebral astrocytoma of childhood		I
	Subependymal giant cell astrocytoma		I
Oligodendrocyte	Oligodendroglioma	Oligodenderoglioma	II
		Anaplastic oligodendroglioma	III
	Mixed glioma	Oligoastrocytoma	II
		Anaplastic oligoastrocytoma	III
Ependymal cells	Ependymoma	Ependymoma	II
		Anaplastic ependymoma	III
		Myxopapillary ependymoma	I
		Subependymoma	I

multiforme where tumour recurrence is usually inevitable within a few months even following radical radiotherapy and chemotherapy. Oligodendrogliomas are much rarer tumours, often relatively well differentiated and even the anaplastic tumours carry a much better prognosis than malignant astrocytomas with median survival in excess of 5–7 years from diagnosis. A large proportion of these tumours also appear to be sensitive to radiation and combination chemotherapy. Ependymomas, the tumours which arise from the epithelial layer which lines the ventricles of the brain and the discontinuous remnants of the central canal of the spinal cord, are rather uncommon and account for only about 5% of all CNS neoplasms. They occur throughout the spinal axis although most commonly in the infratentorial compartment in the first two decades of life and are clinically unpredictable but do show a propensity for local recurrence and in some cases, leptomeningeal spread.

EMBRYONAL NEOPLASMS

The embryonal tumours which arise in the nervous system constitute an extremely important group of tumours which arise predominantly in childhood and present a significant clinical challenge (Table 4.15.2). In the CNS, these small blue cell tumours comprise a diverse group of neoplasms which while largely undifferentiated, usually exhibit sufficient features so that they can be classified into groups with common features. Those embryonal tumours which display some ependymal features are termed ependymoblastoma, those which display neuroblastic features as CNS neuroblastoma and with the term medulloblastoma reserved for those typically undifferentiated neoplasms which occur in the cerebellum. It has been suggested that all of these tumours should be more properly termed 'primitive

Table 4.15.2 Simplified classification of embryonal tumours of the nervous system

Site	Tumour	Subtypes	WHO grade
CNS	Medulloblastoma	Medulloblastoma	IV
		Medullomyoblastoma	IV
		Melanotic medulloblastoma	IV
		Lipomatous medulloblastoma	I or II
	Supratentorial PNET		IV
	Ependymoblastoma		IV
	Neuroblastoma		IV
	Atypical teratoid/rhabdoid tumours		IV
PNS	Neuroblastoma	Classical neuroblastoma	NA
		Ganglioneuroblastoma	NA
		Composite ganglioneuroblastoma	NA
		Gangloneuroma	NA
	Olfactory neuroblastoma		NA

NA = not applicable

neuroectodermal tumours' (PNET), on the grounds that they must have a common origin from primitive, undifferentiated cells and they differ from each other only in their location, cell type and degree of differentiation. However, this is merely an assumption and there is no evidence, as yet, to suggest that there is a common pathogenesis for these diverse tumours. In the current WHO classification, the term supratentorial PNET is reserved for those rare tumours which histologically resemble medulloblastoma, but arise in the cerebral hemispheres rather than the cerebellum.

Although these tumours are less common than glial neoplasms, the clinical outlook for patients with these tumours is poor, as there is a distinct propensity to disseminate throughout the craniospinal axis. However, for the most frequent of these tumours in the CNS, medulloblastoma, a regimen of craniospinal irradiation following complete surgical removal has produced 10-year survival rates in the best series of between 50% and 70%.

Of considerable clinical and biological significance are the highly malignant neuroblastomas that occur in the peripheral nervous system and which are derived from migratory neuronal precursor cells of the neural crest that would ultimately become either the adrenal medulla or elements of the sympathetic nervous system. Because CNS maturity occurs relatively early in embryonic development, most of these neuroblastic tumours arise during childhood with over 80% occurring before the age of 4. The most common site for these tumours are the sympathetic ganglia with about half the tumours occurring in the abdominal, thoracic, cervical or pelvic sympathetic ganglia with most of the remainder in the adrenal glands. Much more rarely, these tumours can arise from small sympathetic ganglia or by transdifferentiation of neural crest derived cells in organs like the orbit, kidney, skin, ovary, spermatic cord or bladder. Neuroblastomas typically present as an abdominal mass usually with some metastic spread, especially in older children, to the bone or bone marrow. The prognosis for patients with disseminated neuroblastoma is poor and even following aggressive chemotherapy, less than 20% of these patients survive 2 years from diagnosis.

Occasionally, neuroblastomas occur within the CNS (CNS neuroblastoma). These neoplasms occur deep in the cerebral hemispheres, most commonly in children under the age of five. They carry a relatively poor prognosis and may give rise to extracranial metastases. Rarely, neuroblastoma may occur in the vault of the nose overlying the cribiform plate. Although these so-called olfactory neuroblastomas may extend to form an intracranial subfrontal mass, it is usually possible to remove them completely and consequently the prognosis is good.

GROWTH OF NERVOUS SYSTEM TUMOURS *IN VITRO*

The laboratory procedures for producing short-term cell lines and eventually established cell lines from surgical biopsy material from nervous system tumours have been published by several authors (Ali-Osman 1998; Darling 1999, 1991; Freshney 1980; Westphal *et al.* 1997; Westphal & Meissner 1998). Basically two main approaches have been employed to produce single-cell suspensions for primary culture using either enzymatic disaggregation with proteolytic enzymes including trypsin or comparatively aggressive enzyme cocktails containing pronase, or its main component neutral protease, combined with DNAase and collagenase (Ali-Osman 1998; Rosenblum *et al.* 1983; Rutka *et al.* 1986, 1987; Westphal *et al.* 1997; Westphal & Meissner 1998). The second main approach has been to use mechanical disaggregation using either energetic pipetting or repeated passage through 14-gauge hypodermic syringe needles (Ponten & Macintyre 1968; Studer *et al.* 1985; Westermark *et al.* 1973). One potential advantage of using mechanical disaggregation is that the loosely attached tumour is removed from the blood vessels leaving a whole intact network which may well reduce endothelial cell contamination (Westphal, 1998).

An alternative approach has been to use explanted biopsy material. It is essential that the small tumour fragments are anchored firmly to the growth substratum for this approach to be successful. Techniques used to achieve this include mechanical methods using coverslips anchored with silicone grease (Thomas *et al.* 1984) or allowing the tumour fragments to dry slightly to the growth substrate (Cravioto 1986; Jacobsen *et al.* 1987; McKeever *et al.* 1987; Studer *et al.* 1985).

The types of culture medium which have been used for the growth of these tumours has been unremarkable, and various authors have used Ham's F-10 or 12, DMEM or RPMI. Any of these media with supplementation with batch-tested foetal bovine serum seem to be satisfactory for the growth of both established cell lines or short-term cultures from malignant astrocytic tumours (see review by Darling 1991) and neuroblastoma (Israel and Thiele 1994). However, there is a suggestion that human umbilical vein serum is more satisfactory for the growth of medulloblastoma *in vitro* (Pietsch *et al.* 1994).

NERVOUS SYSTEM TUMOURS IN CULTURE

It has been recognized for nearly 30 years that it has been relatively easy to establish cell lines from adult malignant astrocytic gliomas and some embryonal tumours like peripheral neuroblastoma. The early work on the culture of CNS tumours has been reviewed in detail by a number of authors (Collins 1983; Nister & Westermark 1994; Ponten & Macintyre 1968; Ponten & Westermark 1978). In contrast, it has been extremely difficult to produce cell lines from gliomas of non-astrocytic lineages and most other types of nervous system embryonal tumours.

Glial tumours in culture

In reviewing the literature on cell lines derived from glial tumours, there are a number of striking features that emerge (Table 4.15.3). While it is possible to produce vigorous short-term cultures from many types of glioma, there is marked variation in the proportion of different types of tumour which will establish in culture. More than 90% of biopsies from malignant astrocytoma (WHO grade III and IV astrocytoma) produce short-term cell lines *in vitro*. Of these, about half will go on to form established cell lines. Consequently, the number of cell lines derived from malignant astrocytoma that have been described in the literature now exceeds 100, although only about 20 are widely used. The vast majority

Table 4.15.3 *In vitro* cultures of nervous system tumours

Tumour type	Subtype	Short-term cultures	Established cell lines	Comments
Astrocytoma	Low-grade astrocytoma	Yes > 50% of biopsies	No	
	Anaplastic astrocytoma	Yes > 50% of biopsies	Yes Uncommon	< 10 lines reported in the literature
	Glioblastoma multiforme	Yes > 90% of biopsies	Yes > 50% of biopsies	More than 100 cell lines reported in the literature. About 20 extensively characterized and in widespread use
Oligodendro-glioma		Yes > 50% of biopsies	Yes Very rarely	Three cell lines reported in the literature. Limited characterization
Ependymoma		Yes > 50% of biopsies	No	No cell lines reported in the literature
Medulloblastoma		Yes > 50% of biopsies	Yes < 20% of biopsies	About 10 well-characterized cell lines reported. Non-adherent. Most successful when cultured from xenografted material
Neuroblastoma		Yes > 50% of biopsies	Yes < 20% of biopsies	> 50 well-characterized cell lines reported in the literature. Most successful from metastatic sites

of these are derived from grade IV astrocytoma (glioblastoma multiforme) and only a handful of cell lines derived from grade III astrocytomas have established *in vitro* (Table 4.15.4). It is quite difficult to determine if the lower grade malignant astrocytomas variously described in the literature as anaplastic astrocytoma, malignant glioma or grade III astrocytoma were really a single type of tumour. It is likely that some of these tumours might now be more properly reclassified as glioblastoma multiforme.

There are no convincing reports of cell lines derived from low-grade astrocytomas of either children or adults although many biopsies from these tumours appear to give rise to vigorous short-term cultures composed predominantly of morphologically well differentiated astrocytic-like cells (Table 4.15.3). It is interesting to note that high grade malignant astrocytomas of childhood do not seem to become established as cell lines with the same ease as their adult counterparts with only three cell lines described in the literature including SF-188 (Rutka *et al.* 1987) and A 382 (Giard *et al.* 1973), neither being in common usage. It may be the relative rarity of these tumours which accounts for this discrepancy, although it seems more likely that it is related to intrinsic biological differences. They appear to be identical, histologically however, these are fundamental differences in their molecular genetic profiles compared with their adult counterparts (Warr *et al.* 2000). It is also clear that cell lines cannot be reliably established from even anaplastic oligodendrogliomas with anything like the same frequency as those from malignant astrocytomas. Only three permanent cell lines have been reported in the literature (Table 4.15.4). There are no reported cell lines from ependymomas, although it has been possible to establish these tumours as serially transplantable xenografts in immune-deprived animals (Horowitz *et al.* 1987).

What factors predispose towards the establishment of cultures from malignant astrocytoma? Although a systematic study has not been carried out on the biological or molecular characteristics of malignant astrocytoma which predispose towards establishment *in vitro*, some clinical features of the tumour appear to correlate with establishment *in vitro*, for example anatomical site or sex of the patient (Westermark *et al.* 1973). A common cytological feature which Ponten and his colleagues observed in their studies was that primary cultures composed of predominantly spindle shaped cells were those which would always go on to eventually establish *in vitro*, while those primary cultures composed of astrocytic-like or polygonal cells always eventually senesced (Ponten & Macintyre 1968; Ponten & Westermark 1978). More recently, it has been shown that these spindle-shape, fibroblastic-like cells, express cell surface fibronectin but not GFAP (Kennedy *et al.* 1987; McKeever *et al.* 1987). It is equally clear that these are not fibroblasts or other adventitious cell types, but display a wide range of neoplastic features and astrocytic features or under appropriate conditions can be induced to show them. FN + GFAP – cells in early passage culture have been shown to be aneuploid by cytometric DNA analysis (McKeever *et al.* 1986) and are able to grow in reduced serum concentrations (McKeever 1984). Both short-term cell lines and established glioma cell lines with this phenotype have been shown to exhibit a wide range of neoplastic features. These include growth on confluent monolayers of normal cells (Freshney 1980; MacDonald *et al.* 1985), expression of plasminogen activator (Frame *et al.* 1984), angiogenic activity on chick chorioallantoic membrane (Frame *et al.* 1984), 'invasive' behaviour in either chick heart fragment (de Ridder *et al.* 1987) or normal rat brain aggregate confrontation assays (Knott *et al.* 1998). Additionally, many, but not all established glioma cell lines grow subcutaneously or intracranially in immune-suppressed mice (Bigner *et al.* 1981a,b; Bullard *et al.* 1981). Short-term cell lines from malignant astrocytoma which do not express GFAP have been shown to display other astrocytic features like glutamine synthetase and β-alanine-sensitive blockade of high-affinity uptake of GABA (Frame *et al.* 1984). Glucocorticoids like dexamethasone or methyl prednisolone inhibit the proliferation of these cells *in vitro*. This inducation of cytostasis produces significant increases in high affinity GABA uptake and also the activity of glutamine synthetase while producing a concomitant and marked depression of neoplastic features like plasminogen activator activity (McLean *et al.* 1986).

Table 4.15.4 Established cell lines derived from malignant astrocytic gliomas

Tumour	Cell line designation	Age/Sex	Antigenic profile	Availability	References
Anaplastic astrocytoma	U87MG	54 yr/F	-/+	ATCC, ECACC, DKFZ	Ponten & Macintyre (1968)
	U373MG	61 yr/M	+/-	ATCC, ECACC, DKFZ	Ponten & Macintyre (1968)
	U118MG	50 yr/M	-/+	ATCC, DKFZ	Ponten & Macintyre (1968)
Mixed glioma	D54MG	39 yr/F	-/-	NK	Bigner et al. (1981a)
Glioblastoma multiforme	A-172	53 yr/M	-/±	ATCC, ECACC, DKFZ, ICLC	Giard et al. (1973)
	D37MG	61 yr/F	-/±	NK	Bigner et al. (1981a)
	D65MG	65 yr/F	-/±	NK	Bigner et al. (1981a)
	D263MG	53 yr/M	+/-	NK	Bigner et al. (1981a)
	DBTRG-05MG	59 yr/F	-/+	ATCC, ECACC, ICLC	Kruse et al. (1992)
	Hs683	76 yr/M	NK	ATCC, DKFZ	Owens et al. (1976)
	M059J	33 yr/M	NK	ATCC	Allalunis-Turner et al. (1993)
	M059K	33 yr/M	NK	ATCC	Allalunis-Turner et al. (1993)
	SF-188	8 yr/M	-/+	NK	Rutka et al. (1987)
	SF-210	72 yr/F	-/+	NK	Rutka et al. (1987)
	SF-268	24 yr/F	-/-	NK	Rutka et al. (1987)
	SF-295	67 yr/F	-/-	NK	Rutka et al. (1987)
	T98G	61 yr/M	NK	ATCC, ECACC, IZSBS	Stein (1979)
	U105MG	62 yr/M	-/+	NK	Ponten & Macintyre (1968)
	U138MG	47 yr/M	-/+	ATCC, DKFZ	Ponten & Macintyre (1968)
	U178MG	56 yr/M	-/+	NK	Ponten & Macintyre (1968)
	U251MG	75 yr/M	+/-	NK	Ponten & Macintyre (1968)
	U343MG	60 yr/M	-/+	DKFZ	Ponten & Macintyre (1968)
	U410MG	42 yr/M	-/+	NK	Ponten & Macintyre (1968)
	G-CCM	NK	+/NK	ECACC	Frame et al. (1984)
	G-UVW	NK	NK	ECACC	McLean et al. (1986)
	LN 405	62 yr/F	+/NK	DSMZ	Bodmer et al. (1989)
	CCF-STTG1	NK	+/NK	ECACC	Barna et al. (1985)
	LI	51 yr/M	-/NK	RMCIS	Zupi et al. (1988)
	DF	48 yr/M	-/NK	RMICS	Zupi et al. (1988)

Table 4.15.5 Established cell lines from embryonal tumours of the nervous system

Tumour	Cell line designation	Age/Sex	Site	Availability	References
Medulloblastoma	D283 Med	6 yr/M	Metastasis	ATCC	Friedman et al. (1985)
	D341 Med	3 yr/M	Primary tumour	ATCC	Friedman et al. (1988)
	D384 Med	1 yr/M	Primary tumour	NK	Bigner et al. (1990)
	D425 Med	5 yr/M	Primary tumour	NK	Bigner et al. (1990)
	Daoy	4 yr/M	Primary tumour	ATCC	Jacobsen et al. (1985)
	MHH-MED1	10 yr/M	CSF seeding	NK	Pietsch et al. (1994)
	MHH-MED2	6 yr/F	Primary tumour	NK	Pietsch et al. (1994)
	MHH-MED3	3 yr/F	Primary tumour	NK	Pietsch et al. (1994)
	MHH-MED4	4 yr/M	Primary tumour	NK	Pietsch et al. (1994)
Neuroblastoma	IMR-32	1 yr/M	Primary tumour	ATCC, ECACC, DSMZ, IZSBS, ICLC	Tumilowicz et al. (1970)
	SK-N-MC	14 yr/F	Metastasis	ATCC, DKFZ	Biedler et al. (1973)
	SK-N-SH	4 yr/F	Metastasis	ATCC, ECACC	Biedler et al. (1973)
	SK-N-AS	6 yr/F	Metastasis	ECACC,ICLC	Helson & Helson (1985)
	SK-N-FI	11 yr/M	Metastasis	ECACC, ICLC	Helson & Helson (1985)
	SK-N-DZ	2 yr/F	Metastasis	ATCC, ECACC	Helson & Helson (1985)
	SK-N-BE (2)	2 yr/M	Metastasis	ECACC, ICLC	Biedler & Spengler (1976)
	CHP-126	1 yr/F	Metastasis	NK	Schlesinger et al. (1976)

Key to Tables 4.15.4 and 4.15.5

NK – not known
Antigenic Phenotype:
+/– indicates GFAP positive, fibronectin negative antigenic phenotype; –/+ indicates GFAP negative, fibronectin positive antigenic phenotype; ± weak positivity

ATCC – American Type Culture Collection
http://www.atcc.org/
Contact: American Type Culture Collection,
10801 University Boulevard,
Manassas, VA 20110–2209, USA
Tel: +1-703-365-2700

DKFZ – Deutsches Krebsforschungszentrum (German Cancer Research Center) Cell Culture Collection
Contact: Heinz Lohrke
Tumorbank-Diagnostik und Experimentelle Therapie
German Cancer Research Center
Im Neuenheimer Feld 280
69120 Heidelberg 1, Germany
Tel: +49-06221-423246 Fax: +49-06221-423265
E-mail: H.Loehrke@DKFZ.Heidelberg.DE

DSMZ – Deutsche Sammlung von Mikroorganismen und Zellkulturen GmbH (German Collection of Microorganisms and Cell Cultures)
http://www.dsmz.de/
Contact: Hans G. Drexler
Human and Animal Cell Cultures
German Collection of Microorganisms and Cell Cultures
Mascheroder Weg 1b
38124 Braunschweig, Germany
Tel: +49–531–2616.161 Fax: +49–531–2616.150
E-mail: dsmzmutz@gbf-braunschweig.de

ECACC – European Collection of Cell Cultures
http://www.camr.org.uk/ecacc.htm
Contact: Sally Warburton
European Collection of Cell Cultures
CAMR Centre for Applied Microbiology and Research
Porton Down, Salisbury, Wiltshire, SP40JG, UK
Tel: +44–1980–612512 Fax: +44–1980–611315
E-mail: ecacc@camr.org.uk

ICLC – Interlab Cell Line Collection
http://www.biotech.ist.unige.it/bio/engl.htm
Contact: Barbara Parodi
Interlab Cell Line Collection
Istituto Nazionale per la Ricerca sul Cancro c/o CBA
Largo Rosanna Benzi, 10
16132 Genova, Italy
Tel: +39-0105737474 Fax: +39-0105737295
E-mail: iclc@ist.unige.it

IZSBS – Cell line collection from the Istituto Zooprofilattico Sperimentale, Brescia, Italy
Contact: Maura Ferrari
Centro Substrati Cellulari
Istituto Zooprofilattico Sperimentale
Via A. Bianchi, 7
25100 Brescia, Italy
Tel: +39-0302290248 Fax: +39-030335613

RMCIS – Istituto Regina Elena, Rome, Italy
Contact: Gabriella Zupi
Chemioterapia Sperimentale Preclinica
Istituto Regina Elena
Via delle messi d'Oro, 156
00158 Roma, Italy
Tel: +39-064985537 Fax: +39-0649852505
E-mail: ZUPI@SYSV.IFO.IT

The extensive molecular genetic changes which occur in cell lines from high-grade astrocytomas have been extensively characterized and reflect the changes seen in tumours *in situ*. For example, gains of chromosome 7 and losses of all or part of one copy of chromosome 10 can readily be identified *in vitro* and *in situ* (Mohapatra *et al.* 1995). Mutations in the *p53* gene are commonly seen in both glioblastoma cell lines and tumours *in situ* (Louis 1994) while allelic loss involving the CDKN2 gene is a very common feature of cell lines derived from glioblastoma multiforme (Dreyling *et al.* 1995).

Embryonal tumours

A limited number of cell lines have been produced from cerebellar medulloblastoma (Table 4.15.5). These lines often show little or no adherence to the growth substrate and grow as cell suspensions of single cells or loosely attached aggregates of cells (Friedman *et al.* 1985; Pietsch *et al.* 1994). Attempts to establish such cell lines of adherent cells from these tumours have proved difficult, and only one cell line, DAOY has been grown in this way (Jacobsen *et al.* 1985). The most successful approach has been to use tumour material from metastatic sites and to initially propagate the cells as xenografts in immune deprived animals (Friedman *et al.* 1985). This strongly suggests that considerable cell selection has occurred and they may be more appropriate models for disseminated medulloblastoma than for the primary disease. More recently, an alternative *in vitro* approach has been to use sequential cycles of incubation in plastic flasks to remove adherent cells thereby selecting for non-adherent cells. This author has used a standard growth medium, DMEM, but supplemented with high glucose levels and human umbilical cord serum rather than foetal bovine serum. Using this approach an unprecedented five cell lines from seven consecutive medulloblastoma specimens were established (Pietsch *et al.* 1994).

Medulloblastoma cell lines *in vitro* do not express GFAP, but some at least express neurofilaments protein types L, M and H (He *et al.* 1989, 1991), while others do not (Pietsch *et al.* 1994). More commonly, these cell lines express neurone-specific enolase and synaptophysin, but not chromogrannin (Pietsch *et al.* 1988). This suggests that cell lines with different levels of neuronal differentiation occur; those with 'mature' neuronal features like neurofilament protein expression and others which have only 'early neuronal' features like synaptophysin. A common and characteristic cytogenetic abnormality found in about half of medulloblastoma biopsies is the presence of an isochromosome 17q with loss of chromosome arm 17p (Biegel *et al.* 1989). As the *p53* gene is virtually never mutated in medulloblatoma *in situ*, this has led to speculation that there is a second tumour supressor gene distal to the *p53* gene, although this has not yet been identified. Double minute chromosomes involving the c-myc proto-oncogene are often present in medulloblastoma *in vitro*, but is less commonly observed *in situ*, suggesting that those rare cells in medulloblastoma *in situ* which have this property are selected *in vitro* (Bigner *et al.* 1990).

In contrast to medulloblastoma, it has been relatively easy to establish cell lines from peripheral neuroblastoma. There is a large portfolio of established cell lines derived from these tumours, which have been extensively characterized

(Table 4.15.5). The literature suggests that between 10% and 20% of tumours will give rise to established cell lines and that the most successful source of material is metastatic cells derived from bone marrow (Israel & Thiele 1994).

Neuroblastoma-derived cell lines have been extensively characterized and clearly retain neuronal or neuro-endocrine/chromaffin-like properties (Cooper *et al.* 1991). They frequently express neurone-specific enolase and neurofilament protein both *in situ* and *in vitro* (Ciccarone *et al.* 1989; Ross *et al.* 1988). Chromogranin-A, a member of a family of secretory proteins present in the dense core vesicles of neuro-endocrine cells, is also commonly expressed in neuroblastoma cell lines (Ou *et al.* 1998). However, other antigens like synaptophysin and ganglioside GD2 which are often expressed *in situ* do not seem to be widely expressed *in vitro*.

Two specific genetic abnormalities found in neuroblastoma *in situ* are useful features which can be used to characterize cell lines *in vitro*. For example, loss of some or all of the short arm of chromosome 1 has been seen in about a third of primary tumours and about three quarters of cell lines (Brodeur & Fong 1989). A second cytogenetic abnormality is over-amplification of the N-myc proto-oncogene located on *2p23–24* which is present either as double minutes or homogeneously staining regions (Christiansen *et al.* 1987; Corvi *et al.* 1994; Schwab *et al.* 1984).

REFERENCES

Ali-Osman F (1998) Human glioma cultures and *in vitro* analysis of therapeutic response. In: Berger M & Wilson CB (eds) *Gliomas*, pp. 134–141.

Allalunis-Turner MJ, Barron GM, Day RS, Dobler KD & Mirzayans R (1993) Isolation of two cell lines from a human malignant glioma specimen differing in sensitivity to radiation and chemotherapeutic drugs. *Radiation Research* 134: 349–354.

Barna BP, Chou SM, Jacobs B, Ransohoff RM, Hahn JF & Bay JW (1985) Enhanced DNA synthesis of human glial cells exposed to human leukocyte products. *Neuroimmunology* 10: 151–158.

Biedler JL, Helson L & Spengler BA (1973) Morphology and growth, tumorigenicity, and cytogenetics of human neuroblastoma cells in continuous culture. *Cancer Research* 33: 2643–2652.

Biedler JL & Spengler BA (1976) A novel chromosome abnormality in human neuro-blastoma and antifolate-resistant Chinese hamster cell lives in culture. *Journal of the National Cancer Institute* 57: 683–695.

Biegel JA, Rorke LB, Packer RJ, Sutton LN, Schut L, Bonner K & Emanuel BS (1989) Isochromosome 17q in primitive neuroectodermal tumors of the central nervous system. *Genes Chromosomes Cancer* 1: 139–147.

Bigner DD, Bigner SH, Ponten J, Westermark B, Mahaley MS, Ruoslahti E, Herschman H, Eng LF & Wikstrand CJ (1981a) Heterogeneity of genotypic and phenotypic characteristics of fifteen permanent cell lines derived from human gliomas. *Journal of Neuropathology and Experimental Neurology* 40: 201–229.

Bigner SH, Bullard DE, Pegram CN, Wikstrand CJ & Bigner DD (1981b) Relationship of *in vitro* morphologic and growth characteristics of established human glioma-derived cell lines to their tumorigenicity in athymic nude mice. *Journal of Neuropathology and Experimental Neurology* 40: 390–409.

Bigner SH, Friedman HS, Vogelstein B, Oakes WJ & Bigner DD (1990) Amplification of the c-myc gene in human medulloblastoma cell lines and xenografts [published erratum appears in *Cancer*

Research 1990 Jun 15;50(12):3809]. *Cancer Research* 50: 2347–2350.

Bodmer S, Strommer K, Frei K, Siepl C, de Tribolet N, Heid I & Fontana A (1989). Immunosuppression and transforming growth factor-beta in glioblastoma. Preferential production of transforming growth factor-beta 2. *Journal of Immunology* 143: 3222–3229.

Brodeur GM & Fong CT (1989) Molecular biology and genetics of human neuroblastoma. *Cancer Genetics and Cytogenetics* 41: 153–174.

Bullard DE, Schold SC, Jr, Bigner SH & Bigner DD (1981) Growth and chemotherapeutic response in athymic mice of tumours arising from human glioma-derived cell lines. *Journal of Neuropathology and Experimental Neurology* 40: 410–427.

Christiansen H, Franke F, Bartram CR, Adolph S, Rudolph B, Harbott J, Reiter A & Lampert F (1987). Evolution of tumor cytogenetic aberrations and N-myc oncogene amplification in a case of disseminated neuroblastoma. *Cancer Genetics and Cytogenetics* 26: 235–244.

Ciccarone V, Spengler BA, Meyers MB, Biedler JL & Ross RA (1989) Phenotypic diversification in human neuroblastoma cells: expression of distinct neural crest lineages. *Cancer Research* 49: 219–225.

Collins VP (1983) Cultured human glial and glioma cells. *International Review of Experimental Pathology* 24: 135–202.

Cooper MJ, Hutchins GM, Cohen PS, Helman LJ & Israel MA (1991) Neuroblastoma cell lines mimic chromaffin neuroblast maturation. *Progress in Clinical and Biological Research* 366: 343–350.

Corvi R, Amler LC, Savelyeva L, Gehring M & Schwab M (1994) MYCN is retained in single copy at chromosome 2 band p23–24 during amplification in human neuroblastoma cells. *Proceedings of the National Academy of Sciences of the USA* 91: 5523–5527.

Cravioto H (1986) Human and experimental gliomas in tissue culture. In: *Progress in Neuropathology* Vol. 6, pp. 166–188.

Darling JL (1990) The biology of human brain tumours. In: Thomas DGT (ed.) *Malignant Brain Tumours*, pp. 1–26. Edward Arnold, London.

Darling JL (1991) Brain. In: Masters JRW (ed.) *Human Cancer in Primary Culture*,

pp. 231–251. Kluwer Academic Publishers, Dordrecht.

de Ridder LI, Laerum OD, Mork SJ & Bigner DD (1987) Invasiveness of human glioma cell lines *in vitro*: relation to tumorigenicity in athymic mice. *Acta Neuropathologica* 72: 207–213.

Dreyling MH, Bohlander SK, Adeyanju MO & Olopade OI (1995) Detection of CDKN2 deletions in tumour cell lines and primary gliomas by interphase fluorescence *in situ* hybridization. *Cancer Research* 55: 984–988.

Frame MC, Freshney RI, Vaughan PF, Graham DI & Shaw R (1984) Interrelationship between differentiation and malignancy-associated properties in glioma. *British Journal of Cancer* 49: 269–280.

Freshney RI (1980) Tissue culture of glioma of the brain. In: Thomas DGT & Graham DI (eds) *Brain Tumours: Scientific basis, clinical investigation and current therapy*, pp. 21–50. Butterworths, London.

Friedman HS, Burger PC, Bigner SH, Trojanowski JQ, Brodeur GM, He XM, Wikstrand CJ, Kurtzberg J, Berens ME, Halperin EC *et al.* (1988). Phenotypic and genotypic analysis of a human medulloblastoma cell line and transplantable xenograft (D341 Med) demonstrating amplification of c-myc. *American Journal of Pathology* 130: 472–484.

Friedman HS, Burger PC, Bigner SH, Trojanowski JQ, Wikstrand CJ, Halperin EC & Bigner DD (1985) Establishment and characterization of the human medulloblastoma cell line and transplantable xenograft D283 Med. *Journal of Neuropathology and Experimental Neurology* 44: 592–605.

Giard DJ, Aaronson SA, Todaro GJ, Arnstein P, Kersey JH, Dosik H & Parks WP (1973) *In vitro* cultivation of human tumors: establishment of cell lines derived from a series of solid tumors. *Journal of the National Cancer Institute* 51: 1417–1423.

He, XM, Skapek SX, Wikstrand CJ, Friedman HS, Trojanowski JQ, Kemshead JT, Coakham HB, Bigner SH & Bigner DD (1989) Phenotypic analysis of four human medulloblastoma cell lines and transplantable xenografts. *Journal of Neuropathology and Experimental Neurology* 48: 48–68.

He XM, Wikstrand CJ, Friedman HS, Bigner SH, Pleasure S, Trojanowski JQ & Bigner DD (1991) Differentiation characteristics of newly established medulloblastoma cell lines (D384 Med, D425 Med, and D458 Med) and their transplantable xenografts. *Laboratory Investigations* 64: 833–843.

Helson L & Helson C (1985) Human neuroblastoma cells and 13-cis-retinoic acid. *Journal of Neurooncology* 3: 39–41.

Horowitz ME, Parham DM, Douglass EC, Kun LE, Houghton JA & Houghton PJ (1987) Development and characterization of human ependymoma xenograft HxBr5. *Cancer Research* 47: 499–504.

Isiuchi S, Nakazato Y, Iino M, Ozawa S, Tamura M & Ohye C (1998) In vitro neuronal and glial production and differentiation of human central neurocytoma cells. *Journal of Neuroscience Research* 51: 526–535.

Ishiuchi S & Tamura M (1997) Central neurocytoma: an immunohistochemical, ultrastructural and cell culture study. *Acta Neuropathologica (Berlin)*, 94: 425–435.

Israel M & Thiele C (1994) Tumor cell lines of the peripheral nervous system. In: Hay R, Park J-G. & Gazdar A (eds) *Atlas of Human Tumor Cell Lines* pp. 45–78. Academic Press, San Diego.

Jacobsen PF, Jenkyn DJ & Papadimitriou JM (1985) Establishment of a human medulloblastoma cell line and its heterotransplantation into nude mice. *Journal of Neuropathology and Experimental Neurology* 44: 472–485.

Jacobsen PF, Jenkyn DJ & Papadimitriou JM (1987) Four permanent cell lines established from human malignant gliomas: three exhibiting striated muscle differentiation. *Journal of Neuropathology and Experimental Neurology* 46: 431–450.

Kennedy PG, Watkins BA, Thomas DG & Noble MD (1987) Antigenic expression by cells derived from human gliomas does not correlate with morphological classification. *Neuropathology and Applied Neurobiology* 13: 327–347.

Knott JC, Mahesparan R, Garcia-Cabrera I, Bolge Tysnes B, Edvardsen K, Ness GO, Mork S, Lund-Johansen M & Bjerkvig R (1998) Stimulation of extracellular matrix components in the normal brain by invading glioma cells. *International Journal of Cancer* 75: 864–872.

Kruse CA, Mitchell DH, Kleinschmidt-DeMasters BK, Franklin WA, Morse HG, Spector F.B & Lillehei KO (1992) Characterization of a continuous human glioma cell line DBTRG-05MG: growth kinetics, karyotype, receptor expression, and tumor suppressor gene analyses. *In Vitro Cell Development Biology* 28A: 609–614.

Louis DN (1994) The p53 gene and protein in human brain tumors. *Journal of Neuropathology and Experimental Neurology* 53: 11–21.

MacDonald CM, Freshney RI, Hart E & Graham DI (1985) Selective control of human glioma cell proliferation by specific cell interaction. *Experimental Cellular Biology* 53: 130–137.

McKeever PE (1984) Persistence of cells from gliomas with glial and mesenchymal markers under conditions which inhibit growth of fibroblasts. *Journal of Neuropathology and Experimental Neurology* 43: 300.

McKeever PE, Fligiel SE, Varani J, Hudson JL, Smith D, Castle RL & McCoy JP (1986) Products of cells cultured from gliomas. IV. Extracellular matrix proteins of gliomas. *International Journal of Cancer* 37: 867–874.

McKeever PE, Smith BH, Taren JA, Wahl RL, Kornblith PL & Chronwall BM (1987) Products of cells cultured from gliomas. IV. Immunofluorescent, morphometric, and ultrastructural characterization of two different cell types growing from explants of human gliomas. *American Journal of Pathology* 127: 358–372.

McLean JS, Frame MC, Freshney RI, Vaughan PF, Mackie AE & Singer I (1986) Phenotypic modification of human glioma and non-small cell lung carcinoma by glucocorticoids and other agents. *Anticancer Research* 6: 1101–1106.

Mohapatra G, Kim DH & Feuerstein BG (1995) Detection of multiple gains and losses of genetic material in ten glioma cell lines by comparative genomic hybridization. *Genes Chromosomes Cancer* 13: 86–93.

Nister M & Westermark B (1994) Human glioma cell lines. In: Hay R, Park J-G & Gazdar A (eds) *Atlas of Human Tumor Cell Lines*, pp. 17–44. Academic Press, San Diego.

Ou XM, Partoens PM, Wang JM, Walker JH, Danks K, Vaughan PF & De Potter WP (1998) The storage of noradrenaline, neuropeptide Y and chromogranins in and stoichiometric release from large dense cored vesicles of the undifferentiated human neuroblastoma cell line SH-SY5Y. *International Journal of Molecular Medicine* 1: 105–112.

Owens RB, Smith HS, Nelson-Rees WA & Springer EL (1976) Epithelial cell cultures from normal and cancerous human tissues. *Journal of the National Cancer Institute* 56: 843–849.

Pietsch T, Gottert E, Meese E, Blin N, Feickert HJ, Riehm H & Kovacs G (1988) Characterization of a continuous cell line (MHH-NB-11) derived from advanced neuroblastoma. *Anticancer Research* 8: 1329–1333.

Pietsch T, Scharmann T, Fonatsch C, Schmidt D, Ockler R, Freihoff D, Albrecht S, Wiestler OD, Zeltzer P & Riehm H (1994) Characterization of five new cell lines derived from human primitive neuroectodermal tumors of the central nervous system. *Cancer Research* 54: 3278–3287.

Ponten J & Macintyre EH 91968) Long term culture of normal and neoplastic human glia. *Acta Pathol. Microbiol. Scand.* 74: 465–486.

Ponten J & Westermark B (1978) Properties of human malignant glioma cells *in vitro*. *Medical Biology* 56: 184–193.

Rosenblum ML, Gerosa MA, Wilson CB, Barger GR, Pertuiset BF, de Tribolet N & Dougherty DV (1983) Stem cell studies of human malignant brain tumors. Part 1: Development of the stem cell assay and its potential. *Journal of Neurosurgery* 58: 170–176.

Ross RA, Ciccarone V, Meyers MB, Spengler BA & Biedler JL (1988) Differential expression of intermediate filaments and fibronectin in human neuroblastoma cells. *Progress in Clinical and Biological Research* 271: 277–289.

Rutka JT, Giblin JR, Dougherty DY, Liu HC, McCulloch JR, Bell CW, Stern RS, Wilson CB & Rosenblum ML (1987) Establishment and characterization of five cell lines derived from human malignant gliomas. *Acta Neuropathologica* 75: 92–103.

Rutka JT, Giblin JR, Hoifodt HK, Dougherty DV, Bell CW, McCulloch JR, Davis RL, Wilson CB & Rosenblum ML (1986) Establishment and characterization of a cell line from a human gliosarcoma.

Cancer Research 46: 5893–5902.

Schlesinger HR, Gerson JM, Moorhead PS, Maguire H & Hummeler K (1976) Establishment and characterization of human neuroblastoma cell lines. *Cancer Research* 36: 3094–3100.

Schwab M, Varmus HE, Bishop JM, Grzeschik KH, Naylor SL, Sakaguchi AY, Brodeur G & Trent J (1984) Chromosome localization in normal human cells and neuroblastomas of a gene related to c-myc. *Nature* 308: 288–291.

Stein GH (1979) T98G: an anchorage-independent human tumor cell line that exhibits stationary phase G1 arrest in vitro. *Journal of Cell Physiology* 99: 43–54.

Studer A, de Tribolet N, Diserens AC, Gaide AC, Matthieu JM, Carrel S & Stavrou D (1985) Characterization of four human malignant glioma cell lines. *Acta Neuropathologica* 66: 208–217.

Tumilowicz JJ, Nichols WW, Cholon JJ & Greene AE (1970) Definition of a continuous human cell line derived from neuroblastoma. *Cancer Research* 30: 2110–2118.

Valdueza JM, Westphal M, Vortmeyer A, Muller D, Padberg B & Herrmann HD (1996) Central neurocytoma: clinical, immunohistologic, and biologic findings of a human neuroglial progenitor tumor. *Surgical Neurology* 45: 49–56.

Warr TJ, Ward SJ, Burrows J, Harding B, Wilkins P, Harkness W, Hayward R, Darling JL & Thomas DGT (2000) Consistent novel regions of genetic loss and gain identified by CGH in paediatric malignant astrocytoma, *Genes Chromosomes Cancer* in press.

Westermark B, Ponten J & Hugosson R (1973) Determinants for the establishment of permanent tissue culture lines from human gliomas. *Acta Pathol. Microbiol. Scand. [A]*, 81: 791–805.

Westphal M, Giese A, Meissner H & Zirkel D (1997) Culture of cells from human tumors of the nervous system on an extracellular matrix derived from bovine corneal endothelial cells. *Methods in Molecular Biology* 75: 185–207.

Westphal M & Meissner H (1998) Establishing human glioma-derived cell lines. *Methods in Cellular Biology* 57: 147–165.

Zupi G, Candiloro A, Laudonio N, Carapella C, Benassi M, Riccio A, Bellocci M & Greco C (1988) Establishment, characterization and chemosensitivity of two human glioma derived cell lines. *Journal of Neurooncology* 6: 169–177.

4.16 COLON ADENOCARCINOMA CELLS (PRIMARY/ESTABLISHED CULTURES OF HUMAN COLON CARCINOMAS)

Cell lines derived from tumours of the colorectum are valuable for studies of the cellular and molecular biology of colorectal cancer. For such studies it is often important to retain a differentiated phenotype in culture. There are three differentiated cell types: (1) mucus-secreting goblet cells; (2) fluid-transporting enterocytes; and (3) endocrine cells. Goblet cells can secrete mucin into the culture medium, and enterocytic cells can give rise to domes, or haemicysts, as fluid transported vectorially accumulates between the cell monolayer and the culture vessel. The following procedure produces colorectal adenocarcinoma cell lines in which these differentiated phenotypes are expressed.

PRELIMINARY PROCEDURE: PREPARATION OF COLLAGEN-COATED FLASKS

Materials and equipment

- Human placental collagen type VI (Sigma), 50 mg

1. Sterilize collagen as powder by gamma irradiation (2.5 Mrad, 25 kgrey).
2. Add 0.05 ml glacial acetic acid to 50 ml warm sterile tissue culture grade water.
3. Add 50 mg collagen using forceps.
4. Leave, stirring at 37°C, for 2–3 h, until completely dissolved.
5. Pipette into four universal containers and centrifuge at 1600 g for 3 min to remove impurities.
6. Transfer collagen supernatant to fresh universal containers and store at 4°C.
7. To coat the flasks, keeping the collagen solution on ice, add about 4 ml to a flask, coat the bottom of the flask, pipette the collagen out and into the next flask and so on. Aspirate off excess collagen solution. This will leave a thin coat of collagen on the bottom of the flasks.
8. Leave the flasks to air-dry with the lids off in a tissue culture cabinet for 2–4 h.
9. Store the collagen-coated flasks at 4°C. The flasks can be stored for up to 4 weeks after coating.

Cell and Tissue Culture for Medical Research, edited by A. Doyle and J.B. Griffiths.
© 2000 John Wiley & Sons, Ltd

PRELIMINARY PROCEDURE: PREPARATION OF SWISS 3T3 FEEDER CELLS

Treat Swiss 3T3 cells with mitomycin C at a concentration of 10 μg ml^{-1} for 2 h to prevent them from growing. An alternative method is to lethally irradiate the cells with 60 kgrey (6 Mrad) gamma irradiation. Although unable to grow, the Swiss 3T3 feeders attach and produce growth factors when added to epithelial cell cultures.

Materials and equipment

- Swiss 3T3 cells (ECACC No. 85022108)
- Mitomycin C (Sigma)
- Trypsin 0.1% in phosphate-buffered saline (PBS)

 Gloves should be worn when handling mitomycin C. Mitomycin C is light sensitive, and therefore tissue culture cabinet lights should be switched off during use.

1. Grow Swiss 3T3 cells on 90 mm Petri dishes in Dulbecco's modified Eagle's medium (DMEM) supplemented with 10% calf serum (calf serum is better for Swiss 3T3 cells than foetal bovine serum) and 2 mM glutamine, plus antibiotics as for standard culture medium, until they are 24 h post-confluent.
2. Dissolve 2 mg mitomycin C in warmed growth medium added using a 10 ml syringe into the vial. Pierce the rubber stopper with another syringe needle to release the pressure before removing the solution from the vial. Make the volume up to 200 ml. A final concentration of 10 μg ml^{-1} is given.
3. Remove the medium from the Swiss 3T3 cells and replace with 10–12 ml of the mitomycin-C-containing medium. Incubate for 2 h (no less).
4. Carefully remove the mitomycin C and rinse the cells with sterile PBS.
5. Wash the cells using 0.1% trypsin and then aspirate off most of the trypsin, leaving only a thin layer covering the cells.
6. Incubate at 37°C for 5–10 min to allow the cells to detach.
7. Suspend the cells in growth medium and centrifuge at 1600 g for 3 min.
8. Resuspend the cell pellet in fresh growth medium.
9. Repeat steps 7 and 8 three times to wash excess mitomycin C off the cells.
10. Count the cells using an improved Neubauer counting chamber and Trypan blue and make up the cell suspension to 1×10^6 viable cells ml^{-1}. Flasks should be seeded to give a surface coverage of 30–50% confluence, i.e. $2–5 \times 10^4$ cells cm^{-2}.
11. Store prepared feeder cell suspension at 4°C and use for up to 1 week.

PROCEDURE: ESTABLISHMENT OF CELL LINES FROM COLORECTAL TUMOURS

Colorectal adenocarcinoma cells are routinely grown on collagen type IV coated 25 cm^2 flasks in the presence of Swiss 3T3 feeder cells (at a density between 1×10^3 and 1×10^4 cells cm^{-2}) at 37°C in a 5% CO^2/air incubator. Standard growth medium is DMEM supplemented with 20% foetal bovine serum (FBS), hydrocortisone sodium succinate (1 μg ml^{-1}), insulin (0.2 units ml^{-1}), glutamine (2 mM), penicillin (100 units ml^{-1}) and streptomycin (100 μg ml^{-1}). For primary cultures, in the first 2 weeks it is advisable to use medium with high antibiotic concentrations. Primary culture medium is as above, but with 200 units ml^{-1} penicillin, 200 μg ml^{-1} streptomycin and 50 μg ml^{-1} gentamicin (ICN Flow).

Tumour specimens collected from surgery must be placed directly in washing medium (DMEM supplemented with 5% FBS, 2 mM glutamine, 200 units ml^{-1} penicillin, 200 μg ml^{-1} streptomycin and 50 μg ml^{-1} gentamicin) to prevent drying out. They should then be kept on ice for transportation to the laboratory.

The disaggregation technique used will depend, in part, on the histology and structure of the tumour specimen. Poorly differentiated adenocarcinomas may only require physical cutting using scalpels, since such tumours have little glandular organization, and small clumps of tumour cells are released into the surrounding washing medium on cutting. A well-differentiated adenocarcinoma may not release small clumps of cells on cutting and in such cases enzyme digestion is necessary. An overnight digestion of 1 mm^3 pieces in collagenase and hyaluronidase also digests associated mesenchymal tissue away from the released epithelial organoids. Organoids are then cleaned by repeated pipetting and by gravity sedimentation. Epithelial organoids will settle out, but the more diffuse mesenchymal tissue will not.

Reagents and solutions

Hyaluronidase Type 1-S (Sigma) solution

Working solution is 100 units ml^{-1}. Stock solution is made up at 1000 units ml^{-1}. Dissolve in DMEM supplemented with penicillin 100 U ml^{-1}, streptomycin 100 μg ml^{-1}. Filter-sterilize and freeze at –20°C in 2 ml aliquots.

Collagenase ClsIII (Worthington Biochemical Group, Freehold, New Jersey, USA) solution

Working solution is 240 units ml^{-1}. Stock solution is made up at 480 units ml^{-1}. Dissolve in DMEM supplemented with penicillin 100 U ml^{-1}, streptomycin 100 μg ml^{-1}. Filter-sterilize and freeze at –20°C in 10 ml aliquots.

Dispase Grade 1 (Boehringer) solution

Working solution is 2 units ml^{-1}. Dissolve in warmed medium containing 10% FBS, 2 mM glutamine, penicillin 100 U ml^{-1}, streptomycin 100 μg ml^{-1}. Filter-sterilize and freeze at –20°C in 5 ml aliquots.

Hydrocortisone (21-hemisuccinate – sodium salt, Sigma) solution

Working concentration is 1 μg ml^{-1}. Stock is made up at 100 μg ml^{-1}. Hydrocortisone is dissolved in warm DMEM containing penicillin 100 U ml^{-1}, streptomycin

100 μg ml⁻¹. Filter-sterilize and aliquot into 5 ml volumes. Hydrocortisone may be stored at –20°C, but must not be refrozen after thawing. Once thawed, any unused solution must be stored at 4°C.

Glutamine solution

Used at 2 mM. Stock is made up at 100 times the working concentration; therefore, dissolve 2.93 g glutamine in 100 ml tissue culture grade water. Filter-sterilize, aliquot and store at –20°C.

Penicillin/streptomycin solution

The antibiotics can be dissolved together when preparing the stock solution. Penicillin is used at 100 units ml⁻¹ and streptomycin at 100 μg ml⁻¹. The stock solution should be made up with each of the antibiotics at ×100 these working concentrations. Dissolve in tissue culture tested water and filter sterilize. Aliquot and store at –20°C.

Materials and equipment

- Collagenase solution
- Hyaluronidase solution
- Washing medium
- Primary culture growth medium
- Swiss 3T3 feeders
- Collagen-coated flasks

 Gloves should be worn for handling human tissue.

Preparation of colorectal adenocarcinoma tissue for culture

1. Wash the tumour biopsy four times in washing medium.
2. A small sample of the tumour specimen should be fixed for histological diagnosis, for subsequent comparison with the tumour histology reported by the hospital pathologists.
3. In a very small volume of washing medium, cut the remaining tissue using crossed scalpel blades until the pieces are 1 mm³ or less. The pieces should be small enough to be pipetted easily with a 10 ml pipette. It is advisable to cut only 1 cm³ of tissue at any one time.
4. Wash the tissue pieces four times by bench centrifugation (1600 *g* for 3 min). If, on cutting the tumour sample, small clumps of tumour cells were released, enzymic digestion will not be necessary; therefore go to step 9.
5. Resuspend the tissue pieces in 8 ml washing medium per 1 cm³ tissue in a universal container.
6. Add 2 ml hyaluronidase solution and 10 ml collagenase solution to the 8 ml suspension.
7. Rotate the samples at 37°C – usually overnight (12–16 h).

8. Centrifuge at 1600 g for 3 min and remove the enzymes. If the sample is very mucinous, this will also remove some of the mucin and make separation easier.
9. Resuspend the sample in washing medium and vigorously pipette up and down to clean mesenchymal tissue from the epithelial organoids. Allow the pieces to settle by gravity sedimentation.
10. Remove the supernatant and examine microscopically. It will contain many fibroblastic cells. If small clumps of epithelial cells are present, transfer to a separate tube and allow to settle. If gravity alone will not bring down the cell clumps, differential centrifugation (1000 rev min^{-1} for 1–2 min) can be used. Retaining cells from the supernatant can have two uses: (a) it safeguards against discarding more dysplastic elements of the tumour (b) it can be a good source of fibroblast cells which can be used for molecular studies as a supply of the normal constitutional DNA.
11. Repeat the gravity sedimentation, resuspension and pipetting four times for each tube.
12. Examine the settled organoid material microscopically. If organoids have been cleaned of fibrous material, centrifuge at 1600 g for 3 min.
13. Remove the supernatant and resuspend in 4 ml primary culture growth medium per tube.
14. Transfer to collagen-coated flasks, and to each flask add 0.1–0.2 ml Swiss 3T3 feeder cells.
15. Incubate at 37°C in 5–10% CO_2 in air.

Maintenance of colorectal epithelial cell cultures

Medium should be changed twice weekly. Until all the epithelial organoids have adhered to the culture vessel, non-adherent material should be centrifuged, resuspended in fresh growth medium and replated.

For the first 2 weeks of culture, the high antibiotic concentraions of the primary culture growth medium should be maintained. After this time, the standard culture medium may be used.

The 3T3 feeder cells must be maintained at 30–40% confluence, and this will normally entail replenishment at the rate of approximately 0.1 ml per week.

Removal of contaminating fibroblastic cells

Well-isolated fibroblastic colonies can be removed by scraping with a cell scraper or the plunger of a sterile syringe. The flasks should then be washed several times to remove the fibroblasts.

Differential trypsinization can be used to remove fibroblasts without the epithelial cells detaching. (Note that if growing benign adenomas, this technique should not be used, since adenoma cells are more sensitive to trypsin than are carcinoma cells.)

The neutral protease dispase can be used to preferentially remove the epithelium. This leaves many fibroblastic cells behind on the flask. The cell suspension removed can be incubated for 2–6 h on tissue culture plastic to allow preferential attachment of fibroblasts before transferring the suspension to a fresh collagen-coated flask.

Subculture

Although carcinoma cells can be subcultured using trypsin, this may be selective. Passaging the cultures as cell clumps using dispase retains cell–cell contacts and permits cell differentiation. From primary culture, several 1 : 1 passages may be required before there are sufficient cells for a 1 : 2 split.

1. Centrifuge dispase solution before use (1600 *g* for 3 min) if any precipitate forms.
2. Pipette 2.5 ml dispase into each 25 cm² flask. Established cell lines should take 30 min to detach, but primary cultures may take 60 min or longer. Monolayers will detach as sheets; less confluent cultures will detach as clumps.
3. Add 7.5 ml standard growth medium to the flask to help wash off the cells and transfer to a universal container.
4. Centrifuge at 1600 *g* for 3 min.
5. Remove the dispase and resuspend in standard growth medium, with repeated pipetting up and down to break up the cell clumps a little.
6. Put into collagen-coated flasks and add 0.1 ml Swiss 3T3 feeder cells. After passaging, the feeder cells remain important.

DISCUSSION

Background information

Development of cancer in the large intestine occurs as a multi-stage process. Most carcinomas are derived from premalignant adenomas in the adenoma–carcinoma sequence. The adenomas are derived from the normal colorectal epithelium. To study the basis of colorectal cancer development experimentally, a major effort has been made to develop cell culture techniques which permit the growth and differentiation of colonic epithelium from normal, premalignant and malignant tissue. The culture of normal colonic epithelium is as yet limited to only short-term cultures of adult tissue or to growth of tissue of foetal origin. Colorectal adenocarcinomas are relatively easy to establish as cell lines, and several methods have been described for their culture (Leibovitz *et al.* 1976; Brattain *et al.* 1983; Paraskeva *et al.* 1984; Kirkland & Bailey 1986). Friedman *et al.* (1981) have described conditions for the primary culture of adenoma-derived cells. The procedure described above for the isolation of colorectal adenocarcinoma cell lines is also suitable for the culture of adenoma cells (Paraskeva *et al.* 1984, 1989).

Carcinoma cell lines can usually be subcultured using trypsin, but some adenoma cell lines cannot be passaged as single cells, and dispase should be used to remove the cells as clumps. Even with carcinomas, it is best to use dispase for the first few passages, as this treatment will be less severe. Cultures can also be grown on tissue culture plastic instead of collagen, and feeder-independent cell lines can be derived. These culture conditions are useful for certain experiments, but the cultures will grow at a reduced rate and the differentiation pattern of the cells may be altered. Removal of collagen and Swiss 3T3 feeder cells results in a significant reduction in the number of mucin-secreting goblet cells and a corresponding increase in brush border columnar-like cells. However, colon cancer cell lines can be grown on a simple medium consisting of DMEM supplemented with 10% FBS and glutamine only, without a collagen substrate and without Swiss 3T3 feeder cells, and still retain the ability to produce some mucin glycoproteins *in vitro*. The standard culture conditions described in the procedure above give cultures of carcinoma or adenoma cells which differentiate into both goblet-like and columnar-like cells.

Once cultures are set up, it is essential to confirm the epithelial nature of the cells. Epithelial cells have a classic cuboidal morphology which gives cultures the appearance of pavement-like sheets, but morphology is only a clue to the nature of the cells since, for example, endothelial cells at high density can have an epithelial-like morphology. Further characterization is therefore necessary. A number of criteria can be used, including the staining of cytokeratin filaments using anti-keratin monoclonal antibodies (e.g. LE61 or Dako-CK1), electron microscopic examination to demonstrate desmosomes, microvilli and goblet-like cells and detection of mucins by histochemical stains (e.g. alcian blue) (Paraskeva *et al.* 1984). Other characteristics of colonic epithelium are the production of carcinoembryonic antigen (CEA), which can be detected using anti-CEA monoclonal antibodies, and the formation of domes due to vectorial fluid transport.

Troubleshooting

The batch of FBS used can greatly affect the growth of colorectal adenocarcinoma cells. It is important to test any batch of serum before purchasing, using a recently established colorectal epithelial cell line, preferably one which exhibits a differentiated phenotype *in vitro*. It is important not to use a well established cell line for batch testing serum, if the aim is to grow primary cultures.

The maintenance of Swiss 3T3 feeder cell support is important. It is important to maintain a high density, but if too many feeder cells are present this may prevent attachment of the epithelial organoids.

The epithelial oranoids may take several weeks to attach, and these should be collected and replated at each medium change.

A major obstacle in the successful establishment of a new cell line is the presence of contaminating fibroblasts. Methods to remove these have been given above. An additional method is to use a conjugate between an anti-Thy-1 antibody and the toxin ricin (Paraskeva *et al.* 1985), which allows specific killing of fibroblasts, but unfortunately this conjugate is not yet commercially available. Each of the methods given has been used with success, but in cases where there is close association between the epithelial and the mesenchymal cells it is often very difficult to eliminate the fibroblasts which grow out from underneath the epithelium. Time spent in cleaning the epithelial cell clumps during the preparation of the tissue before culture is therefore well invested.

Time considerations

The collection of the tumour sample from the hospital can be time-consuming, depending on the arrangements made for collection. The initial preparation for culture will take approximately 2 h before an overnight digest in collagenase and hyaluronidase and a further 1–2 h the following day to separate and clean the epithelial organoids. This will vary according to the number of samples obtained.

Expected results

Once they are in culture, attachment and spreading out of cells should be seen after 24–48 h, although some epithelial organoids may take up to 6 weeks to attach. In some cases the cell population will continue to expand until the culture can be passaged, while in other cases the culture may thrive for a while but then regress and only a few colonies survive. It may be up to 3 months before passaging can be attempted.

After a few trial runs, it should be possible to obtain a success rate of establishing colorectal adenocarcinoma cell lines of 50–80%.

REFERENCES

Brattain MG, Marks ME, McCombs J, Finely W & Brittain DE (1983) Characterization of human colon carcinoma cell lines isolated from a single primary tumour. *British Journal of Cancer* 47: 373–381.

Friedman EA, Higgins PJ, Lipkin M, Shinya H & Gelb AM (1981) Tissue culture of human epithelial cells from benign colonic tumours. *In Vitro* 17: 632–644.

Kirkland SC & Bailey IG (1986) Establishment and characterization of six colorectal adenocarcinoma cell lines. *British Journal of Cancer* 53: 779–785.

Leibovitz A, Stinson JC, McComb WB, McCoy CE, Mazur KC & Mabry ND (1976) Classification of human colorectal adenocarcinoma cell lines. *Cancer Research* 36: 3562–3569.

Paraskeva C, Buckle BG, Sheer D & Wigley CB (1984) The isolation and characterization of colorectal epithelial cell lines at different stages in malignant transformation from familial polyposis coli patients. *International Journal of Cancer* 34: 49–56.

Paraskeva C, Buckle BG & Thorpe PE (1985) Selective killing of contaminating fibroblasts in epithelial cultures derived from colorectal tumours using an anti-Thy-l antibody-ricin conjugate. *British Journal of Cancer* 41: 908–912.

Paraskeva C, Finerty S, Mountford RA & Powell SC (1989) Specific cytogenetic abnormalities in two new human colorectal adenoma-derived epithelial cell lines. *Cancer Research* 49: 1282–1286.

CHAPTER 5

MODERN CLINICAL APPLICATIONS OF CULTURED CELLS

5.1 IMMORTALIZATION METHODS – AN OVERVIEW

Primary cell cultures can be established from a wide variety of mammalian tissues by enzyme digestion at warm or cold temperatures or by outgrowth from primary explants. Such cultures can be maintained *in vitro* for a limited period of time, the length of which is determined in part by whether or not the cells are capable of proliferation. However, it has been shown (Hayflick & Moorhead, 1961) that even those cells which are capable of division have only a limited *in vitro* lifespan. As the cells approach the end of this lifespan an increasing proportion of them become unable to synthesize DNA. The cells then enter a crisis phase, cell division ceases and the culture becomes senescent.

It is possible, however, to obtain cultures of cells with unlimited growth potential. Such immortal cell lines can arise spontaneously (as a rare event) or by transformation either by treatment with carcinogenic chemicals or as a result of exposure to DNA tumour viruses. Permanent lines isolated in this way include the mouse L5178Y lymphoblast cell line isolated after exposing cells to methylcholanthrene (Beer *et al.* 1983) and WI-26 VA4 obtained by infecting the diploid human line WI-26 with SV40 virus (Girardi *et al.* 1965). Another approach, to construct a fusion hybrid between the cell with a limited lifespan and a permanent cell line, has been particularly successful for generating hybridoma lines secreting antibody (Koehler & Milstein 1975). Whichever approach is used, alterations in the properties of the cells occur: changes include the loss of tissue-specific characteristics and chromosomal changes.

Cells can also be immortalized by using genetic engineering techniques to introduce and express viral genes which are capable of overcoming senescence. These genes have been categorized into two groups: those which are capable of immortalizing cells, that is, establishing proliferating cultures from primary cells or increasing the lifespan of cells such as primary embryo fibroblasts; and those which can transform cells, e.g. the established mouse fibroblast cell line, NIH-3T3 (Ruley 1983; Land *et al.* 1983). Thus the oncogenic transformation of rodent cells by DNA viruses occurs as a two-step process as a result of cooperation between two or more oncogenes (Rassoulzadegan *et al.* 1982), in a process which has been compared with the two-stage model of carcinogenesis proposed by Knudson (1986). Examples of oncogenes with immortalizing activity include the adenovirus early region 1A (Ad E1A) and the polyoma large T antigen coding region. Genes with transforming activity include polyoma middle T, the adenovirus early region 1B and the *ras* oncogenes. The situation with human cell immortalization is more complex. In rodent cells immortalization appears to occur as a single step, whereas in human cells it probably occurs in two stages. One of these stages is equivalent to the rodent immortalization step and can be mediated by the same oncogenes.

Cell and Tissue Culture for Medical Research, edited by A. Doyle and J.B. Griffiths.
© 2000 John Wiley & Sons, Ltd

The second step, however, is distinct, occurs at a frequency of around 10^{-7} and appears to involve a mutation event (Shay & Wright 1989). Thus, although differentiated mammalian cell lines have been isolated, by introducing plasmids containing immortalizing genes into a variety of different primary cells (MacDonald 1990), the ease with which such lines can be isolated varies considerably with the tissue involved and the species from which the cells were obtained.

The mechanisms involved in the reversal of senescence by viral oncogenes are not fully understood but are believed to involve the *p53* and *Rb* pathways. The onset of senescence is accompanied by an increase in the cyclin-dependent kinase inhibitors *p16* and *p21* (Alcorta *et al.* 1996; Loughran *et al.* 1997) and telomere shortening (Allsopp *et al.* 1992). However, although expression of the catalytic component of telomerase significantly extends the lifespan of human foreskin fibroblasts and retinal pigment epithelial cells (Bodnar *et al.* 1998), telomerase activity is not sufficient for the immortalization of human keratinocyte or mammary epithelial cells (Kiyono *et al.* 1998).

The nomenclature used in the literature can be confusing. The term primary cell should be restricted to cells which have not been subcultured. After subculture they become secondary cultures. However, this terminology becomes clumsy after a few subcultures and is not strictly adhered to. The term 'limited lifespan culture' is preferable, and can be used for any diploid cells, or cell strain as they are sometimes called. The question of immortality is problematic – how long must a culture last to be considered immortal? One approach is to use the term 'extended lifespan' since this avoids the presumption of indefinite cell division. Transformed cell lines are often considered to be permanent, but SV40-transformed human fibroblasts, for example, normally go through a crisis of senescence, the onset of which is delayed in comparison with untransformed cells, rather than abolished (Huschtscha & Holliday 1983).

As indicated, immortalization can be achieved by viral infection, cell fusion and oncogene transfection. Currently, most laboratories use approaches based upon recombinant DNA techniques, although different techniques are more or less suitable depending on the species or tissue of interest. For example, the best approach for immortalizing murine antibody-secreting cells is hybridoma fusion. Human lymphocytes are best immortalized by transformation with the Epstein Barr Virus. For other cell types, SV40 large T antigen has been used most widely to isolate immortalized lines from a variety of different types of tissue and species (McLean 1999).

REFERENCES

Alcorta D A, Xiong Y, Phelps D, Hannon G, Bleach D & Barrett J C (1996) Involvement of the cyclin-dependent kinase inhibitor p16 (INK4a) in replicative senescence of normal human fibroblasts. *Proceedings of the National Academy of Sciences of the USA* 93: 13742–13747.

Allsopp R C, Vaziri H, Patterson C,

Goldstein S, Younglai E V, Futcher A B, Greider C W & Harley C B (1992) Telomere length predicts replicative capacity of human fibroblasts. *Proceedings of the National Academy of Sciences of the USA* 89: 10114–10118.

Beer J Z, Budzicka E, Niepokojczycka E, Rosiek O, Szumiel I & Walicka M (1983)

Loss of tumorigenicity with simultaneous changes in radiosensitivity and photosensitivity during *in vitro* growth of L5178Y murine lymphoma cells. *Cancer Research* 43: 4736–4742.

Bodnar A G, Ouellette M, Frolkin M, Holt S E, Chiu C-P, Morin G B, Harley C B, Shay J W, Lichtsteiner S & Wright W E (1998) Extension of lifespan by introduction of telomerase into normal human cells. *Science* 279: 349–352.

Girardi A J, Jensen F C & Koprowski H (1965) SV40-induced transformation of human diploid cells: crisis and recovery. *Journal of Cellular and Comparative Physiology* 65: 69–84.

Hayflick L & Moorhead P S (1961) The serial cultivation of human diploid cell strains. *Experimental Cell Research* 25: 585–621.

Huschtscha L I & Holliday R (1983) Limited and unlimited growth of SV40-transformed cells from human diploid MRC-5 fibroblasts. *Journal of Cell Science* 63: 77–99.

Kiyono T, Foster S A, Koop J I, McDougall J K, Galloway D A & Klingelhutz A J (1998) Both Rb/p16^{INK4a} inactivation and telomerase activity are required to immortalize human epithelial cells. *Nature* 396: 84–88.

Knudson A G (1986) Genetics of human cancer. *Annual Review of Genetics* 20: 231–251.

Koehler G & Milstein C (1975) Continuous cultures of fused cells secreting antibody of predefined specificity. *Nature* 256: 495–497.

Land H, Parada L F & Weinberg R A (1983) Tumorigenic conversion of primary embryo fibroblasts requires at least two co-operating oncogenes. *Nature* 304: 596–602.

Loughran O, Clark L J, Bond J, Baker A, Berry I J, Edington K G, Ly I-S, Simmons R, Haw R, Black D M, Newbold R F & Parkinson E K (1997) Evidence for the inactivation of multiple replicative lifespan genes in immortal human squamous cell carcinoma keratinocytes. *Oncogene* 14: 1955–1964.

MacDonald C (1990) Development of new cell lines for animal cell biotechnology. *CRC Critical Reviews in Biotechnology* 10: 155–178.

McLean J S (1999) Immortalization strategies for mammalian cells. In: Jenkins N (ed.) *Methods in Biotechnology*, vol. 8, pp. 61–72. Animal Cell Biotechnology, Humana Press Inc, Totowa, NJ.

Rassoulzadegan M, Cowie A, Carr A, Glaichenhaus N, Kamen R & Cuzin F (1982) The roles of individual polyoma virus early proteins in oncogenic transformation. *Nature* 300: 713–718.

Ruley H E (1983) Adenovirus early region 1A enables viral and cellular transforming genes to transform primary cells in culture. *Nature 304:* 602–606.

Shay J W & Wright W E (1989) Quantitation of the frequency of immortalisation of normal human diploid fibroblasts by SV40 large T-antigen. *Experimental Cell Research* 184: 109–118.

5.2 LONG-TERM MAINTENANCE AND STABILITY OF HYBRIDOMAS

Having identified antibody-positive wells from a fusion experiment it is important to derive as many stable, monoclonal-antibody-secreting clones as possible so that faster-growing, non-secreting clones do not overgrow the desired clones. There are two basic cloning strategies. Using either method it is possible to derive stable clonal cultures from a large number of positive wells within 6 weeks of initial identification (see Figure 5.2.1).

Having established a cloned cell line at the third cloning step, it is important to prepare and freeze a sufficient number of ampoules to ensure that the cells will not be lost at a later date. The established cell line is then expanded from the two 1 ml wells to two 75 cm^2 flasks via growth in two 25 cm^2 small flasks. Once the cells in the 1 ml wells have been used to seed the small flask, they are grown to confluence and are cryopreserved in ampoules using standard techniques (see module 'Cryopreservation'). The cells from the two large flasks are grown to approximately $6-8 \times 10^5$ cells ml^{-1} (viability > 90%), and then cryopreserved for storage.

Having generated an established cloned hybridoma cell line and expanded it through the 1 ml well, small flask to the large flask stage it is important to validate the cells prior to laying down stocks for later study.

There are three tests which are used to confirm the characteristics of the established cell line at the time of completing the initial culture phase:

(a)   Confirmation of antibody secretion
(b)   Mycoplasma testing
(c)   Isotype analysis

PROCEDURE: ENDPOINT CLONING

Materials and equipment

- RPMI 1640 culture medium (with or without 10 mM HEPES) with 20% foetal bovine serum (FBS) and HT (lyophilized hypoxanthine and thymidine)
- Peritoneal feeder layers

1. Count hybridomas to be cloned by the haemocytometer method (see Section 1.3) and dilute in growth medium containing 20% FBS and HT, to give 1000 cells ml^{-1}.

Cell and Tissue Culture for Medical Research, edited by A. Doyle and J.B. Griffiths.
© 2000 John Wiley & Sons, Ltd

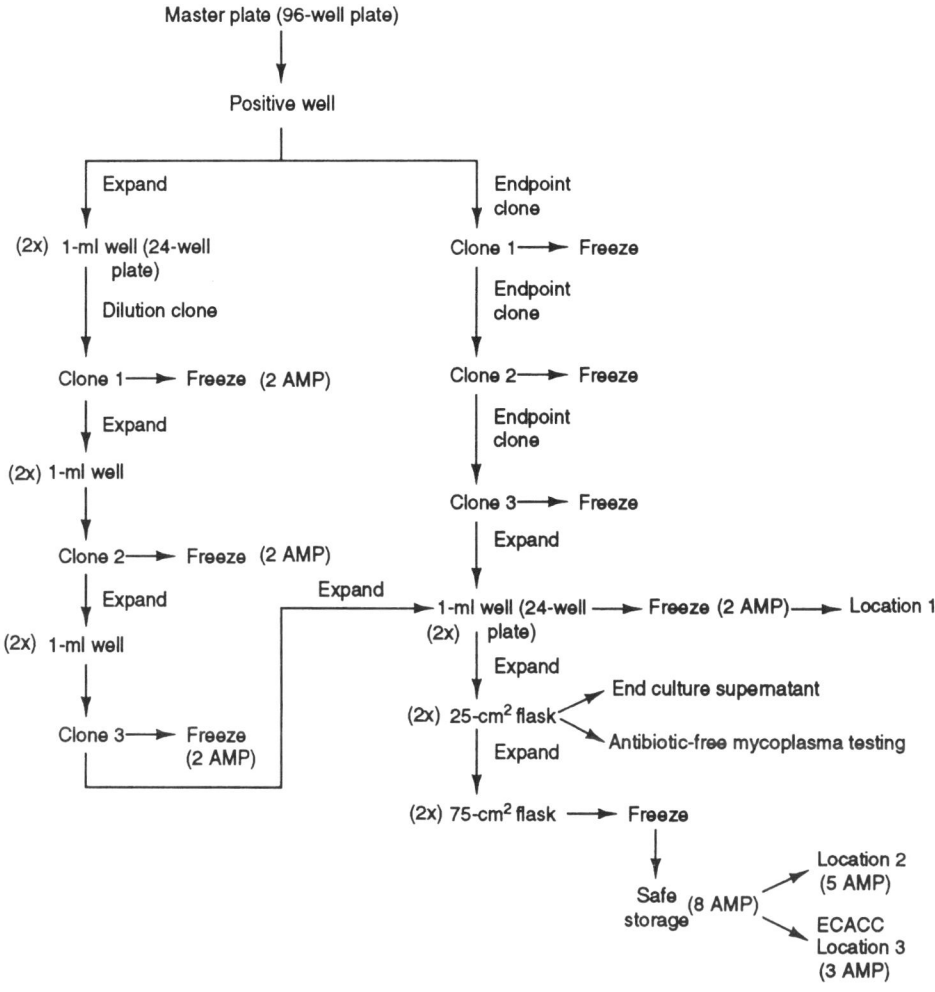

Figure 5.2.1 Idealized flow chart of the process leading to a stable monoclonal-antibody-producing hybridoma. It must be emphasized that failure can occur at virtually any stage and that constant monitoring for antibody production is a prerequisite.

2. Plate the diluted hybridomas into the first column of a mictrotitre plate (100 μl well^{-1}) containing mouse peritoneal macrophages. Alternatively, with experience an appropriate number of cells, or volume of cell suspension, can be taken from the master well.

3. Double dilute the cells to the end of the plate, i.e. 100 μl aliquots are advanced across the microtitre plate from column 1 to column 12, with mixing at each step, using a multichannel automatic pipette. Using this strategy there should be less than 1 cell well^{-1} in columns 8–12.

4. Add 100 μl RPMI 1640 with 20% FBS and HT to each well of the cloning plate.

5. Incubate the cells for 1 week and then check on an inverted microscope for growth. Columns with growth in fewer than one third of the wells should be clonal according to the statistical parameters of Poisson distribution.
6. Check the supernatants for specific monoclonal antibody levels between days 7 and 10, depending on the rate of growth of the cells. At this stage the hybridomas should cover one fifth to one quarter of the base of the microtitre well.
7. Transfer selected clones to 24-well plate wells in 1 ml aliquots for further study and scale up for cyropreservation and storage, or reclone as required.

ALTERNATIVE PROCEDURE: MINI-CLONING

1. As for endpoint cloning, count the hybridoma cells to be cloned, by haemocytometer, and dilute in medium with 20% FBS (200 cells ml^{-1}) and HT.
2. Aliquot the first hybridoma (at 100 μl well^{-1}) into the first three wells of a microtitre plate containing macrophages.
3. Aliquot subsequent hybridomas into groups of three wells in row A, giving a total of four mini-clonings per microtitre plate.
4. Double dilute the cells down the plate (i.e. 100 μl aliquots are advanced down the microtitre plate from row A to row H with mixing at each step) using a multichannel automatic pipette. Using this strategy there should, theoretically, be clonal cells 0.312–0.156 from rows F–H.
5. Add 100 μl 20% FBS growth medium and HT to every well of the cloning plate.
6. Incubate the cells for 1 week in 5% CO_2/95% air at 37°C and then check visually for growth.

Proceed as from step 6 in previous procedure.

This strategy may not select monoclonal cell lines at the first cloning step but has significant advantages in the early stages of establishing stable hybridoma cell lines. In particular, the cloning is not as harsh as that used in the endpoint cloning technique, and usually results in a higher percentage of antibody secretors being taken through to a third stage. The number of colonies which can be taken through from a fusion producing large numbers of positives is greater, as four colonies/plate. Experience indicates that a combination of both methods can be very successful where the first two clonings are mini-clonings and the final cloning uses the endpoint method.

ALTERNATIVE PROCEDURE: SOFT-AGAR CLONING

Additional materials and equipment

• Seaplaque agarose (Miles)
• RPMI 1640 culture medium with 20% FBS and HT
• BALB/c mouse
• Anti-mouse Ig antiserum
• Pressure cooker

- Long-stemmed glass Pasteur pipettes
- Humidified box

1. Keep medium and agarose at 40°C in a water-bath until required.
2. Add Seaplaque agarose (0.17 g) to 5 ml distilled water in a 100 ml bottle. Autoclave the bottle at 15 lb in^{-2} for 15 min, in a pressure cooker. Cool to 43°C.
3. Add warm medium with 20% FBS (95 ml) and HT to 5 ml molten agarose and mix well by shaking.
4. Pipette aliquots of the agarose medium (4 ml) into each of 15 (20 mm diameter) Petri dishes. Keep the remainder of the agarose medium at 43°C in the water-bath.
5. Place a beaker of dry ice, with a few millilitres of sterile distilled water added, in the centre of the open Petri dishes in the tissue culture cabinet.
6. Close the cabinet and switch off the fan. The cold CO_2 vapour provides a cooling effect and maintains the pH of the medium in the Petri dishes as they set over 20–30 min.
7. Anaesthetize a BALB/c mouse and collect 0.1 ml blood by cardiac puncture.
8. Sacrifice the mouse by cervical dislocation before recovery, and then dilute the blood 1 : 200 in RPMI 1640 containing 20% FBS (20 ml) and HT.
9. Centrifuge 1.5×10^5 of the hybridoma cells to be cloned at 150 g for 10 min.
10. Resuspend the hybridoma cells in 15 ml of the diluted mouse blood.
11. Centrifuge the cells as before, and then dilute the cell pellet to 15 ml with the warm agarose/medium.
    The agarose concentration may be too high for a particular cell line so it may be necessary to further dilute the warm agarose medium to as much as half that used for the base layer.
    Beware of diluting the top layer too much as it may become sloppy before the clones are ready to pick.
12. Aliquot 1 ml of the cell suspension, dropwise, to each of the Petri dishes.
13. Allow the top layer to set in the same manner as described earlier.
14. Once the top layer has set, incubate the Petri dishes in a humidified box at 37°C for 4–6 days.
15. After 4–5 days overlay the agarose cultures with 0.2 ml of an appropriate dilution of sterile, filtered rabbit anti-mouse Ig to detect clones secreting immunoglobulin.
16. Check the plates for the next 7 days for antibody precipitates appearing over secreting clones.
17. Using an inverted microscope in a tissue culture cabinet, identify the useful clones. Select the clones using a long-stemmed Pasteur pipette, which has been drawn out by heating in a Bunsen flame, and then sheared to produce a fine bore size. Sterilize the fine-bore pipettes by autoclaving prior to use.
18. Place the selected clones onto mouse macrophage feeder layers in microtitre wells and then culture further for expansion, screening, cryopreservation and/or further cloning steps.

Although this is a more complex method of selection for appropriate clones and requires a great deal more technical competence to perform, especially in maintaining sterility, it has some useful applications. In the selection of new parent lines for murine or human hybridoma technology it is often essential to select a non-secreting clone from a HAT-sensitive cell line. Using the overlay technique described it is relatively easy to rapidly select non-secreting sublines for use as parental cells in fusion experiments.

PROCEDURE: LONG-TERM MAINTENANCE

Growth characteristics

Hybridomas normally grow in suspension. However, it is commonly found that a percentage of these cells, given time, will partially adhere to the surface of the vessel and may require a significant amount of force to make them detach (Figure 5.2.2).

The doubling rate of hybridomas is usually between 16 and 24 h, which means that they require constant attention. In most cases cells would be subcultured three times a week, i.e. Monday/Wednesday/Friday.

A concentration of between 10^5 and 10^6 per ml should be maintained by the passaging regimen, so that the cells are constantly in the log phase of growth (Figure 5.2.3).

Hybridomas passaged below this level may struggle to repopulate the vessel, resulting in a lag period of growth which may be detrimental to the surviving cells. Permitting the hybridomas to exceed the upper limit may result in a decline/ stationary phase which again may affect the performance of surviving hybridomas in subsequent passages, in terms of both monoclonal antibody secretion and growth rate. To monitor the growth of hybridoma cultures it is useful to perform regular viable cell counts by Trypan blue exclusion (see Section 1.3).

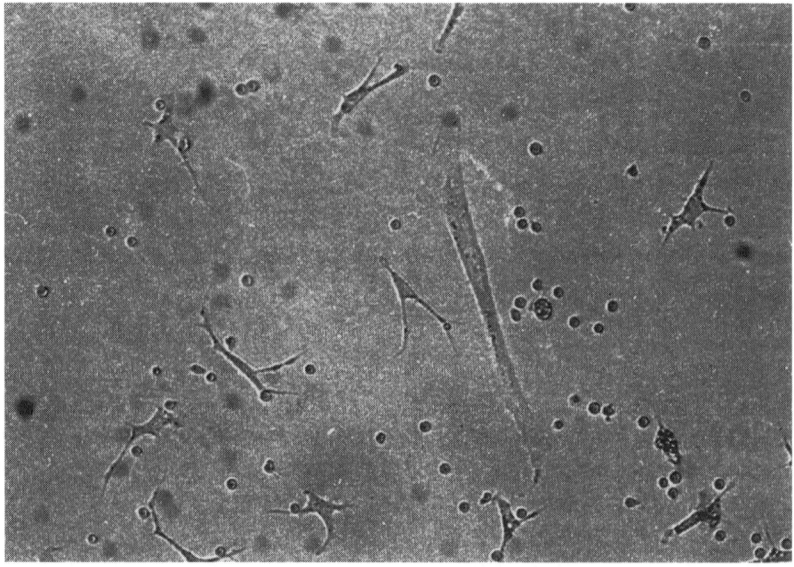

Figure 5.2.2(a) Macrophage feeder layer. Magnification × 4.

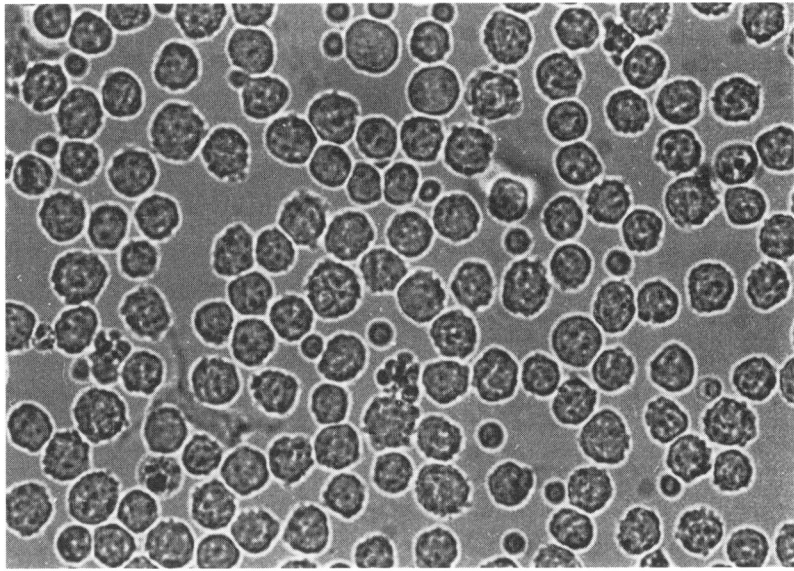

Figure 5.2.2(b) Seven-day culture of hybridomas on thymocyte feeder layer. Magnification ×20.

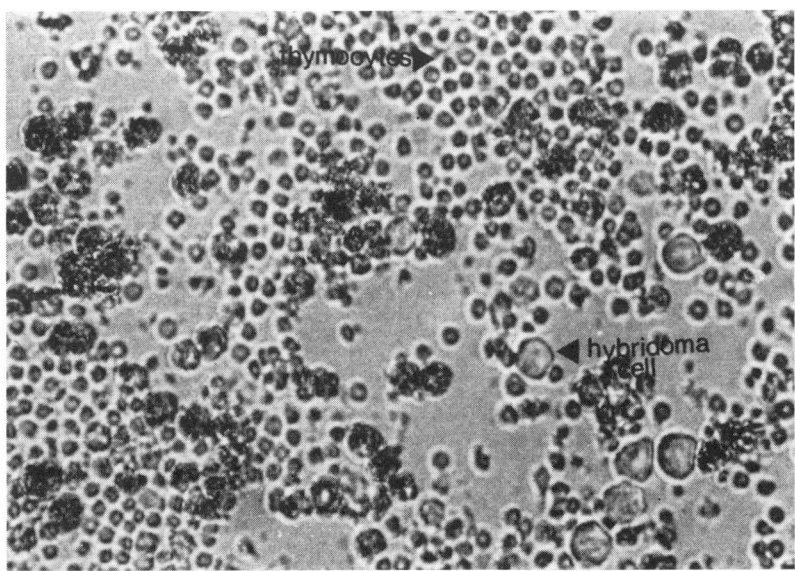

Figure 5.2.2(c) Hybridoma colony ready for expansion on macrophage feeder layer. Magnification ×20.

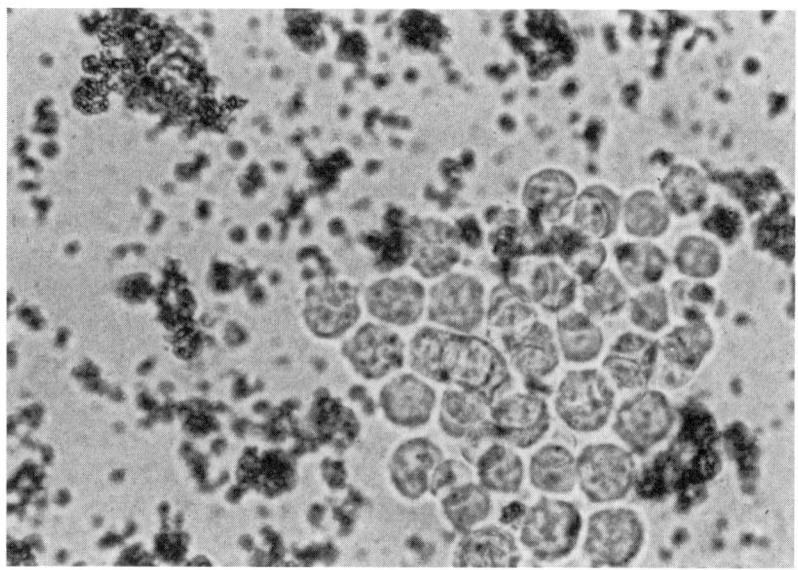

Figure 5.2.2(d) Three-day culture of hybridomas on thymocyte feeder layer. Magnification × 20.

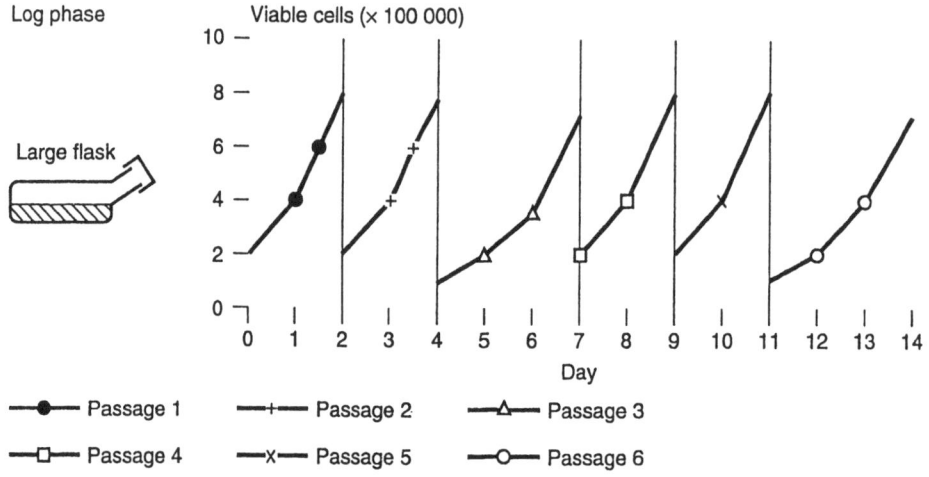

Figure 5.2.3 Routine passage of hybridomas in flasks.

Spinner culture

To maintain hybridomas over long culture periods it is important to keep cells in suspension such that they are continuously in a homogeneous distribution in the culture medium. This ensures the most efficient use of the nutrients available and minimum local concentrations of toxic metabolites encountered by the cells.

A range of small spinner culture vessels are commercially available, which are ideal for studying long-term culture, physiology, growth and secretion characteristics of hybridomas.

Additional materials

- RPMI 1640 culture medium (with or without 10 mM HEPES)
- Heat-activated FBS
- Kanamycin, gentamicin or penicillin/streptomycin
- Spinner culture vessel, 1 litre (Techne, UK)
- Spinner table, 2 or 4 place (Techne, UK)

1. Dilute the hybridoma cells to 1×10^5 cells ml^{-1} in 250 ml medium containing 5% FBS and seed the spinner vessel.
2. Set the stirring speed to a constant 25 rev min^{-1}.
3. Take samples (2 ml) daily using a 5 ml pipette.
4. Perform a viable/total cell count on the sample using Trypan blue exclusion.
5. Centrifuge the remainder of the sample at 150 g for 10 min and store the supernatant, frozen, at $-50°$C until the end of the culture.
6. Continue the culture until the cell viability has dropped below 25%.
7. At the end of the culture period a growth profile will have emerged from the daily cell counts. Assay the frozen samples for total antibody by quantitative ELISA.

By producing a superimposed profile of growth and secretion it is possible to determine whether the hybridoma secretes antibody during the growth phase or enhances secretion at the onset of the stationary/decline phase (Figure 5.2.4). The results from the profile established in the spinner culture can fundamentally affect the choice of bioreactor for any subsequent bulk production work.

Flask culture

The standard method of continuous *in vitro* culture of murine hybridomas is in tissue culture flasks with culture surface of 25 cm^2, 75 cm^2, 125 cm^2 or 225 cm^2. However, as hybridomas are suspension cells, the standard volumes used in these flasks is a more appropriate factor than the surface area. For the four sizes of flasks mentioned, the standard volumes used are 10 ml, 50 ml, 100 ml and 200 ml, respectively.

ALTERNATIVE PROCEDURE: MONITORING OF ANTIBODY SECRETION USING QUANTITATIVE ELISA

Having derived a stable hybridoma secreting a monoclonal antibody, large amounts may be required for diagnostics, therapeutics or antigen purification. When gram quantities are required it is important to monitor antibody production by the cell line, as this will determine the scale of the task. To this end a quantitative ELISA assay is essential as it permits continuous monitoring of antibody secretion.

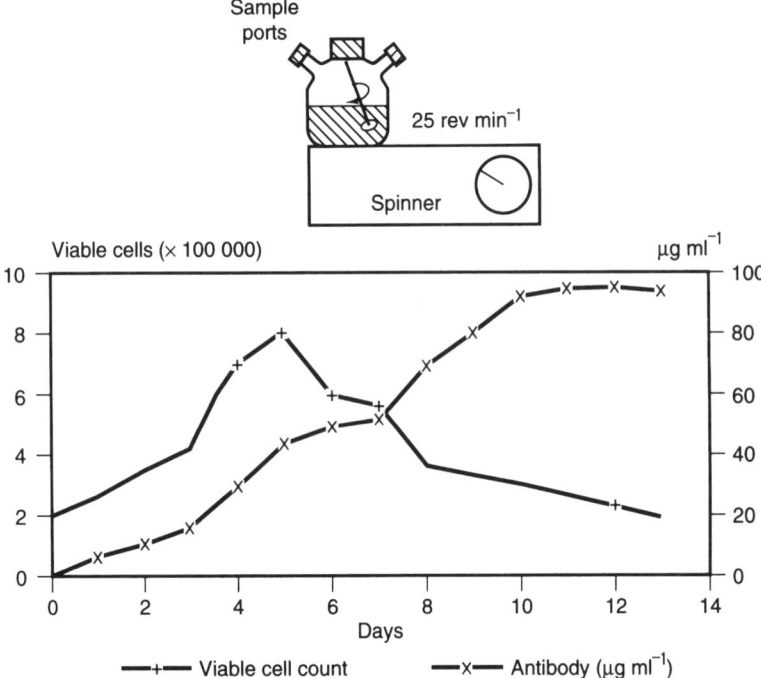

Figure 5.2.4 Hybridoma spinner culture in serum-free medium (growth and secretion profile).

Resultant secretion profiles can be used to assess the type and size of bioreactor required and the time span required to produce the quantity of antibody needed. In this technique the mouse antibody is captured using a rat anti-mouse kappa antibody. The bound mouse immunoglobulin is detected by anti-mouse IgG antibodies coupled to peroxidase.

Reagents and solutions

Coupling buffer

Sodium carbonate 1.59 g and sodium hydrogen carbonate 2.93 g, made up to 1 litre with distilled water (pH 9.6).

ELISA wash buffer

×20 phosphate-buffered saline (PBS) 1 litre, Tween-20, 10 ml, diluted 1 : 20 in distilled water.

Additional materials

- Rat anti-mouse kappa (Serotec)
- Coupling buffer

- ELISA wash buffer
- Sheep anti-mouse Ig-horseradish peroxidase (HRP) conjugate

1. Dilute the rat anti-mouse kappa (4 μg ml⁻¹) in coupling buffer.
2. Incubate the plates at 4°C overnight.
3. Wash the plates four times with ELISA wash buffer.
4. Pat the plates dry on a paper towel and then aliquot the blocking solution (150 μl well⁻¹) into each well of the microELISA plates (optional).
5. Incubate the plates at room temperature for 1 h.
6. Wash the plates as before.
7. Aliquot the supernatants to be tested as follows: 1–7 into column 2, rows A–G respectively (100 μl well⁻¹) at a dilution of 1 : 500.
8. Aliquot the pure antibody standard (ideally the antibody under test, if not the same isotype), 200 ng ml⁻¹ into well H2. No samples are aliquoted into column 1. Double dilute the samples and standard across the plate from columns 2–12 inclusive. Thus the first column is a blank and each well should contain 100 μl.
9. Incubate the plates at 37°C for 1 h wrapped in clingfilm and aluminium foil.
10. Wash the plates as before.

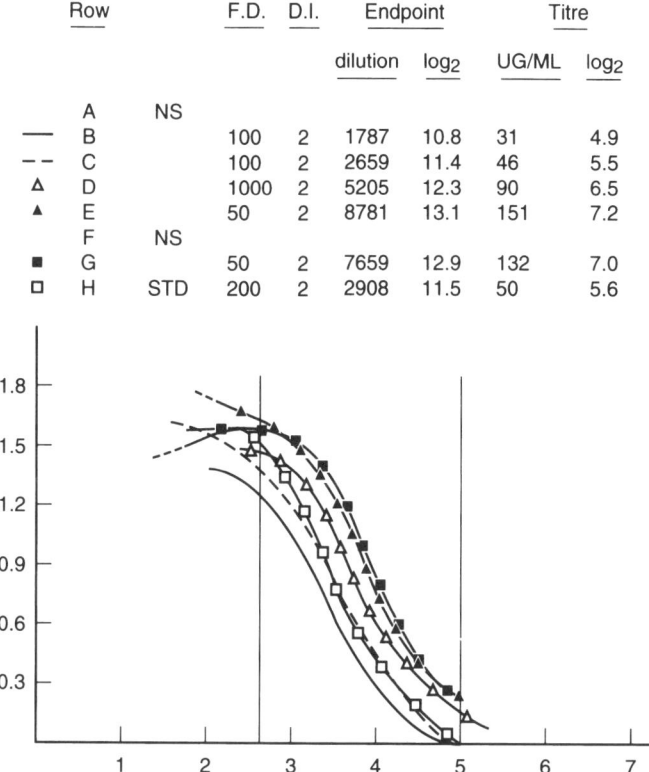

Figure 5.2.5 A sample printout of antibody concentrations in supernatants.

11. Dilute the sheep anti-mouse Ig-HRP conjugate 1 : 1000 and aliquot (100 μl well^{-1}) into all the microtitre plate wells.
12. Incubate the plates at 37°C for 1 h wrapped in clingfilm and aluminium foil.
13. Wash the plate as before.
14. Detect bound peroxidase as previously described.

Note: The concentration/dilutions of standard/sample given refer to IgG$_1$ monoclonal antibodies. The concentrations of the samples used for other isotypes tends to be higher as the conjugate has a greater sensitivity for IgG$_1$.

To calculate the concentration of antibody in the samples, an absorbance reading which corresponds to the steepest part of the dilution curve is selected and the log of the dilution factor at that point for each sample is directly compared to the log of the standard at the same absorbance point. This difference is multiplied by the standard concentrate and a sample concentration is produced. It is relatively simple to generate a computer program which interfaces with the ELISA reader to calculate the antibody concentrations in the sample supernatants (Figure 5.2.5).

5.3 ORGAN CULTURE TECHNIQUES IN TISSUE BANKING

The demand for human tissues for transplantation continues to increase rapidly. Tissue banks play a central role not only in applying standards for donor selection to minimize the risk of disease transmission via tissue grafts but also in ensuring the use of appropriate methods of storage for tissues that maximize both graft availability and successful graft outcome. Tissue banks have at their disposal a range of techniques for tissue storage, including cryopreservation, freeze-drying, hypothermia, and cell, tissue and organ culture methods. The method used will depend on a variety of factors, but of primary concern is whether viable cells or structural integrity or both are required for graft success. Thus, processed bone, which is used for reconstructive surgery, does not require viable cells for a successful graft outcome and may be freeze-dried and terminally sterilized, and skin may be stored in glycerol, which also renders it non-viable; at the other extreme, both cellular viability and structural integrity are essential for successful full-thickness corneal grafts, and storage methods have to be chosen accordingly.

Culture techniques have not typically been exploited purely as a means of storage of tissues for therapeutic use, with the notable exception of cornea. Organ and tissue culture have, of course, been extensively applied to skin, but mainly in the areas of toxicity testing and tissue engineering (Nemecek & Dayan 1999; Damour *et al.* 1998) rather than for actual preservation of intact skin grafts, which are cryopreserved, or stored at 4°C, or stored in glycerol. This section, therefore, will focus on the application of organ culture to the storage of corneas for transplantation.

CORNEAL STRUCTURE AND FUNCTION

Cornea possesses the remarkable property of transparency, yet it also has a protective function, forming part of the tough outer coating of the eye (Armitage 1999a). Human cornea is approximately 11 mm in diameter and has a central thickness of 0.52 mm, which increases towards the periphery. The stroma, which accounts for 90% of the thickness of the cornea, is composed of sheets of collagen fibrils, the lamellae, embedded in proteoglycans. The fibrils in any one lamella run parallel to each other and the transparency of the cornea relies crucially on a highly ordered arrangement of the fibrils, which show a remarkable degree of uniformity both in diameter and in interfibrillar spacing. Scattered throughout the stroma are the keratocytes, which maintain the collagen/proteoglycan matrix. The stroma is bounded

Cell and Tissue Culture for Medical Research, edited by A. Doyle and J.B. Griffiths.

on its outer surface by a multi-layered epithelium and on its inner surface by a monolayer of closely apposed, polygonal (mainly hexagonal) endothelial cells. Since the cornea is avascular, the cells derive all their nutrients from the aqueous humour, and, consequently, there is an essential and constant influx of water and solute into the stroma from the aqueous. This influx of water has to be countered in order to avoid the tissue becoming oedematous and, as a result, losing transparency. It is the function of the endothelium to control stromal hydration by actively pumping ions, including bicarbonate, from the stroma to the aqueous, creating an osmotic gradient that induces a coupled flow of water out of the stroma.

CORNEAL TRANSPLANTATION

Loss of transparency through disease or trauma can be treated by removing a disc, typically 7–8 mm in diameter, of cornea and replacing it with healthy tissue from a cadaveric donor. The operation has a 90% success rate at one year (Vail et al. 1994). Human endothelial cells only rarely undergo mitosis, being arrested in G1 phase (Harris & Joyce 1999; Joyce et al. 1996), and there is a gradual decline in endothelial cell density throughout life (Bourne et al. 1997). The success of a full-thickness corneal graft is therefore crucially dependent on the presence of an adequate number of viable endothelial cells in the transplanted tissue (typically greater than 2000–2500 cells mm^{-2}). Not only does the surgical trauma induce an accelerated loss of endothelial cells post-operatively, but episodes of immunological rejection also cause cell loss (Redmond et al. 1992). If the endothelial cell density falls below about 500 cells mm^{-2}, the endothelium becomes incapable of controlling hydration and the corneal graft fails. The method of corneal storage must, therefore, adequately preserve the endothelium. Although methods for crypreservation of cornea were developed in the 1960s (Capella et al. 1965), they are not used routinely since they can cause considerable damage to the corneal endothelium, with loss of 30–70% of endothelium after grafting. Hypothermic storage is used extensively for corneas, but this technique allows only 1–2 weeks of storage (Lindstrom et al. 1992). Although the reduced temperature slows biochemical reactions and reduces the energy demands of the cells and hence the tissue, there remains a gradual depletion of adenine nucleotides, ionic balance is disrupted, which leads to cellular oedema, and cell membranes undergo temperature-induced changes (Fuller 1991). These changes serve to limit the useful storage period achievable with hypothermia.

CORNEAL STORAGE BY ORGAN CULTURE

Corneas may also be stored by organ culture at 31 to 37°C, and this is the method of choice for many European eye banks. This technique, which was developed in the early 1970s, routinely allows up to four weeks of storage (Summerlin et al. 1973; Doughman et al. 1976), but there have been reports of successful grafts using corneas stored for seven weeks (Ehlers et al. 1999). This technique has been used in the Corneal Transplant Service Eye Bank, Bristol, since 1986 (Armitage et al.

1990) and, at the time of writing, approximately 20 000 organ-cultured corneas
have been supplied from the Bristol eye bank for grafts throughout the UK.
Although the general principles are similar, there are inevitable differences in
detail in the application of organ culture in different eye banks, and the method
as applied in Bristol will be described (Armitage 1999b).

The preparation of corneas for organ culture is carried out in a Class II Biological
Safety Cabinet both to provide a sterile work area and to protect staff from potentially pathogenic material. The major steps in the process include:

- Cleaning of eyes
- Excision of corneoscleral disc
- Suspension of corneoscleral disc in organ culture medium
- Testing medium for bacterial or fungal infection
- Examination of corneal endothelium
- Reversal of stromal oedema before transplantation

Cleaning of eyes

Virtually every eye received in the eye bank will carry bacteria and/or fungi on
the ocular surface. It is important, therefore, to reduce as much as possible this
microbial load in order to minimize the risk of loss of corneas during organ culture
through growth of organisms in the culture medium. Clearly, the tissue cannot be
sterilized as it is essential to have viable cells for a successful graft outcome. Eyes
are therefore taken through several washes of sterile saline and then immersed
in 3% polyvinylpyrrolidone-iodine (PVP-I) for 2 min. The iodine is neutralized
by 0.3% sodium thiosulphate for 1 min followed by a final wash in sterile saline.
The concentration of PVP-I used by different eye banks ranges from 0.5% to 5%.
Even after this cleaning stage, however, up to 20% of eyes may still have viable
bacteria or fungi on the ocular surface (unpublished data).

Excision of corneoscleral discs

An eye is either placed cornea uppermost in a plastic stand or held gently in
sterile gauze. An incision is made in the sclera with a scalpel approximately 3–4 mm
from the limbus (the boundary between the cornea and the sclera) without penetrating the choroid. The incision is completed around the cornea using Castroviejo
corneal section scissors, keeping the blades in the suprachoroidal space and maintaining the 3–4 mm scleral rim. After completing the incision, the corneoscleral
disc should be attached to the ciliary body-choroid only at the scleral spur. The
scleral rim is lifted gently using fine-toothed forceps while detaching the adhe-
sions to the ciliary-body-choroid with a second pair of forceps – extreme care
should be taken during this procedure not to bend or buckle the cornea, and not
to allow the corneal endothelium to come into contact with the iris.

Suspension of corneoscleral disc in organ culture medium

The corneoscleral disc is placed endothelium uppermost in a plastic Petri dish.
The sclera is held with modified cilia forceps, and a sterile suture passed through

the scleral rim, taking care not to fold or crease the sclera. The other end of the suture is passed through the inner base of a silicone rubber stopper, leaving just sufficient suture for the corneoscleral disc to be suspended in organ culture medium when the stopper is placed in the neck of a 100 ml glass DIN bottle containing 80 ml of medium.

The medium comprises Eagle's minimum essential medium with Earle's salts (MEM) buffered with 20 mmol l^{-1} HEPES and containing 26 mmol l^{-1} sodium bicarbonate, 2% foetal bovine serum (FBS), 2 mmol l^{-1} L-glutamine, penicillin (100 U ml^{-1}), streptomycin (0.1 mg ml^{-1}) and amphotericin B (0.25 μg ml^{-1}). In the Bristol eye bank, the corneas are placed in dry incubators at 34°C, which is approximately normothermia for the cornea. The medium is not changed during the storage period, which may last up to four weeks. The silicone rubber stopper allows some gas exchange, allowing oxygen in and the escape of excess carbon dioxide. Although there is clearly nutrient depletion and a build-up of waste products during the culture period, the benefits of changing the medium, which some eye banks do every 1–2 weeks, are not conclusive (Hjordtal et al. 1989).

Testing medium for bacterial or fungal infection

After 7 days, a sample of the culture medium is taken to test for bacterial and fungal infection using brain-heart-infusion broth (BHI) and tryptic soy-agar (TSA) plates. In the case of positive cultures, the cornea is withdrawn from the eye bank and sent to a medical microbiology laboratory for identification of the contaminating organisms. The percentage of corneas withdrawn because of microbial infection is between 1% and 3% of those stored. Two thirds of the infections are bacterial, most frequently *Staphylococcus epidermidis*, with *Candida spp.* being the overwhelming cause of the fungal infections. The incidence of infection shows some seasonal variation, being highest in summer, and increases slightly with the interval between the death of a donor and the removal of the eyes (Armitage & Easty 1997).

Postoperative endophthalmitis caused by bacterial or fungal infection is a rare but extremely serious complication of intraocular surgery that is sight threatening and can even lead to the loss of an eye. In the case of corneal transplantation, there is a risk that microorganisms may be transferred into the eye from the grafted tissue. One of the main advantages of organ culture over other storage methods for corneas is the readiness with which bacteria and fungi may be detected during the storage period, allowing the tissue to be discarded rather than transplanted. Moreover, the antibiotics present in the organ culture medium will, because of the higher temperature, be far more effective against microorganisms than those included in hypothermic storage solutions. Although the sterility of a graft cannot be guaranteed, the likelihood of transplanting organ-cultured corneas that are carrying bacteria or fungi is clearly very much less than for corneas stored at 4°C where there is no chance to screen for contaminating organisms.

Examination of corneal endothelium

To ensure that the endothelium is intact and contains sufficient numbers of cells, organ-cultured corneas are always examined by light microscopy before they are issued for transplantation, usually 2–3 days before the scheduled date of graft. The endothelium is stained with 0.2% Trypan blue, to reveal areas of damaged or missing cells, and a hypotonic sucrose solution is used to induce slight cell swelling, which renders the cells borders visible. A cell count is made to provide an estimate of cell density, and other morphological features are evaluated, such as the degree of folding in the endothelial basement membrane and the regularity of the endothelial cell pattern.

Since endothelial cell density is inversely related to donor age, some eye banks that use hypothermic storage methods do not routinely examine the endothelial layer but simply exclude donors above 60–65 years old. In the UK, this would rule out half of our eye donors. Donor age does undoubtedly have a strong influence on whether corneas are suitable for transplantation (Armitage & Easty 1997); but provided that corneas meet or exceed the minimum criteria (i.e. endothelial cell density > 2200 cells mm^{-2}), donor age does not adversely affect graft survival (Vail et al. 1994). Similarly, storage times of up to 4 weeks do cause a decline in the percentage of corneas suitable for transplantation from > 80% to 70%; but again, provided that the minimum criteria are met, storage time has no influence on graft survival (Armitage & Easty 1997; Vail et al. 1994).

Reversal of stromal oedema before transplantation

During organ culture, the cornea is completely immersed in culture medium, and the endothelium is unable to control hydration adequately; consequently, the stroma becomes oedematous with the result that the cornea approximately doubles in thickness. After examination of the endothelium, corneas that are judged suitable for transplantation are placed into organ culture medium that contains 5% dextran M_r 500 000). This reverses the oedema and thins the cornea towards normal thickness, but corneas should not remain in this dextran medium for more than a few days (a 4-day limit is set in the Bristol eye bank). Early attempts to prevent the occurrence of oedema by including dextran in the medium throughout the organ culture period proved to be detrimental to the endothelium.

GRAFT OUTCOME

Clinical follow-up studies of organ-cultured grafts have shown that overall graft survival at one year is 90% (Vail et al. 1994), and the accelerated postoperative loss of endothelial cells has an exponential half-time of 3.4 years (Redmond et al. 1992). Both of these values are similar to those for corneas stored for much shorter periods at 4°C. The advantages, therefore, of organ culture for corneal preservation are extended storage time and greater microbiological safety.

FUTURE DEVELOPMENTS

There is clearly room for improvement in the organ culture system. The development of a fully defined medium that maintains endothelium, epithelium and stromal keratocytes would be a significant advance. Furthermore, ways to control stromal hydration and thus prevent oedema during organ culture would simplify the method. Dextran and chondroitin sulphate are both currently incorporated into hypothermic storage solutions to prevent stromal oedema; but the reduced temperature clearly obviates the detrimental effects of long-term exposure to dextran seen during organ culture.

Low endothelial cell density is one of the main reasons for judging corneas to be unsuitable for transplantation. Attempts have been made over the years to increase endothelial cell density by seeding corneas in organ culture with extra cells (Engelmann *et al.* 1999). An alternative would be to expand the existing population of cells by stimulating cell division. The key to these approaches lies in understanding why human corneal endothelial cells are arrested in G_1 phase. It is not simply due to contact inhibition – wound healing studies show that deficits in the endothelial monolayer are repaired by cells migrating and expanding to fill gaps rather than by cells at the wound margin undergoing mitotic division to replace missing cells. It has been suggested that the TGF-beta pathway may be involved in preventing cells progressing to S-phase (Harris and Joyce 1999). Corneal endothelium of other species, such as rabbit and pig, do readily divide and *in vitro* cultures can be established. Cultures of human endothelium can be established but with much greater difficulty (Engelmann *et al.* 1988).

REFERENCES

Armitage WJ (1999a) Anatomy and physiology of the cornea. In: Easty DL & Sparrow JM (eds) *Oxford Textbook of Ophthalmology*. Oxford University Press, Oxford.

Armitage WJ (1999b) Eye banking. In: Easty DL & Sparrow JM (eds) *Oxford Textbook of Ophthalmology*. Oxford University Press, Oxford.

Armitage WJ & Easty DL (1997) Factors influencing the suitability of organ-cultured corneas for transplantation. *Investigative Opthalmology & Visual Science* 38: 16.

Armitage WJ, Moss SJ, Easty DL & Bradley BA (1990) Supply of corneal tissue in the United Kingdom. *British Journal of Ophthalmology* 74: 685.

Bourne WM, Nelson LR & Hodge DO (1997) Central corneal endothelial cell changes over a ten-year period. *Investigative Ophthalmology & Visual Science* 38: 779.

Capella JA, Kaufman HE & Robbins JE (1965) Preservation of viable corneal tissue. *Cryobiology* 2: 116.

Damour O, Augustin C & Black AF (1998) Applications of reconstructed skin models in pharmaco-toxicological trials. *Med. Biol. Eng. Comp.* 36: 825.

Doughman DJ, Harris JE & Schmitt MK (1976) Penetrating keratoplasty using 37 C organ cultured cornea. *Transactions of the American Academy of Ophthalmology and Otolaryngology* 81: 778.

Ehlers H, Ehlers N & Hjortdal JO (1999) Corneal transplantation with donor tissue kept in organ culture for 7 weeks. *Acta Ophthalmologica Scandinavica* 77: 277.

Engelmann K, Böhnke M & Friedl P. (1988) Isolation and long-term cultivation of human corneal endothelial cells. *Investigative Ophthalmology & Visual Science* 29: 1656.

Engelmann K, Drexler D & Böhnke M (1999) Transplantation of adult human or porcine corneal endothelial cells onto human recipients in vitro. I. Cell culturing and transplantation. *Cornea* 18: 199.

Fuller BJ (1991) The effects of cooling on mammalian cells. In: Fuller BJ & Grout BWW (eds) *Clinical Applications of Cryobiology*. CRC press, Boca Raton, FL.

Harris DL & Joyce NC (1999) Transforming growth factor-beta suppresses proliferation of rabbit corneal endothelial cells in vitro. *Journal of Interferon and Cytokinetic Research* 19: 327.

Hjortdal JO, Ehlers N & Andersen CU (1989) Some metabolic changes during human corneal organ culture. *Acta Ophthalmolica Scandinavia.* 67: 295.

Joyce NC, Meklir B, Joyce JJ & Zieske JD (1996) Cell cycle protein expression and proliferative status in human corneal cells. *Investigative Ophthalmology & Visual Science* 37: 645.

Lindstrom RL, Kaufman HE, Skelnik DL, Laing RA, Lass JH, Musch DC, Trousdale MD, Reinhart WJ, Burris TE, Sugar A, Davis RM, Hirokawa, K, Smith T & Gordon JF (1992) Optisol corneal storage medium. *American Journal of Ophthalmology* 114: 345.

Nemecek GM & Dayan AD (1999) Safety evaluation of human living skin equivalents. *Toxicology and Pathology* 27: 101.

Redmond RM, Armitage WJ, Whittle J, Moss SJ & Easty DL (1992) Long-term survival of endothelium following transplantation of corneas stored by organ culture. *British Journal of Ophthalmology* 76: 479.

Summerlin WT, Miller GE, Harris JE & Good RA (1973) The organ-cultured cornea: an in vitro study. *Investigative Ophthalmology* 12: 176.

Vail A, Gore SM, Bradley BA, Easty DL, Rogers CA & Armitage WJ (1996) Influence of donor and histocompatibility factors on corneal graft outcome. *Transplantation* 58: 121.

5.4 CELL AND TISSUE ENGINEERING

Cell therapy is the replacement, repair, or enhancement of biological function of damaged tissue or organs. This is achieved by transplantation of cells to a target organ by injection (e.g. foetal cells into the brain of patients with Parkinson's or Alzheimer's disease), or by implantation of cells selected/engineered to secrete missing gene products. The capability of growing large quantities of differentiated human cells (either xenograft or allograft) on complex matrices has opened up the possibility of tissue engineering by providing human recipients with replacement transplantable tissues and, in the not too distant future, organs. This technology has developed from advances made in three-dimensional cell culture in which cells are given the correct physical and physiological environment to ensure continued differentiated function. This has required the correct medium conditions (with complex growth factors) as well as physical substratum and low shear stress conditions to be in place (Mueller-Kleiser 1997).

The key technical issues that remain are maintenance of the differentiated state of the cells (it must be remembered that proliferation and differentiation are often at the opposite ends of a cellular spectrum) and the ability to provide adequate levels of nutrients to cells assembled in tissue-like structures (Ferber 1999). Tissue engineering requires first, the reproduction of cells (in conventional culture dishes), second, the induction of differentiation, and third, maintenance of differentiation (Minuth *et al.* 1998). The latter two steps require novel tissue carriers and perfusion culture containers as described by Minuth *et al.* (1998) in Table 5.4.1, and in the following section. In addition, there are in some cases safety considerations to be borne in mind especially in the use of xenotransplanted cells with issues raised over cross-species transmission of porcine retroviruses (Stoye *et al.* 1998). The US FDA have been particularly active in this matter (Fox 1998)

THREE-DIMENSIONAL CELL CULTURE SYSTEMS

Tissue culture cells grown in two-dimensional (2D) monolayers in traditional glass or plastic tissue culture flasks have been used successfully for many purposes in research and industrial production. However, such cultures may lose key phenotypic characteristics (e.g. virus susceptibility, morphology, surface markers/receptors) after repeated passage. *In vivo* the presence of three-dimensional (3D) cellular structures is critical to the correct development, function and stability of cells, tissues and organs. The characteristics that the researcher or technologist wishes to utilize are often a feature of the tissue and not individual cells, e.g. a

Cell and Tissue Culture for Medical Research, edited by A. Doyle and J.B. Griffiths.
© 2000 John Wiley & Sons, Ltd

Table 5.4.1 Differentiation of cells and tissues is triggered at different levels. Tissue-specific differentiation is obtained by a synergistic action

Degree of differentiation	Necessities	Realizations
	Optimal anchorage for differentiation	Filters, fleeces, biomatrices
	Optimizing the extra-cellular matrix	Coating with extra-cellular proteins
	Cell and tissue transportation	Tissue carrier for optimal handling
	Elimination of metabolites/ maintaining paracrine factors on a constant level	Perfusion culture containers
	Growth/differentiation factors	Additives to the medium, binding to the artificial matrix
	Proliferation versus contact inhibitation	Avoiding mitotic stress in serum-free medium
	Long-term culture	Optimizing electrolytes in the medium for human cells

From: Tissue engineering: generation of differentiated artificial tissue for biomedical applications. *Journal of Cell and Tissue Research* 291(1): 1–11, January 1998. © Springer-Verlag.

functional bladder epithelium or crypt structures of the gut. In this section we describe some of the approaches that can be used to simulate certain features of the *in vivo* environment in an attempt to promote natural gene expression and tissue function in cultured cells. The described technologies address these features from two aspects:

1. Endogenous features (i.e. within 3D cell structures): autocrine and paracrine factors; mass transfer characteristics; establishment of appropriate cell–cell and cell–matrix contacts; cell signalling pathways; and pressure and tensile forces.
2. Exogenous features (i.e. in culture medium): hormones; nutrition, including conditioned media; pO_2; pH; pCO_2; and temperature.

Many of these specialized functions of cells are lost or expressed at low levels when they are growing in monolayers and this is due, in part, to the lack of appropriate cell–cell or cell–matrix interactions. Cultivation in a 3D system can promote or improve characteristic cell functions such as hormone secretion, production of extracellular matrix components and expression of differentiation markers.

The resulting cellular structures can be used as models for investigating development, drug metabolism, toxicity, biotransformations, pathogenesis and microorganism replication.

In industrial (e.g. recombinant protein) and medical (e.g. bioartificial organ) fields 3D cultures can be used to improve the surface area/volume ratio compared with 2D cultures, which is a useful feature where cells are used as the machinery for biological production. Such approaches promote high cell yield and increased production of cellular or recombinant proteins.

Diverse techniques for 3D culture are now available and here we will primarily consider:

- 3D multilayers or spheroids
- 3D supports, e.g. microcarriers
- 3D matrices, such as gels, sponges and porous microcarriers

Usually a single cell type is cultured in isolation, as in 3D systems to simulate liver function where hepatocytes are propagated to reach high organ-like densities (e.g. $> 10^7$ cells ml^{-1}) in a 3D matrix, thereby enabling the reproduction of specific tissue-like function. Furthermore, other organ-like structures can be reproduced using co-cultures, e.g. mesenchymal and epithelial cells of intestinal origin (Goodwin *et al.* 1993) and foetal rat spinal cord co-cultured with human muscle cells (Mariotti *et al.* 1993). A further technique used to reproduce tissue-like structures is re-aggregation of primary chick cell suspensions (Funk *et al.* 1994). All of these approaches are very useful for the study of differentiation, cell–cell interactions and tissue function, and can benefit enormously from a 3D culture approach. The applications can range from cell biology and medical research to industrial-scale production of biologicals. However, there is no single technique applicable for all purposes and in the following the key features of a number of 3D culture techniques are reviewed.

Spheroids

Spheroids are suspended multicullular aggregates where cells adhere to each other instead of attaching to an artificial substrate. They are applied as models for embryogenesis of different tissues, in tumour biology as therapeutic models and also used in developmental studies. Spheroids can be generated efficiently by culture of cells on the normal tissue culture surfaces (glass and plastic) coated with agarose (Sussman & Sussman 1961) or poly(hydroxyethyl methacrylate) (polyHEMA) (Folkman & Moscona 1978). Short-term spheroid cultures provide useful spherical structures and in some cases spheroid culture methods significantly improve the cultivation of primary cells (Ijima *et al.* 1997) and the retention of differentiated characteristics.

Early research by Moscona (1961) on embryonic cells cultured as spheroids showed tissue-like structural properties and *in vivo*-like growth. A high differentiation capacity, which is not observed in monolayers, can be found in multicellular spheroids of hepatocytes (Tong *et al.* 1994), adult human glioma cells (Glimelius *et al.* 1988), avian foetal brain cells (Funk *et al.* 1994) and outer root sheath cells of human hair follicles (Limat *et al.* 1994). This is an especially important consideration in drug tests, where cells with a full range of natural metabolic activities are required to achieve a biochemical turnover that would be observed in tissues.

In particular, malignant human ovary and mammary carcinomas can form solid aggregates with histological similarities to the primary tumours (Schleich, 1967). In addition, the production of extracellular matrix components similar to tumours *in vivo* has been demonstrated by Glimelius *et al.* (1988) with multicellular spheroids of human glioma cells.

A series of examples reveal that cell systems like these can be used in wide ranging studies of, for example, cancer sensitivity to radiation and chemotherapy and the analysis of penetration by cytotoxic drugs in targeted tumour therapies (Lindstrom & Carlsson 1993; Rotmensch *et al.* 1994). Spheroid-like hollow bodies with a multicellular epithelial morphology have also been utilized as a model of pathogenesis of infection as a result of *Neisseria* infection. Numerous cell–cell contacts representative of cells *in vivo* have been demonstrated in spheroids from nasopharyngeal cells, e.g. junctional complexes, desmosomes, specific orientation of the cytoskeleton and cellular organelles (Boxberger *et al.* 1993).

Microcarriers

An important feature of 3D cultures for industrial purposes is the improved surface area/volume ratio allowing high cell densities in bioreactors with relatively low costs of production and maintenance. A system often applied for 3D large-scale productions is the microcarrier bead. The use of porous carriers (e.g. Verax Microsphere, Cultispher G, Immobisil or Siran Porous Beads has not only enabled increased unit productivity (reviewed by Looby & Griffiths 1989; Griffiths 1990) but allows cells to form a 3D organization within the porous microsphere.

A number of microcarrier systems have been established with primary and secondary cells from birds and mammals, permanent cell lines from fish and mammals as well as diploid human cells (Reuveny 1985). A further possibility is the mass production of cells with the retention of differentiation potential, as can be seen with bone cells (Sautier *et al.* 1992) and human retinal pigment epithelial cells (Kuriyama *et al.* 1992). In addition to the widely used continuous cell lines for the production of biologicals, freshly harvested cells such as endothelial cells have been applied for the production of endothelium-derived relaxing factor (Bing *et al.* 1991).

Hepatocytes growing in microcarrier systems are used for the study of liver failure and drug metabolism. Transplantation experiments of hepatocytes grown on microcarriers showed detoxification of ammonium and reduction of bilirubin concentrations after induced acute liver failure (Nagaki *et al.* 1990). Furthermore, co-cultivation assays using hepatocytes and Balb/c 3T3 fibroblasts as target cells have been applied to analyse the metabolism-mediated toxicity of xenobiotics *in vitro* (Voss & Seibert 1992).

Other examples for the successful employment of co-cultures are: the interaction between muscle and nerve cells (Shahar *et al.* 1985) and the co-cultivation of vascular endothelial and smooth-muscle cells using microcarrier techniques (Davies & Kerr 1982); co-cultivation of cerebellar granule cells, cerebral cortical neurons and cortical astrocytes on collagen-coated dextran beads (Cytodex3) and the production and release of specific neurotransmitters and enzyme synthesis resembling *in vivo* interactions between neurons and astrocytes (Westergaard *et al.* 1991).

Microcarrier technology has provided valuable systems for industrial-scale use of animal cells. However, microcarriers have also proved to be a highly adaptable and very successful approach to studying a wide range of cell types, both in mono- and co-culture.

Filterwells

In the filterwell or 'Transwell®', systems, cells are grown on a permeable membrane between two separated liquid phases (McCall *et al.* 1981). In this condition confluent epithelial cells can achieve their full polarized functional state without the stresses induced by 'doming' of epithelial monolayers on glass or plastic surfaces (Rabito *et al.* 1980). Filterwell culture has proved valuable for modelling the epithelium of the human intestine (Hidalgo *et al.* 1989). This technique is rapidly becoming the preferred culture system for many studies using epithelial cultures.

Matrix sponges or 3D gels and matrix sandwiches

Sponges made of cellulose or collagen gels can be used as matrices in Petri dishes where cells penetrate the inert material. A collagen-coating can be applied to improve cell adherence and growth. Analysis of cell characteristics in this system is readily achieved by histological analysis of sections. The influence of culture conditions on differentiation has been analysed by Bruns *et al.* (1994), comparing the formation of matrix substances by sheep rib perichondrium cultured on collagen sponges, fibrin gels and cellulose acetate filters. An *in vitro* model simulating normal human secretory endometrium was successfully established by Bentin-Ley *et al.* (1994). In this system the polarized epithelial cells were grown on 'Matrigel®', which separated the underlying collagen-embedded endometrial stromal cells. Mechanical stress-induced orientation has been studied using arterial smooth-muscle cells in 3D collagen lattices (Kanda & Matsuda 1994). Sandwiches of collagen matrix to simulate the space of Disse in the liver have proved highly successful for the extended and improved culture of primary hepatocytes (e.g. Bader *et al.* 1996). Such changes in physical characteristics of culture conditions may offer valuable developments for tissue models in the future.

Microcontainers

The microcontainer technique is a recently developed method for 3D cultures (Weibezahn *et al.* 1994). In this patented system cells grow on the vertical walls until a multicellular layer is formed. The containers are then transferred into a vessel and perfused continuously. A controlled supply of growth media of different compositions on each side of the layer allows the culture of highly differentiating cell types and hence the establishment of tissue-like models. The full range of applications for this technique remains to be determined and it may well provide some useful *in vitro* tissue models in the future.

Simulated microgravity

In order to maintain microgravity conditions achieved in space flight when cell culture experiments returned to Earth's surface, the US National Aeronautical Space Agency (NASA) laboratories developed a device called the rotating wall vessel (RWV). In this system cells are grown in a rotating body of culture medium with no direct air/medium interface. This provides for conditions of very low shear stress in which cells of different buoyant density can aggregate and proliferate together. This technique has been used for a wide range of cell cultures, including cartilage (Freed *et al.* 1993), ovarian tumour cells (Becker *et al.* 1993), hepatocytes (Battle *et al.* 1999) and colorectal carcinoma (Goodwin *et al.* 1992). Co-culture systems have revealed some very useful and novel characteristics that mimic *in vivo* tissue (Goodwin *et al.* 1993). Thus, low shear stress technology holds great promise for the development of new *in vitro* models.

Conclusion

Whilst the use of standard tissue culture flasks in glass or plastic has provided much of the data on which our understanding of *in vitro* cell biology is based, a very wide range of novel culture formats are now available for the benefit of those in basic research and biotechnology. Some of these approaches, notably filterwell culture, are now fairly commonplace but others, such as microcontainers and the RWV system, remain to have their full potential identified. Recent developments in the area of 3D cell culture have been rapid. It is evident that there is still much to be learned about the way cell function is modulated in 3D cell–cell and cell–matrix interactions.

TISSUE ENGINEERED SKIN

The first recorded application over a decade ago was to grow keratinocytes from a small skin biopsy into large cell sheets which were then surgically grafted onto burns patients (Auger *et al.* 1998; Rheinwald & Green 1975; Burt & McGrouther 1992). This has now been extended to treating traumatic injuries and ulceration and has advanced to commercial production of dermal replacement products (e.g. Dermagraft).

Generally these skin substitutes are produced by the culture of keratinocytes on a matured dermal equivalent including fibroblasts in a collagen gel matrix. The introduction of more sophisticated biomaterials, such as biodegradable microspheres (La France & Armstrong 1999) permitted the delivery of known numbers of cells which can migrate freely out of the matrix and allow the biomaterial to be resorbed. This type of technology will further improve the ready availability of replacement tissue in providing easy storage and transportation in a form that retains the essential three-dimensional structure in an 'off-the-shelf' format.

APPLICATIONS OF TISSUE ENGINEERING

To avoid destruction of implants by the host's immune system encapsulation of the transplant cells in semi-permeable devices is widely used (Scharp *et al.* 1994). Examples include pancreatic islet cells for diabetes (Joseph *et al.* 1994), chromaffin cells for chronic pain (Aebischer *et al.* 1986), and genetically engineered BHK cells secreting neurotrophic factors for neuro-degenerative diseases (Aebischer, 1997; Peshwa *et al.* 1994). It has not yet been possible to replace the liver or kidney but artificial organs situated outside the patient containing primary or recombinant cells through which the patient's blood is perfused have been developed (Iijma et al. 1997; Gage 1998). Dialysis techniques only remove the toxic products whereas the cells in the artificial organs perform biotransformations, i.e. as well as degrading toxic products they additionally regenerate many essential metabolites which are returned to the body.

Thus greater sophistication has been achieved in the application of this cell culture technology to particular clinical problems. In addition, however, there is the increased likelihood that manipulation of matrix composition and medium conditions will provide organotypic co-cultures which would provide a suitable basis for study of the molecular mechanisms of tissue homeostasis and in vitro pharmacotoxicology (Stark et al. 1999).

Summary of potential medical applications of tissue/cell engineering

- Diabetes (pancreatic islet cells)
- Parkinson's disease (foetal dopamine cells)
- Duchenne's muscular dystrophy (myoblasts)
- Liver disease (parenchymal hepatocytes)
- Burns patients (keratinocytes and fibroblasts)
- Cartilage damage (chondrocytes)
- Pain (chromaffin cells)
- Cardiovascular disease (endothelial cells)
- Brain and spinal cord (neurotrophic factor-secreting cells)
- Cancer (haemopoietic cells, bone marrow, adoptive cellular therapy)
- Retinal pigmented epithelium
- Huntingdon's disease

(Reviewed by Gage 1998)

CURRENT ADVANCES

The current developments in the field are summarized in Table 5.4.2. They incorporate the reports that outline new approaches to therapy in experimental animals as well as human patients. In the main they constitute proof of principle rather than routine practices and are not yet in everyday use. There are also a number of commercial ventures that are established to exploit the new technology and offer for sale readily available biomaterials. These include Advanced Tissue Sciences, Creative Biomolecules, Genzyme Tissue Repair and Hepatix. Of these companies, Genzyme Tissue Repair has received US FDA approval for engineered tissues derived from a patient's own cells in the repair of traumatic knee cartilage tissue damage. There is still a considerable distance to go before these techniques are widely applied and receive full regulatory approval in every type of application.

An important potential development for tissue engineering is expected to be based on stem cells (self-renewing cells that give rise to phenotypically and

Table 5.4.2 Advances in development of tissue engineered tissues and organs. Included here are studies that have involved an *in vivo* step as well as *in vitro* culture. Most studies have been conducted using animal models.

Tissue/organ	Cell type	Matrix/biomaterial	References
Bladder	Canine urothelial and smooth muscle cells	Bladder-shaped polymers	Oberpinning *et al.* (1999)
Urothelial	Human urothelial and smooth muscle cells	Polyglycolic acid	Review: Atala (1998)
Cardiovascular	Rat cardiac myocytes	Cardiogel (fibroblast derived)	Bick *et al.* (1998)
	Human fibroblasts	Resorbable Polyglycolic acid mesh	Hoerstrup *et al.* (1998)
	Porcine/Bovine endotholial cells	Gelfoam materials	Nugent *et al.* (1999)
	Bovine smooth muscle and endothelial cells	Biomimetic system Polyglycolic acid mesh	Niklason *et al.* (1999)
	Human vascular smooth muscle (not endothelial)	None. Ascorbic acid promoted cell sheet formation	L'Heureux *et al.* (1998)
Intestine	Rat epithelial organ explants	Biodegradable tubular scaffold of Polyglycolic acid and Polylactic acid	Choi *et al.* (1998)
			Kaihara *et al.* (1999)
			Kim *et al.* (1999)
Bone/cartiledge	Bovine periosteum cells	Bioresorbable polymer fibres	Puelacher *et al.* (1996)
	Rabbit chondrocytes	Collagen gel	Wakitani *et al.* (1998)
	Human auricular chondrocytes	Polyglycolic acid	Rodriguez *et al.* (1999)
	Porcine chondrocytes	Fibrinogen and thrombin 'glue'	Silverman *et al.* (1999)
Kidney	Porcine renal proximal tubule cells	Haemofiltration cartridge with polysulphone fibres	Humes *et al.* (1999)
			Colton (1999)
	Human/porcine	Microcarriers/encapsulation/ biodegradable polymer scaffold	Review: Davis and Vacanti (1996)
			Amiel *et al.* (1999)
Liver	Rat hepatocytes	Poly-l-lactic acid materials	Cusick *et al.* (1997)
	Porcine hepatocytes	Bioartificial liver support system; microencapsulation/hollow fibres	Dixit and Gitnick (1998)
Muscle	Rat smooth muscle cells	Polyglycolic acid fibres	Kim *et al.* (1998)
Nerve	Rat and Human Schwann cells	Gel matrix in a Polyacrylonitrile/ Polyvinylchloride conduit.	Heath and Rutkowski (1998)
Cornea	Human epithelial cells	Polyglycolic acid matrix	
		Collagen and fibroblasts	Germain *et al.* (1999)
Pancreas	Mouse B cell line MIN6 plus human insulin promoter gene	Microencapsulation in 3 layerbeads. Agarose/Polystyrene sulfonic acid/carboxymethylcellulose	Kawakami *et al.* (1997)
	Canine islets of Langerhans	Microencapsulation in agarose	Tashiro *et al.* (1997)
	Porcine islets of Langerhans	Hollow fibres/hydrogel	Delaunay *et al.* (1998)
	Rat islets of Langerhans	Microencapsulation	Zekorn *et al.* (1999)

genotypically identical daughter cells). Stem cells develop via a 'committed progenitor stage' to a terminally differentiated cell. They are multi-potent, i.e. able to develop into a wide range of tissues and organs, but only fertilized germ cells are toti-potent, i.e. able to give rise to all cell tissues in the body. Control of the development of stem cells into the required tissue, or to stimulate quiescent 'committed progenitor cells' of the required tissue, with the relevant growth factors and hormones would allow the most effective cell therapy possible. This approach is causing some ethical controversy as the most suitable source of stem cells is to clone from the human embryo. The technique is to extract the genetic material from an adult patient needing transplantation, introduce it into a human egg with its nucleus removed, grow the embryo *in vitro* for 8 divisions until stem cells can be treated with growth factors to form the required tissue (e.g. Pancreas, nerve etc.).

REFERENCES

Aebischer P *et al.* (1986) A bioartificial parathyroid. *Transactions of the American Society for Artificial Internal Organs* 32: 134–137.

Aebischer P (1997) Polymer encapsulated xenogenec cell lines: a novel approach for gene therapy In: Carrondo M (ed.) *Animal Cell Technology:From vaccines to genetic medicine*, pp. 29–31. Kluwer Academic, Dordrecht.

Amiel GE & Atala A (1999) Current and future modalities for functional renal replacement. *Urol. Clin. North Am.* 26: 235–246.

Atala A (1998) Tissue engineering in urologic surgery. *Urol. Clin. North. Am.* 25: 39–50.

Auger FA, Rouabhia M, Goulet F, Berthod F, Moulin V & Germain L (1998) Tissue engineered human skin substitutes developed from collagen populated hydrated gels: clinical and fundamental applications. *Med. Biol. Eng. Comput.* 36: 801–812

Bader A, Knop E, Kern A, Boker K, Fruhaul N, Crome O, Esselman H, Pape C, Kempka G & Searing KF (1996) 3D co-culture of hepatic sinusoidal cells with primary hepatocytes – design of an organotypical node. *Experimental Cell Research*, 226(1): 223–233.

Battle T, *et. al.* (1999) Progressive maturation resistance to microcystin-LR cytotoxicity in two different hepatospheroidal models. *Cell Biology and Toxicology* 15: 3–12.

Becker JL, Prewett TL, Spaulding GF & Goodwin TJ (1993) Three-dimensional growth and differentiation of ovarian tumour cell lines in high aspect rotating wall vessel. Morphologic and embryonic considerations. *Journal of Cellular Biochemistry* 51: 283.

Bentine-Ley U, Pedersen B, Lindenberg S, Falck Larsen J, Hamberger L & Horn T (1994) Isolation and culture of human endometrial cells in a three-dimensional culture system. *Journal of Reproduction and Fertility* 101: 327–332.

Bick R. J, Snuggs M. B, Poindexter B. J, Buja LM & Van Winkle WB (1998) Physical, contractile and calcium handling properties of neonatal cardiac myocytes cultured on different matrices. *Cell Adhesion Communications* 6: 301–310.

Bing RJ, Binder T, Pataricza J, Kibira S & Narayan KS (1991) The use of micro-carrier beads in the production of endothelium-derived relaxing factor by freshly harvested endothelial cells. *Tissue & Cell* 23: 151–159.

Boxberger HJ, Sessler MJ, Maetzel B & Meyer TF (1993) Highly polarized primary epithelial cells from human nasopharynx grown as spheroid-like vesicles. *European Journal of Cellular Biology* 62: 140–151.

Bruns J, Kersten P, Lierse W, Weiss A & Sibermann M (1994) The *in vitro* influence of different culture conditions on the potential of sheep rib perichondrium to form hyaline-like cartilage. *Virchows Archiv* 424: 169–175.

Burt AM & McGrouther DA (1992) The production and use of skin cells cultures in therapeutic situations In: *Animal Cell Biotechnology*, Vol. 5, pp. 151–168. Academic Press, London.

Choi RS, Riegler M, Pothoulakis C, Kim BS, Mooney D, Vacanti M & Vacanti JP (1998) Studies of brush border enzymes, basement membrane components and electrophysiology of tissue engineered neointestine. *Journal of Pediatric Surgery* 33: 991–996.

Colton CK (1999) Engineering a bioartificial kidney. *Nature Biotechnology* 17: 421–422.

Cusick RA, Lee H, Sano K, Pollock JM, Utsonomiya H, Ma PX, Langer R, & Vacanti J.P (1997) The effect of donor and recipient age on engraftment of tissue engineered liver, *Journal of Pediatric Surgery* 32: 357–360.

Davies PF & Kerr C (1982) Co-cultivation of vascular endothelial and smooth muscle cells using microcarrier techniques. *Experimental Cell Research* 141: 455–459.

Davis MW, & Vacanti JP (1996) Toward development of an implantable tissue engineered liver. *Biomaterials* 17: 365–372.

Delaunay C, Darquay S, Honiger J, Capron F, Rouault, C & Reach G (1998) Glucose-insulin kinetics of bioartificial pancreas made of an AN 69 hydrogel hollow fiber containing porcine islets and implanted in diabetic mice. *Artificial Organs* 22: 291–299.

Dixit V & Gitnick G (1998) The bioartificial liver: state of the art. *European Journal of Surgery* 582: 71–76.

Ferber D (1999) Lab-grown organs begin to take shape. *Science* 284: 422–423.

Folkman J & Moscona A (1978) Role of cell shape in growth control. *Nature (London)* 273: 345–349.

Fox JL (1998) FDA seeks 'comfort factors' before removing hold on porcine xeno-transplantation trials (news item). *Nature Biotechnology* 16: 224.

Freed LE, Vunjak-Novakovic G & Langer R (1993) Cultivation of cell–polymer cartilage implants in bioreactors. *Journal of Cellular Biochemistry* 51: 257.

Freshney RJ (1986) *Animal Cell Culture.* IRL Press, Oxford.

Funk KA, Liu CH, Wilson BW & Higgins RJ (1994) Avian embryonic brain reaggregate culture system. *Toxicology and Applied Pharmacology* 124: 149–158.

Gage FH (1998) Cell therapy. *Nature (Supplement)* 392: 18–24.

Germain L, Auger FA, Grandbois E, Guignard R, Giasson M, Biosjoly H & Guerin S.L (1999) Reconstructed human cornea produced *in vitro* by tissue engineering. 67: 140–147.

Glimelius B, Noring B, Nedermann T & Carlsson J (1988) Extracellular matrices in multicellular spheroids of human glioma origin: increased incorporation of proteoglycans and fibronectin as compared to monolayer cultures. *Acta Pathologica et Microbiologica Scandinavica* 96: 433–444.

Goodwin TJ, Jessup JM & Wolf DA (1992) Morphological differentiation of human colorectal carcinoma in rotating wall vessels. *In Vitro Cell Development Biology* 28A: 47–60.

Goodwin TJ, Schroeder WF, Wolf DA & Moyer MP (1993) Rotating-wall vessel coculture of small intestine as a prelude to tissue modeling: aspects of simulated microgravity. *Proceedings of Society for Experimental Biology and Medicine* 202: 181–192.

Griffiths JB (1990) Advances in animal cell immobilization technology. *Animal Cell Biotechnology* 4: 149–166.

Heath CA & Rutkowski GE (1998) The development of bioartificial nerve grafts for peripheral nerve regeneration. *Trends in Biotechnology* 16: 163–168.

Hildago IJ, Raub TJ & Borchardt RT (1989) Characterisation of the human colon carcinoma cell line (Caco-2) as a model system for intestinal epithelial permeability. *Gastroenterology* 96: 736–749.

Humes DH, Buffington DA, Mackay SM, Funke AJ & Weitzel WF (1999) Replacement of renal function in uremic animals with a tissue-engineered kidney. *Nature Biotechnology* 17: 451–455.

Ijima H, Matsushita T, Nakazawa N, Koyama S, Gion T, Shirabe K, Shimada M, Takenaka K, Sugimachi K & Funatsu K (1997) Spheroid formation of primary dog hepatocytes using polyurethane foam and its application to hybrid artificial liver. In Carrondo M (ed.) *Animal Cell Technology From vaccines to genetic medicine*, pp. 577–583. Kluwer Academic, Dordrecht.

Joseph JM et al.(1994) Transplantation of encapsulated chromaffin in the sheep subarachnoid space: a preclinical study for treatment of cancer pain. *Cell Transplantation,* 3: 355–364.

Kanda K & Matsuda T (1994) Mechanical stress-induced orientation and ultrastructural change of smooth muscle cells cultured in three-dimensional collagen lattices. *Cell Transplantation* 3: 481–492.

Kaihara S, Kim SS, Benvenuto M, Choi R, Kim BS, Mooney D, Tanaka K & Vacanti JP (1999) Succesful anastamosis between tissue-engineered intestine and native small bowel. *Transplantation* 67: 241–245.

Kawakami Y, Inoue K, Hayashi H, Wang WJ, Setoyama H, Gu YJ, Imamwa M, Iwata H, Ikada Y, Nozawa M & Miyazaki J (1997) Subcutaneous xenotransplantation of hybrid artificial pancreas encapsulating B cell line (MIN6): functional and histological study. *Cell Transplant* 6: 541–545.

Kim BS, Putnam AJ, Kulik TJ & Mooney DJ (1998) Optimising sending and culture methods to engineer smooth muscle tissue on biodegradable polymer matrices. *Biotechnology and Bioengineering* 57: 46–54.

Kim SS, Kaihara S, Benvenuto MS, Choi RS, Kim BS, Mooney DJ, Taylor GA & Vacanti JP (1999) Regenerative signals for intestinal epithelial organoid units transplanted on biodegradable polymer scaffolds for tissue engineering of small intestine. *Transplantation* 67: 227–233.

Kuriyama S, Nakano T, Yoshimura N, Ohuchi T, Moritera T & Honda Y (1992) Mass cultivation of human retinal pigment epithelial cells with microcarrier. *Ophthalmologica* 205: 89–95.

Leist CH, Meyer H-P & Fiechter A (1990) Potential and problems of animal cells in suspension culture. *Journal of Biotechnology* 15: 1–46.

La France ML & Armstrong DW (1999) Novel living skin replacement: biotherapy approach for wounded skin tissues. *Tissue Engineering* 5: 153–170.

L'Hereux N, Paquet S, Labbe R, Germain L & Auger FA (1998) A completely biological tissue-engineered human blood vessel. *FASEB Journal* J 12: 47–56.

Limat A, Breitkreutz D, Hunziker T, Klein CE, Nozer F, Fusenig NE & Braathen LR (1994) Outer root sheath (ORS) cells organize into epidermoid cyst-like spheroids when cultured inside Matrigel: a light-microscope and immunohistological comparison between human ORS cells and interfollicular keratinocytes. *Cell and Tissue Research* 275: 169–176.

Lindstrom A & Carlsson J (1993) Penetration and binding of epidermal growth factor–dextran conjugates in spheroids of human glioma origin. *Cancer Biotherapy* 8: 145–158.

Looby D & Griffiths JB (1989) Immobilization of animal cells in fixed and fluidized porous glass sphere reactors. In: Spier RE, Griffith JB, Stephenne J & Crooy PJ (eds) *Advances in Animal Cell Biology and Technology for Bioprocesses*, pp. 336–343.

Mariotti C, Askanas V & King Engel W (1993) New organotypic model to culture the entire fetal rat spinal cord. *Journal of Neuroscience Methods* 48: 157–167.

McCall E, Povey J & Dumonde DC (1981) The culture of vascular endothelial cells on microporous membranes. *Thrombosis Research* 24: 417–431.

Minuth MW, Sittinger, M & Kloth S (1998) Tissue engineering: generation of differentiated artificial tissues for biomedical applications. *Cell Tissue Research* 291: 1–11.

Moscona A (1961) Rotation-mediated histogenetic aggregation of dissociated cells. A quantifiable approach to cell interaction *in vitro*. *Experimental Cell Research* 22: 455–475.

Mueller–Kleiser W (1997) Three dimensional cell cultures: from molecular mechanisms to clinical applications. *American Journal of Physiology* 273: C1109-C1123.

Nagaki M, Kano T, Muto Y, Yamada T, Ohnishi H & Moriwaki H (1990) Effects of intraperitoneal transplanation of microcarrier-attached hepatocytes on D-galactosamine-induced acute liver failure in rats. *Gastroenterologia Japonica* 25: 78–87.

Niklason LE, Gao J, Abbott WM, Hirschi KK, Houser S, Marini R & Langer R (1999) Functional arteries grown *in vitro*. *Science* 284: 489–493.

Nugent HM, Rogers C & Edelman ER (1999) Endothelial implants inhibit intimal hyperplasia after porcine angioplasty. *Circulation Research* 84: 384–391.

Oberpinning F, Meng J, Yoo JJ & Ataea A (1999) De novo reconstitution of a functional mammalian urinary bladder by tissue engineering. *Nature Biotechnology* 17: 149–155.

Peshwa MV, Nyberg SL, Wu FJ, Amiot B, Cerra FB & Hu W-S (1994) A novel hepatocyte entrapment, hollow fiber bioreactor as a bio-artificial liver. In: *Animal Cell Biotechnology: Products of today, prospects for tomorrow*, pp. 273–277. Butterworth-Heinemann, Oxford,

Puelacher WC, Vacanti JP, Ferraro NF, Schloo B & Vacanti CA (1996) Femoral shaft reconstruction using tissue-engineered growth of bone. *International Journal of Oral Macillofax Surgery* 25: 223–22.

Rabito CA, Tchao R, Valentich J & Leighton J (1980) Effect of cell substratum interaction of hemicyst formation by MDCK cells. *In Vitro* 16: 461–468.

Reuveny S (1985) Microcarriers in cell culture: structure and applications. *Advances in Cell Culture* 4: 213–247.

Rheinwald JG & Green H (1975) Serial cultivation of strains of human epidermal keratinocytes: the formation of keratinizing colonies from single cells. *Cell* 6: 331–344.

Rodriguez A, Cao YL, Ibarra C, Pap S, Vacanti M, Eavey RD & Vancanti CA (1999) Charactersitics of cartilage engineered from human pediatric auricular cartilage. *Plastic Reconstrustion Surgury* 103: 1111–1119.

Rotmensch J, Whitlock JL, Culbertson S, Atcher RW & Schwartz JL (1994) Comparison of sensitivities of cells to X-ray therapy, chemotherapy, and isotope therapy using a tumor spheroid model. *Gynecologic Oncology* 55: 290–293.

Sautier JM, Nefussi JR & Forest N (1992) Mineralization and bone formation on microcarrier beads with isolated rat calvaria cell population. *Calcified Tissue International* 50: 527–532.

Scharp *DW et al.* (1994) Production of encapsulated human islets implanted without immunosuppression in patients with type I or type II diabetes in non-diabetic control subjects. *Diabetes* 43: 1167–1170.

Schleich, A (1967) Studies on aggregation of human ascites tumor cells. *European Journal of Cancer* 3: 243–246.

Shahar A, Mizrahi A, Reuveny S, Zinman T & Shainberg A (1985) Differentiation of myoblasts with nerve cells on microcarriers in culture. *Developments in Biological Standardization* 60: 263–268.

Silverman RP, Passaretti D, Huang W, Randolph MA & Yaremchuk MJ (1999) Injectable tissue-engineered cartilage using a fibrin glue polymer. *Plastic Reconstruction Surgery* 103: 1809–1818.

Stark HJ, Baur M, Breitkreutz D, Mirancea N & Fusenig NE (1999) Organotypic keratinocyte cocultures in defined medium with regular epidermal morphogenesis and differentiation *Journal of Investigative Dermatology* 112: 681–691.

Stoye JP, Le Tissier P, Takeuchi Y, Patience C & Weiss RA (1998) Endogenous retroviruses: a potential problem for xenotransplantation? *Annals of the New York Academy of Sciences* 862: 67–74.

Sussman M & Sussman RR (1961) Aggregative performance. *Experimental Cell Research* 8: 91–106.

Tashiro H, Iwata H, Warnock GL, Takagi T, Machida H, Ikada Y & Tsuji, O (1997) Characterization and transplantation of agarose microencapsulated canine islets of Langerhans. *Annals of Transplants* 2: 33–39.

Tong JZ, Sarrazin S, Cassio D, Gauthier F & Alvarez F (1994) Application of spheroid culture to human hepatocytes and maintenance of their differentiation. *Biologie Cellulaire* 81: 77–81.

Voss JU & Seibert H (1991) Microcarrier-attached rat hepatocytes as a xenobiotic-metabolizing system in cocultures. *Cell Biology and Toxicology* 7: 387–399.

Wakitani S, Goto T, Young RG, Mansour JM Goldberg VM & Caplan AI (1998) Repair of large full-thickness articular cartilage defects with allograft articular chodrocytes embedded in a collagen gel. *Tissue Engineering* 4: 429–444.

Weibezahn KF, Knedlitschek G, Dertinger H, Bier W, Schaller Th & Schubert K (1994) An *in vitro* tissue model using mechanically processed microstructures. Presented at the 41st ETCS Congress, Verona.

Westergaard N, Sonnewald U, Peterson SB & Schousboe A (1991) Characterization of microcarrier cultures of neurons and astrocytes from cerebral cortex and cerebellum. *Neurochemistry Research* 16: 919–923.

Zekorn TD, Horcher A, Siebers U, Federlin K & Bretzel RG (1999) Synergistic effect of microencapsulation and immunoalteration on islet allograft survival in bio-artificial pancreas. *Journal of Molecular Medicine* 77: 193–198.

5.5 GENE THERAPY AND RECOMBINANT CELLS

5.5A Recombinant DNA Cell Lines

OVERVIEW

Mammalian cells can be genetically modified by the direct introduction of DNA which is taken up by the cell and may be retained and expressed transiently in the cytoplasm. Alternatively, the DNA may be transported to the nucleus and integrated into the cellular genome. Transient expression is advantageous when there is a requirement for high levels of protein production over a rapid time scale. Expression levels of 1–10 mg L^{-1} of product have been reported for secreted proteins, corresponding to specific productivities of around 0.1–1 pg cell day^{-1}, by transfecting large numbers of cells and maintaining them under production conditions for 5–10 days (Wurm & Bernard 1999). Expression can be achieved by introducing DNA vectors which contain a viral origin of replication, into cells which already contain other viral sequences, e.g. COS or HEK293EBNA. Alternatively, cells can be infected with recombinant viruses based on vaccinia, adenovirus, Sindbis virus, Semliki Forest virus or bovine papillomavirus which are capable of autonomous replication and transmission to daughter cells.

In the case of stable gene transfer systems the level of expression depends, not only on the activity of the promoter used to express the gene but also on the number of copies of the gene of interest present in each cell. The copy number is dependent both on the method used to introduce the DNA and on whether the gene is amplified post-transfection: techniques exist for amplifying the vector sequences to over 2000 copies cell^{-1} (Crouse *et al.* 1983). The most widely used systems are based on dihydrofolate reductase (DHFR) (Kaufman *et al.* 1985) and on glutamine synthetase (GS) (Cockett *et al.* 1990). These systems rely on the fact that the cell amplifies DNA in order to overcome the toxic effect of the selective agent – methotrexate in the case of DHFR and methionine sulphoximine in the case of GS – resulting in an increase in copy number of the gene of interest along with that of the enzyme of approximately 10-fold per round of selection (Bebbington 1993).

A variety of different techniques exist for introducing the DNA based both on physical methods and on viral vectors. The key features of these different techniques are summarized below.

Cell and Tissue Culture for Medical Research, edited by A. Doyle and J.B. Griffiths.
© 2000 John Wiley & Sons, Ltd

DNA transformation/transfection

The original and most commonly used method for introducing genes into mammalian cells was to mix the DNA with $CaCl_2$ in the presence of HEPES buffer (Graham & van der Eb 1973; Wigler *et al.* 1977). The calcium phosphate-DNA complexes which are formed adsorb to the surface of the cell and are subsequently endocytosed. Many modifications exist, e.g. the addition of DEAE-dextran (McCutchan & Pagano 1968); the use of a modified buffer at a slightly higher pH (Chen & Okayama 1987) to allow the DNA-$CaPO_4$ complex to form more slowly; the use of chloroquine to prevent endosomal acidification (Luthman & Magnusson 1983); the use of glycerol or DMSO to osmotically shock cells (Sussman & Milman 1984; Lopato *et al.* 1984) and the use of butyrate (Gorman & Howard 1983). Strontium phosphate has been used as an alternative for cells such as bronchial epithelial cells which are sensitive to calcium (Brash *et al.* 1987).

Liposome-mediated transfection/lipofection

When DNA is mixed with unilamellar liposome vesicles (which have a net positive charge due to the presence of amine groups) they associate in such a way as to neutralize the negative charge of the DNA phosphate backbone. The lipid–DNA complex retains a net positive charge which can associate with the negatively charged cell membranes and allow the DNA to be delivered into the cell. The technique was developed by Felgner and colleagues (1987) but a variety of different lipid reagents are now available as commercial kits and these offer a technically simple, reproducible and efficient method for getting DNA into cells.

Electroporation

Cells can be permeabilized by the application of an electrical voltage across the membrane. The holes in the membrane which are produced are large enough to permit the entry of DNA or other large molecules into the cell if it is incubated in a DNA solution during permeabilization. The high voltages used can have a toxic effect on cells, efficiencies are low and normally only one copy of the vector enters any given cell. However, this method is particularly useful for non-adherent cell lines (Neumann *et al.* 1982; Chu *et al.* 1987).

Microinjection

Small volumes of DNA can be directly injected into cells, either the nucleus or cytoplasm, using fine glass capillary micropipettes (Graessman 1970; Capecchi 1980). This approach has the advantage of being very efficient and consequently is the method of choice if the number of recipient cells is limited (e.g. in the case of embryos) or if there is a mixture of cell types present. However, the procedure needs highly trained and experienced staff, is labour-intensive and requires specialized micromanipulation equipment. A modification of this method was developed by Yamamoto and Furusawa (1978) in which cells are incubated in culture medium in which DNA is present, then pricked with a needle so that the

DNA can enter the cells when the needle is withdrawn. A finely focused laser beam has also been used to perforate the cell in order to introduce DNA (Kurata *et al.* 1986).

Particle bombardment

DNA-coated gold particles have been introduced into cells, both *in vivo* and *in vitro*, using a controlled electric discharge to generate a shock wave which accelerates the particles at high speed into target cells. Transient expression was detected in liver, skin and muscle tissue in rats and mice, and both transient and stable expression was observed in a variety of cell lines (Yang *et al.* 1990).

Protoplast/spheroplast fusion

DNA which has been cloned into bacterial plasmids can be introduced into mammalian cells by direct fusion with bacterial protoplasts (Rassoulzadegan *et al.* 1982). These protoplasts (also known as spheroplasts) are produced by removing the cell wall with lysozyme but care must be taken to ensure that this digestion is complete otherwise the mammalian cells will rapidly be overgrown with bacteria. The protoplasts are then fused to the mammalian cells with polyethylene glycol in a method similar to that used for mammalian cell fusion.

Transferrinfection

In an attempt to find an alternative method which is less toxic to hematopoietic cells, Zenke *et al.* (1990) have investigated receptor-mediated endocytosis as a method for introducing DNA into cells. Transferrin-polycation conjugates efficiently bind DNA and have been used to introduce DNA into cells via the transferrin cycle in a process termed 'transferrinfection'. However, endocytosed DNA is trapped in intracellular vesicles and is readily destroyed by lysosomal action. The addition of replication-defective adenovirus during transfection can enhance the efficiency of gene transfer by disrupting the endosomes and allowing the DNA to get into the cytoplasm and subsequently into the nucleus (Wagner *et al.* 1992).

Viruses play a direct role in delivering DNA to cells by the use of vectors based on recombinant eukaryotic viruses. The viruses attach to receptors on the surface of the cell thus providing a very cell-specific system. Two of these commonly used viral vector systems based on adenoviral vectors and retroviral vectors are described in detail in subsequent sections. Adenoviral vectors have a broad host range, are stable and can be produced at high titre, which explains why they are of such interest as vectors for gene therapy. Retroviral vectors are also widely used although they infect non-dividing cells only and there is always a risk from the production of replication-competent retroviruses (RCR) due to recombination between retroviral vector and endogenous retroviral sequences. Their host range is limited by the choice of packaging cell line which determines the envelope protein of the viral vector. Other viral delivery systems include those based on the bovine papillomavirus type 1 (BPV-1). BPV vectors replicate episomally in the cell and can be present at up to 100 vector copies per cell (Di Maio 1987). However, they can also be found integrated into the chromosome in single form, tandem repeats or in rearranged or deleted form (Matthias *et al.* 1983). Less frequently used now are SV40 pseudovirions in which the SV40 DNA has been removed and plasmids of up to 5.4 kb are encapsidated and transmitted to the target cells by viral infection (Oppenheim 1993). More recently recombinant baculoviruses have been used for both stable and transient expression of recombinant proteins in mammalian cells (Condreay *et al.* 1999).

There are two main disadvantages to the current methods used for gene transfer. First, the DNA is integrated randomly within the chromosome of the host cell and may, therefore, integrate into another gene and abolish the activity. Alternatively, the DNA may integrate into a region of the genome where gene expression is shut down. Gene targeting techniques are being developed to overcome these limitations, and indeed the possibility of ablation of activity can be regarded positively and used to produce loss of function mutants (Mansour *et al.* 1988), e.g. the so-called 'knock-out' phenotype in mice (Robertson 1991).

The second disadvantage with most methods is that they are very inefficient. The proportion of cells which fail to take up DNA depends on the cell type and the method used, but with the notable exception of microinjection, often no more than 0.1% of cells express the DNA in a stable fashion. Because of this inefficiency of uptake, it is vital to have screening and selection systems, and since most genes cannot be selected for directly, these are usually based upon an antibiotic resistance gene encoded on the plasmid. Examples of such markers include the *neo* gene from the bacterial transposon Tn5 which encodes aminoglycoside phosphotransferase. Expression of this gene confers resistance to the antibiotic Geneticin (G418), an aminoglycoside which is similar to kanamycin and blocks protein synthesis (Jimenez & Davies 1980; Colbere-Garapin *et al.* 1981). Other selectable markers include the *hyg* gene from *E. coli* which encodes hygromycin-B-phosphotransferase and allows cells to grow in hygromycin B an aminocyclitol antibiotic which inhibits protein synthesis (Blochlinger & Diggelmann 1984); the *pac* gene from *Streptomyces alboniger* which allows cells to grow in the protein synthesis inhibitor puromycin (Vara *et al.* 1986); and the *E. coli gpt* gene which codes for xanthine-guanine phosphoribosyltransferase and can allow cells to grow in HAT medium (Mulligan & Berg 1981). Full details of these and other selection systems are described in Jaggar (1993).

REFERENCES

Bebbington C (1993) Selection for gene amplification in CHO cells. In: Doyle A & Griffiths J B (eds) *Cell & Tissue Culture: Laboratory Procedures*, 27D: 3.1–3.8, Wiley, Chichester.

Blochlinger K & Diggelmann H (1984) Hygromycin B phosphotransferase as a selectable marker for DNA transfer experiments with higher eukaryotic cells. *Molecular Cell Biology* 4: 2929–2931.

Brash D E, Reddel R R, Quanrad M, Yang K, Farrell M P, & Harris C C (1987) Strontium phosphate transfection of human cells in primary culture: stable expression of the simian virus 40 large T antigen gene in primary human bronchial epithelial cells. *Molecular Cell Biology* 7: 2031–2034.

Capecchi M R (1980) High efficiency transformation by direct microinjection of DNA into cultured mammalian cells. *Cell* 22: 479–488.

Chen C & Okayama H (1987) High efficiency transformation of mammalian cells by plasmid DNA. *Molecular Cell Biology* 7: 2745–2752.

Chu G, Hayakawa H & Berg P (1987) Electroporation for the efficient transfection of mammalian cells with DNA. *Nucleic Acids Research* 15: 1311–1326.

Cockett M I, Bebbington C R & Yarranton G T (1990) High level expression of tissue inhibitor of metalloproteinases in Chinese hamster ovary cells using glutamine synthetase gene amplification. *Bio/Technology* 8: 662–667.

Colbere-Garapin F, Horodniceanu F, Kourilsky P & Garapin A-C (1981) A new dominant hybrid selective marker for higher eukaryotic cells. *Journal of Molecular Biology* 150: 1–14.

Condreay, J P, Witherspoon S M, Clay W C & Kost T A (1999) Transient and stable gene expression in mammalian cells transduced with a recombinant baculovirus vector. *Proceedings of the National Academy of Sciences of the USA* 96: 127–132.

Crouse G F, McEwan R N & Pearson M L (1983) Expression and amplification of engineered mouse dihydrofolate reductase minigenes. *Molecular and Cellular Biology* 3: 257–266.

Di Maio D (1987) Papilloma virus cloning vectors. In: Salzman N P & Howley P M (eds) *The Papovaviridae 2. The Papillomaviruses*, pp. 293–319 Plenum Press, New York.

Felgner P L, Gadek T R, Holm M, Roman R, Chan H W, Wenz M, Northrop J P, Ringold G M & Danielson M (1987) Lipofection: A highly efficient, lipid-mediated DNA transfection procedure. *Proceedings of the National Academy of Sciences of the USA* 84: 7413–7417.

Gorman C M & Howard B H (1983) Expression of recombinant plasmids in mammalian cells is enhanced by sodium butyrate. *Nucleic Acids Research* 11: 7631–7648.

Graessman O A (1970) Mikrochirurgische Zellkerntransplantation bei Saeugetierzellen. *Experimental Cell Research* 60: 373–382.

Graham F L & van der Eb A J (1973) A new technique for the assay of infectivity of human adenovirus 5 DNA. *Virology* 52: 456–467.

Jaggar R T (1993) Drug selection including stable gene transfer. In: Doyle A & Griffiths J B (eds) *Cell & Tissue Culture: Laboratory Procedures*, 27D: 2.1–2.11, Wiley, Chichester.

Jimenez A & Davies J (1980) Expression of a transposable antibiotic resistance element in Saccharomyces. *Nature* 287: 869–871.

Kaufman R J, Wasley L C, Spiliotes A J, Gossels S D, Latt S A, Larsen G A & Kay R M (1985) Coamplification and coexpression of human tissue-type plasminogen activator and murine dihydrofolate reductase sequences in Chinese hamster ovary cells. *Molecular and Cellular Biology* 5: 1750–1759.

Kurata S-I, Tsukakoshi M, Kasuya T & Ikawa Y (1986) The laser method for efficient introduction of foreign DNA into cultured cells. *Experimental Cell Research* 162: 372–378.

Lopato M A, Cleveland D W & Sollner-Webb B (1984) High level transient expression of a chloramphenicol acetyl transferase gene by DEAE-dextran mediated transfection coupled with a dimethyl sulfoxide or glycerol shock treatment. *Nucleic Acids Research* 12: 5707–5717.

Luthman H & Magnusson G (1983) High efficiency polyoma DNA transfection of chloroquine treated cells. *Nucleic Acids Research* 11: 1295–1308.

Mansour S L, Thomas K R & Capecchi M R (1988) Disruption of the proto-oncogene int-2 in mouse embryo-derived stem cells: a general strategy for targeting mutations to non-selectable genes. *Nature* 336: 348–352.

Matthias P D, Bernard H U, Scott A, Brady G, Hashimoto-Gotoh T & Schuetz G (1983) A bovine papilloma virus vector with a dominant resistance marker replicates extrachromosomally in mouse and E coli cells. *The EMBO Journal* 2: 1487–1492.

McCutchan J H & Pagano J S (1968) Enhancement of the infectivity of simian virus 40 deoxyribonucleic acid with diethyl-aminoethyl dextran. *Journal of the National Cancer Institute* 41: 351–357.

Mulligan R C & Berg P (1981) Selection for animal cells that express the Escherichia coli gene coding for xanthine-guanine phosphoribosyltransferase. *Proceedings of the National Academy of Sciences of the USA* 78: 2072–2076.

Neumann E, Schaefer-Ridder M, Wang Y & Hofschneider P H (1982) Gene transfer into mouse lyoma cells by electroporation in high electric fields. *The EMBO Journal* 1: 841–845.

Oppenheim A (1993) SV40 pseudovirions. In: Doyle A & Griffiths J B (eds) *Cell & Tissue Culture: Laboratory Procedures,* 27B: 7.1–7.11, Wiley, Chichester.

Rassoulzadegan M, Binetruy B and Cuzin F (1982) High frequency of gene transfer after fusion between bacteria and eukaryotic cells. *Nature* 295: 257–259.

Robertson E J (1991) Using embryonic stem cells to introduce mutations into the mouse germ line. *Biology of Reproduction* 44: 238–245.

Sussman D J & Milman G (1984) Short term high efficiency expression of transfected DNA. *Molecular and Cellular Biology* 4: 1641–1643.

Vara J A, Portela A, Ortin J & Jimenez A (1986) Expression in mammalian cells of a gene from Streptomyces alboniger conferring puromycin resistance. *Nucleic Acids Research* 14: 4617–4624.

Wagner E, Zatloukal K, Cotton M, Kirlappos H, Mechtler K, Curiel D T & Birnstiel M L (1992) Coupling of adenovirus to transferrin-polylysine/DNA complexes greatly enhances receptor-mediated gene delivery and expression of transfected genes. *Proceedings of the National Academy of Sciences of the USA* 89: 6099–6103.

Wigler M, Silverstein S, Lee L-S, Pellicer A, Cheng Y-C & Axel R (1977) Transfer of purified herpes virus thymidine kinase gene to cultured mouse cells. *Cell* 11: 223–232.

Wurm F & Bernard A (1999) Large-scale transient expression in mammalian cells for recombinant protein production. *Current Opinion in Biotechnology* 10: 156–159.

Yamamoto F & Furusawa M (1978) A simple microinjection technique not employing a micromanipulator. *Experimental Cell Research* 117: 441–445.

Yang N-S, Burkholder J, Roberts B, Martinell B & McCabe D (1990) In vivo and in vitro gene transfer to mammalian somatic cells by particle bombardment. *Proceedings of the National Academy of Sciences of the USA* 87: 9568–9572.

Zenke M, Steinlein P, Wagner E, Cotton M, Beug H & Birnstiel M L (1990) Receptor-mediated endocytosis of transferrin-polycation conjugates: an efficient way to introduce DNA into hematopoietic cells. *Proceedings of the National Academy of Sciences of the USA* 87: 3655–3659.

5.5B Adenovirus Vectors

OVERVIEW

Viruses are the subject of much investigation as vectors for the treatment of acquired and inherited diseases and in the transduction of genes for basic research. The retroviruses, adenoviruses, herpesviruses and adeno-associated viruses are being developed for gene transfer and each of these has their own advantages and disadvantages making them suited to specific gene transfer applications. Table 5.5B.1 gives a summary of the advantages and disadvantages of each vector.

Viral vectors is an area with very considerable range and depth and a complete review of the area is beyond the scope of this section. The two most used viral vectors at present are those derived from retroviruses and adenoviruses and these are dealt with in considerable detail within the chapter. For more detailed information on herpesviruses and adeno-associated viruses see Robbins *et al.* (1998) and references therein.

The study of adenoviruses as transforming agents and as regulators of cellular gene expression has, to a large extent, defined the function and properties of the adenovirus genome. This has subsequently permitted the development of the virus as a vector for gene transfer.

THE ADENOVIRUS GENOME

The adenovirus genome comprises a double-stranded linear DNA molecule of 34–36 kb (depending on the serotype), which is transcribed in both directions.

Table 5.5B.1 Summary comparison of viral vectors

Virus	Advantages	Disadvantages
Retrovirus	• Relatively simple genetics • Low immunogenic response • Integrative – maintenance of expression of gene of interest	• Low viral titres 10^6 ml^{-1} • Only infects dividing cells. • Integrative – activate or damage cellular genes
Adenovirus	• Relatively simple genetics • Can accommodate large inserts – allows controllable expression • High viral titres • Infect non-dividing cells	• Non-integrative • Transient expression • Immunogenic responses observed
Herpes-simplex	• Broad host range • Infect specialized cell types e.g. neurones	• Complicated molecular biology • Toxicity
Adeno-associated	• Site-specific integration	• Limited host range

During the adenovirus lytic cycle there is an ordered expression of viral genes. Early genes are expressed immediately after infection and continue to be expressed throughout the cycle, whilst late genes are expressed during the later stages of infection. The late phase is marked by the onset of viral replication and is required for manufacture of the viral capsid proteins. Figure 5.5B.1 illustrates the adenovirus transcription map.

Early genes

The E1 region

The E1 region consists of two genes, namely E1A and E1B. E1A is the first gene to be transcribed after viral infection. It encodes two products which differ only by a few central amino acids due to differential splicing of the mRNA. The E1A products are transcriptional activators of a variety of both viral and cellular promoters (Berk 1986) and as such are absolutely required for viral replication

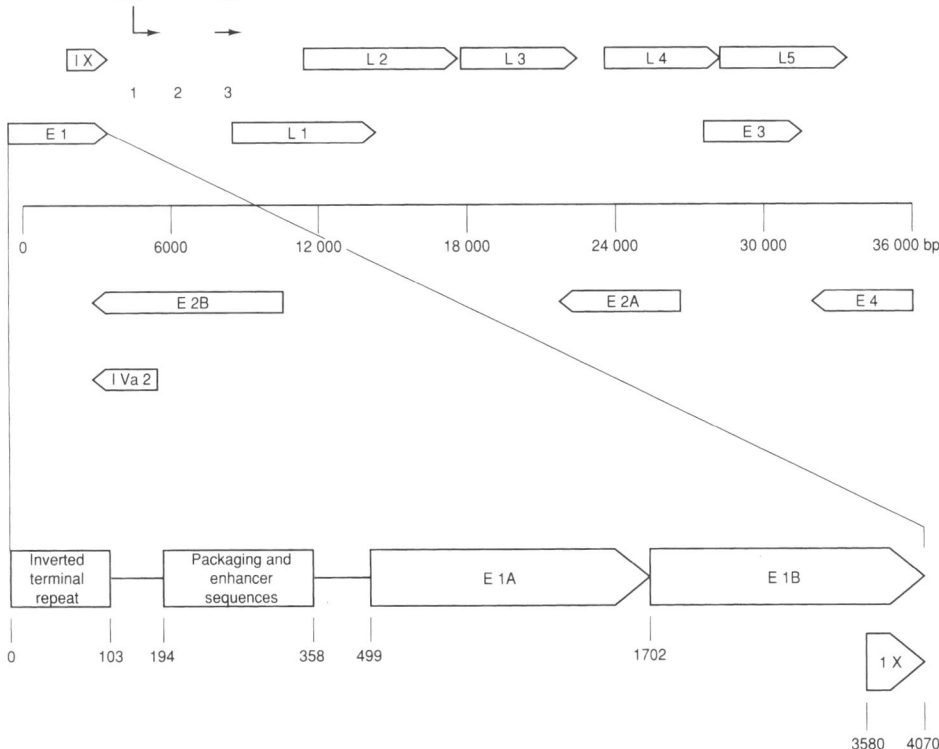

Figure 5.5B.1 Adenovirus type 5 genome: Transcriptional pattern. The locations of the major early (E) and late (L) transcription units are shown. The splicing patterns of the mRNAs are not shown. The structural proteins IX and Iva2 are expressed at late times in infection. 1, 2 and 3 represent the tripartite leader sequences and MLP the major late promoter. VA encodes an RNA which prevents downregulation of translation.

and maintenance. The regulatory elements of E1A are located upstream of the mRNA cap site. These include an inverted terminal repeat required for viral replication (Hay & McDougall 1986), an enhancer sequence important for regulation of E1A and the other early genes, E1B, E2, E3 and E4 (Hearing & Shenk 1986), and a packaging signal required for encapsidation of viral DNA (Grable & Hearing 1990). There are two main E1B proteins encoded early in infection, a 19 kDa protein which is important for protection of viral DNA from degradation by cellular nucleases during infection (Subramanian et al. 1984), and a 55 kDa protein which is required for shut-off of host protein synthesis (Babiss & Ginsberg 1984), and for accumulation of late viral mRNA which probably leads to the downregulation of translation of cellular mRNA seen at late times (Babiss et al. 1985). It is evident, therefore, that the E1B region influences the progression of the adenovirus cycle to late events (see Figure 5.5B.1)

The E2 region

The E2 region encodes proteins which are important for viral replication, namely an 80 kDa protein which attaches to both ends of the viral genome, acting as a primer for initiation of DNA replication, a 72 kDa DNA-binding protein, and a 140 kDa DNA polymerase (review: Stillman 1985).

The E3 region

The E3 region is thought to help the virus evade the immune response, but is otherwise totally dispensable for replication. A 19 kDa E3 protein downregulates expression of class I MHC molecules in cells infected or transformed by most serotypes of adenovirus (Burgert & Kvist 1985). In addition, a 14.5 kDa E3 protein inhibits the cytotoxic activity of tumour necrosis factor (TNF) (Gooding et al. 1988).

The E4 region

The E4 region has seven open reading frames and may encode up to 16 proteins. Products of this region are necessary for viral DNA replication, late mRNA synthesis and shut-off of host protein synthesis (Halbert et al. 1985), and, in addition, for virus assembly (Falgout & Ketner 1987). An E4 34 kDa protein can be found as a complex with the E1B 55 kDa protein (Sarnow et al. 1984). The fact that these two proteins have many of the same properties has led to the suggestion that the early to late progression of the lytic cycle is regulated by this E1B–E4 protein complex.

Late genes

The late phase of the lytic cycle is marked by a decrease in host protein translation and the onset of viral replication. It is at this time that the late mRNAs encoding the capsid proteins are transcribed as a result of an increase in the activity of the major late promoter (MLP). There are five families of late mRNAs.

Variations within each family occur as a result of differential splicing. All the mRNAs are spliced onto the tripartite leader, which is thought to be important for efficient translation of these messages (Berkner & Sharp 1985).

Late stage events also include a change in early promoter activities leading to production of protein IX (pIX) and protein Iva2 (Iva2). pIX is absolutely required for the packaging of full-length genomic recombinants and increases the thermostability of the virion (Ghosh-Choudhury *et al.* 1987). Another sequence which is important at late stages is the VA sequence encoding RNA polymerase III which is required for the translation of adenovirus messages.

ADENOVIRUSES AS EXPRESSION VECTORS

In the last decade, adenovirus vectors have been extensively used for the expression of a large number of foreign genes (Haj-Ahmad & Graham 1986; Graham *et al.* 1988; Prevec *et al.* 1989; Yoo *et al.* 1992). The biology and genome structure of adenoviruses offer a number of advantages for the development of recombinant vectors. Thus high-level expression of heterologous genes may be obtained through use of adenovirus promoters, whilst shut-off of host protein synthesis at late stages allows over-expression of the foreign protein. In addition, adenoviruses have a broad host range, are highly stable, and can be amplified to high titres, typically 10^{10}–10^{14} PFU ml^{-1}. It is these properties which have generated interest in the use of adenoviruses as vectors for gene therapy.

Vector development has been carried out with adenovirus types 2, 4 and 7; however, the use of adenovirus type 5 is more common. Adenovirus type 5 has been the most extensively studied of the adenovirus serotypes, both biochemically and genetically. DNA packaging constraints allow insertion of a maximum of 2 kb of foreign DNA into the wild-type adenovirus genome. Expression of larger sequences of foreign DNA requires that deletions first be made in the adenovirus vector. The E3 gene is dispensable for virus growth, and therefore foreign genes can be inserted in place of the E3 gene without affecting the ability of the virus to propagate. Most of the E1 region may also be deleted. This region is necessary for viral growth, but since the E1A and E1B proteins are *trans*-acting, they can be provided exogenously if the virus is grown in helper cells which express these proteins. 293 cells are human kidney cells which have been transformed using the E1 region and as a consequence may be used for the growth of E1-deletions in E1 must not encompass the region encoding pIX. In addition the inverted terminal repeat, packaging and enhancer regions are required in *cis* and must be maintained within the virus genome. With these deletions it is now theoretically possible to insert up to 7.5 kb of heterologous sequences into the adenovirus genome to produce a helper-independent recombinant adenovirus, although to date the largest reported insert is of 6 kb (Ghosh-Choudhury *et al.* 1987).

There are two main problems associated with the use of adenoviral vectors in gene therapy. The first is the reduced duration of gene expression due to an immune response. The most successful approach to in the generation of more-attenuated adenovirus is the deletion of all or some of the E4 open reading frames. However, there is evidence of reduced long-term gene expression in these modified vectors (Wang & Finer 1996). The removal of all the viral coding sequences other than the terminal repeats, needed for viral replication, produces more defective viruses. These adenoviral vectors are described as being 'gutted' or 'gutless' and can be grown to high titres (Parks *et al.* 1996; parks & Graham 1997; Wang *et al.* 1995). However, purification from the gutted vector from the helper virus is difficult.

VECTOR CONSTRUCTION AND RESCUE

The large size of the adenovirus genome means that there are few unique restriction sites, which makes *in vitro* genetic manipulation of the virus difficult. Nevertheless a number of adenoviruses with genomes with reduced numbers of restriction sites have been isolated (Jones & Shenk 1979). The most convenient way of rescuing adenovirus recombinants is by homologous recombination between the wild-type adenovirus genome and a bacterial plasmid (early replacement plasmid) encoding either a modified E1 or E3 region. A large part of these adenovirus sequences can be substituted but, as previously described, it is essential to maintain the inverted terminal repeat, packaging and enhancer regions and coding region for protein IX. Foreign gene expression can be driven by the resident E1 promoter or by a duplicated copy of the MLP. Alternatively, a heterologous promoter can be inserted, allowing strong or tissue-specific expression. For more details on the design of recombinant adenoviruses the reader is referred to Berkner (1992).

Once the replacement plasmid has been constructed there are three possible ways of generating recombinant adenovirus stocks.

First, Ad309 DNA can be rescued by homologous recombination with the replacement plasmid. The Ad309 adenovirus 5 mutant was selected because it has single ClaI and XbaI restriction sites at positions 916 bp and 1336 bp in the early region (Jones & Shenk 1979). Ad309 DNA can therefore be dually digested with ClaI and XbaI and then co-transfected with the early replacement plasmid (Figure 5.5B.2(a). A problem with this method, however, is that any residual uncut viral DNA will replicate after transfection to give a high-wild-type background.

Second, a bacterial plasmid such as PJM17 can be used (Graham *et al.* 1988). PJM17 contains the entire adenovirus 5 genome with the prokaryotic vector pBRX inserted into the early region. As a result of the pBRX insert, PJM17 exceeds the packaging capacity of the adenovirus capsid and only upon recombination between the PJM17 and the early replacement plasmid can infectious virus result (Figure 5.5B.2(b)).

A third possible approach is to ligate the early replacement plasmid directly onto the adenovirus genome via appropriate restriction sites. Again this method requires a mutant adenovirus genome with single restriction sites in the E1 or E3 region where the foreign gene is to be inserted (Figure 5.5B.2(c)).

Figure 5.5B.3 outlines the stages and the timescale for the rescue of recombinant adenoviruses.

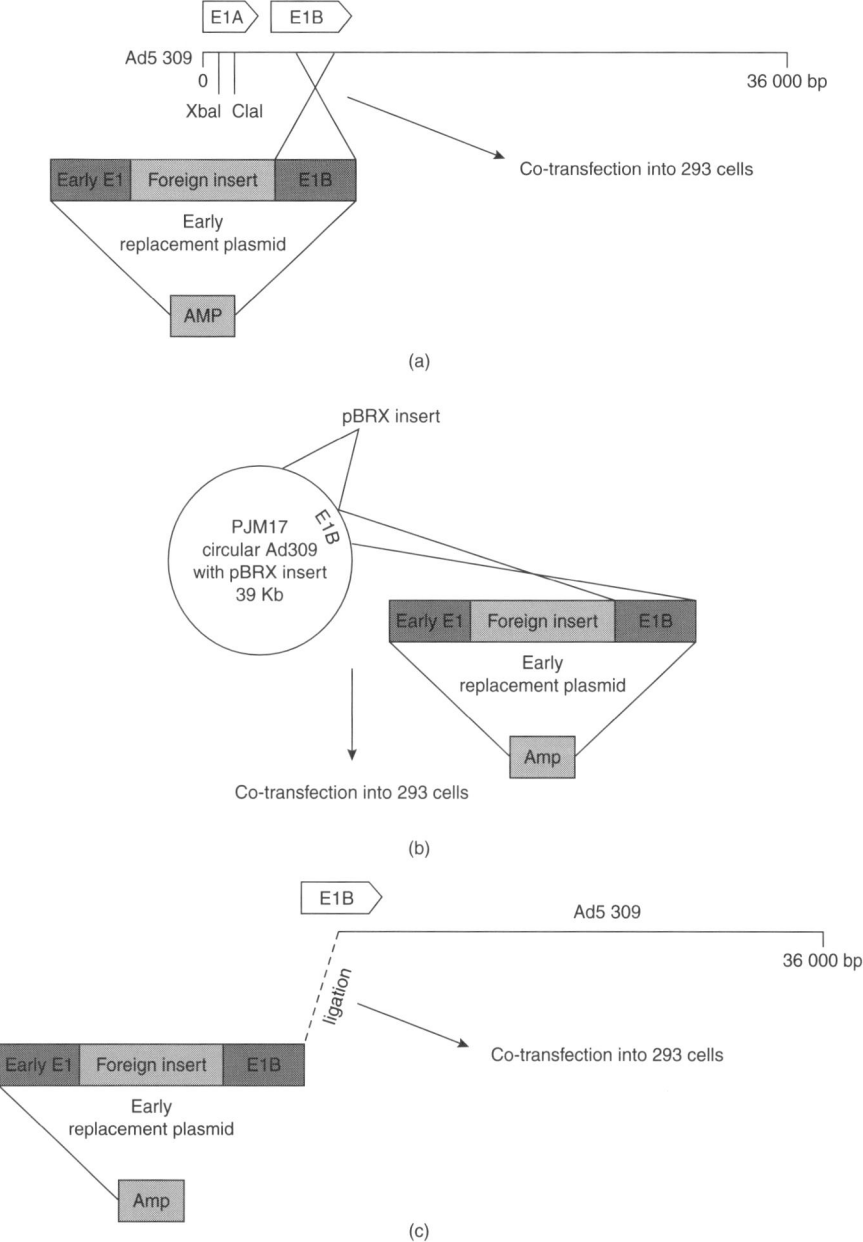

Figure 5.5B.2 Methods for the rescue of recombinant adenoviruses.

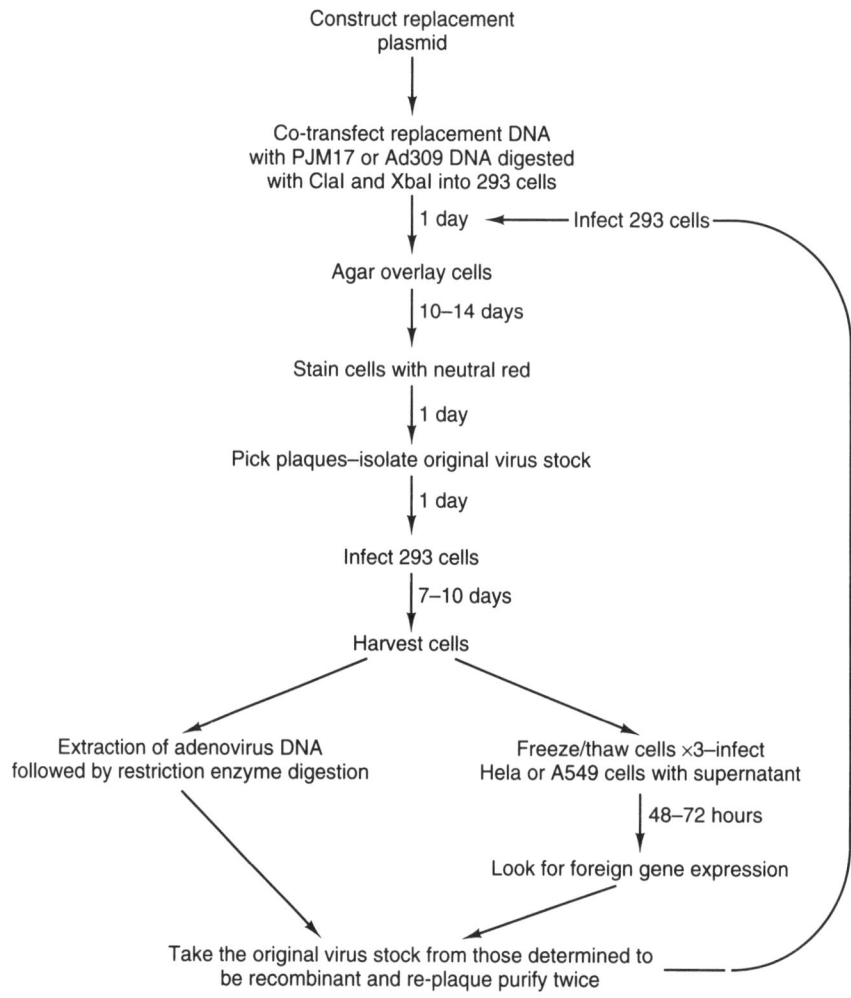

Figure 5.5B.3 Flow chart outlining the stages and the timescale when rescuing recombinant adenoviruses.

PROCEDURE: INFECTION OF CELLS WITH ADENOVIRUS AND ISOLATION OF ADENOVIRUS DNA FROM INFECTED CELLS

In order to rescue recombinant adenoviruses using the Ad309 DNA, it is necessary to grow up a large stock of virus from which the DNA is then isolated. Work using wild-type and recombinant adenoviruses should be done in a tissue culture cabinet. The category of hood used should depend on the risk assessment of the individual project.

Materials and equipment

- Tissue culture medium: DMEM-HEPES buffered pH 7.4, 2 mM glutamine, 8% foetal bovine serum (FBS)
- Adenovirus 309 stock (Jones & Shenk 1979)
- Twenty (10 cm dishes of 293 cells (ATCC and ECACC)
- Arcton (1,1,2-trichlorotrifluoroethane)
- Caesium chloride
- 0.02 M Tris pH 7.9, 1 mM EDTA
- Phenol (buffered with 10 mM Tris, pH 8.0)
- Chloroform/isoamyl alcohol (24 : 1)
- 5 M sodium acetate, pH 5.2
- 70% ethanol
- TE (10 mM Tris, 1 mM EDTA, pH 8.0)
- Cell scrapers
- Ultracentrifuge tubes for Beckman SW60 (11×61 mm)
- Dialysis tubing (boiled for 10 min in 2% (w/v) sodium bicarbonate, 1 mM EDTA, pH 8.0, followed by washing in distilled water and boiling for a further 10 min in 1 mM EDTA, pH 8.0)

Infection of 293 cells with virus

1. Split two confluent dishes of 293 cells between 20×10 cm dishes.
2. When the cells reach about 80% confluence, remove the medium and infect with adenovirus 309 stock by adding 100–200 µl stock virus solution of at least 10^9 PFU ml^{-1} (i.e. multiplicity of infection (MOI) > 100).
3. Place the cells in an incubator at 37°C. Rock the dishes every 20 min for 4 h, and then add fresh medium.

Isolation of high titre virus from infected 293 cells

1. After 5–7 days viral lytic infection should be apparent. Cells should appear larger and more rounded and in places may detach from the dish.
2. Harvest 20×10 cm dishes of infected cells by removing the medium and scraping the cells from the dish using a cell scraper.
3. Add 10 ml medium to each dish of detached cells and resuspend the cells in the medium.
4. Combine the cell suspension, divide between 4×50 ml centrifuge tubes and centrifuge to pellet the cells. Resuspend each cell pellet in 10 ml medium.
5. Freeze and thaw the cell suspension three times by placing the suspension at –70°C until frozen and then at 37°C. This breaks open the cells, resulting in release of the virus.
6. Add an equal volume of Arcton in a fume cupboard and mix with a pipette. Arcton is a fluorocarbon which solubilizes the cell membranes. Centrifuge at 400 g for 4 min.
7. Carefully remove the top layer, avoiding the interface, and place in a fresh universal. Accurately measure the volume and add 0.5 g caesium chloride/ml. Mix by inversion until the caesium chloride is completely dissolved.

8. Place in 61×11 mm ultracentrifuge tubes (Beckman), filling each tube to 2–3 mm from the top and balancing the tubes exactly. Centrifuge in a swing-out rotor (e.g. Beckman SW60) at 45 000 rev min^{-1} (equivalent to 208 000 g) at 4°C overnight with the brake off.

9. The following day, remove the tubes from the centrifuge, keeping them in a vertical position. In the middle of each tube a white adenovirus band should be visible. A higher band may also be visible which consists of empty virion capsids.

   In a Class II tissue culture cabinet pierce the bottom of the tube with a 21 G needle and let the gradient drip from the hole into a waste container. When the viral band nears the bottom of the tube, collect approximately 1 ml into a microcentrifuge tube.

10. Place the pooled virus into boiled dialysis tubing and dialyze against 5 litre dialysis buffer (0.02 M Tris pH 7.9, 1 mM EDTA) at 4°C. After 2 h replace the dialysis buffer with another 5 litre fresh buffer. Dialyse overnight at 4°C.

Isolation of viral DNA

1. The following day, remove the dialysed virus from the dialysis tubing and place 500 μl aliquots in microcentrifuge tubes.

2. Extract once with phenol buffered with 10 mM Tris pH 8.0 by adding an equal volume of phenol, vortexing and centrifuging at 13 000 rev min^{-1} for 5 min in a microcentrifuge. Remove the top layer and place in fresh microcentrifuge tubes.

3. Add equal volumes of chloroform/isoamyl alcohol (24 : 1), vortex and centrifuge at 13 000 rev min^{-1} for 5 min in a microcentrifuge.

4. Remove 400 μl of the top layer from each tube and place in a fresh micro-centrifuge tube. Add 40 μl 3 M sodium acetate pH 5.2 and 1 ml absolute ethanol. Place at –20°C for 30 min and centrifuge at 13 000 rev min^{-1} for 10 min in a microcentrifuge.

5. After removal of the ethanol with a Pasteur pipette, wash the precipitated DNA with 1 ml 70% ethanol and centrifuge at 13 000 rev min^{-1} for 2 min in a micro-centrifuge. Again remove the ethanol using a drawn-out Pasteur pipette so as not to disturb the DNA, and leave the pellet to dry in a sterile environment.

6. Resuspend each pellet in 100 μl TE. Measure the concentration of DNA by reading the absorbance of diluted aliquots at 260 nm. An optical density of 1 at 260 nm, with a 1 cm light path, is taken to equal 50 mg ml^{-1} DNA.

PROCEDURE: RESCUE OF ADENOVIRUS EARLY REPLACEMENT VECTORS

As mentioned in the discussion below there are three methods of rescuing recombinant adenoviruses, but only the two recombination methods will be given. The first involves co-transfection of the uncut early replacement plasmid and the Ad309 DNA which has been digested with XbaI and ClaI using standard molecular biology techniques (Maniatis et al. 1982). This ensures that the first 1336 bp of the left-hand end of the genome is removed. Nevertheless digestion should be verified by subjecting a small amount (2–5 µg) of the digested DNA to agarose gel electrophoresis. The amount of each DNA co-transfected can vary and is often down to the personal choice of the worker. Ideally several ratios of Ad5 309 to replacement plasmid should be used, for example 1 : 10, 1 : 5 and 1 : 1 respectively (values in micrograms). A total of 10–20 µg per 5 cm dish of 293 cells should be transfected (see below). In theory, the higher the ratio of replacement plasmid to Ad5 309 DNA, the greater the chance of obtaining recombinant plaques. One microgram of uncut Ad309 DNA should serve as a positive control and 1 µg of cut Ad309 DNA should be used as a negative control.

The second method of rescuing recombinant adenoviruses is to co-transfect uncut early replacement vector with an uncut plasmid encoding the whole Ad309 genome such as pJM17. Again the ratio of early replacement plasmid to pJM17 can vary from 1 : 1 to 10 : 1. A total of 10–20 µg per 5 cm dish of 293 cells should be transfected (see below). Control pJM17 should also be transfected as a negative control and 1 µg of uncut Ad309 DNA should serve as a positive control.

Calcium phosphate transfection and agar overlaying of transfected cells

The calcium phosphate transfection (Graham & van der Eb 1973) method works well for the transfection of adenovirus DNA and has been used efficiently to generate recombinant adenovirus stocks. Twenty-four hours after post-transfection, the cells are overlaid with agar. This prevents virus spread throughout the dish so that separate regions of adenovirus infection can be isolated individually.

Reagents and solutions

×2 HEPES-buffered saline solution

16.4 g l^{-1} NaCl, 11.9 g l^{-1} HEPES acid, 0.21 g l^{-1} NaHPO$_4$. Increase pH to exactly 7.05 with 5 M NaOH, filter sterilize through a 0.2 µm filter, and store at –20°C as 50 ml aliquots.

Phosphate-buffered saline (PBS, Oxoid)

PBS comprises 0.8 g l^{-1} NaCl, 0.02 g l^{-1} KCl, 0.115 g l^{-1} disodium hydrogen phosphate, and 0.02 g l^{-1} potassium dihydrogen phosphate, and has a pH of 7.3.

Additional materials and equipment

- 2.5 M CaCl$_2$ (Analar) made fresh. Filter-sterilize through 0.2 µm filter
- 10% glycerol (1 ml glycerol in 9 ml tissue culture medium)
- ×2 tissue culture medium (88 ml 2 × DMEM, 8 ml FBS, 4 ml 0.57 M NaHCO$_3$, 2 ml 0.2 M glutamine)
- 4% low melting point agar (4 g low melting point agar in 100 ml distilled water and autoclaved at 15 lb in^{-2}, 121°C, for 10 min)
- 293 cells (ATCC and ECACC)

1. Seed 293 cells 24–48 h prior to transfection onto 5 cm dishes. Use approximately 10% of cells from a 10 cm confluent dish to seed each 5 cm dish. Ideally on the day of transfection the cells should be 80% confluent. Two to four hours before transfection, refeed cells with 5 ml medium.
2. Ethanol precipitate the DNA to be transfected and air-dry. Resuspend the DNA to be transfected in a total of 225 μl distilled water. Add 25 μl 2.5 M CaCl$_2$.
3. Add 250 μl × 2 HEPES-buffered saline solution to a 5 ml plastic Bijoux. Then add the DNA/CaCl$_2$ solution dropwise with a Pasteur pipette. As the DNA/CaCl$_2$ is added, bubble air through the HEPES-buffered saline solution with a Pasteur pipette to produce a fine, cloudy precipitate.
4. Allow the complete formation of the precipitate at room temperature for 10–20 min. The precipitate should make the solution slightly cloudy; a very heavy, almost milky, precipitate should not be used.
5. Remove the medium from the 5 cm dish of cells and replace with 2 ml fresh medium. Add the precipitate in a dropwise fashion evenly around the dish.
6. Incubate the cells for 6 h.
7. Remove medium. Add 2 ml 10% glycerol solution carefully to the edge of the dish. Tilt the dish so that the glycerol solution covers the cell surface. Leave for 90–120 s. (Glycerol shocking is used when transfecting some cells, facilitating more efficient uptake of the precipitated DNA.)
8. Gently add 5 ml medium at the edge of the dish and remove immediately. Follow this with another two 5 ml washes with medium, being careful to add medium at the edge of the dish so as not to disrupt the cell layer. Finally, add another 5 ml medium and place the cells in the incubator overnight.
9. The following day overlay the cells with agar. Melt the low melting point agar in a microwave or a pan of boiling water with the cap loosened and then place in a 37°C water-bath. Place the × 2 medium in a 37°C water-bath. When both media are at 37°C, mix the two at a ratio of 1 : 1, avoiding the creation of air bubbles, and leave at 37°C before use. Remove medium and overlay each dish with 5 ml agar/medium mixture.

    It is important that the temperature of the mixture is close to 37°C when it is added to the cells: too cold and the agar will solidify upon pipetting; too hot (>39°C) and the cells will be killed.
10. When the agar has solidified, place the cells back in the incubator. Refeed the cells twice weekly by overlaying with 2 ml agar/medium.

Isolation of virus from cell sheet plaques

Viral replication within the 293 cells leads to local cell death, usually forming a hole in the cell sheet. Visible sites of adenovirus infection should be observed at 10–14 days post-transfection.

Where 1 μg uncut adenovirus DNA has been transfected you should expect a large number of plaques covering the cell sheet; very few plaques on this plate would suggest that the original transfection was not efficient.

XbaI and ClaI digested Ad309 and plasmid DNA co-transfection

Where 5 μg of cut Ad309 DNA alone has been transfected onto the 293 cells, the presence of plaques would indicate inefficient restriction enzyme digestion of the adenovirus DNA leading to a high background of wild-type plaques. If there are similar numbers of plaques on the dishes transfected with cut Ad309 DNA alone and dishes co-transfected with the cut Ad309 and early replacement DNA, this would suggest that the majority of the plaques are wild type. In such circumstances it would be preferable to repeat the restriction enzyme digestion of the Ad309 DNA and to repeat the co-transfection. However, if there are more plaques on the dishes where the adenovirus and early replacement DNA have been transfected together, individual plaques then need to be picked, and the virus from each plaque grown up.

PJM17 and plasmid DNA co-transfection

Where PJM17 DNA alone has been transfected the presence of plaques would indicate that the PHM17 DNA had internally recombined to produce a smaller plasmid which is able to be packaged into the adenovirus capsid. Internal recombination nevertheless occurs relatively infrequently and whre the PJM17 and early replacement DNA have been co-transfected it is not uncommon for most of the plaques present to be recombinant.

Neutral red staining

Whichever method is used, once plaques can be observed microscopically, it is best to stain the cells to allow easier identification of infected regions. To achieve this, neutral red stain is used which is selectively absorbed by the live cells and will not be taken up by the adenovirus-infected cells.

Additional materials and equipment

- Neutral red solution (3.3 g l⁻¹). Dilute 1 : 10 in PBS for use in PBS (Sigma)

1. Add 2 ml 10% neutral red solution to each dish, and return the cells back to the incubator for 2–4 h. Remove the stain and reincubate overnight.
2. The following day, the healthy cells will have taken up the stain, and plaques where cells have not taken up the red dye will be easily observed. With a wide bore Pasteur pipette, remove plugs of agar above each plaque and place each in a separate Bijou containing 1 ml medium.
3. Extract the virus as described above ('Isolation of high titre virus from infected 293 cells', steps 1–5), freeze-thawing the 1 ml medium and agar plug at least three times by placing at −70°C until frozen and then at 37°C. This solution will from now on be referred to as the original virus solution.

Screening for recombinant virus

Plaques which have been picked may contain wild-type virus for recombinant virus and it is essential to establish which plaques contain recombinant virus before moving on. This can be achieved using two different procedures, which should be carried out in parallel. For each plaque, 100 μl of the original virus solution should be used to infect two 5 cm dishes of 293 cells as described above ('Infection of 293 cells with virus', steps 1–3). Once lytic infection is evident, both dishes should be harvested. One dish of cells should be used to isolate adenovirus DNA, which can then be analysed by restriction enzyme digestion or polymerase chain reaction (PCR). The second dish should be used to isolate high-titre virus. This virus should then be used to infect cells, such as HeLa or A549, which should be checked for foreign gene expression at 48–72 h post-infection by western blotting, indirect immunofluorescence, or northern blotting.

Additional materials and equipment

- 0.5% (w/v) trypsin, pH 7.3
- 0.04% versene
- 10 mM spermine (Sigma)
- TE (10 mM Tris, 1 mM EDTA, pH 9.0)
- Lysis buffer (20% ethanol, 100 mM Tris pH 9.0, 0.4% sodium deoxycholate)
- 10% sodium dodecylsulfate (SDS)
- 0.25 M EDTA, pH 8.0
- Proteinase K (20 mg ml^{-1}) (Merck)
- HeLa cells (ATCC and ECACC)
- A549 cells (ATCC and ECACC)

Isolation of viral DNA

1. Harvest 293 cells by trypsinization. Remove the medium from the cells and wash once with saline to remove any residual FBS. Add 1 ml trypsin/versene (1 : 1); after 2 min the cells will begin to detach from the plastic of the dish. Add 10 ml medium to inactivate the trypsin and pellet the cells by centrifugation at 400 g for 5 min.
2. Remove the supernatant with a narrow bore Pasteur pipette and resuspend the cell pellet in 400 μl TE pH 9.0, 10 mM spermine. Mix, transfer to a microcentrifuge tube, and then add 400 μl lysis buffer. Mix gently; do not vortex.
3. Centrifuge at 13 000 rev min^{-1} for 15 min in a microcentrifuge.
4. Pipette out the supernatant into a new microcentrifuge tube and add 60 μl 10% SDS, 40 μl 0.25 M EDTA pH 8.0 and 20 μl proteinase K. Incubate at 37°C for 1 h.
5. Extract once with an equal volume of phenol/chloroform/isoamyl alcohol, 25 : 24 : 1, followed by vortexing and microcentrifugation.
6. Transfer the aqueous phase to a fresh microcentrifuge tube.
7. Repeat extraction twice with chloroform/isoamyl alcohol (24 : 1). To the final aqueous phase add a 10% (v/v) 3 M sodium acetate pH 5.2 and 1 ml absolute ethanol. Place at –20°C for 30 min and centrifuge at 13 000 rev min^{-1} for 10 min in a microcentrifuge.
8. Remove the ethanol with a Pasteur pipette, add 1 ml 70% ethanol and centrifuge at 13 000 rev min^{-1} for 2 min in a microcentrifuge.

9. Again remove the ethanol with a drawn-out Pasteur pipette and leave the pellet to dry.
10. Resuspend each pellet in 50 µl TE. A typical yield should be between 5 µg and 15µg DNA which is suitable in this form for restriction enzyme digestion.

Infection of HeLa/A549 cells with concentrated virus

1. Take one of the dishes of infected cells. Remove medium and scrape the cells from the dish using a cell scraper. Add 1 ml medium to resuspend the cells and place in microcentrifuge tubes.
2. To isolate the virus, freeze–thaw the 1 ml cell suspension at least three times as described above ('Isolation of high-titre virus from infected 293 cells', steps 1–5).
3. Pellet the cells by centrifugation at 13 000 rev min^{-1} in a micorcentrifuge tube for 5 min and then use 200 µl supernatant to infect one 10 cm dish of 80% confluent HeLa or A549 cells as described above ('Infection of 293 cells with virus', steps 1–3). Use uninfected cells as a negative control.
4. Harvest the cells 48–72 h later and look for the expression of the foreign gene by western blotting, indirect immunofluorescence, PCR or northern blotting. This should allow the detection of plaques generated by recombinant adenovirus replication.

Plaque purification of recombinant virus

Wild-type virus, if present, will very often overgrow recombinant virus, so it is essential that the final adenovirus stock is completely free of wild-type virus. When recombinant virus has been identified, it is essential to purify this virus by plaque purification, which serves to eliminate any wild-type virus. Ideally the plaque purification procedure should be repeated at least twice to ensure a final pure virus stock.

Additional materials and equipment

* 293 cells (ATCC and ECACC)

1. Plate out 16×5 cm dishes of 293 cells at a 20% plating density 24–48 h before infection so that on the day of infection they are approximately 80% confluent.
2. Make serial 10-fold dilutions of virus in microcentrifuge tubes (e.g. 10^{-1}–10^{-15}) in 1 ml medium.
3. On the day of infection remove the medium and to each dish of cells add 200 µl diluted virus solution. Reincubate and rock the dishes briefly every 20 min to ensure that virus covers the cells.
4. Overlay the cells 5–6 h later as described above ('Calcium phosphate transfection and agar overlaying of transfected cells', step 9) and 5–7 days post-infection repeat the plaque picking as described above ('Isolation of virus from cell sheet plaques'). Once recombinant plaques have been identified, the plaque purification should be repeated at least once more.

Stock production of an adenovirus stock and its titration

Once a pure recombinant virus stock is obtained, it is preferable to make a large volume of this virus which should then be titrated. There are two methods which can be used. Plaque assay is the most accurate and the one traditionally used; however, the simpler 50% endpoint method can also be used.

Additional materials and equipment

- 293 cells (ATCC and ECACC)
- ×2 tissue culture medium (88 ml ×2 DMEM, 8 ml FBS, 4 ml 0.57 M NaHCO$_3$, 2 ml 0.2 M glutamine)
- 4% low melting point agar (4 g low melting point agar in 100 ml distilled water, autoclaved at 15 lb in^{-2}, 121°C, for 10 min)
- Neutral red solution (3.3 g l^{-1}). Dilute 1 : 10 in PBS for use

1. Infect 20 × 10 cm dishes of 293 cells with plaque purified recombinant virus stock. 20 μl of purified recombinant virus diluted in a 180 μl medium should be used to infect each dish. When virus infection is evident 5–7 days post-infection, the cells should be harvested in 40 ml medium.
2. Freeze–thaw the cell suspension at least three times as described above ('Isolation of high-titre virus from infected 293 cells', steps 1–5) and centrifuge to pellet the cells.
3. The supernatant containing the recombinant adenovirus should then be frozen at –70°C in 1 ml aliquots. Once frozen, one of these aliquots should then be thawed out for titration.

Titration using the plaque assay

1. Make serial 10-fold dilutions of the plaque purified virus solution from 10^{-1} to 10^{-15} in medium. Use 200 μl of each dilution to infect one 5 cm dish of 293 cells as in 'Plaque purification of recombinant virus', steps 1–3, keeping one 5 cm dish of 293 cells as an uninfected control.
2. After infection, overlay the cells with agar immediately. Refeed by overlaying with 2 ml fresh agar and medium twice weekly for 14 days.
3. At 7–10 days post-infection, when plaques are visible microscopically, the cells should be stained with neutral red solution as described in 'Isolation of virus from cell sheet plaques', step 1.
4. The following day count the number of plaques on each plate. One isolated plaque is generated by a single virus particle infecting a single cell. At high virus dilution all the 293 cells would be expected to be infected with virus. As the virus is diluted out 10-fold, 10-fold decreases in the number of viruses, and therefore plaques, will be evident, allowing the titer to be determined.

Titration by the 50% endpoint method

1. Two to three days preceding the titration, plate out 293 cells in a 96-well plate at a concentration of 1×10^3 293 cells well^{-1} so that on the day of titration each 200 μl well is approximately 80% confluent.

2. Make serial 10-fold dilutions of the plaque purified virus solution from 10^{-1} to 10^{-15} in medium. Remove the medium from each well of the multiwell plate and add 100 μl of each 10-fold dilution of the plaque purified virus to 12 wells. Reincubate and 5–6 h later carefully add a further 100 μl medium to each well, ensuring no cross-contamination between wells.

3. Refeed the cells in each well every 4 days for 10–14 days, after which adenovirus infection should become apparent in certain wells. Titres expected should be in the range of 10^8–10^{14} infectious units ml^{-1}. By scoring wells for adenovirus infection it is possible to estimate the concentration of the virus. For example, if all wells are infected where 100 μl of the 10^{-10} virus dilution was added, and only two out of the six wells are infected where 100 μl of the 10^{-11} virus dilution was added, and no wells are infected where 100 μl of the 10^{-12} virus dilution was added, it can be concluded that the virus titre is in the range of 10^{12} PFU ml^{-1}. To more accurately calculate the 50% endpoint, the method of Reed & Muench (1938) should be used.

DISCUSSION

The advantages of using adenoviruses as vectors for gene transfer have become widely appreciated in recent years. It is now recognized that adenoviruses have the potential to carry large segments of foreign DNA, and that high-titre stocks of recombinant adenoviruses can be generated in the laboratory (approximately 10^{11}–10^{14} PFU ml^{-1}, a million-fold higher than typical retrovirus titres). Adenoviruses can be used to infect a wide range of cell types, and unlike retroviruses, can also infect slowly proliferating or even non-dividing cells. However, the frequency of integration into cellular DNA is low compared with that of retroviruses, reducing the likelihood of stable gene expression in an individual cell. Viral vectors, like plasmids, can of course be 'fine tuned' for particular requirements. For example, varying levels of gene expression could be accomplished by incorporating different promoters into the vector, or tissue-specific promoters could be used to achieve expression of a gene in a particular cell type. One of the most exiting prospects for adenovirus vectors in the future is their potential use in gene therapy. Nevertheless, difficulties involving the delivery of the gene to dividing cells in order to achieve stable integration, and the possibility of destruction of infected cells by the host immune system, remain to be overcome.

REFERENCES

Babiss LE & Ginsberg HS (1984) Adenovirus type 5 early region 1b gene product is required for efficient shut off of host protein synthesis. *Journal of Virology* 50: 202–212.

Babiss LE, Ginsberg HS & Darnell JE (1985) Adenovirus E1B proteins are required for accumulation of late viral mRNA and for effects on cellular mRNA translation and transport. *Molecular and Cellular Biology* 5: 2552–2558.

Berk AJ (1986) Adenovirus promoters and E1A transcription. *Annual Review of Genetics* 20: 45–79.

Berkner KL (1992) Expression of heterologous sequences in adenoviral vectors.

Current Topics in Microbiology and Immunology 158: 39–66.

Berkner KL & Sharp PA (1985) Effect of the tripartite leader on synthesis of a non-viral protein in an adenovirus 5 recombinant. *Nucleic Acids Research* 13: 841–857.

Burgert HG & Kvist S (1985) An adenovirus type 2 glycoprotein blocks cell surface expression of human histocompatibility class I antigens. *Cell* 41: 987–997.

Falgout B & Ketner G (1987) Adenovirus early region 4 is required for efficient virus particle assembly. *Journal of Virology* 61: 3759–3768.

Ghosh-Choudhury G, Haj-Ahmad Y & Graham FL (1987) Protein IX, a minor component of the human adenovirus capsid, is essential for the packaging of full length genomes, *EMBO Journal* 6: 1733–1739.

Gooding LR, Elmore LW, Tollefson AE, Brady HA & Wold WSM (1988) A 14 700 MW protein from the E3 region of Adenovirus inhibits cytolysis by tumor necrosis factor. *Cell* 53: 341–346.

Grable M & Hearing P (1990) Adenovirus type 5 packaging domain is composed of a repeated element that is functionally redundant. *Journal of Virology* 64: 2047–2056.

Graham FL & van der Eb AJ (1973) A new technique for the assay of infectivity of adenovirus 5 DNA. *Virology* 36: 59–72.

Graham FL, Prevec LA, Schneider M, Ghosh-Choudhury G, McDermott M & Johnson DC (1988) Cloning and expression of glycoprotein genes in human adenovirus vectors. In: Lasky L (ed.) *Technological Advances in Vaccine Development, UCLA Symposia on Molecular and Cellular Biology*, Vol. 84, pp. 243–253. Alan R Liss, Inc., New York.

Haj-Ahmad Y & Graham FL (1986) Development of a helper independent human adenovirus vector and its use in the transfer of the Herpes Simplex Virus Thymidine Kinase gene. *Journal of Virology* 57: 267–274.

Halbert DN, Cutt JR & Shenk T (1985) Adenovirus early region 4 encodes functions required for efficient DNA replication, late gene expression, and host cell shut off. *Journal of Virology* 56: 250–257.

Hay R & McDougall IM (1986) Viable viruses with deletions in the left inverted terminal repeat define the adenovirus origin of DNA replication. *Journal of General Virology* 67: 321–332.

Hearing P & Shenk T (1986) The adenovirus type 5 E1A enhancer contains two functionally distinct domains: one is specific for E1A and the other modulates all early units in Cis. *Cell* 45: 229–236.

Jones N & Shenk T (1979) Isolation of adenovirus type 5 host range deletion mutants defective for transformation of rat embryo cells. *Cell* 16: 683–689.

Maniatis T, Fritsch EF & Sambrook J (1982) *Molecular Cloning – a Laboratory Manual*. Cold Spring Harbor Laboratory, Cold Spring Harbor, New York.

Parks RJ, Chen L, Anton M, Sankur U, Rudnicki MA & Graham FLA (1996) Helper-dependent adenovirus vector system: removal of helper virus by Cre-mediated excision of the viral packaging system. *Proceedings of the National Academy of Sciences of the USA* 93: 13565–13570.

Parks RJ & Graham FL (1997) A helper-dependent system for adenovirus vector production helps define a lower limit for efficient DNA packaging. *Journal of Virology* 71: 3293–3298.

Prevec L, Schneider M, Rosenthal KL, Belbeck LW, Derbyshire JB & Graham FL (1989) Use of human adenovirus based vectors for antigen expression in animals. *Journal of General Virology* 70: 429–434.

Reed L & Muench H (1938) A simple method of measuring 50% end points. *American Journal of Hygiene* 127: 493–497.

Robbins PD, Thara H & Ghivizzani SC (1998) Viral vectors for gene therapy. *Trends in Biotechnology* 16: 35–40.

Sarnow P, Hearing P, Andersen CW, Halbert DN, Shenk T & Levine A (1984) Advenovirus early region 1b 58 000 dalton tumor antigen is physically associated with an early region 4 25 000 dalton protein in productively infected cells. *Journal of Virology* 49: 692–700.

Stillman BW (1985) Biochemical and genetic analysis of adenovirus DNA replication *in vitro*. In: Setlow JK & Hollaender A (eds) *Genetic Engineering: Principles and Methods*, pp. 1–27. Plenum Press, New York.

Subramanian T, Kuppuswamy M, Gysbers J, Mak S & Chinnadurai G (1984) 19kDa tumor antigen coded by the early region E1b of adenovirus 2 is required for efficient synthesis and for protection of viral DNA. *Journal of Biological Chemistry* 259: 11777–11783.

Wang Q, Jia XC & Finer MH (1995) A packaging cell line for propagation of recombinant adenovirus containing lethal gene-region deletions. *Gene Therapy* 2, 775–783.

Wang Q & Finer MH (1996) Second-generation adenovirus vectors. *Nature Medicine* 2: 714–716.

Yoo D, Graham FL, Prevec L, Parker MD, Benko M, Zamb T & Babiuk LA (1992) Synthesis and processing of the haemagglutinin-esterase glycoprotein of bovine coronavirus encoded in the E3 region of adenovirus. *Journal of General Virology* 73: 2591–2600.

5.5C Retroviruses and Retroviral Vectors

The *Retroviridae* family of animal viruses are RNA viruses that replicate through a DNA intermediate. The DNA intermediate, the provirus, is stably integrated into the cellular DNA of the host cell. Like most enveloped viruses, retroviruses efficiently enter host cells (Figure 5.5C.1) via cell fusion or by receptor-mediated endocytosis. Following virus entry, the core particle is released, and the process of reverse transcription occurs, leading to double-stranded linear DNA. This DNA is then transported to the nucleus, where it is integrated to form the provirus. Transcription of the provirus by cellular RNA polymerase II leads to the production of viral RNA transcripts, genomic length or spliced, which are transported to the cytoplasm. The genomic length RNA can act either as messenger RNA for translation into viral proteins or as genomic RNA to be packaged into virions; the spliced RNA acts as messenger RNA. Assembled virus particles are then released from the infected cell to produce progeny virus.

Many retroviruses isolated from animals also include sequences (oncogenes) modified from certain cellular genes, the proto-oncogenes. These retroviruses, the highly oncogenic retroviruses, integrate and express oncogenes. The genetic organization of retroviruses allows for insertion of non-retroviral genes (such as oncogenes) without interfering with any steps in the normal virus life cycle. Typically, retroviral infection and replication does not kill the infected cells. Thus,

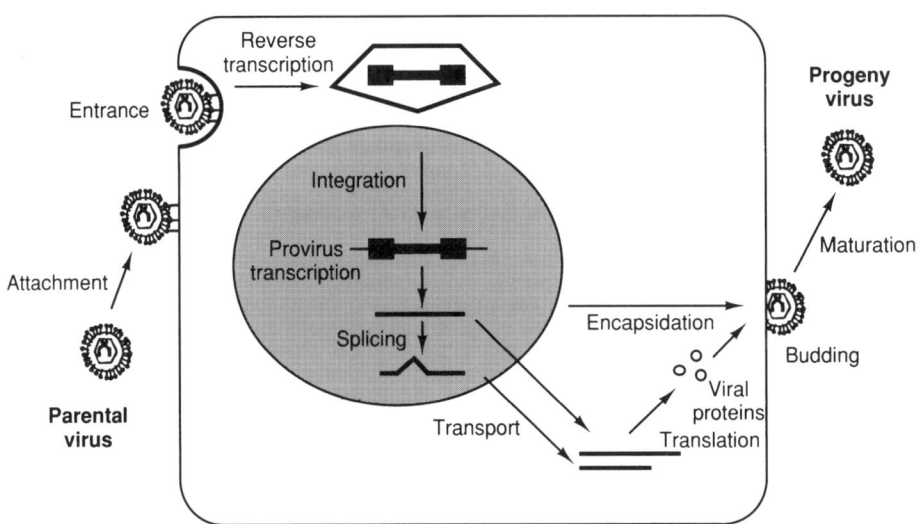

Figure 5.5C.1 Replication cycle of a simple retrovirus. A detailed description of the replication cycle is given in the text.

in nature, retroviruses have evolved highly efficient mechanisms for gene transfer. Retroviruses that contain oncogenes or other types of non-retroviral sequences are called retroviral vectors (Temin 1986).

The existence of naturally occurring retroviral vectors has led, with the advent of genetic engineering technology, to the construction of a large number of retroviral vectors, which are usually not replication-competent (Temin 1986; Dornburg & Temin 1991; Miller 1992). The construction of such vectors has been aided tremendously by the convenient location of the *cis*- and *trans*-acting elements necessary for retrovirus replication. A retroviral vector (Figure 5.5C.2(b)) derived from a simple retrovirus (Figure 5.5C.2(a)) contains all the *cis*-acting sequences necessary for retrovirus replication and for virus production, but is deficient in the production of some or all of the viral proteins necessary for replication and virus production. The *cis*-acting sequences, located primarily at the ends of the retroviral genome, include the long terminal repeats (LTRs) which contain the regions U3 (containing the promoter and enhancer sequences necessary for transcription to RNA), U5 and R. Located in the 5' untranslated region of the genome are the encapsidation (packaging) signal (E or ψ) for packaging of two identical copies of viral RNA into virus particles, and the primer binding site (PBS) for initiation of minus-strand DNA synthesis. Located in the 3' untranslated region of the genome is the polypurine tract (PPT), a region important for initiation of plus-strand viral DNA synthesis. At the ends of the genome are the attachment sites (attR and attL), which are necessary for integration of the double-stranded linear DNA into the cellular DNA.

The viral proteins, located in the internal region of a simple retroviral genome, include *gag*, *pol* and *env* (Figure 5.5C.2(a)). The *gag* (group-specific *a*ntigen) gene usually encodes three proteins: matrix, capsid and nucleocapsid. The *pol* gene encodes the viral protease, reverse transcriptase (RNA-dependent DNA polymerase) and the viral integrase. The *env* gene codes for the envelope glycoprotein, which is synthesized as a precursor and is subsequently cleaved by host cell proteases into the surface protein (SU) and a transmembrane protein (TM).

Originally, retrovirus vectors were produced as virus particles by using a replication-competent virus as a helper. This results in production of both vector virus and helper virus. The use of such vector virus stocks for gene transfer leads to the spread of the vector and the helper virus in the susceptible cell types. To obviate the possibility of virus spread, helper (or packaging) cells were first constructed by Mann *et al.* (1983) for murine leukaemia virus (MLV) and by Watanabe & Temin (1983) for avian reticuloendotheliosis virus. These cells contain the coding sequences for the viral proteins necessary for virus production, but do not produce virus due to a partial or complete removal of the viral *cis*-acting sequences. In more recently constructed helper cell lines, the viral coding sequences have been separated into different expression plasmids (Miller 1992; Temin 1990). The introduction of a retroviral vector into helper cells results in the production of vector virus that is free of helper virus (Figure 5.5C.3).

There are several desirable qualities to consider in the choice of a retroviral vector. Non-replication-competent retroviral vectors are generally more desirable because larger amounts of foreign sequences (up to 7–8 kb in length) may be inserted. Retroviral vectors usually contain a dominant-selectable marker (e.g.

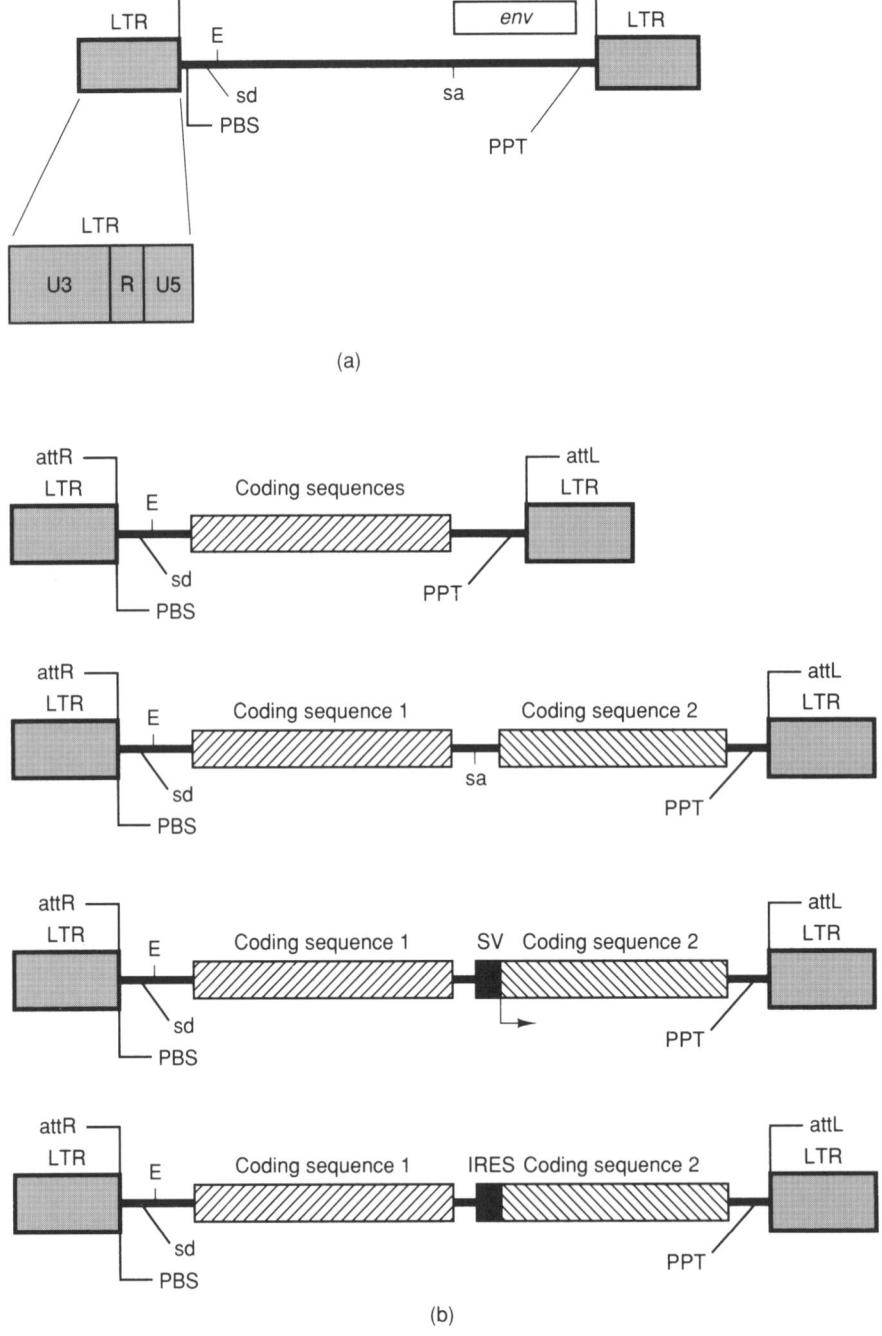

Figure 5.5C.2 A simple retrovirus (a) and simple retrovirus vectors (b) shown in proviral DNA form. Grey rectangular boxes represent the retroviral long terminal repeat (LTR); the unique 3′ (U3), repeat (R), and unique 5′ (U5) regions (shown in the enlarged LTR)

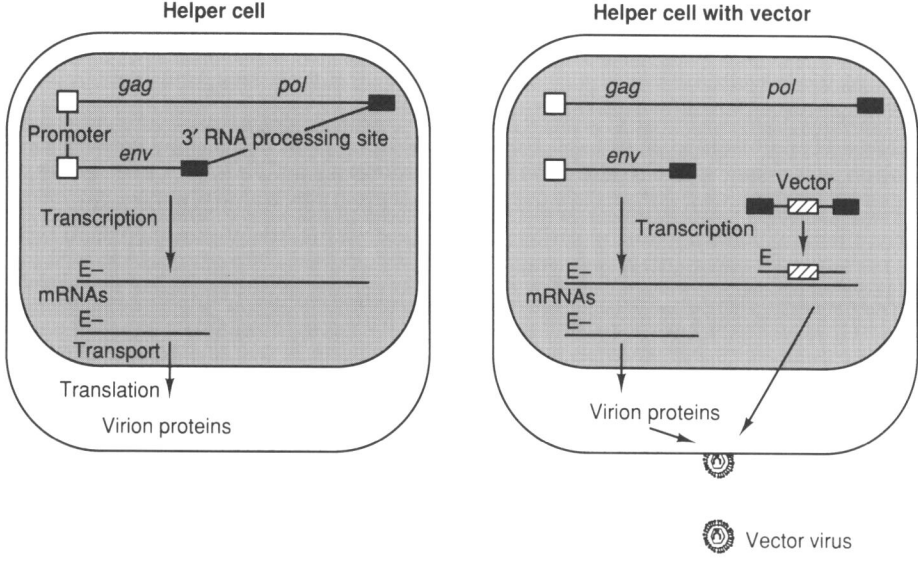

Figure 5.5C.3 A retrovirus helper cell and vector virus production from a helper cell.

neomycin phosphotransferase, *neo*, or hygromycin B phosphotransferase, *hyg*) to aid in the screening for vector transduction. In addition, many retroviral vectors contain an internal promoter or a *cis*-acting sequence that promotes cap-independent translation initiation at internal initiation codons (Boris-Lawrie & Temin 1993). Such *cis*-acting sequences are from the 5′ untranslated region of picornaviruses (called an internal ribosome entry site (IRES)) which allows expression of a marker or some inserted DNA.

There are two important considerations with regard to the choice of helper cells. The first is the host range of the vector virus, which is mostly determined by the helper cells used. A major determinant of host range is the envelope protein. It is desirable to use an envelope protein that confers a relatively wide host range (or the host range of choice). The envelope protein from the amphotropic subgroup of MLV is most often used. Production of pseudotyped MLV-based retroviral vectors where the retroviral envelope glycoprotein is completely replaced by the G glycoprotein of vesicular stomatitis virus can yield titres of > 10^9 colony-forming units per ml and can also increase the host range (Burns *et al.* 1993; Cossett *et al.* 1995). A second consideration is to use helper cells that express the *gag-pol* and

---

of the viral RNA are shown. The remaining viral sequences are indicated by a thick black line. The open rectangular boxes shown indicate the locations and translation reading frames of the viral genes *gag*, *pol* and *env*. The attachment sites relative to the polarity of the resulting provirus in the host chromosome (attL and attR for left and right, respectively) are shown, as is the primer binding site (PBS), the splice donor and acceptor sites (sd and sa, respectively), the *cis*-acting encapsidation sequence (E), and the polypurine tract (PPT). The inserted coding sequences are indicated. The SV40 early promoter (SV), with the direction of transcription indicated by the arrow, and a picornavirus internal ribosome entry site sequence (IRES), are shown as labelled black boxes.

the *env* genes on separate expression plasmids (Figure 5.5C.3). This is done to minimize the spontaneous generation of replication-competent helper virus by limiting the possibility for recombination in regions of sequence homology between the expression plasmids and the retroviral vector.

Specific limitations for the use of retroviruses include the lack of specific integration at a particular chromosome locus, insertional activation of host genes by integration of the retrovirus vector into the host cell DNA, and the titre of vector virus stock, which may not be sufficiently high for transducing particular cell types or for certain transduction experiments.

A severe limitation of retroviral vectors is their inability to infect non-dividing cells. However, a solution to this problem is the use of non-MLV derived retroviral vectors. There are a number of recent reports based on the development of lentivirus vectors (e.g. human immunodeficiency virus (HIV) and simian immunodeficiency virus (SIV) being used to infect non-dividing cells (Yee *et al.* 1994; Naldini *et al.* 1996a,b).

REFERENCES

Boris-Lawrie KA & Temin HM (1993) Recent advances in retrovirus vector technology. *Current Opinion in Genetics and Development* 3: 102–109.

Burns JC, Friedmann T, Driever W, Burrascano M & Yee J-K (1993) Vesicular stomatitis virus G glycoprotein pseudotyped retroviral vectors: concentration to very high titer and efficient gene transfer into mammalian and non-mammalian cells. *Proceedings of the National Academy of Sciences of the USA* 90: 8033–8037.

Cossett FL, Takeuchi Y, Battini JL, Weiss RA & Collins MK (1995) High-titer packaging cells producing recombinant retroviruses resistant to human serum. *Journal of Virology* 69: 7430–7436.

Dornburg R & Temin HM (1991) Retroviruses as vectors for gene transfer. In: Dulbecco R (ed.) *Encyclopedia of Human Biology*, Vol. 6, pp. 653–659. Academic Press, New York.

Mann R, Mulligan RC & Baltimore D (1983) Construction of a retrovirus packaging mutant and its use to produce helper-free defective retrovirus. *Cell* 33: 153–159.

Miller AD (1992) Retrovial vectors. *Current Topics in Microbiology and Immunology* 158: 1–24.

Naldini L, Blomer U, Gallay P, Ory D, Mulligan R, Gage FH, Verma IM & Trono D (1996a) In vivo gene delivery and stable transfection of nondividing cells by lentivirial vector. *Science* 272: 263–267.

Naldini L, Blomer U, Gage FH, Trono D & Verma IM (1996b) Efficient transfer, integration and sustained long-term expression of the transgene in adult rat brains injected with a lentiviral vector. *Proceedings of the National Academy of Sciences of the USA* 93: 11382–11388.

Temin HM (1986) Retrovirus vectors for gene transfer: efficient integration into and expression of exogenous DNA in vertebrate cell genomes. In: Kucherlapati R (ed.) *Gene Transfer*, pp. 149–187. Plenum Press, New York.

Temin HM (1990) Safety considerations in somatic gene therapy of human disease with retrovirus vectors. *Human Gene Therapy* 1: 111–123.

Watanabe S & Temin HM (1983) Construction of a helper cell line for avian reticuloendotheliosis virus cloning vectors. *Molecular and Cellular Biology* 3: 2241–2249.

Yee JK, Miyanohara A, LaPorte P, Bouic K, Burns JC & Friedmann T (1994) A general method for the generation of high-titer, pantropic retroviral vectors: highly efficient infection of primary hepatocytes. *Proceedings of the National Academy of Sciences of the USA* 91: 9564–9568.

5.5D Retroviral-vector-mediated Gene Transfer

Gene transfer using retroviruses first involves the introduction of the retroviral vector (as a plasmid) into retrovirus helper or packaging cells via transfection. Next, virus is harvested from the transfected cells and used to infect fresh helper cells in order to obtain helper cells that stably produce the vector virus, which can then be used to infect susceptible target cells (the transfection process is known to be mutagenic; this extra step helps reduce the chance of using a vector that has undergone mutation) (Boris-Lawrie & Temin 1993; Miller 1990, 1992; Miller *et al.* 1993). In this section, common methods used for the transfection of retroviral vectors into helper cells will be detailed along with methods for harvesting and testing vector virus produced from helper cells.

SELECTION OF RETROVIRUS HELPER CELLS AND DESIGN OF RETROVIRAL VECTORS

A large number of retroviral vectors (Figure 5.5D.1) and retrovirus helper cell lines (Table 5.5D.1) are currently available. These vectors and cell lines can commonly be obtained directly from the investigators who constructed them. In some cases, these reagents can also be obtained from the American Type Culture Collection (ATCC, Manassas, VA) and the European Collection of Cell Cultures (ECACC, Salisbury, Wiltshire, UK).

In choosing a particular retroviral vector (when a drug-resistance marker(s) is to be expressed from the vector), it is important to consider the drug-resistance markers used to construct the helper cell line. For retroviral vectors that contain the same drug-resistance marker as the chosen helper cell line, it is not possible to select for the presence of the vector in these cells. The drug-resistance markers present in many of the commonly used retroviral helper cell lines are indicated in Table 5.5D.1. Some of the selectable markers that have been included in retroviral vectors are: neomycin phosphotransferase (*neo*), hygromycin B phosphotransferase (*hyg*), dihydrofolate reductase (*dhfr*), guanine phosphoribosyltransferase (*gpt*), histidine (*hisD*), puromycin (*pac*) and ouabain (*oua*) (Miller 1992).

 Most of the drugs used for selection with these markers (e.g. neomycin sulphate, hygromycin B, methotrexate, puromycin and ouabain) are highly toxic and should be handled with care; instructions provided by the manufacturer should be consulted regarding proper use and handling.

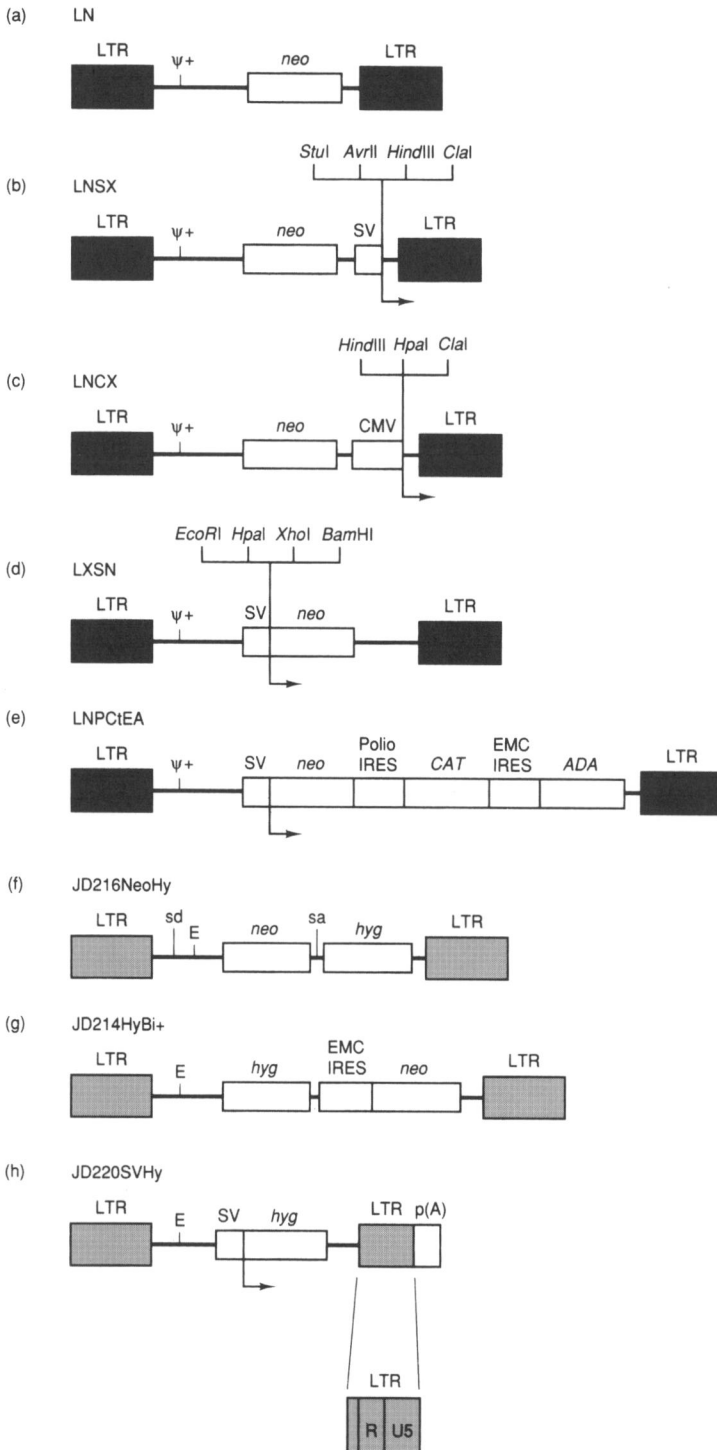

CONSTRUCTION OF RETROVIRAL VECTORS WITH INSERTED GENE SEQUENCES

Several things must be considered prior to inserting a cDNA sequence or other sequences of interest into a chosen retroviral vector. First, when expression of this sequence is to be transcribed from a promoter sequence already present in the vector (either in the LTR or a promoter sequence previously engineered into the vector), it is important to remove promoter sequences of the inserted gene in order to avoid promoter interference. Second, polyadenylation (poly A) signals

Table 5.5D.1 Commonly used retrovirus helper cell lines

Name	Host range[a]	Drug resistance markers[b]	Titre[c] (CFU ml^{-1})	Reference
ΨCRE	Ecotropic	*hyg, gpt*	10^6	Danos & Mulligan (1988)
ΨCRIP	Amphotropic	*hyg, gpt*	10^6	Danos & Mulligan (1988)
GP + E-86	Ecotropic	*gpt*	10^6	Markowitz *et al.* (1988a)
GP + envAm12	Amphotropic	*hyg, gpt*	10^6	Markowitz *et al.* (1988b)
DSDh	SNV	*dhfr*	10^5	Hu & Temin (1990)
DSN	SNV	*neo*	10^5	Dougherty *et al.* (1989)
DAN	Amphotropic	*neo*	10^4	Dougherty *et al.* (1989)
ΩE	Ecotropic	*gpt*	10^5	Morgenstern & Land (1990)
Isolde	ALV	*hyg, ble*	10^5	Cosset *et al.* (1990)
PG13	GALV	*tk, dhfr*	10^6	Miller *et al.* (1991)
PG53	GALV	*tk, hyg*	10^5	Miller *et al.* (1991)
ampli-GPE	Amphotropic	*neo*	10^6	Takahara *et al.* (1992)

[a]The host range of vectors produced from these helper cell lines are of the ecotropic or amphotropic subgroup of murine leukaemia virus (MLV), spleen necrosis virus (SNV), avian leukosis virus (ALV), or gibbon ape leukaemia virus (GALV).

[b]The drug-resistance genes that are already present in the helper cells from their construction by co-transfection with the retroviral gene expression constructs. The drug markers in each helper cell line cannot be used for selection of vectors introduced by transfection or infection in these cells. *hyg*, Hygromycin B phosphotransferase; *neo*, neomycin phosphotransferase; *gpt*, xanthine-guanine phos-phoribosyltransferase; *ble*, a bacterial gene conferring resistance in mammalian cells to bleomycin and phleomycin; *tk*, thymidine kinase.

[c]The titres presented represent the typical titre expected of vector virus from the helper cell line.

Figure 5.5D.1 Examples of retroviral vectors based on murine leukaemia virus (MLV) (a–e) and spleen necrosis virus (SNV) (f–h). Black or grey rectangular boxes represent the long terminal repeat (LTR) of MLV or SNV, respectively, Ψ or E represent the MLV or SNV encapsidation sequence, respectively; Ψ+ indicates the MLV encapsidation sequence and a portion of the MLV *gag* coding sequence. Open boxes representing the neomycin phosphotransferase gene (*neo*), the SV40 early promoter (SV), the cytomegalovirus imme-diate early promoter (CMV), the poliovirus and encephalomyocarditis virus internal ribosome entry site sequences (polio IRES and EMC IRES, respectively), the chloram-phenicol transacetylase gene (*CAT*), the adenosine deaminase gene (*ADA*), the hygromycin B phosphotransferase gene (*hyg*) and the SV40 polyadenylation signal (p(A)) are indi-cated. The splice donor and acceptor (sd and sa, respectively) are indicated. Locations of useful restriction enzyme sites are indicated. For more details regarding vectors (a)–(d), see Miller & Rosman (1989); for (e), see Morgan *et al.* (1992); for (f), see Dougherty & Temin (1986); for (g), see Koo *et al.* (1992); and for (h), see Dougherty *et al.* (1989).

should also be removed from the inserted gene, as they would cause premature termination of transcription and reduce the level of full-length vector RNA that is necessary for vector virus production. Third, sequences that may cause RNA instability and lead to lower RNA levels of the inserted gene(s) and/or of the full-length vector RNA (Schwartz *et al.* 1992) should also be removed.

PROCEDURE: TRANSFECTION OF MAMMALIAN CELLS WITH RETROVIRAL VECTORS

Retroviral vectors are first introduced into helper cell lines via transfection. Two methods of transfection, calcium phosphate precipitation and dimethylsulphoxide (DMSO)/Polybrene, are commonly used. In each method, the retroviral vector may be either transiently transfected into helper cells for rapad virus production, or stably transfected by drug selection into helper cells with individual clones selected and expanded for production of a relatively homogeneous vector virus, usually at a high titre.

Reagents and solutions

Reagents, solutions and media for maintenance of helper cell lines

The maintenance of cell lines varies depending on the cell line used to construct the helper cells. For many of the helper cell lines for MLV vectors (e.g. PA317 and PG13), Dulbecco's modified Eagle's medium (DMEM) with high glucose (4.5 g l^{-1}) supplemented with 10% (v/v) foetal bovine serum (FBS) (Sigma Chemical Co., St. Louis, MO) is used as a growth medium. For helper cell lines for SNV-based vectors (e.g. DSDh, DAN and DSN), Temin's modified Eagle's medium (American Bioorganics, Inc., Niagara Falls, NY) supplemented with 6% (v/v) calf serum (Biologos, Inc., Naperville, IL) is used (Temin 1968).

Precipitation buffer

100 μl 500 Mm HEPES-NaOH (pH 7.1), 125 μl 2.0 M NaCl, and 10 μl 150 mM Na$_2$HPO$_4$-NaH$_2$PO$_4$ (pH 7.0), mixed and diluted to 1 ml with double distilled water.

Materials and equipment

- Retroviral vector plasmid DNA
- Retrovirus helper cell line
- ×2 HEPES-buffered saline (HBS)
- 2 M CaCl$_2$
- Precipitation buffer
- ×0.1 TE (10 mM Tris, pH 7.5, 10 mM EDTA)
- 1 mg ml^{-1} Polybrene in TD (25 mM Tris, pH 7.5; 0.4 M NaCl; 5 mM KCl; 0.7 mM Na$_2$HPO$_4$)
- 12 × 75 mm polystyrene tubes

Calcium phosphate precipitation method

All the reagents used are sterilized prior to use by filtration through 0.22 μm filters (Nalgene Co., Rochester, NY).

1. Plate retrovirus helper cells at 5×10^5 cells (for helper cells based on NIH-3T3 cells; this number varies depending upon the cell type) per 60 mm dish 1 day prior to transfection.
2. The next day, replace medium with 5 ml fresh medium. The cells are then ready to be transfected.
3. Prepare each precipitate by mixing 25 μl 2 M $CaCl_2$, 10 μg plasmid DNA (in double distilled water or 10 mM Tris-HCl, pH 7.5, 1 mM EDTA) and double distilled water to a final volume of 200 μl.
4. To precipitate the DNA, add dropwise 200 μl of the DNA-$CaCl_2$ with constant agitation into a 12×75 mm polystyrene tube (Falcon 2054 (Becton Dickinson, Oxnard, CA)) containing 200 μl of freshly made precipitation buffer. The mixture should immediately become slightly cloudy; if the mixture remains clear or if a clumpy precipitate develops, the precipitation is not optimal, and the mixture should be discarded.
5. After about 30 min at 20°C, add the fine precipitate to the medium in each tissue culture dish and mix by swirling to distribute the precipitate evenly.
6. The next day, aspirate the medium, and add 5 ml fresh medium to each tissue culture dish.
7. To harvest the transiently produced vector virus, remove the virus-containing medium the following day, and centrifuge the medium at 3000 g for 5 min at 20°C to remove cells and debris. The virus-containing medium can be used immediately for infection of target cells or can be frozen at –70°C (virus is stable for years at –70°C but titers are generally lower than that of freshly harvested virus). In order to obtain stably transfected helper cell lines, replace medium in each culture dish 48 h post-transfection with medium containing selective agents and maintain in this medium, which should be changed every 2–3 days, until the formation of large, drug-resistant colonies (approximately 10–20 days post-transfection, depending on cell type and selective agent used).

DMSO/Polybrene method

1. Plate retrovirus helper cells at 5×10^5 cells (again, this number varies depending upon the cell type) per 60 mm dish 1 day prior to transfection.
2. Replace medium with 1 ml of a DNA/medium mixture (using 1–5 μg DNA per tissue culture dish). Add 30 μg of a 1 μg μl^{-1} solution of Polybrene (Sigma Chemical Co., St Louis, MO) to each 60 mm dish. Incubate the cells at 37°C for 6 h.
3. Swirl dishes periodically during this period in order to distribute the DNA mixture evenly.
4. Remove the DNA/Polybrene mixture by aspiration and replace with 2 ml of a 25% (v/v) solution of DMSO/medium which has been premixed and cooled to 37°C. The incubation time must be optimized for each cell type. The incuation time for D-17 cells and their derivatives (e.g. DSN, DAN and DSDh) is

3.5 min, and for NIH-3T3 cells and their derivatives (e.g. PG13 and PG53) is 1 min.

5. Remove this solution by aspiration, and wash the cells three times with 2 ml of medium quickly and very gently.

6. Place cells in 5 ml of medium per 60 mm dish for 24 h.

7. Harvest transiently produced vector virus at 48 h post-transfection by removing the virus-containing medium and placing in a centrifuge at 3000 g for 5 min at 20°C to remove cells and debris.

Obtain stably transfected helper cell lines by replacing the medium with medium containing selective agents 24 h post-transfection, and maintaining cells in this medium (adding fresh medium every 2–3 days) until drug-resistant colonies form (approximately 10–20 days post-transfection, depending on cell type and selective agent used).

PROCEDURE: GENERATION OF HELPER CELLS STABLY PRODUCING A RELATIVELY HOMOGENEOUS SUPPLY OF VECTOR VIRUS

In order to obtain stable cell lines that contain a single provirus and that produce relatively homogeneous vector virus, helper cells must first be transfected transiently with retroviral vector DNA. Virus harvested 2 days post-transfection is then used to infect fresh helper cells. The infected helper cells are placed under drug selection, and drug-resistant cell clones are isolated and screened (by Southern analysis and/or by polymerase chain reaction (PCR)), for the presence of an integrated, non-arranged vector. In addition, the clones are screened for the expression of the inserted gene of interest, for the production of high-titre vector virus, and for the absence of helper virus. The effect of integration sites on the level of expression of the gene of interest can be very dramatic and may need to be more closely analysed. The vector virus produced from these cell clones is relatively homogeneous because the virion RNA is transcribed from a single integrated provirus, in contrast to virus produced from transfected cells, where multiple integrated DNA copies containing various rearrangements may be present. The provirus is relatively stable and is not easily lost during cell passage. These virus-producing helper cells can be stored in liquid nitrogen to provide a long-term source of vector virus.

Materials and equipment

- Retrovirus vector plasmid DNA
- Retrovirus helper cell line
- 1 mg ml^{-1} Polybrene in TD (25 mM Tris, pH 7.5; 0.4 M NaCl; 5 mM KCl; 0.6 mM Na$_2$HPO$_4$)
- 12 × 75 mm polystyrene tubes
- Cloning rings (4 mm inner diameter)
- Silicone grease

1. Assay plates with 5×10^5 helper cells (depending on the cell type) per 60 mm dish are transfected with the retroviral vector plasmid DNA using one of the methods described above.
2. After 2 days, harvest medium, centrifuge at 3000 g at room temperature for 5min, and use the supernatant for infection of fresh helper cells.
   Add, to each 60 mm dish, 0.2 ml virus supernatant and 0.2 ml 100 μg ml^{-1} Polybrene (for most mammalian cell types; 15 μg ml^{-1} for avian cells) and incubate for 40 min. Swirl dishes periodically during this period in order to distribute the virus evenly. Then replace the medium with 5 ml fresh medium.
3. Place infected helper cells under selection with the appropriate medium 24 h after infection.
4. After large (around 2–3 mm in diameter), drug-resistant colonies are formed (10–20 days depending upon cell type and selective agent), isolate individual clones using cloning rings (Bellco Glass Inc., Vineland, NJ).
   To prepare the cloning rings, apply a thin coat of silicone grease (Dow Corning high-vacuum grease (Midland, MI)) evenly to one open end of each ring. (*Note*: the grease used is placed in a glass 60 mm Petri dish and autoclaved prior to use.) To isolate clones, well-isolated colonies are located by drawing a circle around each colony on the bottom of the dish with a felt-tip pen. Colonies can be most easily visualized by holding the dish up to the light, taking care not to spill the medium.
5. Aspirate the medium and place the cloning rings over the colonies to be isolated and press down with a pair of forceps. Add a drop of trypsin-EDTA to each cylinder, and monitor the extent of trypsinization through a microscope.
   When the cells have rounded up, neutralize the trypsin by adding bovine serum to each ring (one at a time) and force the solution vigorously in and out of a glass Pasteur pipette to dislodge the cells. Depending upon the application, 10 colonies or more are isolated for analysis.
   Note: as an alternative to using cloning rings, individual colonies can be picked by suction suing a P-1000 pipetter (Gilson) with a cotton-barrier tip (VWR Scientific). After marking the location of the drug-resistant colony with a felt-tip pen and aspirating the medium from the Petri dish, place the pipette tip, with the plunger fully depressed, directly over the colony until tight contact is made. The plunger is slowly released, and the colony is dislodged from the plate into the pipette tip. The dislodged colony is then placed in trypsin first; after a few minutes, add serum and medium in 2 ml of medium in one well of a 24-well cluster dish (Costar).
6. After expansion of each cell clone, analyse the clonal lines as previously discussed.

SUPPLEMENTARY PROCEDURE: INFECTION OF SUSCEPTIBLE TARGET CELLS WITH VECTOR VIRUS PRODUCED FROM HELPER CELL LINES

Materials and equipment

- Target cell line susceptible to infection
- 1 mg ml^{-1} Polybrene in TD (25 mM Tris, pH 7.5; 0.4 M NaCl; 5 mM KCl; 0.7 mM Na$_2$HPO$_4$)

1. Remove medium from virus-producing cells and centrifuge at 3000 g at room temperature for 5 min.
2. Carefully remove the top portion of supernatant and use for infection. To each 60 mm dish, add 0.2 ml supernatant and 0.2 ml 100 μg ml^{-1} Polybrene (for most mammalian cell types; 15 μg ml^{-1} for avian cells) and incubate for 40 min.
   Note: in some instances, co-cultivation of virus-producing cells with target cells can also be used to increase the rate of infection (Mansky & Temin 1994).
3. Place infected cells under selection with the appropriate medium 2 h after infection. Resistant colonies are typically observed in 10–20 days.

REFERENCES

Boris-Lawrie KA & Temin HM (1993) Recent advances in retrovirus vector technology. *Current Opionion in Genetics and Development* 3: 102–109.

Cosset F-L, Legras C, Chebloune Y, Savatier P, Thoraval P, Thomas JL, Samarut J, Nigon VM & Verdier G (1990) A new avian leukosis virus-based packaging cell line that uses two separate transcomplementing helper genomes. *Journal of Virology* 64: 1070–1078.

Danos O & Mulligan RC (1988) Safe and efficient generation of recombinant retroviruses with amphotropic and ecotropic host ranges. *Proceedings of the National Academy of Sciences of the USA* 85: 6460–6464.

Dougherty JP & Temin HM (1986) High mutation rate of a spleen necrosis virus-based retrovirus vector. *Molecular and Cellular Biology* 6: 4387–4395.

Dougherty JP, Wisniewski R, Yang S, Rhose BW & Temin HM (1989) New retrovirus helper cells with almost no nucleotide sequence homology to retrovirus vectors. *Journal of Virology* 63: 3209–3213.

Hu W-S & Temin HM (1990) Genetic consequence of packaging two RNA genomes in one retroviral particle: pseudodiploidy and high rate of genetic recombination. *Proceedings of the National Academy of Sciences of the USA* 87: 1556–1560.

Koo H-M, Brown AMC, Kaufman RJ, Prorock CM, Ron Y & Dougherty JP (1992) A spleen necrosis virus-based retroviral vector which expresses two genes from a dicistronic mRNA. *Virology* 186: 669–675.

Mansky LM & Temin HM (1994) Lower mutation rate of bovine leukemia virus relative to that of spleen necrosis virus. *Journal of Virology* 68: 494–499.

Markowitz D, Goff S & Bank A (1988a) A safe packaging line for gene transfer: separating viral genes on two different plasmids. *Journal of Virology* 62: 1120–1124.

Markowitz D, Goff S & Bank A (1988b) Construction and use of a safe and efficient amphotropic packaging cell line. *Virology* 167: 400–406.

Miller AD (1990) Retrovirus packaging cells. *Human Gene Therapy* 1: 5–14.

Miller Ad (1992) Retroviral vectors. *Current Topics in Microbiology and Immunology* 158: 1–24.

Miller AD & Rosman GJ (1989) Improved retroviral vectors for gene transfer and expression. *Biotechniques* 7: 980–986.

Miller AD, Garcia JV, von Suhr N, Lynch CM, Wilson C & Eiden MV (1991) Construction and properties of retrovirus packaging cells based on gibbon ape leukemia virus. *Journal of Virology* 65: 2220–2224.

Miller AD, Miller DG, Garcia JV & Lynch CM (1993) Use of retroviral vectors for gene transfer and expression. *Methods in Enzymology* 217: 581–599.

Morgan RA, Couture L, Elroy-Stein O, Ragheb J, Moss B & Anderson WF (1992) Retroviral vectors containing putative internal ribosome entry sites: development of a polycistronic gene transfer system and applications to human gene therapy. *Nucleic Acids Research* 20: 1293–1299.

Morgenstern JP & Land H (1990) Advanced mammalian gene transfer: high titre retroviral vectors with multiple drug selection markers and a complementary helper-free packaging cell line. *Nucleic Acids Research* 18: 3587–3596.

Schwartz S, Campbell M, Nasioulas G, Harrison J, Felber BK & Pavlakis GN (1992) Mutational inactivation of an inhibitory sequence in human immunodeficiency virus type 1 results in rev-independent gag expression. *Journal of Virology* 66: 7176–7182.

Takahara Y, Hamada K & Housman DE (1992) A new retrovirus packaging cell for gene transfer constructed from amplified long terminal repeat-free chimeric proviral genes. *Journal of Virology* 66: 3725–3732.

Temin HM (1968) Studies on carcinogenesis by avian sarcoma viruses. VIII. Glycolysis and cell multiplication. *International Journal of Cancer* 3: 273–282.

5.6 *IN VITRO* TOXICITY TESTING

TOXICITY OF XENOBIOTICS

Humans are habitually exposed to a large variety of foreign substances (food additives, cosmetics, pollutants, chemicals, pharmaceuticals, etc.) which are potentially toxic and harmful to different organs and tissues. Substances capable of producing cell damage are known as toxins and are classified according to whether they exert their effects in all individuals, in a dose-dependent and hence *predictable* manner (intrinsic toxins), or do so only in some individuals, usually after several contacts, in a non-dose-dependent and therefore *unpredictable* way (idiosyncratic toxins). Intrinsic toxins may act directly on cellular systems (active toxins) or after biotransformation by hepatocytes (latent toxins). Idiosyncratic toxicity may be the consequence of an unusual metabolism of the drug (metabolic idiosincrasy) or be mediated by the immune system after repeated previous contacts (sensitization) (Pessayre 1986; Benford *et al.* 1987).

THE USE OF CULTURED CELLS IN PHARMACO-TOXICOLOGICAL RESEARCH

The three Rs (*replacement, reduction* and *refinement*) concept developed by Russel and Burch (1959) is at the root of the development of *in vitro* alternatives to animal experimentation (Balls *et al.* 1995). They defined *replacement* as any scientific method employing non-sentient material which can replace methods using conscious living vertebrates; *reduction* as a means of lowering the number of animals needed to obtain information of a given amount and precision; and *refinement* as the elimination of any incidence or severity of inhumane procedures applied to those animals which have to be used. It is therefore desirable to replace the use of animals and whenever possible, *in vitro* methods should be used instead of *in vivo* methods.

Testing for the toxicity of new drugs forms part of the normal battery of assays to which new compounds are routinely subjected, and at present it is performed on experimental animals, despite the difficulties involved in extrapolating results obtained from animals to humans (Anderson *et al.* 1998). The development of *in vitro* methods as alternatives to animal experimentation is of great relevance in biomedical research aimed at detecting the potential toxicity of xenobiotics in humans. *In vitro* methods offer a series of advantages: (1) they can be used in the early stages of drug development; (2) only a small amount of a compound is needed for the assays; (3) they drastically reduce the use of laboratory animals; and (4) for many purposes the use of human primary cultured cells from different

Cell and Tissue Culture for Medical Research, edited by A. Doyle and J.B. Griffiths.
© 2000 John Wiley & Sons, Ltd

target organs (liver, kidney, lung, skin, nervous system, etc.), can provide direct information about the potential effects on the target cells in humans, which would be more scientifically relevant. Therefore, the need for rapid and cheaper methods for screening toxicity at a very early stages of drug development, the necessary reduction in the number of animals used, and the possibility of experimenting with human-derived cells are factors that have promoted the development of *in vitro* models that could be used as alternatives to anticipate the potential toxicity of xenobiotics in humans.

The quality and specificity of the data generated by *in vitro* models depends on several factors (Castell & Gómez-Lechón 1992; Castell *et al.* 1997): (1) the use of a biological system that reproduces, to a large extent, the metabolic behaviour of the target organ for the toxic effect of the xenobiotic; (2) the choice of appropriate parameters for evaluating the toxic effect *in vitro*; and (3) a correct experimental design so that the *in vitro* data are predictive of the potential *in vivo* effects.

THE CHOICE OF THE *IN VITRO* BIOLOGICAL MODEL: CELL LINES VERSUS PRIMARY CULTURES

We may have different *in vitro* conditions for maintaining cells outside the organism, which can be chosen according to the aim and experimental design of our research. In addition, the possibility of experimenting with human-derived cells has made the search for *in vitro* methods able to detect the potential toxicity of drugs to humans very attractive (Anderson *et al.* 1998).

Primary cultures

Cells derive directly from organs or tissues and are obtained by methods that frequently involve enzymatic or mechanical means and cultured in hormonally defined conditions. Cells in primary culture have a limited lifespan, are proliferating (lung, keratinocytes, kidney cells, etc.) or non-proliferating (hepatocytes), and express most of the specialized functions typical of the tissue or organ of origin. These differentiated cells are the model of choice for organo-specific toxicity studies.

Cell lines

Cell lines are subcultures derived from a primary culture, and they may have a limited (finite, diploid) or continuous life-span (established, permanent, heteroploid). The lifespan of a diploid cell line is approximately 20–30 cell cycles, and the cells are usually normal ones that are undergoing the process of senescence *in vitro*. Established cell lines derive from primary cultures of diploid cell lines by transformation processes which are either spontaneous or induced by viruses, chemical or physical agents, or they derive from tumoral tissues. Although a few established cell lines are able to express specialized functions of the tissue or organ of origin, most of them are undifferentiated. Established cell lines are used for basal cytotoxicity studies (Ekwall and Ekwall 1988; Ekwall *et al.* 1990; Zucco 1992; Barile 1997).

TOXICITY RISK ASSESSMENT IN CULTURED CELLS

When developing screening protocols for toxicity, many researchers have focused their attention on the so-called basal cytotoxicity end-point parameters (cell viability, cell survival, enzyme leakage, etc.). This certainly represents a first step towards evaluating toxicity, but evaluation of these parameters alone may not consider xenobiotics that impair target cell function without causing cell death. This may not be critical for the target cell itself but is of toxicological significance for the whole organism. Therefore, an *in vitro* model attempting to evaluate the toxicity of xenobiotics has to take this possibility into account (Flint, 1990; Balls & Fentem 1992; Castell & Gómez-Lechón 1992; Castell *et al.* 1997). Figure 5.6.1 shows an integrated screening protocol for evaluating the effect of xenobiotics on cells, in which both cytotoxic and metabolic effects are investigated (Ekwall *et al.* 1990; Castell & Gómez-Lechón 1992; Castell *et al.* 1997). The cytotoxicity end-points give information on the maximum drug concentration compatible with cell survival. The target-organ toxicity end-points provide direct information about the extent to which a cell's specific functions are altered (Flint 1990; Shrivastava *et al.* 1991; O'Hare & Atterwill 1995; Castell *et al.* 1997). The organ-specific or basal acute toxic effects of a particular compound can be assessed by combining meta-bolically competent cells from the target organ with non-differentiated cell lines. For example, by comparing the concentration–toxicity curves of the compound (1) in fully competent primary cultured hepatocytes; (2) in non-hepatic cells (i.e. fibroblasts); and (3) in non-metabolizing hepatocytes (i.e. well-differentiated human hepatoma HepG2 that lack cytochrome P-450), it is possible to ascertain whether the compound elicits toxic effects preferentially on hepatocytes or whether bioactivation (metabolization) of the xenobiotic is required in order to produce cellular damage (Castell and Gómez-Lechón 1992; Castell *et al.* 1997). The species-specific effect of a chemical can also be easily evaluated using *in vitro* cellular systems from different species.

The ultimate goal of *in vitro* experiments is to generate the type of scientific information needed to identify compounds that are potentially toxic to humans. For this purpose, not only the design of experiments but also the interpretation of results are essential. Even simple parameters for assessing cell toxicity have

---

Figure 5.6.1 Screening protocol for *in vitro* evaluation of the potential risk of toxicity of a drug. *In vitro* models can be used for different goals. *Cytotoxicity* evaluation using several parameters gives a first indication about the toxic potential (IC50) of a certain compound and *toxic mechanisms* underlying these effects can be evaluated. The maximal non-toxic concentration (IC10) of the drug is also determined and only at concentrations compatible with cell survival makes sense to asses drug effects on *functionality* of target-cells. Not only the design of experiments but also the interpretation of results obtained *in vitro* are essen-tial An important point is the *metabolic relevance* of the observed alteration and its *reversibility* upon withdrawal of the xenobiotic. The investigation of *drug metabolic profile* for anticipating the metabolism of a new compound in humans is also a key point.

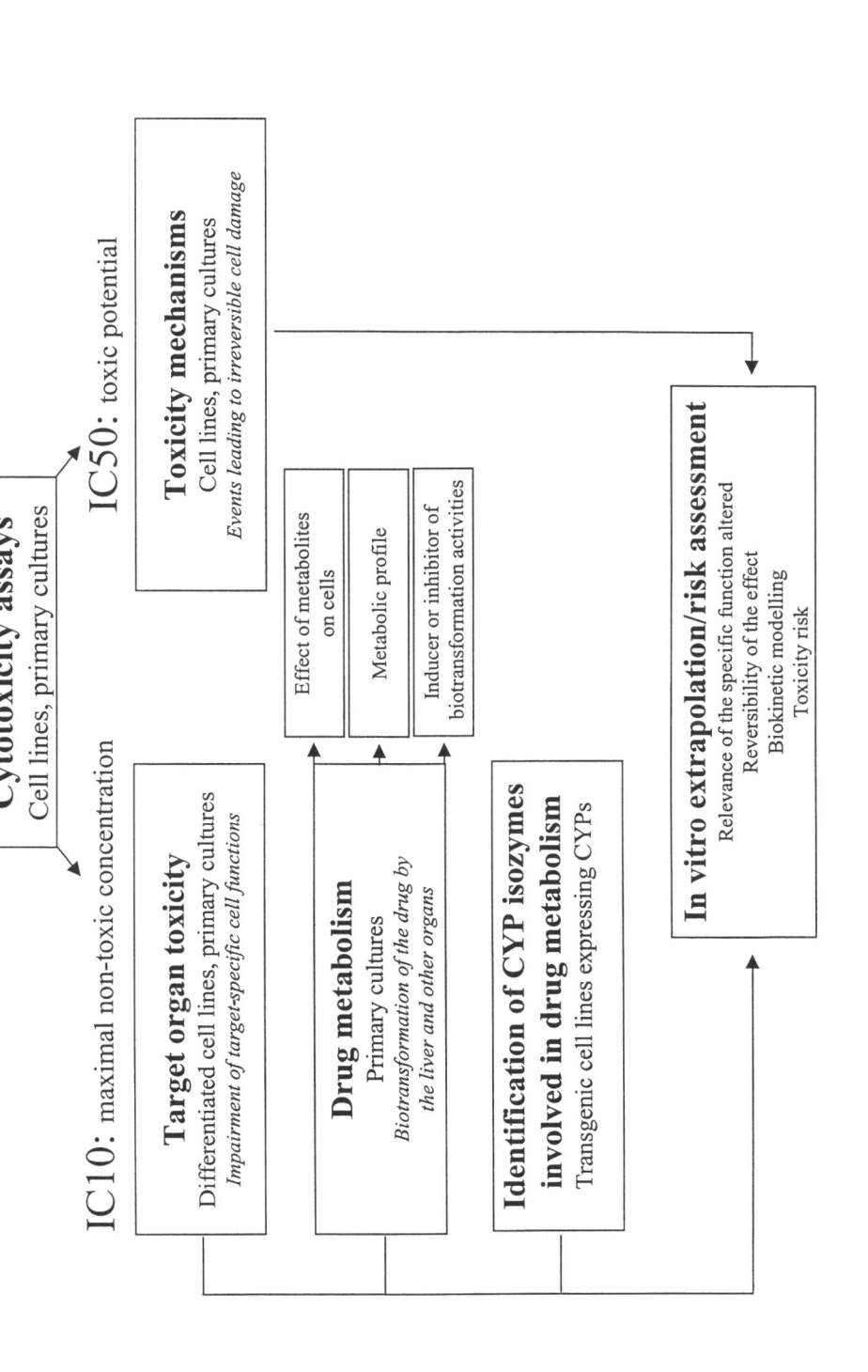

yielded promising results when comparing *in vitro* effects with human toxicity. False negatives are infrequent in compounds that are toxic without biotransformation. However, the *in vitro* models may lack sensitivity for xenobiotics that require a high degree of biotransformation or prolonged exposures to exert their toxic effect.

A second point is the metabolic relevance of the observed alteration and its reversion upon withdrawal of the xenobiotic from incubation media (Castell & Gómez-Lechón 1992; Castell *et al.* 1997). Certain cell functions can be transiently altered by a xenobiotic, yet this might lack *in vivo* significance if the cell rapidly recovers upon elimination of the xenobiotic.

A final point is the influence of *in vivo* pharmacokinetics on the toxicity of the compound. In *in vitro* experiments, xenobiotics are kept in culture plates at a constant concentration along the incubation time. This contrasts with what occurs *in vivo* where the concentration of the drug reaches a maximum and decreases thereafter in a characteristic concentration/time curve. Whenever possible, this circumstance should be taken into account in the experimental design of *in vitro* experiments. This is, however, technically complex *in vitro*. A simplified procedure to bring the *in vivo* situation closer to *in vitro* experimental conditions is to incubate cells for a time and concentration equivalent to the *in vivo* AUC (area under the concentration/time curve), experimentally determined or estimated by PBPK models (Krishnan & Andersen 1994).

A simple way to rank the relative potential toxicity of a drug within a homologous series of compounds is to compare the plasmatic concentration of the drug *in vivo* with the concentration causing toxic effects *in vitro*. The *toxicity risk* (TR) is thus defined as the quotient of both magnitudes. The larger the values of TR are (closer to 1 or even greater), the greater the toxicity risk will be for a given drug. It can be reasonably assumed that if a drug reaches a plasmatic concentration at a concentration that is toxic *in vitro* and stays there for a period of time that is also conducive to toxicity *in vitro*, the compound will show toxic effects *in vivo* (Ponsoda *et al.* 1991b; Castell *et al.* 1997).

Cytotoxicity evaluation

Damage to cells by xenobiotics commonly results in an early alteration of cell membrane permeability. Consequently, leakage of cytoplasmic enzymes into the culture medium is a first-choice parameter for a rapid and sensitive evaluation of cytotoxicity. Among cytosolic enzymes, LDH is usually a very good marker. However, for *in vitro* assays, intracellular measurement is recommended because different stability of isozymes once outside the cells may affect the results obtained. In these assay conditions, viable cells will show higher activities while affected cells only will contain a part of the activity (Ponsoda *et al.* 1991a; Castell *et al.* 1997). The procedure described below consists of two enzymatic reactions by which cellular LDH produces NADH and pyruvate from NAD and lactate. Pyruvate is then metabolized further by GPT, and the NADH together with MTT is metabolized by NADH:dye-DH to render the formazan that is going to be measured. The avantage of this colourimetric method over the classic UV method that it uses a single reaction is that its activity can easily be tracked by the colour changing from yellow to blue. Leakage of enzymes associated with organelles (mitochondrial activities GOT and GPT) is also used.

The MTT and XTT tests are based on the conversion of tetrazolium salts into coloured products by the mitochondrial enzyme succinate dehydrogenase with the requirement of cellular NADH and are the ones most often used for cytotoxicity assessment in cells. The MTT test is probably one of the easiest cytotoxicity assays to perform. When using MTT salts the formazan formed is water insoluble, precipitates into the cells and should be extracted with organic solvents (Castell *et al.* 1997). XTT salt yields a soluble product that is secreted to the culture medium. This method is especially useful when a cell culture is being monitored several times, i.e. to measure cell proliferation, since this product is non-toxic and therefore allows repetitive testing on a single cell preparation (Ponsoda *et al.* 1998). The most frequently used cytotoxicity end-point parameters are listed in Table 5.6.1.

Table 5.6.1 Biological endpoint parameters for detecting cytotoxicity

Endpoint	Experimental parameter
Cell morphology	Cell size Cell–cell contacts Nuclear number, size, shape and inclusions Nucleolar vacuole formation Cytoplasmic vacuole formation
Cell viability	Vital dye uptake (e.g. fluorescein diacetate and derivatives) Trypan blue exclusion test Cell number counting Replating efficiency
Cell adhesion	Attachment to culture surface Detachment from culture surface Cell–cell adhesion
Cell proliferation	Increase in cell number Increase in total DNA Increase in total RNA Increase in total protein Colony formation
Membrane damage	Loss of cytosolic enzymes (e.g. LDH, GOT, GPT) Loss of ions or co-factors (e.g. Ca^{2+}, K^+. NADPH) Leakage from pre-loaded cells (e.g. vital dye, ^{51}Cr) Leakage across cellular membrane (e.g. fluorescein)
Uptake/incorporation	Thymidine and DNA synthesis Uridine and RNA synthesis Amino acids and protein synthesis
Metabolic effects	Inhibition of metabolic cooperation Co-factor depletion (e.g. ATP) Impairment of mitochondrial function (e. g. MTT and XTT tests) Lisosomal alterations (e.g. Neutral red uptake test)

Experiments on cytotoxicity are designed to determine the maximal non-toxic concentration (MNTC) of a drug, i.e. the highest concentration compatible with cell survival. This is roughly estimated for each parameter as the concentration causing only 10% of the maximal cytotoxic effect. Toxicity data are expressed as IC10 and IC50 (concentration causing 10% or 50% of cell death; Figure 5.6.2). To calculate these IC values, a Logit transformation of results is recommended, by which the results of concentration (x) and viability (y) are expressed as $x'=-\log(x)$ and $y'= \text{LOGIT } (y)= \ln (y - y_{00}) / (y_0 - (y - y_{00}))$

Here y_{00} is the asyndote of an infinite concentration (almost 0) and y_0 is the asyndote of a concentration of 0 (100%). The transformed results fit a linear regression, and to obtain a given IC value one should interpolate the desired value in the transformed equation.

The question of how many toxicity end-point parameters need to be assessed to identify a potentially cytotoxic compound is worth discussing. In most cases, the different quantitative cytotoxic parameters currently used give equivalent information on the toxicity of a compound. For a first screening of cytotoxicity it normally suffices to determine only one parameter. However, the use of several markers can provide additional information, for example as to whether the damage affects only cell membranes (i.e. LDH leakage) or also involves subcellular structures (GPT, succinate dehydrogenase or ATP production for mitochondrial damage; ATP-mediated neutral red uptake for lysosomes).

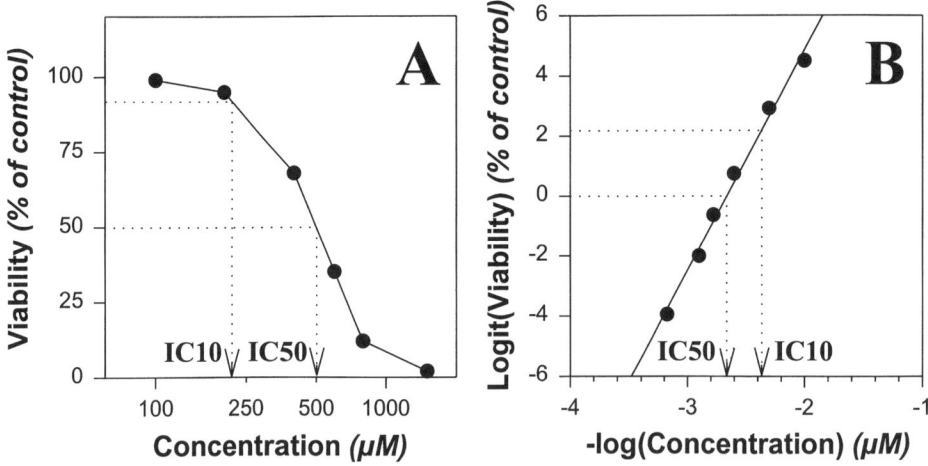

Figure 5.6.2 Estimation of IC10 and IC50. Experiments on cytotoxicity are designed to determine the highest concentration compatible with cell survival. This is roughly estimated for each parameter as the concentration causing only 10% of the maximal cytotoxic effect (IC10). The toxic potential of a given compound is expressed as the concentration causing 50% of cell death (IC50). Using the IC values, compounds can be ranked according their toxic potential. Both parameters can be estimated either graphically from the dose/effect curves (IC10 ≈ 220 μM, IC50 ≈ 500 μM ; Panel A) or mathematically in the Logit transformed curve (IC10 = 231.8 μM and IC50 = 461.0 μM; Panel B).

MTT test

Materials and equipment

- 5 mg ml^{-1} MTT (3-[4,5-dimethylthiazol-2-yl]-2,5-diphenyl tetrazolium bromide) solution in PBS. Sterilized by filtration. Stored in dark at 4°C. Stable for a month.
- 96-well plates
- Multichannel pipettes
- Microplate reader equipped with a 490 nm filter

Experimental procedure

1. After treatment of cells with the xenobiotics, the cell monolayers are gently washed with warm PBS using a multichannel pipette.
2. 100 µl of culture medium containing MTT solution (10 : 1) is added to each well. It is recommended to have some wells in the plate without cells but incubated with MTT solution in order to have blanks of the readings.
3. Cells are kept for 1–3 h (depending on cell density and activity) in the cell incubator.
4. Incubation medium is carefully removed.
5. 100 µl of dimethyl sulphoxide is added and the plate is gently shaken to resuspend formed formazan. You must now wait until a homogenized colour is formed.

Reading of results and calculations

1. Absorbance is read at 490 nm of wells containing cells and blanks.
2. The mean of the absorbance of wells is calculated with the same treatment after subtracting of blank absorbance.
3. The results are normalized considering control wells as 100% (maximum absorbance obtained), expressing then the results as percentage of controls.
4. Using the appropriate method of calculation, normally Logit transformation, IC50 and IC10 can be estimated.

XTT Test

Materials and equipment

- A 1 mg ml^{-1} solution of XTT (3'-[(phenylamino)-carbonyl]-3,4-tetrazolium-bis(4-methoxy-6-nitro)benzene-sulfonic acid hydrate in culture medium supplemented with HEPES 15 mM (to avoid alkalinization of the medium during readings). This solution is prepared sterile daily, and the medium should be warmed to 50°C for better XTT solubilization.
- 5 mM phenazine methosulphate in PBS, stored sterile in the dark at 4°C. Stable for a month. This solution has expired when the initial light yellow coloration of the solution changes to dark-yellow.
- 96-well plates.

- Multichannel pipettes.
- Microplate reader equipped with a 450 nm filter for reading and 630 nm as reference wavelength.

Experimental procedure

1. After treatment of cells with xenobiotics, cell monolayers are gently washed with sterile warm PBS using a multichannel pipette.
2. 100 μl of culture medium supplemented with 15 mM HEPES is added.
3. 25 μl of a mixture of the XTT and PMS solutions (1 ml XTT solution + 5 μl PMS solution) is added. It is recommended to leave some wells without cells but incubated with XTT solution in order to have blank readings.
4. The cells are kept for 3–6 h (depending on cell density and activity) in the incubator.
5. 100 μl of incubation medium is carefully removed and transferred to a replicate 96-well plate that will be read immediately.
6. Cell monolayers are washed with sterile warm PBS. Culture medium is added to wells and cultures are kept for a next incubation.

Reading of results and calculations

1. Absorbance is read at 450 nm of wells, using 630 nm as a reference wavelength containing cells and blanks.
2. The mean of the absorbance of wells is calculated with the same treatment after subtracting the blank absorbance.
3. The results are normalized considering control wells as 100% (maximum absorbance obtained), expressing the results as percentage of controls.
4. Using the appropriate method of calculation, normally by Logit transformation, IC50 and IC10 can be estimated.

Intracellular LDH leakage

Materials and equipment

- Main reagent solution containing: 17.36 g l⁻¹ Tris; 5.5 g l⁻¹ L-lactate, lithium salt; 0.84 g L⁻¹ glutamic acid; 2.8 g L⁻¹ N-Cetyl-N,N,N-trimethyl-ammoniumbromid; 9.6 ml L⁻¹ Triton X-100; 118.8 mg L⁻¹ MTT. This solution is adjusted to pH 8 before addition of MTT, then stored in the dark at 4°C. Stable for a month.
- NAD solution: 55 mg ml⁻¹ in distilled H_2O. Prepared daily.
- Enzyme solution: 88 U ml⁻¹ GPT from pig heart and 16 U ml⁻¹ NADH-dye-DH. This enzyme solution is prepared daily in Tris-citrate buffer, 50 mM, pH 7.6.
- Working solution for a 96-well plate is prepared with 14 ml of main reagent solution, 0.5 ml of NAD solution and 0.5 ml of enzyme solution.
- 96-well plates.
- Multichannel pipettes.
- Ultrasound bath suitable for a 96-well plate.
- Microplate reader equipped with a 570 nm filter.

Experimental procedure

1. After treatment of cells with xenobiotics, cell monolayers are gently washed with warm PBS using a multichannel pipette.
2. If analysis is not going to be performed immediately, cell monolayers are frozen in liquid N_2 and stored at $-20°C$ until analysis.
3. 50 µl of saline is added to each well and the plate is sonicated with ten short pulses to allow complete intracellular LDH leakage to medium. Cells do not usually detach from the bottom of the plate and there is no need to centrifuge the plates.
4. Supernatant is diluted in saline, if necessary, depending on the expected activity. The range of units/assay is limited within 0.5 and 12 U L^{-1}. For example, primary cultured hepatocytes are normally diluted 5–10 times.
5. 50 µl of diluted sample is put in a replicate 96-well plate.
6. 150 µl of working solution is added. It is very important to avoid bubble formation during pipetting. This solution contains detergents and easily forms many bubbles.
7. The plate is sonicated for better homogenization of mixture.
8. Wait 3 min before first reading the plate to allow all reactions to reach maximum.

Reading of results and calculations

1. Absorbance is read at 570 nm at different times, subtracting the value of the blank (incubation of saline with working solution).
2. Slope of the reaction is calculated (absorbance/time).
3. The following formula is applied to obtain the equivalent U/L of the sample:

$$\text{Activity (U L}^{-1}) = \frac{\Delta\text{absorbance}}{\Delta\text{time}} \times \frac{\text{Assay volume}}{\text{Sample volume}} \times \frac{1000}{\epsilon \times d} \quad \mu\text{mol} \times \text{min}^{-1} \text{ L}^{-1}$$

Since $d = 5$ mm, $\epsilon = 1.72 \times 10^4$ L mol^{-1} cm^{-1}, assay volume = 200, sample volume = 50, the calculations can be simplified to:

$$\text{Activity (U L}^{-1}) = \frac{\Delta\text{absorbance}}{\Delta\text{time}} \times 465.11 \quad \mu\text{mol min}^{-1} \text{ L}^{-1}$$

4. The value of activity obtained should be corrected with the dilution used in assay incubations.
5. The results are normalized considering control wells as 100%, expressing all results as a percentage of controls.
6. Using the appropriate method of calculation, normally by Logit transformation, IC50 and IC10 can be estimated.

Target-organ toxicity

Primary cultures of cells derived from different organs or tissues that retain specialized functions *in vitro* or that maintain specialized structures are also widely used in toxicology for organ-specific toxicity evaluation (Figure 5.6.1). At subcytotoxic concentrations (they should be below the IC10), that should not cause perceptible cell death, it is possible to design experiments to examine the interferences that a given xenobiotic can cause in the specialized functions of a target cell in culture (Jolles and Cordier 1992; Castell *et al.* 1997).

Drug metabolism

Biotransformation of xenobiotics is an evolutionary acquisition of higher organisms that enables them to eliminate lipophilic substances that otherwise might accumulate in tissues, thereby causing toxic effects. This process occurs at different levels in the organism, but the liver is the most active organ in metabolizing foreign compounds (Donato *et al.* 1992). Biotransformation of xenobiotics involves chemical modification of the compounds (Figure 5.6.3). Most of such processes are redox processes catalysed by a family of hemoproteins, namely, cytochrome P450-dependent monooxygenases (phase I reactions). The result is a new metabolite or metabolites that usually are more polar and reactive and are further conjugated by hepatocytes with endogenous molecules (i.e. glucuronic acid, glutathione, sulphate, amino acids, etc.). The new molecule renders N-bond or O-bond derivatives that are much more soluble (phase II reactions), thus facilitating the elimination of lipophilic substances. Although a biotransformation sequence gener-

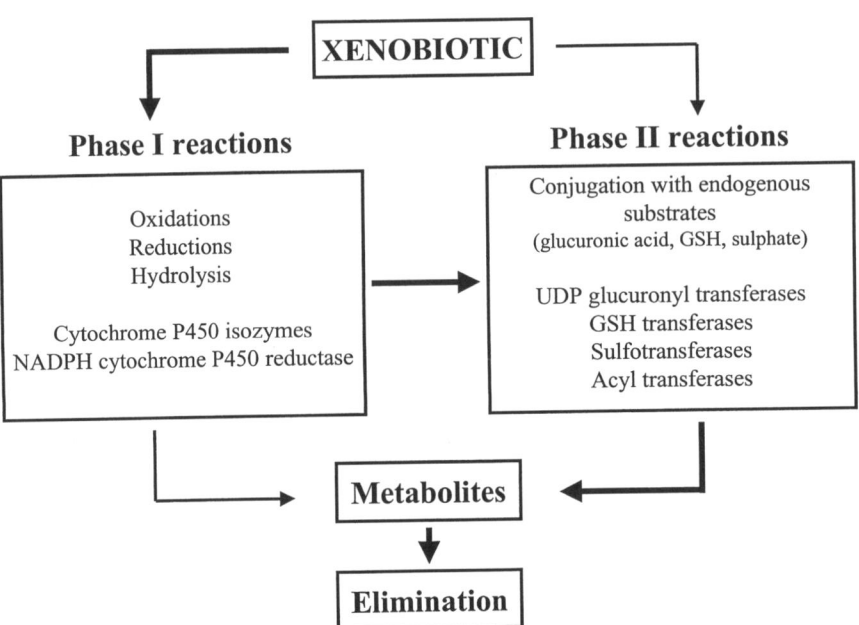

Figure 5.6.3 Phase I and Phase II reactions of drug-metabolism. Biotransformation reactions can be divided into two phases. In phase I the xenobiotic structure is modified by oxidation, reduction or hydrolysis. The cytochrome P450 system is responsible for the majority of the phase I reactions. phase II reactions consist of conjugation of xenobiotic or its Phase I metabolites with endogenous substrates such as glucuronic acid, glutathione or sulfate, thus facilitating the elimination of lipophilic substances.

ally parallels a detoxification process, there are many cases in which the metabolites formed after phase I reactions can cause deleterious effects on cells (Figure 5.6.4) (Pessayre 1986; King 1987). Hepatic metabolism of drugs is very frequently the cause of drug adverse reactions. Drug metabolism often shows significant species differences. In the case of pharmaceuticals, the relevance of such differences may be very important in terms of drug pharmacokinetics and the risk/benefits balance of a particular drug. Unfortunately, no animal model can accurately predict the human metabolism of drugs, and in some cases toxic effects on humans were not discovered until the first clinical trials. This has stimulated the use of human hepatocytes for investigating and anticipating the hepatic metabolism of new compounds before using them in humans (Guillouzo *et al.* 1993; Castell & Gómez-Lechón 1994; Skett *et al.* 1995). Using hepatocytes from different species, including human, allows comparative studies of the metabolism of a particular drug that are of great value in selecting the animal model closest to humans for use at a very early stage of the development of new drugs to investigate their effects (Bort *et al.* 1996a). Moreover, these studies make it possible to anticipate the hepatic metabolism of a compound.

The basic reason for using hepatocytes for drug metabolism studies is the assumption that these cells retain in culture their characteristic *in vivo* drug-metabolizing activities (Donato *et al.* 1992; Gómez-Lechón *et al.* 1997). Consequently, the level

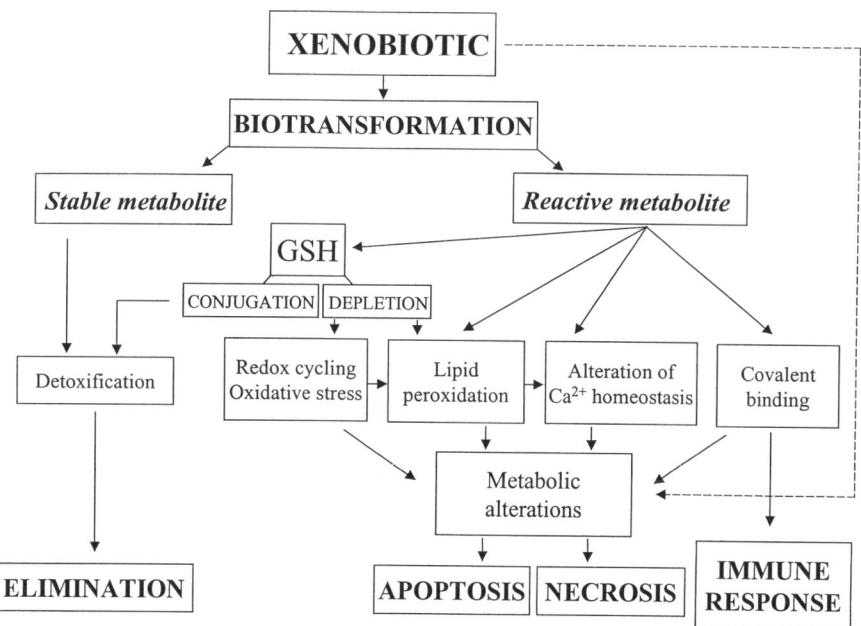

Figure 5.6.4 Mechanisms involved in toxicity of xenobiotics. Toxins may act directly on cellular systems or after biotransformation by hepatocytes. In the latter case, toxicity is ultimately the balance between bioactivation and detoxification which determines whether a reactive metabolite elicits a toxic effect. There are several processes known to play a role in the molecular events leading to irreversible cell damage and cell death by either necrosis or apoptosis.

of these enzymes is of critical concern, and although these activities gradually decrease in culture, they are expressed and can be induced by drugs in primary cultured cells for several days. Human, dog, and pig hepatocytes retain the expression of biotransformation enzymes longer than rat, mouse and rabbit hepatocytes (Donato *et al.* 1999).

For metabolic profile studies, drugs are incubated with intact hepatocytes, and after incubation the unaltered drug and the metabolites formed are recovered and identified by HPLC, GC/ME (Bort *et al.* 1996b, 1999a; Castell *et al.* 1997, 1998). When intact cells are used, no extra cofactors or NADPH-regenerating systems are required because living cells generate their own NADPH (Donato *et al.* 1992).

The use of primary cultures of human cells is restricted by the availability of human tissue. Moreover, as human liver cells do not grow in culture, the number of cells is also limited which makes it difficult to use human hepatocytes for routine testing. To overcome this limitation, genetically manipulated metabolic-competent cell lines have recently been developed by using expression vectors encoding full-length human cytochrome genes. Several groups succeeded in expressing fully active cytochrome P450 isozymes in non-hepatic cells (epithelial, lymphoid, yeasts) that constitutively express such activities under the control of a strong promoter (Crespi *et al.* 1991; Doehmer *et al.* 1992; Langebach *et al.* 1992; Macé *et al.* 1997; Castell *et al.* 1998). These cell lines will probably be a very useful tool for investigating the effect of single isozymes on the metabolism of a particular compounds, but in turn, it will be difficult to predict the *in vivo* metabolic profile of such compounds and further characterization of endogenous phase II enzyme expression. The use of human cell lines with an indefinite lifespan could serve as an alternative to human liver and primary hepatocyte cultures. In particular, highly differentiated hepatocellular carcinomas have been proposed as *in vitro* strategies for routine experiments on drug metabolism (Schuetz *et al.* 1993). However, hepatoma cells show very limited metabolic capacity, primarily due to the reduced expression of CYP activities (Donato *et al.* 1994; Castell *et al.* 1998), and, hence, they do not constitute a real alternative to primary cultured hepatocytes.

Another important issue is the assessment of systemic toxicity of drugs on different organs, and the possibility of ascertaining whether the toxic effects are elicited by the parent compound or by the metabolites produced in the liver. By co-culturing primary cultures of hepatocytes and target-organ cells in two compartment chambers but in the same culture medium it may be possible to investigate the effect of metabolites released to the medium by hepatocytes on the other cell type.

Molecular mechanisms involved in the toxicity of xenobiotics

The molecular mechanisms involved in the toxicity of xenobiotics are of major concern to toxicologists. While it is quite easy to determine the *in vivo* doses that produce toxicity, it is more difficult to find out why cell death occurs or what event leads to an irreversible change in the living system that in the end is responsible for cell death. Some xenobiotics are electrophilic in nature, and others are biotransformed by the liver or other tissues to highly reactive metabolites generally more toxic than the parent compound. This activation process is the key to many toxic phenomena. Some of the metabolites formed are potent electrophiles or carbon-centered radicals capable of reacting with nucleophiles. There are several processes known to play a role in the molecular events leading to irreversible cell damage (Figure 5.6.4):

1. *Lipid peroxidation.* As a consequence of the toxic action of many xenobiotics, lipids of cellular membranes are the target site of action and peroxides are formed in their aliphatic chain. These peroxides produce the increasing degradation of lipids that finally may disrupt the structure and functionality of the membranes leading to cell death.

2. *Alteration of intracellular Ca^{2+} concentration.* Intracellular calcium participates in many cellular functions and its levels should be perfectly regulated to obtain a proper cell function; many substances can interfere in intracellular calcium homeostasis control leading to cell malfunction and death.

3. *Oxidative stress* is produced by compounds able to undergo repeated oxidation and reduction cycles within the cell. A representative example of this type of toxin are quinones that can be partially reduced by reductases to semiquinones with NADPH consumption. The semiquinone can react readily with molecular oxygen in a one electron oxidation, thereby regenerating the parent quinone. This redox cycling causes the continuous production of reactive oxygen species (superoxide anions), and depletion of GSH and nicotinamide nucleotide pools, with a concomitant increase in lipid peroxidation and Ca^{2+} release. The ability of cells to reduce all radicals with GSH to avoid modification of basic macromolecules (proteins, lipids, etc.) protects the cell against the xenobiotic. When GSH is depleted, cellular malfunction and death can be promoted. It is important to measure intracellular GSH levels to determine the toxic mechanism of a given compound.

4. *Covalent binding* of xenobiotics or their metabolites to biomolecules alters cell metabolism and eventually causes DNA damage. Against these potential hazards, cells have their own defence mechanisms (GSH, DNA-repair, suicide inactivation, etc.). Ultimately, toxicity is the balance between bioactivation and detoxification which determines whether a reactive metabolite elicits a toxic effect or not (Acosta 1990; Jover *et al.* 1992; Castell *et al.* 1997).

Measurement of GSH levels

Materials and equipment

- Sodium phosphate buffer, 0.1 M, pH 8, supplemented with 5 mM EDTA. Stored at 4°C. Stable.
- Homogenization solution: 5% Trichloroacetic acid, 2 mM EDTA in H_2O. Stored at room temperature. Stable.
- NaOH 1 M. Stored at room temperature. Stable.
- GSH Standard solution: 10 mM GSH in homogenization solution. Prepared daily.
- 10 mg ml^{-1} *o*-phtaldialdehyde solution in methanol for spectroscopy. Prepared daily, protected from light.
- Centrifuge, for 24-well plates.

- 96-well plates.
- Multichannel pipettes.
- Ultrasound bath suitable for a 96-well plate.
- Water bath.
- Microplate fluorescence reader equipped with filters; 360 nm for excitation and 450 nm for emission.

Experimental procedure

1. After treatment of cells with xenobiotics, cell monolayers are washed with PBS and culture plates are stored at –20°C until analysis. The following steps are described for a 24-well plate.
2. 150 μl of homogenization solution is added and the plate is sonicated 10 times during short pulses of 2 seconds.
3. The plate is centrifuged for 20 min at 3000 g to allow sedimentation of precipitated protein.
4. 50 μl of supernatant is carefully taken in duplicate and transferred to a 96-well plate. The remaining homogenization solution can be discarded and the protein content of monolayers can be measured in the monolayers that should still be attached to wells.
5. A calibration curve of GSH is prepared with 50 μl of sample in the range 0–5 nmols well^{-1}.
6. 190 μl of sodium phosphate buffer supplemented with 1 M NaOH (175 : 15) is added to wells.
7. 10 μl of *o*-phtaldialdehyde solution is finally added.
8. The solution is homogenized in an ultrasound bath, 4–5 short pulses of 2 s.
9. The plate is kept in the dark for 30 min.

Reading of results and calculations

1. Fluorescence is read in the microplate fluorescence reader at 360 nm for excitation and 450 nm for emission.
2. With the standards the linear regression curve is calculated and the fluorescence of the test samples is interpolated. After protein measurement of wells, results can be expressed as nmol of GSH mg^{-1} cellular protein.

Lipid peroxidation

Materials and equipment

- MDA Standard solution: 2 mM malondialdehyde in H_2O. A previous dilution of MDA in methanol (MDA : methanol; 1 : 5) is recommended to enhance miscibility of MDA in water. Stored a 4°C. Stable for a month.
- 1% Phosphotungstic acid in H_2O. Stored at room temperature. Stable.
- 0.67 % Thiobarbituric acid (TBA) in H_2O. Stored at 4°C. If it precipitates, it can be redissolved in a water bath at 37°C. Stable for a month.
- 7% Sodium dodecyl sulphate (SDS) in H_2O. Stored at room temperature. Stable.

- 0.1 N HCl. Stored at room temperature, protected from light. Stable.
- Butanol.
- 96-well plates.
- Centrifuge for tubes.
- Water bath.
- Microplate fluorescence reader equipped with filters; 530 nm for excitation and 595 nm for emission.

Experimental procedure

1. After treatment of cells with xenobiotics, culture medium is collected and centrifuged, if necessary, to remove cells. The medium is then stored at –20°C until analysis. An aliquot of culture medium used for cell incubations should be taken to prepare calibration curve of MDA.
2. Since lipid peroxidation is hardly detectable on cell monolayers, plates may either be washed and continue the culture or the plates can be kept for other assays like GSH or LDH.
3. A calibration curve is prepared by diluting the standard in culture medium in the range 0–1.6 nmol MDA/assay (250 μl of sample)
4. Pipette in a tube: 250 μl sample or standard ; 100 μl of 7% SDS; 1 ml 0.1 N HCl; 150 μl 1% phosphotungstic acid and 500 μl 0.67 % TBA.
5. The tubes are mixed and incubated for 1 h in a boiling bath, protected from light.
6. The tubes are coded in an ice bath.
7. 1 ml of butanol is added and vigorously mixed to allow complete extraction of reaction product.
8. The mixture is centrifuged at 3000 rpm for 10 min.
9. The organic phase of each tube is collected and 250 μl is placed in the wells of a 96-well plate.

Reading of results and calculations

1. Fluorescence is read in the microplate fluorescence reader at 530 nm for excitation and 595 nm for emission.
2. With the standards, the linear regression curve is calculated and the fluorescence of the problem samples interpolated. Results are given as pmol of Thiobarbituric acid reacting substances (TBA-RS) per plate.

Alteration of intracellular Ca^{2+} levels

Materials and equipment

- 5 μM Fluo-3 solution: prepared daily in culture medium containing 5 μM Fluo-3-AM (acetomethyl ester, cell permeant), supplemented with 0.0075% pluronic acid (helps Fluo-3 to enter the cell)
- 8 mM Digitonin in DMSO. Stable at room temperature. This solution is diluted daily at 270 μM in HEPES saline solution.

- 100 mM EGTA in 0.25 M NaOH. Stable at room temperature.
- A HEPES saline solution, containing 2 mM $CaCl_2$; 10 mM HEPES; 137 mM NaCl; 5 mM KCl; 1 mM; Na_2HPO_4; 0.5 mM $MgCl_2$; 10 mM glucose and 1% albumin. Store at $-20°C$ and add $CaCl_2$ daily.
- 96-well plates.
- Multichannel pipettes.
- Microplate fluorescence reader equipped with filters; 485 nm for excitation and 530 nm for emission . It is highly recommended that optical components are under the plate since cell monolayers are at the bottom of the wells.

Experimental procedure

1 After treatment of cells with xenobiotics, culture medium is collected and 40 μl well^{-1} (in a 96-well plate) of 5 μM Fluo-3 solution is added.
2 The plate is incubated for 30 min in the incubator.
3 Cell monolayers are washed three times with HEPES saline solution.
4 40 μl of HEPES saline solution is added and fluorescence (F) is read with the filters mentioned above.
5 5 μl well^{-1} of 270 μM Digitonin, final concentration 30 μM is added. This solution is added to disrupt all cellular membranes, allowing all Fluo-3 captured by cells reach calcium present in HEPES saline solution.
6 After 5 min fluorescence (F_{max}) is read.
7 5 μl well^{-1} of 100 mM EGTA solution, final concentration 10 mM is added. This solution chelates all calcium present in medium, reducing fluorescence to a minimum.
8 After 5 min fluorescence (F_{min}) is read.

Reading of results and calculations

1 Three fluorescence values are obtained for each well: F, F_{max} and F_{min}, corresponding to the fluorescence after Fluo-3 incubation, maximum fluorescence and minimum fluorescence.
2 The calculations should be performed according to the following formula:

$$[Ca^{2+}]_i \text{ (nM)} = 400 \times \frac{(F - F_{min})}{(F_{max} - F)},$$

where 400 is the value of the dissociation constant for calcium and Fluo-3.

APPLICATIONS OF *IN VITRO* METHODS IN BIOMEDICAL TOXICOLOGY

An *in vitro* assay cannot give a straightforward answer as to whether a certain compound will or will not be toxic for humans. However, skilful interpretation of the scientific information generated in experiments using human-relevant cells can be of great value in making decisions during the development of new pharmaceuticals. Several representative examples of the use of *in vitro* methods for evaluating the potential risk of toxicity of xenobiotics in humans can be mentioned:

1. Classification and labeling of chemicals on the basis of their increasing molar toxicity *in vitro* (toxic potential) (Seibert *et al.* 1996; Ponsoda *et al.* 1997; Bottrill 1998; Clemedson *et al.* 1998).
2. Risk assessment for target-organ toxicity and the major metabolic alteration to the target-cell (cardiotoxicity, immunotoxicity, nephrotoxicity, neurotoxicity, hepatotoxicity, etc.) caused by a given drug can be foreseen (Massey 1989; Nelson & Brenneman 1994; Hawksworth *et al.* 1995; Gad 1994; Gribaldo *et al.* 1996; Castell *et al.* 1997; Meredith & Miller 1997; Bort *et al.* 1999b).
3. Developmental toxicology (Whittaker & Faustman 1994).
4. Evaluation of the suitability and toxicity of biomaterials used in odontology, traumatology, etc. (Cortina *et al.* 1997; Fubini *et al.* 1998).
5. Local irritation tests (ocular and cutaneous) which replace the conventional *in vivo* tests (Loprieno 1995; Botham *et al.* 1998; Balls *et al.* 1999).
6. Systemic toxicity, evaluating the toxic effects on different target-organs, helping to predict acute toxicity of xenobiotics to humans (Barratt *et al.* 1995).
7. An upper limit of the AUC curve (concentration × time) beyond which a drug is likely to have toxic effects on human cells *in vivo* can be estimated (Krishnan and Andersen 1994).
8. Detection of chemical mutagens by evaluating their potential carcinogenic activity on cells (Jolles and Cordier 1992; Macé *et al.* 1997; Mitchell & Combes 1997).
9. Phototoxicity of drugs evaluated on cellular models (Spielmann *et al.* 1994; Boscá *et al.* 1995; Miranda *et al.* 1997).
10. The metabolic profile of a new drug in humans can be predicted by using human hepatocytes. Comparative studies of drug metabolism in hepatocytes from different species can be carried out to help select the animal model closest to humans for investigating drug effects (Bort *et al.* 1996a,b, 1999a,b).
11. The molecular mechanisms involved in the toxicological effect can be investigated (Acosta 1990; Castell *et al.* 1997).
12. The evaluation of drug-drug interactions (Jover *et al.* 1991; Li *et al.* 1997).
13. Ecotoxicology (Walker 1998).

REFERENCES

Acosta D (1990) *Cellular and Molecular Toxicology and In Vitro Toxicology*. CRC Press, Boston, MA.

Anderson R, O'Hare M, Balls M, Brady M, Brahams D, Burt A, Chesné C, Combes R, Dennison A, Garthoff B, Hawksworth G, Kalter E, Lechat A, Mayer D, Rogiers V, Sladowski D, Southee J, Trafford J, van der Valk J & van Zeller AM (1998) The availability of human tissue for biomedical research. *ATLA – Alternatives to Laboratory Animals* 26: 763–777.

Balls M, Berg N, Bruner LH, Curren RD, de Silva O, Earl LK, Esdaile DJ, Fentem JH, Liebsch M, Ohno Y, Prinsen MK Spielmann H & Worth AP (1999) Eye irritation testing: The way forward. *ATLA – Alternatives to Laboratory Animals* 27: 53–71.

Balls M & Fentem JH (1992) The use of basal cytotoxicity and target organ toxicity tests in hazard identification and risk assessment. *ATLA – Alternatives to Laboratory Animals* 20: 368–388.

Balls M, Goldberg AM, Fentem J, Broadhead CL, Burch RL, Festing MFW, Frazier JM, Hendriksen CFM, Jennings M, van der Kamp MDO, Morton DB, Rowan AN, Russell C, Russell WMS, Spielmann H, Stephens ML, Stokes WS, Straughan DW, Yager JD, Zurlo J & van Zutphen BFM (1995) The three Rs : The way forward. *ATLA – Alternatives to Laboratory Animals* 23: 838–866.

Barile F (1997) Continuous cell lines as a model for drug toxicity assessment. In: Castell JV and Gómez-Lechón MJ (eds) *In Vitro Methods in Pharmaceutical Research*, pp. 33–54. Academic Press, London.

Barrat MD, Castell JV, Chamberlain M, Combes RD, Dearden JC, Fentem JH, Gerner I, Giuliani A, Gray TJ, Livingstone DJ, Provan WN, Rutten FAJJL, Verhaar HJM & Zbinden P (1995) The integrated use of alternative approaches for predicting toxic hazard. *ATLA – Alternatives to Laboratory Animals* 23: 410–429.

Benford DJ, Gibson GG & Bridges TW (1987) *Drug Metabolism from Molecules to Man*. Taylor and Francis, London.

Bort R, Macé K, Boobis A, Gómez-Lechón MJ, Pfeifer A, & Castell JV (1999a) Hepatic metabolism of diclofenac. Role of human CYPs in the minor oxidative pathways. *Biochemical Pharmacology* 58: 787–796.

Bort R, Ponsoda X, Carrasco E, Gómez-Lechón MJ & Castell JV (1996a). Comparative metabolism of non-steroidal anti-inflamatory drug, aceclofenac, in the rat, monkey, and human. *Drug Metabolism and Disposition* 24: 969–975.

Bort R, Ponsoda X, Carrasco E, Gómez-Lechón MJ & Castell JV (1996b) Metabolism of aceclofenac in humans. *Drug Metabolism and Disposition* 24: 834–841.

Bort R, Ponsoda X, Jover R, Gómez-Lechón MJ & Castell JV (1999b) Diclofenac toxicity to hepatocytes: A role for drug metabolism in cell toxicity. *Journal of Pharmacology and Experimental Therapeutics* 288: 65–72.

Boscá F, Carganico G, Castell JV, Gómez-Lechón MJ, Hernández D, Mauleón D, Martínez LA & Miranda MA (1995) Evaluation of ketoprofen (R,S and R/S) phototoxicity by a battery of *in vitro* assays. *Journal of Photochemistry and Photobiology* 31: 133–138.

Botham PA, Earl LK, Fentem JH, Roguet R & van de Sandt JJM (1998) Alternative methods for skin irritation testing : the current status. *ATLA – Alternatives to Laboratory Animals* 26: 195–211.

Bottrill K (1998) The use of biomarkers as alternatives to current animal tests on food chemicals. *ATLA – Alternatives to Laboratory Animals* 26: 421–480.

Castell JV & Gómez-Lechón MJ (1992) The in vitro evaluation of the potential risk of hepatotoxicity of drugs. In: Castell JV and Gómez-Lechón MJ (eds) *In Vitro Alternatives to Animal Pharmaco-toxicology*, pp. 179–204 Farmaindustria, Barcelona.

Castell JV & Gómez-Lechón MJ (1994) Hepatocytes in vitro pharmacotoxicology, toxicology and drug metabolism. In: Spier RE, Griffiths JB and Berthold W (eds) *Animal Cell Technology: Products for today, prospects for tomorrow*, pp. 711–721. Butterworth-Heinemann, Oxford.

Castell JV, Gómez-Lechón MJ, Ponsoda X & Bort R (1997) *In vitro* investigation of the molecular mechanisms of hepatotoxicity. In: Castell JV & Gómez-Lechón MJ (eds) *In Vitro Methods in Pharmaceutical Research*, pp. 375–410. Academic Press, London.

Castell JV, Jover R, Bort R & Gómez-Lechón MJ (1998) The challenge of using hepatic cell lines for drug metabolism. In: Boobis AR, Kremers P, Pelkonen O & Pithan K (eds) *Proceedings of COST B1 European Symposium on the Prediction of Drug Metabolism in Man : Progress and problems*, pp. 77–92. European Commission, Liege.

Clemedson C, Andersson M, Aoki Y, Barile FA, Bassi AM, Calleja MC, Castaño A, Clothier RH, Dierickx P, Ekwall B, Ferro M, Fikesjo G, Garza-Ocañas L, Gómez-Lechón MJ, Gulden M, Hall T, Imai K, Isomaa B, Kahru A, Kerszman G, Kjellstrand P, Kristen U, Kunimoto M, Karenlampi S, Lewan L, Lilius H, Loukianov A, Monaco F, Ohno T, Persoone G, Romert L, Sawyer TW, Segner H, Seibert H, Shrivastava R, Sjostrom M, Stammati A, Tanaka N, Thuvander A, Torres Alanis O, Valentino M, Wakuri S, Walum E, Wang XH, Wieslander A, Zucco F & Ekwall B (1998) MEIC Evaluation of acute systemic toxicity. Part IV. *In vitro* results from 67 toxicity assays used to test reference chemicals

31–50 and comparative cytotoxicity analysis. *ATLA – Alternatives to Laboratory Animals* 26: 131–183.

Cortina P, Gómez-Lechón MJ, Navea A, Menezo JL, Terencio MC & Díaz-Llopis M (1997) Diclofenac sodium and cyclosporin A inhibit human lens epithelial cell proliferation in culture. *Graefe's Archives of Clinical and Experimental Ophtalmology* 235: 180–185.

Crespi CL, Gonzalez FJ, Steimel DT, Turner TR, Gelboin HV, Penman BW & Langenbach R (1991) A metabolically competent human cell line expressing five cDNAs encoding procarcinogen-activating enzymes: Application to mutagenicity testing. *Chemical Research in Toxicology* 4: 566–572.

Doehmer J, Wolfel C, Dogra S, Doehmer C, Seidel A, Platt KL, Oesch F & Glatt HR (1992) Applications of stable V79-derived cell lines expressing rat cytochrmes P4501A1, 1A2, and 2B1. *Xenobiotica* 22: 1093–1099.

Donato MT, Bassi AM, Gómez-Lechón MJ, Penco S, Herrero E, Adamo D, Castell JV & Ferro M (1994) Evaluation of the xenobiotic biotransformation capability of six rodent hepatoma cell lines in comparison with rat hepatocytes. *In Vitro Cellular and Developmental Biology* 30A: 574–580.

Donato MT, Castell JV & Gómez-Lechón MJ (1999) Characterization of drug metabolizing activities in pig hepatocytes for use in bioartificial liver devices: Comparison with other hepatic cellular models. *Journal of Hepatology* (in press).

Donato MT, Gómez-Lechón MJ & Castell JV (1992) Biotransformation of drugs by cultured hepatocytes. In: Castell JV & Gómez-Lechón MJ (eds) *In Vitro Alternatives to Animal Pharmaco-toxicology*, pp. 149–178. Farmaindustria, Barcelona.

Ekwall B & Ekwall K (1988) Comments on the use of diverse cell systems in toxicity testing. *ATLA – Alternatives to Laboratory Animals* 15: 193–201.

Ekwall B, Silano V, Paganuzzi-Stammati A & Zucco F (1990) Toxicity tests with mammalian cell cultures. In: Bourdeau P (ed.) *Short-Term Toxicity Tests for Non-Genotoxic Effects*, pp. 75–97. Wiley, New York.

Flint PO (1990) *In vitro* toxicity testing: Purpose, validation and strategy. *ATLA – Alternatives to Laboratory Animals* 18: 11–18.

Fubini B, Aust AE, Bolton RE, Borm PJA, Bruch J, Ciapetti G, Donaldson K, Elias Z, Gold J, Jaurand MC, Kane AB, Lison D & Muhle H (1998) Non-animal tests for evaluating the toxicity of solid xenobiotics. *ATLA – Alternatives to Laboratory Animals* 26: 579–617.

Gad SC (1994) Gastrointestinal toxicology: *In Vitro* test systems. In: Gad SC (ed.) *In Vitro Toxicology*, pp. 231–238. Raven Press, New York.

Gómez-Lechón MJ, Donato T, Ponsoda X, Fabra R, Trullenque R & Castell JV (1997) Isolation, culture and use of human hepatocytes in drug research. In: Castell JV & Gómez-Lechón MJ (eds) *In Vitro Methods in Pharmaceutical Research*, pp. 129–153. Academic Press, London.

Gribaldo L, Bueren J, Deldar A, Hokland P, Meredith C, Moneta D, Mosesso P, Parchment R, Parent-Massin A, Pessina A, San Roman J & Schoeters G (1996) The use of *in vitro* systems for evaluating haematotoxicity. *ATLA – Alternatives to Laboratory Animals* 24: 211–231.

Guillouzo A, Morel F, Fardel O & Meunier B (1993) Use of human hepatocyte cultures for drug metabolism studies. *Toxicology* 82: 209–219.

Hawksworth GM, Bach PH, Nagelkerke JF, Dekant W, Diezi JE, Harpur E, Lock EA, MacDonald C, Morin JP, Pfaller W, Rutten AA, Ryan MP, Toutain HJ & Trevison A (1995) Nephrotoxicity testing *in vitro*. *ATLA – Alternatives to Laboratory Animals* 23: 713–727.

Jolles G & Cordier A (1992) *In Vitro Methods in Toxicology*, Academic Press, London.

Jover R, Ponsoda X, Castell JV & Gómez-Lechón MJ (1992) Evaluation of the cytotoxicity of ten chemicals on human cultured hepatocytes: Predictability of human toxicity and comparison with rodent cellular systems. *Toxicology in Vitro* 6: 47–52.

Jover R, Ponsoda X, Gómez-Lechón MJ, Guerrero C, del Pino J, & Castell JV (1991) Potentiation of cocaine hepatotoxicity by ethanol in human hepatocytes. *Toxicology and Applied Pharmacology* 107: 526–534.

King LJ (1987) Metabolism and mechanisms of toxicity: an overview. In: Benford D J, Bridges JW & Gibson GG (eds) *Drug Metabolism from Molecules to Man*, pp. 657–668. Taylor and Francis, London.

Krishnan K & Andersen ME (1994) Physiologically based pharmacokinetic modeling in toxicology. In: Hayes AW (ed.) *Principles and Methods of Toxicology*, pp. 149–188. Raven Press, New York.

Langebach R, Smith PB & Crespi C (1992) Recombinant DNA approaches for the development of metabolic systems used in in vitro toxicology. *Mutation Research* 277: 251–275.

Li AP, Maurel P, Gómez-Lechón MJ, Cheng LC & Jurima-Romert M (1997) Preclinical evaluation of drug-drug interaction potential: present status of the application of primary human hepatocytes in the evaluation of cytochrome P450 induction. *Chemico-Biological Interactions* 107: 5–16.

Loprieno N (1995) *Alternative Methodologies for the Safety Evaluation of the Chemicals in the Cosmetic Industry*. CRC Press, New York.

Macé K, Offord EA & Pfeifer AMA (1997) Drug metabolism and carcinogen activation studies with human genetically engineered cells. In: Castell JV and Gómez-Lechón MJ (eds) *In Vitro Methods in Pharmaceutical Research*, pp 433–456. Academic Press, London.

Massey TE (1989) Isolation and use of lung cells in toxicology. In: McQueen ChA (ed.) *In Vitro Toxicology : Model Systems and Methods*, pp. 35–72. The Telford Press, New Jersey.

Meredith C & Miller K (1997) Immunotoxicology *in vitro*. In: Castell JV and Gómez-Lechón MJ (eds) *In Vitro Methods in Pharmaceutical Research*, pp. 225–240. Academic Press, London.

Miranda MA, Castell JV, Sarabia Z, Hernández D, Puertes I, Morera MI & Gómez-Lechón MJ (1997) Mechanisms of photosensitization of drugs : Involvement of tyrosines in the photomodification of proteins mediated by tiaprofenic acid *in vitro*. *Toxicology In Vitro* 11: 653–659.

Mitchell IG & Combes RD (1997) *In vitro* genotoxicity and cell transformation assessment. In: Castell JV and Gómez-Lechón MJ (eds) *In Vitro Methods in Pharmaceutical Research*, pp. 317–352. Academic Press, London.

Nelson PG & Brenneman DE (1994) Neurotoxicology *in vitro*. In: Gad SC (ed.) *In vitro Toxicology*, pp. 123–148. Raven Press, New York.

O'Hare S & Atterwill CK (1995) *In Vitro Toxicity Testing Protocols*. Humana Press Inc., New Jersey.

Pessayre D (1986) In: Fillastre JP (ed.) *Hepatotoxicity of Drugs*, pp. 39–62. Editions INSERM, Rouen.

Ponsoda X, Gómez-Lechón MJ & Castell JV (1998) Toxicity and cell density monitoring in monolayer and three-dimensional cultures with the XTT assay. *ATLA – Alternatives to Laboratory Animals* 26: 331–342.

Ponsoda X, Jover R, Castell JV & Gómez-Lechón MJ (1991a) Measurement of intracellular LDH activity in 96-well cultures: A rapid and automated assay for cytotoxicity studies. *Journal of Tissue Culture Methods* 13: 21–24.

Ponsoda X, Jover R, Gómez-Lechón MJ, Fabra R, Trullenque R & Castell JV (1991b) The effects of buprenorphine on the metabolism of human hepatocytes. *Toxicology in Vitro* 5: 219–224.

Ponsoda X, Núñez C, Castell JV & Gómez-Lechón MJ. (1997) Evaluation of the cytotoxic effects of MEIC chemicals 31–50 on primary culture of rat hepatocytes and hepatic and non- hepatic cell lines. *ATLA – Alternatives to Laboratory Animals* 25: 423–436.

Russell WMS & Burch RL (1959) *The Principles of Humane Experimental Technique*. Methuen, London.

Schuetz EG, Schuetz JD, Strom SC, Thompson MT, Fisher RA, Molowa DT, Li D & Guzelian PS (1993) Regulation of human liver cytochromes P-450 in family 3A in primary and continuous culture of human hepatocytes. *Hepatology* 18: 1254–1262.

Seibert H, Balls M, Fentem J, Bianchi V, Clothier RH, Dierickx P J, Elwall B, Garle MJ, Gomez-Lechón MJ, Gribalbo L, Gulden M, Liebsh M, Ramussen E, Roguet R, Shivrastava R & Walum E (1996) Acute Toxicity testing *in vitro* and classification and labeling of chemicals. *ATLA – Alternatives to Laboratory Animals* 24: 499–510.

Shrivastava R, John GW, Rispat G, Chevalier A & Massingham R (1991) Can the *in vivo* maximum tolerated dose be predicted using *in vitro* techniques? A working hypothesis. *ATLA – Alternatives to Laboratory Animals* 19: 393–402.

Skett P, Tyson C, Guillouzo A & Maier P (1995) Report of the International Workshop on the use of human *in vitro* liver preparations to study drug metabolism in drug development. *Biochemical Pharmacology* 17: 280–285.

Spielmann H, Lowell WW, Holzle E, Johnson BE, Maurer T, Miranda MA, Pape WJW, Sapora O & Sladowski D (1994) *In vitro* phototoxicity testing. *ATLA – Alternatives to Laboratory Animals* 22: 314–348.

Walker CH (1998) Alternative approaches and tests in ecotoxicology: A review of the present position and the prospects for change, taking into account ECVAM's duties, topic selection and test criteria. *ATLA – Alternatives to Laboratory Animals* 26: 649–677.

Whittaker SG & Faustman EM (1994) *In vitro* assays for developmental toxicity. In: Gad SC (ed.) *In Vitro Toxicology*, pp. 97–122. Raven Press, New York.

Zucco F (1992) Use of continuous cell lines for toxicological studies. In: Castell JV and Gómez-Lechón MJ (eds) *In Vitro Alternatives to Animal Pharmaco-toxicology*, pp. 43–68. Farmaindustria, Barcelona.

APPENDIX 1: TERMINOLOGY

Many people wonder why one should spend time worrying about scientific terminology when, it is said, people will continue to use terms with which they are comfortable in spite of any communication difficulties that occur or action taken by standardization committees. They often point out that major discoveries have been, and continue to be, made in spite of any problems with communication. On the contrary, it should be a major concern of scientists that their ideas be properly conveyed to others. According to the dictionary, to communicate is to 'impart or give information'. Many feel that scientists have not done particularly well in communicating with the lay community (Schaeffer 1984; Iglewski 1989). That this has become a major problem is exemplified by the difficulties that the scientific community encounters when attempting to influence legislation dealing with funding for research. There has been a continual erosion of the money allocated to biomedical research, on a worldwide scale, and this has created a virtual crisis in research and training of future scientists. It may also be argued that a significant aspect of the poor communication with the lay community stems from the fact that scientists do not communicate as well as they should with each other. This can be seen in the microcosm of the larger scientific community constituted by researchers using the techniques of cell and tissue culture.

SOME ASPECTS OF THE PROBLEM

Communication within scientific disciplines has long been efficient via a system of jargon. Individuals within particular fields communicate efficiently but individuals outside the field or discipline find that jargon is simply, as defined in the dictionary, 'unintelligible talk'. Some feel that this is not a major problem in fairly circumscribed fields because there is little necessity for 'outsiders' to understand technical presentations. This may be true for such fields of research. However, difficulties arise when areas of research become interdisciplinary. Then the jargon, techniques and body of information must be understood widely and in diverse disciplines. Terminology adopted by individuals who have not previously used it is often changed and misused, causing confusion. That the above is so is confirmed by the confusion in the fields of vertebrate, invertebrate and plant cell culture (hereafter referred to only as cell culture), molecular biology and molecular genetics over the use and abuse of terminology.

There is hardly a field of biological investigation in which cell cultures or the techniques of molecular biology and molecular genetics are not used. These technologies are being employed by an ever-widening group of researchers who are, therefore, communicating more globally via scientific presentations, publications and research

Cell and Tissue Culture for Medical Research, edited by A. Doyle and J.B. Griffiths.
© 2000 John Wiley & Sons, Ltd

proposals. Unfortunately, often the communicator and recipient represent different areas of specialization and have been brought together by the common technology used in their work. In such situations, misuse of terminology can prove unfortunate indeed, leading to inability to repeat a piece of research, or the notion of relative incompetence in completing a task proposed for a research project, or problems in publishing a paper describing a completed research project. There needs to be a major standardization of the terminology associated with cell culture, molecular biology and molecular genetics lest the situation becomes even worse.

SOLUTIONS TO THE PROBLEM

Because our goal is effective communication, we should be striving to make our communications as lucid as possible. We should make every effort to describe phenomena so that others can repeat and, hopefully, reproduce our findings.

It is important for a commission, such as the Society for *In Vitro* Biology (formerly the Tissue Culture Association Terminology Committee), wherein collective wisdom is brought to bear on an issue, to come forth with definitions that can be adopted by the scientific community. Because these experts come to agreements on terms and phrases after much deliberation, their definitions should carry greater weight than any used in an individual publication. Misuse of terms is widespread and is probably a result of *individual vis-à-vis collective* use of terms. In oral communication, it is easy to explain ambiguities. However, with written communication, where there is no immediate dialogue, it is more difficult to do so. For a manuscript, a grant proposal or a report to a special interest group, it is important not to be misunderstood. Therefore, only precisely defined terms, which are universally accepted, should be used.

In the interest of helping to alleviate the problems mentioned above, the following compendium of terms, compiled and defined by the Tissue Culture Association Terminology Committee (Mueller *et al.* 1990), is reprinted. The glossary is reprinted with the permission of the Society for *In Vitro* Biology (formerly the Tissue Culture Association), Largo, Maryland, USA.

TERMINOLOGY ASSOCIATED WITH CELL, TISSUE AND ORGAN CULTURE, MOLECULAR BIOLOGY AND MOLECULAR GENETICS

Adventitious: Developing from unusual points of origin, such as shoots or root tissues from callus or embryos from sources other than zygotes. This term can also be used to describe agents that contaminate cell cultures.

Anchorage-dependent cells or cultures: Cells, or cultures derived from them, that will grow, survive or maintain function only when attached to a surface such as glass or plastic. The use of this term does not imply that the cells are normal or that they are or are not transformed neoplastically.

Aneuploid: The situation that exists when the nucleus of a cell does not contain an exact multiple of the haploid number of chromosomes; one or more chromosomes being present in greater or lesser number than the rest. The chromosomes may or may not show rearrangements.

Apoptosis: A naturally occurring process of cell death that plays a complementary, but opposite, role to mitosis in regulation of animal cell populations. This is an active cellular process that provides a means for precisely regulating cell numbers and, therefore, biological activity. Unlike simple degeneration, cellular death that is dependent on active participation of cellular components can, potentially, be suppressed. Inhibition of cell death may contribute to oncogenesis. Characteristics of apoptosis may include: volume reduction through blebbing of the cell surface (membranes remain intact; there is no cell lysis but there is a selective loss of intracellular fluids); chromatin condensation (activation of nuclear endonucleases that cleave chromatin at internucleosomal sites); and cell surface changes that allow recognition and disposition by phagocytic cells before they autolyse.

Attachment efficiency: The percentage of cells plated (seeded, inoculated) that attach to the surface of the culture vessel within a specified period of time. The conditions under which such a determination is made should always be stated.

Autocrine cell: In animals, a cell that produces hormones, growth factors or other signalling substances for which it also expresses the corresponding receptors. See also 'Endocrine' and 'Paracrine'.

Axenic culture: A culture without foreign or undesired life forms. An axenic culture may include the purposeful co-cultivation of different types of cells, tissues or organisms.

Cell culture: Term used to denote the maintenance or cultivation of cells *in vitro*, including the culture of single cells. In cell cultures, the cells are no longer organized into tissues.

Cell generation time: The interval between consecutive divisions of a cell. This interval can best be determined, at present, with the aid of cinephotomicrography. *This term is not synonymous with 'population doubling time'.*

Cell hybridization: The fusion of two or more dissimilar cells leading to the formation of a synkaryon.

Cell line: A cell line arises from a primary culture at the time of the first successful subculture. The term cell line implies that cultures from it consist of lineages of cells originally present in the primary culture. The terms **finite** or **continuous** are used as prefixes if the status of the culture is known. If not, the term **line** will suffice. The term **continuous line** replaces the term **established line**. In any published description of a culture, one must make every attempt to publish the characterization or history of the culture. If such has already been published, a reference to the original publication must be made. In obtaining a culture from another laboratory, the proper designation of the culture, as originally named and described, must be maintained and any deviations in cultivation from the original must be reported in any publication.

Cell strain: A cell strain is derived either from a primary culture or a cell line by the selection or cloning of cells having specific properties or markers. In describing a cell strain, its specific features must be defined. The terms **finite** or **continuous** are to be used as prefixes if the status of the culture is known. If not, the term **strain** will suffice. In any published description of a cell strain, one must make every attempt to publish the characterization or history of the strain. If such has already been published, a reference to the original publication must be made. In obtaining a culture from another laboratory, the proper designation of the culture, as originally named and described, must be maintained and any deviations in cultivation from the original must be reported in any publication.

Chemically defined medium: A nutritive solution for culturing cells in which each component is specifiable and, ideally, of known chemical structure.

Clone: In animal cell culture terminology, a population of cells derived from a single cell by mitoses. A clone is not necessarily homogeneous and, therefore, the terms **clone** and **cloned** do not indicate homogeneity in a cell population, genetic or otherwise. In plant culture terminology, the term may refer to a culture derived as above or it may refer to a group of plants propagated only by vegetative and asexual means, all members of which have been derived by repeated propagation from a single individual.

Cloning efficiency: The percentage of cells plated (seeded, inoculated) that form a clone. One must be certain that the colonies formed arose from single cells in order to use this term properly. See 'Colony forming efficiency'.

Colony-forming efficiency: The percentage of cells plated (seeded, inoculated) that form a colony.

Complementation: The ability of two different genetic defects to compensate for one another.

Contact inhibition of locomotion: A phenomenon characterizing certain cells in which two cells meet, locomotory activity diminishes and the forward motion of one cell over the surface of the other is stopped.

Continuous cell culture: A culture that is apparently capable of an unlimited number of population doublings; often referred to as an immortal cell culture. Such cells may or may not express the characteristics of *in vitro* neoplastic or malignant transformation. See also 'Immortalization'.

Crisis: A stage of the *in vitro* transformation of cells. It is characterized by reduced proliferation of the culture, abnormal mitotic figures, detachment of cells from the culture substrate and the formation of multinucleated or giant cells. During this massive cultural degeneration, a small number of colonies usually, but not always, survive and give rise to a culture with an apparent unlimited *in vitro* lifespan. This process was first described in human cells following infection with an oncogenic virus (SV40). See also 'Cell line', '*In vitro* transformation' and '*In vitro* senescence'.

Cryopreservation: Ultra-low temperature storage of cells, tissues, embryos or seeds. This storage is usually carried out using temperatures below –100°C.

Cumulative population doublings: See 'Population doubling level'.

Cybrid: The viable cell resulting from the fusion of a cytoplast with a whole cell, thus creating a cytoplasmic hybrid.

Cytoplast: The intact cytoplasm remaining following the enucleation of a cell.

Cytoplasmic hybrid: Synonymous with 'cybrid'.

Cytoplasmic inheritance: Inheritance attributable to extranuclear genes, e.g. genes in cytoplasmic organelles such as mitchondria or chloroplasts, or in plasmids, etc.

Density-dependent inhibition of growth: Mitotic inhibition correlated with increased cell density.

Differentiated: Cells that maintain, in culture, all or much of the specialized structure and function typical of the cell type *in vivo*.

Diploid: The state of the cell in which all chromosomes, except sex chromosomes, are two in number and are structurally identical with those of the species from which the culture was derived. Where there is a Commission Report available, the experimenter should adhere to the convention for reporting the karyotype of the donor. Commission Reports have been published for mouse (Committee on Standardized Genetic Nomenclature for Mice 1972), human (Paris Conference 1971) and rat (Committee for a Standardized Karyotype of *Rattus norvegicus* 1973). In defining a diploid culture, one should present a graph depicting the chromosome number distribution leading to the modal number determination along with representative karyotypes.

Electroporation: Creation, by means of an electrical current, of transient pores in the plasmalemma usually for the purpose of introducing exogenous material, especially DNA, from the medium.

Embryo culture: *In vitro* development or maintenance of isolated mature or immature embryos.

Embryogenesis: The process of embryo initiation and development.

Endocrine cell: In animals, a cell that produces hormones, growth factors or other signalling substances for which the target cells, expressing the corresponding receptors, are located at a distance. See also 'Autocrine' and 'Paracrine'.

Epigenetic event: Any change in a phenotype that does not result from an alteration in DNA sequence. This change may be stable and heritable and includes alteration in DNA methylation, transcriptional activation, translational control and post-translational modifications.

Epigenetic variation: Phenotypic variability that has a non-genetic basis.

Epithelial-like: Resembling or characteristic of, having the form or appearance of epithelial cells. In order to define a cell as an epithelial cell, it must possess characteristics typical of epithelial cells. Often one can be certain of the histological origin and/or function of the cells placed into culture and, under these conditions, one can be reasonably confident in designating the cells as epithelial. It is incumbent upon the individual reporting on such cells to use as many parameters as possible in assigning this term to a culture. Until such time as a rigorous definition is possible, it would be most correct to use the term epithelial-like.

Euploid: The situation that exists when the nucleus of a cell contains exact multiples of the haploid number of chromosomes.

Explant: Tissue taken from its original site and transferred to an artificial medium for growth or maintenance.

Explant culture: The maintenance or growth of an explant in culture.

Feeder layer: A layer of cells (usually lethally irradiated for animal cell culture) upon which are cultured a fastidious cell type.

Fibroblast-like: Resembling or characteristic of, having the form or appearance of fibroblast cells. In order to define a cell as a fibroblast cell, it must possess characteristics typical of fibroblast cells. Often one can be certain of the histological origin and/or function of the cells placed into culture and, under these conditions, one can be reasonably confident in designating the cells as fibroblast. It is incumbent upon the individual reporting on such cells to use as many parameters as possible in assigning this term to a culture. Until such time as a rigorous definition is possible, it would be most correct to use the term fibroblast-like.

Finite cell culture: A culture that is capable of only a limited number of population doublings after which the culture ceases proliferation. See also '*In vitro* senescence'.

Habituation: The acquired ability of a population of cells to grow and divide independently of exogenously supplied growth regulators.

Heterokaryon: A cell possessing two or more genetically different nuclei in a common cytoplasm, usually derived as a result of cell-to-cell fusion.

Heteroploid: The term given to a cell culture when the cells comprising the culture possess nuclei containing chromosome numbers other than the diploid number. This is a term used only to describe a culture and is not used to describe individual cells. Thus, a heteroploid culture would be one that contains aneuploid cells.

Histiotypic: The *in vitro* resemblance, of cells in culture, to a tissue in form or function or both. For example, a suspension of fibroblast-like cells may secrete a glycosaminoglycan-collagen matrix and the result is a structure resembling fibrous connective tissue, which is, therefore, histiotypic. This term is not meant to be used along with the word 'culture'. Thus, a tissue culture system demonstrating form and function typical of cells *in vivo* would be said to be histiotypic.

Homokaryon: A cell possessing two or more genetically identical nuclei in a common cytoplasm derived as a result of cell-to-cell fusion.

Hybrid cell: The term used to describe the mononucleate cell that results from the fusion of two different cells, leading to the formation of a synkaryon.

Hybridoma: The cell that results from the fusion of an antibody-producing tumour cell (myeloma) and an antigenically-stimulated normal plasma cell. Such cells are constructed because they produce a single antibody directed against the antigen epitope that stimulated the plasma cell. This antibody is referred to as a monoclonal antibody.

Immortalization: The attainment by a finite cell culture, whether by perturbation or intrinsically, of the attributes of a continuous cell line. An immortalized cell is not necessarily one that is neoplastically or malignantly transformed.

Immortal cell culture: See 'Continuous cell culture'.

Induction: Initiation of a structure, organ or process *in vitro*.

***In vitro* neoplastic transformation:** The acquisition, by cultured cells, of the property to form neoplasms, benign or malignant, when inoculated into animals. Many transformed cell populations that arise *in vitro* intrinsically or through deliberate manipulation by the investigator produce only benign tumours that show no local invasion or metastasis following animal inoculation. If there is supporting evidence, the term *in vitro* malignant neoplastic transformation or *in vitro* malignant transformation can be used to indicate that an injected cell line does, indeed, invade or metastasize.

***In vitro* senescence:** In vertebrate cell cultures, the property attributable to finite cell culture; namely, their inability to grow beyond a finite number of population doublings. Neither invertebrate nor plant cell cultures exhibit this property.

***In vitro* transformation:** A heritable change, occurring in cells in culture, either intrinsically or from treatment with chemical carcinogens, oncogenic viruses, irradiation, transfection with oncogenes, etc., and leading to the acquisition of altered morphological, antigenic, neoplastic, proliferative or other properties. This expression is distinguished from the term *in vitro* neoplastic transformation in that the alterations occurring in the cell population may not always include the ability of the cells to produce tumours in appropriate hosts. The type of transformation should always be specified in any description.

Karyoplast: A cell nucleus, obtained from the cell by enucleation, surrounded by a narrow rim of cytoplasm and a plasma membrane.

Liposome: A closed lipid vesicle surrounding an aqueous interior; it may be used to encapsulate exogenous materials for their ultimate delivery into cells by fusion with the cell.

Microcell: A cell fragment, containing one to a few chromosomes, formed by the enucleation or disruption of a micronucleated cell.

Micronucleated cell: A cell that has been mitotically arrested and in which small groups of chromosomes function as foci for the reassembly of the nuclear membrane, thus forming micronuclei, the maximum of which would be equal to the total number of chromosomes.

Morphogenesis: (a) The evolution of a structure from an undifferentiated to a differentiated state. (b) The process of growth and development of differentiated structures.

Mutant: A phenotype variant resulting from a changed or new gene.

Organ culture: The maintenance or growth of organ primordia or the whole or parts of an organ *in vitro* in a way that may allow differentiation and preservation of the architecture and/or function.

Organogenesis: In animal cell cultures, the evolution, from dissociated cells, of a structure that shows natural organ form or function or both. In plant tissue culture, a process of differentiation by which plant organs are formed *de novo* or from pre-existing structures. In developmental biology, this term refers to differentiation of an organ system from stem or precursor cells.

Organotypic: Resembling an organ *in vivo* in three-dimensional form or function or both. For example, a rudimentary organ in culture may differentiate in an *organotypic* manner, or a population of dispersed cells may become rearranged into an *organotypic* structure and may also function in an *organotypic* manner. This term is not meant to be used along with the word 'culture' but is meant to be used as a descriptive term.

Paracrine: In animals, a cell that produces hormones, growth factors or other signalling substances for which the target cells, expressing the corresponding receptors, are located in its vicinity or in a group adjacent to it. See also 'Autocrine' and 'Endocrine'.

Passage: The transfer or transplantation of cells, with or without dilution, from one culture vessel to another. It is understood that any time cells are transferred from one vessel to another, a certain portion of the cells may be lost and, therefore, dilution of cells, whether deliberate or not, may occur. This term is synonymous with the term subculture.

Passage number: The number of times the cells in the culture have been subcultured or passaged. In descriptions of this process, the ratio or dilution of the cells should be stated so that the relative cultural age can be ascertained.

Pathogen free: Free from specific organisms based on specific tests for the designated organisms.

Plating efficiency: This is a term that originally encompassed the terms Attachment (Seeding) efficiency, Cloning efficiency and Colony-forming efficiency but is now better described by using one or more of the them in its place because the term plating is not sufficiently descriptive of what is taking place. See also 'Attachment', 'Seeding', 'Cloning' and 'Colony forming efficiency'.

Population density: The number of cells per unit area or volume of a culture vessel; also, the number of cells per unit volume of medium in a suspension culture.

Population doubling level: The total number of population doublings of a cell line or strain since its initiation *in vitro*. A formula to use for the calculation of 'population doublings' in a single passage is:

$$\text{Number of population doublings} = \frac{\log}{10 \; (N/N_o) \times 3.33}$$

where N = number of cells in the growth vessel at the end of a period of growth and N_o = number of cells plates in the growth vessel. *It is best to use the number of viable cells or number of attached cells for this determination.* Population doubling level is synonymous with the term cumulative population doublings.

Population doubling time: The interval calculated during the logarithmic phase of growth in which, for example, 1.0×10^6 cells increase to 2.0×10^6 cells. This term is not synonymous with the term cell generation time.

Primary culture: A culture started from cells, tissues or organs taken directly from organisms. A primary culture may be regarded as such until it is successfully subcultured for the first time. It then becomes a cell line.

Pseudodiploid: This describes the condition where the number of chromosomes in a cell is diploid but, as a result of chromosomal rearrangements, the karyotype is abnormal and linkage relationships may be disrupted.

Recon: The viable cell reconstructed by the fusion of a karyoplast with a cytoplast.

Saturation density: The maximum cell number attainable, under specified culture conditions, in a culture vessel. This term is usually expressed as the number of cells per square centimetre in a monolayer culture or the number of cells per cubic centimetre in a suspension culture.

Seeding efficiency: See 'Attachment efficiency'.

Senescence: See '*In vitro* senescence'.

Somatic cell genetics: The study of genetic phenomena of somatic cells. The cells under study are most often cells grown in culture.

Somatic cell hybrid: The cell resulting from the fusion of animal cells derived from somatic cells that differ genetically.

Somatic cell hybridization: The *in vitro* fusion of animal cells derived from somatic cells that differ genetically.

Subculture: See 'Passage'. With plant cultures, this is the process by which the tissue or explant is first subdivided and then transferred into fresh culture medium.

Substrain: A substrain can be derived from a strain by isolating a single cell or groups of cells having properties or markers not shared by all cells of the parent strain.

Suspension culture: A type of culture in which cells, or aggregates of cells, multiply while suspended in liquid medium.

Synkaryon: A hybrid cell that results from the fusion of the nuclei it carries.

Tissue culture: The maintenance or growth of tissues, *in vitro*, in a way that may allow differentiation and preservation of their architecture and/or function.

Totipotency: A cell characteristic in which the potential for forming all the cell types in the adult organism is retained.

Transfection: The transfer, for the purposes of genomic integration, of naked, foreign DNA into cells in culture. The traditional *microbiological* usage of this term implied that the DNA being transferred was derived from a virus. The definition as stated here is that which is in use to describe the general transfer of DNA irrespective of its source. See also 'Transformation'.

Transformation: In plant cell culture, the introduction and stable genomic integration of foreign DNA into a plant cell by any means, resulting in a genetic modification. This definition is the traditional microbiological definition. For animal cell culture, see '*In vitro* transformation', '*In vitro* neoplastic transformation' and 'Transfection'.

Variant: A culture exhibiting a stable phenotypic change, whether genetic or epigenetic in origin.

Virus-free: Free from specified viruses based on tests designed to detect the presence of the organisms in question.

REFERENCES

Committee for a Standardized Karyotype of *Rattus norvegicus* (1973) Standard karyotype of Norway rat, *Rattus norvegicus*. *Cytogenetics and Cell Genetics* 12: 199–205.

Committee on Standardized Genetic Nomenclature for Mice (1972) Standard karyotype of the mouse, *Mus musculus. Journal of Heredity* 63: 69–72.

Iglewski B (1989) Communicating with the public: a new scientific imperative. *ASM News* 55(6): 306.

Mueller S, Renfroe M, Schaeffer WI, Shay JW, Vaughn J & Wright M (Tissue Culture Association Terminology Committee Members) (1990) Terminology associated with cell, tissue and organ culture, molecular biology and molecular genetics. *In Vitro Cellular and Developmental Biology* 26: 97–101.

Paris Conference (1971), Supplement (1975) Standardization in Human Cytogenetics. Birth Defects: Original Article Series, XI, 9, 1975. The National Foundation, New York, Reprinted in: *Cytogenetics and Cell Genetics* (1975) 15: 201–238.

Schaeffer WI (1984) In the interest of clear communication. *In Vitro Cellular and Developmental Biology* 25: 389–390.

APPENDIX 2: Company addresses

Amersham Pharmacia Biotech
Bjorkgatan 30
75184
Uppsala
Sweden

American Type Culture Collection
(ATCC)
10801 University Blvd
Manassas
VA 20110
USA

Amersham International
Amersham Place
Little Chalfont
Bucks HP7 9NA
UK

Ashby Scientific Ltd
Unit 11
Atlas Court
Coalville
Leicestershire
LE67 3FL
Tel: (01530) 832590
Fax: (01530) 832591
Email: info@ashby-scientific.co.uk
Web site: ashby-scientific.co.uk

B. Braun Biotech International GmbH
Schwarzenberger Weg 73-79
D-34212 Melsungen
Germany
Tel: (49) 5661 71 3400
Fax: (49) 5661 71 3702
Email: bbi.info@bbraun.com
Web site: www.bbraunbiotech.com

Bayer Diagnostics
Bayer House
Strawberry Hill
Newbury
Berkshire
RG14 1JA
UK

Beckman Coulter (UK) Ltd
Oakley Court
Kingsmead Business Park
London Road
High Wycombe
Bucks HP11 1JU
UK

Becton Dickenson & Co.
Towns Road
Cowley
Oxford
OX4 3LY
UK

Bellco Glass Inc
340 Elrudo Road
Vineland
NJ 08360
USA

Bibby Sterlin Ltd.
Tiling Drive
Stone
Staffordshire ST15 OSA
UK

BioReliance (formerly Microbiological
Associates)
14920 Broschart Road
Rockville
MD 20850
USA
Tel: (0) 301 738 1000
Fax: (0) 301 610 2590
Email: info@bioreliance.com
Website: www.bioreliance.com

BioWhittaker UK Ltd
ISC House
Progress Business Centre
5 Whittle Parkway
Slough SL1 6DQ
UK

Cellex Biosciences Inc.
8500 Evergreen Blvd.
Minneapolis
MN 55433
USA
Tel: (1) 612 786 0302
Fax: (1) 612 786 0915
Web site: www.cellexbio.com

Cellmark Diagnostics
PO Box 265
Abingdon
Oxfordshire OX14 1YX
UK
Tel: (01235) 528609
Fax: (01235) 528141
Email: cellmark@astrazeneca.com
Web site: www.cellmark.co.uk

Cellon
22 rue Dernier Sol
L-2543 Luxembourg
Grand Duchy of Luxembourg

Coulter Electonics Ltd
Northwell Drive
Luton
Bedfordshire LU3 3RH
UK

Cryomed
51529 Birch Street
New Balitmore
MI 48047
USA

DuPont UK
International Centre
Boulton Road
Stevenage
Herts SG1 4QE
UK

Envair Ltd
York Avenue
Haslingden
Rossendale
Lancashire
BB4 4HX
UK
Email: carolev@envair.co.uk

European Collection of Cell Cultures
(ECACC)
CAMR (Centre for Applied
Microbiology Research)
Salisbury
Wiltshire SP4 OJG
UK
Tel: (01980) 612512
Fax: (01980) 611315
Email: ecacc@camr.org.uk
Web site: www.camr.org.uk

Fisher Scientific UK
Bishop Meadow Road
Loughborough
Leicestershire LE11 5RG
UK

Gilson Medical Electronics Inc
Anachen House
20 Charles Street
Luton
Beds LU2 OEB
UK

Hepaire Manufacturing Ltd
Aire Cool House
Spring Gardens
London Road
Romford
Essex RM7 9LY
UK

Heraeus Instruments GmbH
PO Box 15 63
63405 Hanau
Germany

Hitachi Scientific Instruments
Nissei Sangyo America Ltd
460 East Middlefield Road
Mountain View
CA 94043
USA

HyClone
1725 South HyClone Road
Logan
UT 84321
USA
Tel: (1) 435 753 4584
Fax: (1) 435 753 4589
Email: info@hyclone.com
Web site: www.hyclone.com

ICN Biomedicals, Inc
3300 Hyland Avenue
Costa Mesa
CA 92626
USA

Integra-Biosciences AG
(Headquarters)
Industriestrasse 44
P.O. Box 74
8304 Wallisellen
Switzerland
Tel: (41) 1 877 4646
Fax: (41) 1 877 4600
Email: info-ch@integra-
biosciences.com
Web site: www.integra-biosciences.com

Jencons Scientific Ltd
Cherrycourt Way Industrial Estate
Stanbridge Road
Leighton Buzzard
Beds LU7 8UA
UK

Jouan Inc
Rue Bobby Sands CP 3203
44805 Saint-Herblain
Cedex
France

JRH Biosciences
13804 West 107th Street
Lenexa
KS 66215
USA
Tel: (01264) 333311
Fax: (01264) 332412
Email: info@Jhreurope.com
Web site: www. Jrhbio.com

JRH Biosciences
Smeaton Road
West Portway
Andover
Hampshire
SP10 3LF
UK

L' Aire Liquide
57 av Carnot
BP13 94503 Champigny Cedex
France

Life Sciences International
Unit 5
The Ringway Centre
Edison Road
Basingstoke
Hants
RG21 6YH
UK

Life Technologies
3 Fountain Drive
Inchinnan Business Park
Paisley PA4 9RF
Scotland
UK
Tel: (0141) 814 6100
Fax: (0141) 814 6287
Email: euroinfo@lifetech.com
Web site: www.lifetech.com

Media-Cult A/s
Symbion Science Park
Haraldgade 68
DK2100 Copenhagen
Denmark

Merck Ltd (BDH Laboratory Supplies)
Merk House
Poole
Dorset
BH15 1TD
UK

Millipore (UK) Ltd
The Boulevard
Blackmoor Lane
Watford
WD1 8YD
UK

Nikon Corporation
Fuji Building
2-3 Marunouchi 3-chome
Chiyoda-ku
Tokyo 100
Japan

Nucleopore Corp.
7035 Commerce Circus
Pleasanton
CA 94566-3294
USA

NUNC A/S
Kamstrupvej 90
PO Box 280
DK -4000 Roskilde
Denmark
Tel: (45) 4631 2000
Fax: (45) 4631 2175
Email: infociety@nunc.dk
Web site: www. nunc.nalgenunc.com

Nycomed Pharma
Box 4284
Torshov
N-0401 Oslo 4
Norway

Olympus Co. (UK) Ltd
2–8 Honduras Street
London
EC1U OTX
UK

PAA Labor-und
Forchungsgesellschaft mBh
Weiner Strasse 131
A-4020 Linz
Austria

Pall Gelman Laboratory
Europa House
Havant Street
Portsmouth P01 3PD
UK
Tel: (01705) 302600
Email: stephen_profit@pall.com
Wed site: www.pall.com

Planar Products Ltd
110 Windmill Road
Sunbury on Thames
TW16 7HD
UK

Sartorius AG
PO Box 32 43
Weender Landstrasse 94–108
3400 Goettingen
Germany

Schott Glaswerke
Hattenbergstrasse
10 Postfach 2480
D-6500 Mainz 1
Germany

Sigma-Aldrich Company Ltd
The Old Brickyard
New Road
Gillingham
Dorset SP8 4XT
UK
Tel: (0800) 717181
Fax: (0800) 378785
Email: ukcustsv@euronotes.sial.com
Web site: www.sigma-aldrich.com

Techne (Cambridge) Ltd.
Duxford
Cambridge
CB2 4PZ
UK
Tel: (01223) 832401
Fax: (01223) 836838
Email: sales@techneuk.co.uk
Web site: www.techneuk.co.uk

Thermo Optek UK
Sussex Manor Park
Gatwick Road
Crawley
West Sussex
RH10 2QQ
UK
Tel: (01293) 561222
Fax: (01293) 561980

ThermoQuest/Forma Scientific Division
Millcreek Road
PO Box 649
Marietta
OH 45750
USA
Tel: (1) 740 373 4763
Fax: (1) 740 374 1817
Email: forma.marketing@tmquest.com
Website: www.forma.com

TPC Microflow
Walworth Road
Andover
Hampshire SP10 5AA
UK
Tel: (01264) 835853
Fax: (01264) 835854
Email: enquiries@tpc.microflow.co.uk
Web site: www.tpc-microflow.co.uk

Whatman LabSales
Whatman House
St Leonards Road
20/20 Maidstone
Kent ME16 OLS
UK

Wheaton Scientific
1000 North Tenth Street
Millville
NJ 08332
USA

Worthington Biomedical Corp
Hall Mill Road
Freehold
NJ 07728
USA

Zeiss (Carl) Germany
PO Box 1380
D-7082 Oberkochen
Germany

Zeiss (Carl) Inc.
Microscope Division
1 Zeiss Drive
Thornwood
NY 10594
USA

INDEX